Lecture Notes on Data Engineering and Communications Technologies **184**

Series Editor

Fatos Xhafa, *Technical University of Catalonia, Barcelona, Spain*

The aim of the book series is to present cutting edge engineering approaches to data technologies and communications. It will publish latest advances on the engineering task of building and deploying distributed, scalable and reliable data infrastructures and communication systems.

The series will have a prominent applied focus on data technologies and communications with aim to promote the bridging from fundamental research on data science and networking to data engineering and communications that lead to industry products, business knowledge and standardisation.

Indexed by SCOPUS, INSPEC, EI Compendex.

All books published in the series are submitted for consideration in Web of Science.

AboulElla Hassanien · Rawya Y. Rizk ·
Dragan Pamucar · Ashraf Darwish ·
Kuo-Chi Chang
Editors

Proceedings of the 9th International Conference on Advanced Intelligent Systems and Informatics 2023

Springer

Editors
AboulElla Hassanien
Faculty of Computers and Information
Cairo University
Giza, Egypt

Rawya Y. Rizk
Port Said University
Port Fouad City, Egypt

Dragan Pamucar
Department of Operations Research and Statistics
The University of Belgrade, Faculty of Organizational Sciences
Belgrade, Serbia

Ashraf Darwish
Faculty of Science
Helwan University
Cairo, Egypt

Kuo-Chi Chang
Shulin District
Fujian University of Technology
New Taipei, Taiwan

ISSN 2367-4512 ISSN 2367-4520 (electronic)
Lecture Notes on Data Engineering and Communications Technologies
ISBN 978-3-031-43246-0 ISBN 978-3-031-43247-7 (eBook)
https://doi.org/10.1007/978-3-031-43247-7

This Springer imprint is published by the registered company Springer Nature Switzerland AG
The registered company address is: Gewerbestrasse 11, 6330 Cham, Switzerland

Paper in this product is recyclable.

Preface

This volume constitutes the refereed proceedings of the 9th International Conference on Advanced Intelligent Systems and Informatics (AISI 2023), which took place in Port Said, Egypt, from Sept. 20–22, 2023. It is an international interdisciplinary conference covering research and development in informatics and intelligent systems. In response to the call for papers for AISI 2023, we have accepted 49 papers, from the main conference. The papers cover current research in deep learning, machine learning-based applications, real-time system, the Internet of things and fog computing, and smart systems and data science. We sincerely thank the plenary speakers, workshop chairs, and international program committee members for helping us formulate a rich technical program. The 9th edition of AISI is organized by the Scientific Research Group in Egypt (SRGE) in collaboration with Port Said University, Egypt.

We would like to extend our sincere appreciation for the outstanding work contributed over many months by the local organization chair and publicity chair of the organizing committee. We also wish to express our appreciation to the SRGE members for their assistance and the team from the Faculty of Engineering at Port Said University. We would like to emphasize that the success of AISI 2023 would not have been possible without the support of many committed volunteers who generously contributed their time, expertise, and resources toward making the conference an unqualified success. Finally, thanks to the Springer team for their support in all stages of the production of the proceedings. We hope that you will enjoy the conference program.

Organization

Honorary Chairs

Ayman Mohamed	President of Port Said University, Egypt

International Advisory Board

Xiao-Zhi Gao	University of Eastern Finland, Finland
Yu-Dong Zhang	University of Leicester, UK
M. F. Tolba	Ain Shams University, Egypt
Siddhartha Bhattacharyya	Christ University, Bangalore, India
Nagwa Bader	Ain Shams University, Egypt
Hala El Moshier	Ain Shams University, Egypt
Mrutyunjaya Panda	Utkal University, Vani Vihar Bhubaneswar-4, Odisha, India
Abdelkrim Haqiq	FST, Hassan 1st University, Settat, Morocco
Janusz Kacprzyk	Polish Academy of Sciences, Poland
Kamel Hussein Rahouma	Faculty of Computers, Nahda University, Beni Suef, Egypt
Tien-Wen Sung	Fujian University of Technology, Fuzhou, China
Fuquan Zhang	Minjiang University, Fujian, China
Mohammad Hammoudeh	Manchester Metropolitan University, UK
Sunil Vadera	University of Salford, UK
Narayan C. Debnath	Eastern International University, Vietnam
Vaclav Snasel	VSB-Technical University of Ostrava, Czech Republic

General Chairs

Rawya Y. Rizk	Port Said University, Egypt
Aboul Ella Hassanien	Cairo University, Egypt

Conference Co-chairs

Ashraf Darwish	Helwan University, Egypt
Rehab F. Abdel-Kader	Port Said University, Egypt

Technical Program Chairs

Kuo-Chi Chang	Department of Applied Intelligent Mechanical and Electrical Engineering, Preparatory Chair, Yu Da University of Science and Technology, Taiwan, (R.O.C)
Dragan Pamucar	The University of Belgrade, Faculty of Organizational Sciences, Department of Operations Research and Statistics, Belgrade, Serbia
Farid Meziane	University of Derby, UK

Publicity Chairs

Mohamed Abdelfatah	Egypt
Randa Atta	Port Said University, Egypt
Heba Nashaat	Port Said University, Egypt

International Program Committee

Maytham Alabbas	University of Basrah in Iraq
Heba M. Abdel-Atty	Port Said University, Egypt
Heba Y. M. Soliman	Port Said University, Egypt
Walaa Saber	Port Said University, Egypt
Mohamed F. AbdelKader	Port Said University, Egypt
Rabab F. Abdel-Kader	Port Said University, Egypt
Islam Shaalan	Port Said University, Egypt
Joshua A. Abolarinwa	Namibia University of Science and Technology, Namibia
Tolga Ensari	Department of Computer and Information Science, Arkansas Tech University, USA
Arezou Harraf	Box Hill College Kuwait, Kuwait
Amina Buallay	Secretariat General of the Higher Education Council, Bahrain

Araby Madbouly	Muscat College, Oman
Abdul Nasser Nour	An-Najah National University, Palestine
Qasim Zureigat	Sulaiman Al Rajhi School of Business, Saudi Arabia
Basheer Al-Haimi	Hebei University, China
Talal Al-Hayale	University of Windsor, Canada
Shakir Ullah	Stratford University, USA
Chin-Shiuh Shieh	National Kaohsiung University of Science and Technology, Taiwan
Chris Hughes	University of Salford, UK
Essam Rashed	Nagoya Institute of Technology, Japan
Pavel Krömer	VSB Technical University of Ostrava, Czech Republic
Ibrahima Faye	Universiti Teknologi PETRONAS, Malaysia
Ahmed Hamed Attia	Suez Canal University, Egypt
Yannis Korkontzelos	Edge Hill University, UK
Amna Eleyan	Manchester Metropolitan University, UK
Eman H. Alkhammash	Taif University, Saudi Arabia
Rania Hodhod	Columbus State University, USA
Djamel Saba	Renewable Energy Research Unit in Saharan Region Adrar, Algeria

Local Arrangement Committee Chairs

Mohamed Abd ElFatah	Misr Higher Institute for Commerce and Computers, Mansoura, Egypt
Mourad Raffat	Egypt
Heba Aboul Ella	Egypt

Contents

Energy and Power Control

Data Mining, Decision Making, and Intelligent Systems

Health Informatics

Security and Communications

Image and Video Processing

Deep Learning Technology and Applications

A Robust Deep Learning Techniques for No-Show Prediction in Hospital Appointments

Phat Tien Nguyen[1,2], Doan Thai Dang[3], and Vinh Dinh Nguyen[3](✉)

[1] Becamex Business School, Eastern International University, Thủ Dầu Một, Vietnam
phat.nguyen@eiu.edu.vn

[2] Becamex International Hospital, Thuận An, Binh Duong, Vietnam
phat.nguyent@bih.vn

[3] School of Computing and Information Technology, Eastern International University, Thủ Dầu Một, Vietnam
{doan.dang,vinh.nguyen}@eiu.edu.vn

Abstract. Machine learning (ML) has been widely adopted in the healthcare industry for improving patient outcomes and operational efficiency. Predicting no-show appointments is one of the areas where ML has been applied to optimize appointment scheduling and management. However, there are still challenges to be addressed when applying ML to predict no-show appointments, including data quality and potential biases in the models. In recent years, deep learning (DL) has emerged as a powerful tool for solving complex problems, including those in the healthcare industry. This paper proposes a DL technique to predict no-show appointments in the hospital sector using a dataset of historical patient appointments and their attendance. The study aims to accurately predict which patients are at risk of not attending their appointments and explore ways to improve appointment scheduling and management using these predictions. The paper also discusses the benefits and challenges of using ML and DL methods in healthcare services and provides theoretical and managerial implications for future research and practice.

Keywords: No-show Appointments · Healthcare Management · Machine Learning · Deep Learning · RNN

1 Introduction

Machine learning (ML) has been increasingly used in the healthcare industry to improve patient outcomes and streamline operations [1, 2]. The healthcare and hospital sector is a complex and dynamic environment that generates large amounts of data, including patient information, medical records, and images [3]. The ability of ML to process and analyze this data can provide valuable insights and support decision-making [4, 5]. One of the main areas where ML has been applied in healthcare is diagnosing and treating diseases [6]. For instance, ML algorithms have been used to support interpreting medical images, such as X-rays, CT scans, and MRIs, to detect and diagnose cancer, cardiovascular disease, and other conditions [1, 7–9].

A. Hassanien et al. (Eds.): AISI 2023, LNDECT 184, pp. 3–18, 2023.
https://doi.org/10.1007/978-3-031-43247-7_1

Additionally, ML algorithms have been used to predict patient outcomes, such as the risk of readmission, survival, and response to treatment [10]. Another area where ML has been applied in healthcare is managing and optimizing healthcare systems [3, 11]. For instance, ML algorithms have been used to predict patient demand, schedule resources, and optimize the allocation of beds and staff fields [11, 12]. Additionally, ML algorithms have been used to predict no-show appointments [13], leading to wasted resources, longer wait times for other patients, and decreased revenue for healthcare providers [14]. The application of ML to predict no-show appointments is based on the idea of training a model on historical data of patients' appointments and their attendance [2]. The model uses various features, such as demographics, medical history, and appointment details, to predict the likelihood of a patient not showing up for their appointment [2, 15–18]. The predictions can then be used to improve appointment scheduling and management and to provide targeted interventions to patients at high risk of not showing up [1, 13, 19]. Several studies have been conducted to apply ML to predict no-show appointments, and the results have been promising. These include ensemble learning techniques such as some ensemble methods, including Random Forest (RF) [18, 20], Gradient Boosting (GB) [13], ADA Boots [21], decision tree algorithms [22], and neural networks [18]. Several studies have reported high accuracy rates in their predictions, with some models achieving over 80% accuracy [15, 23]. In recent years, Deep Learning (DL) has emerged as a powerful tool for solving complex problems in various fields, including healthcare [24, 25]. However, challenges must still be addressed when applying ML and DL to predict no-show appointments [24, 26]. One of the main challenges is the quality and availability of data. The healthcare sector generates a large amount of data, but it is often unstructured, incomplete, and heterogeneous, which makes it challenging to use for ML [27]. Additionally, there is a need for more research to investigate the potential biases in the models and how to mitigate them [1, 10].

This paper introduces a new DL technique to predict no-show appointments in the hospital sector. We will use a dataset of historical patient appointments and their attendance from the Becamex International Hospital (BIH) database to train and evaluate our DL models. Our goal is to accurately predict which patients are at risk of not attending their appointments and explore how these predictions can be used to improve appointment scheduling and management. Moreover, we also profoundly discussed the benefits and challenges of using DL and ML methods in the healthcare service sector.

2 Literature Review

2.1 Online Outpatient Appointment Scheduling

In early 2022, the Vietnamese Ministry of Health approved a plan to promote developing and utilizing digital medical platforms as part of the national digital transformation program through 2025, with an orientation towards 2030 [28]. Advances in information and communication technologies have led to the growth of the online medical services industry in Vietnam. These services include online consultations, outpatient appointment scheduling, electronic medical prescriptions, and online payments [28–30].

Among these services, outpatient appointment scheduling is the most widely used and is the focus of our study. An appointment system can significantly benefit the patient experience and hospital operations during the COVID-19 pandemic and beyond [31, 32]. There are a few ways an appointment system can improve patient journeys: (1) By spreading out patient visits throughout the day, an appointment system can help maintain social distancing and reduce crowding, making the patient journey safer [32, 33]; (2) Long wait times can be reduced by allowing non-urgent patients to schedule their consultation in advance, improving the patient experience [34]; (3) With an appointment system, patients can better plan their day as it eliminates uncertainty and stress caused by long wait times and minimal information [35]; (4) By controlling and steering the flow of patients away from peak hours, an appointment system can create a happier work environment for staff by reducing anxiety and allowing them to serve more patients per day [11, 36, 37]; (5) Automated confirmations and reminders sent through an appointment system can reduce outpatient cancellations [34, 36]; (6) Finally, an appointment system can also reduce the time needed for administrative tasks by automating appointment handling and allowing patients to book online or in an app. [16, 23, 38].

2.2 Patient No-Show Behavior

Patient no-show behavior is a phenomenon that can be influenced by a variety of factors [2, 23, 39]. Previous studies have found that demographic characteristics, such as age and gender, can affect no-show rates [23, 39]. For example, younger patients are more likely to miss outpatient appointments [40], but some studies have suggested that age has no significant impact on no-show rates [41]. Furthermore, some studies found that women have a lower risk of no-shows than men, while others have found the opposite [19, 23, 39]. Patients without health insurance are likelier not to attend a clinic visit than those with private or public health insurance [42]. In addition to demographic factors, other characteristics such as income, marital status, and race have also been found to impact no-show behavior [15, 34, 42, 43]. The appointment time also plays a role, with patients with a long lead time of appointment being at a higher risk of being no-shows [39, 40, 42]. The day of the week and the appointment time can also affect no-show rates, with Mondays and afternoon appointments having higher no-show rates [44]. Weather conditions on the appointment day have also been found to predict no-show behavior [45]. Patient distance from the hospital, previous visit experience, medical insurance, appointment channels, and appointment specialties are also factors that may impact no-show behavior [15, 39, 46]. It is important to note that these findings may vary depending on the population and setting studied, and more research is needed to understand the complex factors contributing to patient no-show behavior fully.

2.3 Classification Algorithms

Several classification algorithms can be used to predict no-show appointments in a hospital setting. These include:

Logistic Regression: Logistic regression is a linear model commonly used for binary classification problems such as no-show prediction. It estimates the probability of an outcome based on the input features [20, 47].

Decision Trees: Decision trees are a popular non-parametric algorithm that can be used for both classification and regression tasks. They work by recursively partitioning the feature space into smaller regions to minimize the impurity or error within each region [22, 48].

Support Vector Machines: Support vector machines (SVMs) are a class of algorithms that can be used for classification and regression tasks [16]. They work by finding the best decision boundary, or hyperplane, that maximizes the margin between different classes [21].

Neural Networks: Neural networks are a class of ML algorithms inspired by the human brain's structure and function [45]. They can be used for various tasks, including image recognition, natural language processing, and no-show prediction [4].

Gradient Boosting: Gradient boosting is an ML technique that creates an ensemble of weak decision trees and combines them to produce a robust model [13]. It iteratively fits the new trees to correct the errors in the previous trees.

DL - Recurrent Neural Networks (RNN): is a type of DL architecture that is designed to process sequential data. It is a type of RNN with multiple layers, allowing it to learn complex relationships between inputs and outputs [49]. The hidden state of the RNN at each time step is passed as input to the next time step, allowing the network to maintain a contextual understanding of the input sequence. DL RNNs are trained using large amounts of data and powerful computational resources, such as GPUs. They have been applied to various tasks, including classify classification, language translation, speech recognition, image captioning, and sequential decision-making.

Long Short-Term Memory (LSTM): Long short-term memory (LSTM) is a type of recurrent neural network that is specifically designed to overcome the problem of vanishing gradients in traditional RNNs [49]. LSTM uses memory cells and gates to selectively retain or discard information from previous time steps, which allows them to model long-term dependencies in the data [49].

All these algorithms have been used to predict no-show appointment appointments in hospitals and have shown promising results. The optimal algorithm depends on the dataset size, nature, and hospital requirements. In this study, the RNN algorithm is well suited for predicting no-show appointments in BIH for several reasons. Firstly, RNNs are designed to handle sequential data, making them well-suited for processing time-series data such as appointment schedules. Additionally, RNNs could maintain an internal state influenced by previous inputs, allowing the network to contextualize the input sequence. This is important when predicting no-show appointments, as the likelihood of a patient missing an appointment may depend on factors such as medical conditions and patient demographic information.

Furthermore, RNNs can learn complex relationships between inputs and outputs, making accurate predictions even when dealing with multiple and interrelated factors. These characteristics make RNNs well-suited for predicting no-show appointments in a hospital setting. Moreover, we also compare and evaluate the performance metrics of the RNN algorithm with other popular classification algorithms to provide more information.

2.4 Related Work

There has been a growing body of research on applying ML (ML) to predict no-show appointments in the healthcare sector. Some studies demonstrate that ML models can be trained to predict no-show appointments with high accuracy and that the performance of the models can vary depending on the dataset and the algorithm used [15, 39, 47]. Factors necessary for prediction include demographics, medical history, and appointment details. Details of recent studies can see in Table 1.

Table 1. Summary and comparison of key reviewed studies related to no-show prediction (last 5 years)

ID	Articles	Dataset Size	No-Show Rate	Application Area	Total Features	Algorithms	Evaluation & Performance Measures
1	(Liu et al., 2022) [45]	161,822 (Jan 2015 - Sept 2016)	20.3%	Predicting no-shows in pediatric medical appointment at Children's Hospital's Primary Care	387	ANN, LR	AUC: 0.97 Precision: 0.81 Recall: 0.83 F1 score: 0.82
2	(Periyathambi et al., 2022) [50]	607 (Jan 2016 – Dec 2019)	35%	Identify the factors associated with non-attendance at National Health Service	17	LR	F1 score: 0.6 Accuracy: 0.67 AUC: 0.7174
3	(Joseph et al., 2022) [16]	110,527 (N/A)	20%	Prediction of Clinical Appointment No-Shows at Public Hospital	13	LR, DT, RF, Linear SVM, GB	Average Precision: 0.91 - 0.94
4	Elkhider et al., 2022) [43]	71,178 (Jul 2013 -Sept 2018)	16%	Identify predictors of a no-show in neurology clinics	9	Multivariate LR	Odds ratio: 0.49–1.62
5	(Dashtban & Li, 2021) [25]	4,046,097 (Apr 2015 – Sept 2018)	8%	Predicting non-attendance in hospital outpatient appointments at National Health Service	32	SDAEs	AUC: 0.70
6	(Qureshi et al., 2021) [21]	6,000,000 (Otc 2007 – Feb 2022)	8%	Prediction of Missed Clinical Appointment at Primary Care Provider	7	RF, DT, LR, XGBoost, GB, AdaBoost, MB, SVM	Accuracy: 0.86 Precision: 0.83 Recall: 0.94 F1 score: 0.87 AUC: 0.92
7	Salazar et al., 2021) [18]	5000 (2017 – 2019)	12%	Predict a Patient's No-Show in the Healthcare Sector at Children's Hospital's Primary Care	22	LR, DT, RF	Recall: 0.91 AUC: 0.97
8	(Krishnan & Sangar, 2021) [22]	110,527 (N/A)	21%	Predict the no-shows and removes the bias towards minority class	14	DT	Precision: 0.79 Recall: 0.99 F1 score:0.88 AUCROC: 0.50

(*continued*)

Table 1. (*continued*)

ID	Articles	Dataset Size	No-Show Rate	Application Area	Total Features	Algorithms	Evaluation & Performance Measures
9	(Suk et al., 2021) [38]	1,252,127 (Sept 2018 – Nov 2018)	6.12%	Evaluate patient no-shows based on modes of appointment-making and types of appointments	14	Multivariate LR	Odds ratio: 0.39 – 12.59

It is important to note that the performance of these ML algorithms can vary depending on the specific characteristics of the data and the specific evaluation metric used (e.g., accuracy, precision, recall, F1-score, ROC-AUC) [4, 5, 18]. Therefore, it is essential to carefully evaluate the performance of different algorithms using appropriate evaluation metrics before making a final decision [4]. However, more research is needed to improve the models' performance and generalizability and investigate the models' potential biases and how to mitigate them [10]. In addition, there are several limitations to previous studies on applying ML to predict no-show appointments in healthcare. (1) Data availability: Many studies have used small, limited datasets that may not represent the general population. This can lead to models that perform well on the specific dataset but may not generalize well to other settings; (2) Feature engineering: Feature engineering is the process of selecting and creating relevant features from the dataset; it can be a limitation as the features selected may not be the optimal ones for the model; (3) Model evaluation: Some studies have used simple evaluation metrics, such as accuracy, which may not fully capture the performance of the model. Additionally, the lack of the use of appropriate validation techniques such as k-fold cross-validation might lead to overfitting of the model; (4) Limited application: Many studies have focused on predicting no-show appointments at a single clinic or within a specific healthcare system, which may not be generalizable to other settings; (5) Ethical concerns: healthcare data is sensitive and private, and the way the data is collected, stored, and used should be compliant with data privacy regulations, not all studies might have considered this.

Overall, while previous studies have shown that ML can be an effective tool for predicting no-show appointments in healthcare, more research is needed to improve the generalizability and robustness of the models and to ensure that they are ethically sound [4, 19, 39, 45].

3 Methods and Techniques

3.1 Data Description

The data were extracted from the BIH located in Binh Duong Province. We retrieved the information for each patient who scheduled an appointment between June 01, 2022, and July 30, 2022. In order to enhance the ability of the proposed techniques to extract useful information from the dataset, data must go through a series of preprocessing steps.

Pseudo-code for data preprocessing steps
Step 1: Clean and organize the data *1.1 Remove missing or duplicate values, correct errors, and ensure all data is in the correct format.* *1.2 Use the RandomForestClassifier algorithm to handle missing data:* *1.2.1 Import RandomForestClassifier class from the sci-kit-learn library.* *1.2.2 Create an instance of the RandomForestClassifier and train the algorithm on the available data to predict the missing values.* *1.2.3 Repeat the process to obtain multiply imputed datasets to obtain robust estimates of the missing values.* ***Step 2: Feature Engineering*** *2.1 Extract features from the date and time of the appointment.* *2.2 Create new features or transform existing ones that better represent the underlying data and improve the model's performance.* ***Step 3: Feature Scaling or Normalization:*** *3.1 Use StandardScaler to scale the features:* *3.1 Import StandardScaler class from the sci-kit-learn library.* *3.2 Create an instance of the StandardScaler and fit it into the training data.* *3.3 Apply the transform method to the training and testing data to scale the features.* ***Step 4: Feature Importance*** *4.1 Use Recursive Feature Elimination with Cross-Validation (RFECV) to identify essential features for a predictive model.* *4.2 Recursively remove the least important features and evaluate the model performance using cross-validation.* *4.3 Determine the optimal number of features resulting in a feature ranking that can be used to identify the most impactful variables in a predictive model.* ***Step 5: Split Data into Training and Test Set*** *5.1 Use the train_test_split function to split the data into training and test sets.* *5.2 Use a 75/25 split, 75% of the data is used for training, and the remaining 25% is used for testing.* *5.3 Ensure the split is representative of the overall dataset and that the test set is not too small.* *5.4 Use stratified sampling, which ensures that the split is representative of the distribution of the target variable.* *5.5 The test set evaluates the model's performance and estimates its accuracy in making predictions on new, unseen data.*

After the preprocessing steps, the final dataset contained 94212 outpatient appointments and seven features. The features included in the final dataset can be categorized as follows: Age, Gender, Patient location, DoA (Date of Appointment), NoD (Name of department), ToP (Type of Patient), and target variable name Status (show or no-show). After the feature extraction, the completed dataset will be selected and transformed into 85 attributes (independence variables) that predict the no-show appointment (dependence variables) in the further training, testing, and evaluation steps (Table 2).

Table 2. Main features and attributes

Features	Data description
Age	0–97
Gender	Male, female
Patient location	58 provinces
Date of appointment	Month, Day of Week, Time of Day
Type of patient	4 types of patients
Department	9 clinical departments
Status	Show/No-show

3.2 Implementation

Use ML and DL to predict no-show appointments in a hospital with the given data explained in Fig. 1.

- Data Collection: Collect and preprocess the data, including the six attributes: Age, Gender, Patient location, DoA (Date of Appointment), NoD (Name of department), and ToP (Type of Patient).
- Data Cleaning: Clean and prepare the data by handling missing or incorrect values, converting categorical variables into numerical values, and standardizing the data.
- Feature Engineering: Create new features or extract relevant information from existing features to improve the model's predictive power. For example, extract information such as the day of the week or time of day from the appointment date.
- Model Selection: Select an appropriate ML or DL model for the task—for example, decision trees, support vector machines, neural networks, etc.
- Model Training: Train the selected model using preprocessed and feature-engineered data.
- Model Evaluation: Evaluate the model's performance using metrics such as accuracy, precision, recall, and F1-score, AUC.

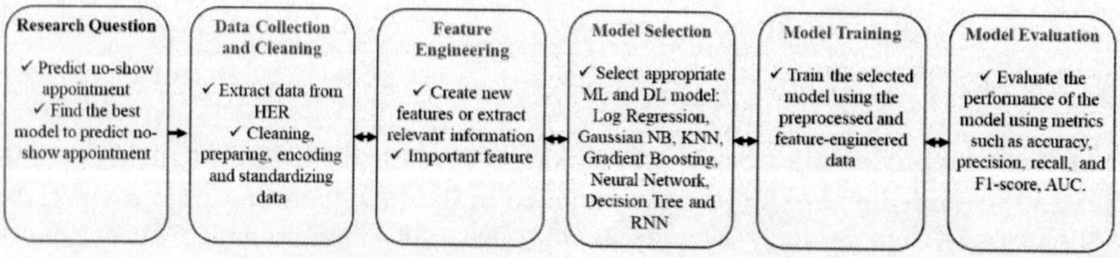

Fig. 1. Summary of the research procedure

The experiment used Python 3.9 on a Windows 10 machine with an Intel Core I9 processor, 32 GB of RAM, and a 10 GB GeForce RTX 3080 graphics card. The data preprocessing, training and evaluation of ML algorithms were executed utilizing the Pandas, Matplotlib libraries and Scikit-learn, a high-level framework designed for supervised and unsupervised ML algorithms. Especially to use RNN with the Keras library from the TensorFlow framework, we followed these steps:

- Install the Keras library and its dependencies.
- Import the necessary modules and classes from the Keras library, such as the Sequential class, LSTM layer, and Dense layer.
- Define the architecture of the RNN model using the Sequential class and add LSTM and Dense layers to it. We can also specify the number of units in each LSTM layer and the activation function used by each Dense layer.
- Compile the RNN model by specifying the optimizer, loss function, and metrics we want to use during training.
- Prepare data for training and testing, including splitting it into training and test sets and normalizing the input features.
- Train the RNN model using the fit method and provide it with the training data and the number of epochs to run.
- Evaluate the performance of the trained RNN model on the test data using the evaluate method.
- Finally, we can use the prediction method to predict new data.

3.3 Evaluation Models

To compare and evaluate the performance of RNN with other popular classification algorithms using various performance metrics (details in Table 3). Some standard performance metrics are used include [4, 5, 18]:

(1) Confusion matrix: This is typically a 2x2 table that includes four metrics:

In a confusion matrix, True Positive (TP): refers to correctly classified positive observations, False Positive (FP): refers to incorrectly classified positive observations, False Negative (FN): refers to incorrectly classified negative observations, and True Negative (TN): refers to correctly classified negative observations. The confusion matrix can be represented as the matrix below.

	Predicted Positive	Predicted Negative
Actual Positive	True Positive (TP)	False Negative (FN)
Actual Negative	False Positive (FP)	True Negative (TN)

Different metrics from the confusion matrix can be calculated to evaluate the model's performance, including accuracy, precision, recall, and F1 score.

(2) Accuracy: This measures the proportion of correct predictions made by the model and is calculated as the number of correct predictions divided by the total number of predictions.

$$\text{Accuracy} = (\text{TP} + \text{TN})/(\text{TP} + \text{TN} + \text{FP} + \text{FN})$$

(3) Precision: This measures the fraction of positive predictions and is calculated as the number of true positive predictions divided by the sum of the true positive and false positive predictions.

$$\text{Precision} = \text{True Positives}/(\text{True Positives} + \text{False Positives})$$

(4) Recall: This measures the fraction of actual positive instances correctly predicted by the model and is calculated as the number of true positive predictions divided by the sum of the true positive and false negative predictions.

$$\text{Recall} = \text{true positives}/(\text{true positives} + \text{false negatives})$$

(5) F1-score: This is the harmonic mean of precision and recall and is a commonly used metric for evaluating the performance of binary classification models.

$$\text{F1} - \text{score} = 2 * (\text{precision} * \text{recall})/(\text{precision} + \text{recall})$$

(6) Receiver Operating Characteristic (ROC) curve: The Receiver Operating Characteristic (ROC) curve is a graphical representation of the performance of a binary classifier. It is created by plotting the true positive rate (TPR) against the false positive rate (FPR) for different classification thresholds. The true positive rate is also known as sensitivity or recall, and it is defined as the ratio of true positives to the sum of true positives and false negatives:

$$\text{TPR} = \text{TP}/(\text{TP} + \text{FN})$$

The false positive rate, on the other hand, is defined as the ratio of false positives to the sum of false positives and true negatives:

$$\text{FPR} = \text{FP}/(\text{FP} + \text{TN})$$

4 Results

In this study, we aimed to investigate using DL and other classification algorithms to predict no-show appointments in a hospital setting. To this end, we used several classification algorithms, including logistic regression, decision trees, support vector machines, gradient boosting, and DL algorithms, such as recurrent neural networks (RNN) to predict no-show appointments. We used a dataset that contains patient demographic information, appointment information, and others information extracted from HER.

4.1 Model 1

The results showed that the DL algorithm (RNN) performed better than the other ML classification algorithms. Specifically, the RNN model achieved the highest AUC of 91% on the test set, outperforming the LogisticRegression, GaussianNB, K-Nearest Neighbors, Gradient Boosting, Neural Network, and Decision Tree algorithms. This

Table 3. Performance measures of prediction full models

Modeling Method	AUC	Overall Accuracy	Precision		Recall		F1-Score	
			1	0	1	0	1	0
Gaussian NB	0.65	0.12	0.96	0.09	0.03	0.99	0.06	0.17
Log-Regression	0.73	0.91	0.91	0.60	1.00	0.00	0.95	0.00
KNN	0.82	0.91	0.93	0.53	0.98	0.22	0.95	0.31
Gradient Boosting	0.82	0.91	0.91	0.74	1.00	0.02	0.95	0.04
Neural Network	0.82	0.91	0.92	0.49	0.99	0.11	0.95	0.18
Decision Tree	0.90	0.96	0.98	0.79	0.98	0.82	0.98	0.80
DL (RNN)	0.91	0.95	0.96	0.74	0.98	0.63	0.97	0.68

Note: 1: Show rates, 0: No-show rates, value in percent (%)

Table 4. Performance measures of light prediction models

Modeling Method	AUC	Overall Accuracy	Precision		Recall		F1-Score	
			1	0	1	0	1	0
Gaussian NB	0.67	0.77	0.93	0.17	0.81	0.40	0.87	0.24
Log-Regression	0.73	0.91	0.91	0.00	1.00	0.00	0.95	0.00
KNN	0.83	0.91	0.93	0.54	0.98	0.21	0.95	0.31
Gradient Boosting	0.81	0.91	0.91	0.78	1.00	0.01	0.95	0.03
Neural Network	0.80	0.91	0.92	0.42	0.99	0.10	0.95	0.16
Decision Tree	0.89	0.96	0.98	0.77	0.98	0.80	0.98	0.78
DL (RNN)	0.92	0.93	0.95	0.67	0.97	0.53	0.96	0.59

Note: 1: Show rates, 0: No-show rates, value in percent (%)

suggests that DL models, particularly RNN, are powerful tools for no-show prediction in a hospital setting.

Based on the table provided, the RNN model achieved the highest AUC score of 91%, indicating that it had the best overall performance in accurately predicting whether patients would show up for their appointments. This means that the RNN model had the highest true positive rate and the lowest false positive rate, both crucial metrics in predicting no-shows in a hospital setting. Furthermore, RNNs are particularly well-suited for time-series data like this, where there is a sequential order to the data points (e.g., appointment times). RNNs are designed to capture dependencies and patterns across data sequences, which can be beneficial in predicting future events based on past events. This may have given the RNN model an advantage over other models that do not consider the sequential nature of the data.

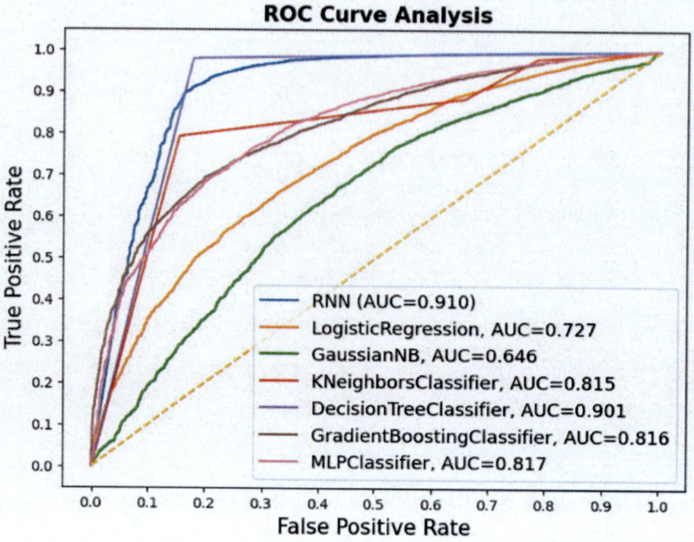

Fig. 2. Full models Receiver operating characteristic curve comparison

4.2 Model 2

Applying feature importance methods to a model can help identify the most significant features in the model and provide insights into which features are driving the model's predictions. This information can clarify the model's behavior and identify any potential problems or biases in the data or the model. By identifying the most critical features, we can create a more straightforward and more interpretable model, which will help us to understand how each feature contributes to the prediction and how we can improve the performance of the proposed model in the future.

The approach to achieving this is by identifying and only using the most important features in the model. In this study, we would like to implement feature importance to identify the features that have the most significant impact on the predictions. The study applies the RFECV method for the entire dataset, including 85 attributes after the RFECV process; the model proposed to keep 26 independent attributes. Then, we apply the training, test, and evaluate processes for the new dataset. The details resulting from this implementation can be found in Table 4 and Fig. 2.

The performance metrics of the RNN model show that by reducing attributes, the ability to predict reduction but not too much and can be accepted to use (ROC = 0.92, accuracy = 0.93, F1-Score = 0.59) (Fig. 3).

In this study, the time to make predictions on a new case was measured for two different models: The full Model and the Light Model for RNN algorithms to evaluate the performance of these algorithms. The models were trained on the same dataset and were tested using the same new data case. The elapsed time for making predictions on the new data case was calculated using the "time module" in Python. The results showed that Light Model took significantly less time to predict than the Full model. The

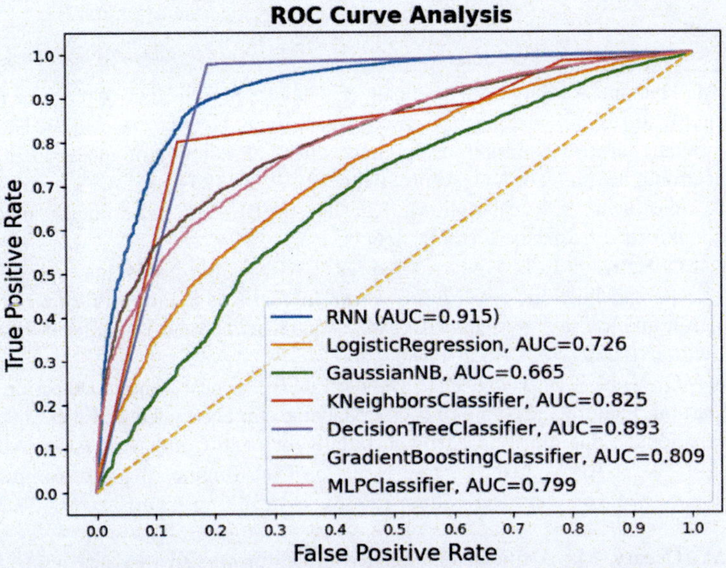

Fig. 3. Light models Receiver operating characteristic curve comparison

lightweight model had an average elapsed time of 0.037 seconds, while the Full model had an average elapsed time of 0.045 seconds.

5 Conclusion

In conclusion, this study has demonstrated the effectiveness of deep learning algorithms, specifically RNN, in predicting no-show appointments in a hospital setting. Utilizing patient demographic and appointment-related features, the RNN algorithm could accurately capture the complex relationships and dependencies in the data, resulting in high-performance metrics such as ROC of 0.93, the precision of 0.95, and an F1-score of 0.85. The comparison with other studies also suggests that the RNN algorithm used in this study outperformed other machine learning algorithms regarding accuracy and AUC. Furthermore, the study highlights the importance of considering multiple evaluation metrics when assessing the performance of prediction algorithms and the need for careful consideration of feature-importance methods in certain situations. The theoretical implications of this study are significant as it demonstrates the potential of DL techniques to analyze complex and sequential data in healthcare settings. These findings have practical implications for healthcare organizations, as the accurate prediction of no-show appointments can lead to improved efficiency and resource management.

Acknowledgment. The authors would like to thank Eastern International University (EIU) and Becamex International Hospital (BIH) Vietnam for funding this research.

References

1. Javaid, M., Haleem, A., Singh, R.P., Suman, R., Rab, S.: Significance of machine learning in healthcare: features, pillars and applications. Int. J. Intell. Netw. **3**, 58–73 (2022)
2. Lin, Q., Betancourt, B., Goldstein, B.A., Steorts, R.C.: Prediction of appointment no-shows using electronic health records. J. Appl. Stat. **47**(7), 1220–1234 (2020)
3. Dash, S., Shakyawar, S.K., Sharma, M., Kaushik, S.: Big data in healthcare: management, analysis and future prospects. J. Big Data **6**(1), 1–25 (2019)
4. Nguyen, M., Nguyen, T.-T., Dinh, V., Jeon, J.W.: Real-time pedestrian detection using a support vector machine and stixel information. In: 2017 17th International Conference on Control, Automation and Systems (ICCAS), Jeju, Korea (South), pp. 1350–1355 (2017). https://doi.org/10.23919/ICCAS.2017.8204203
5. Nguyen, V.D., Tran, T.H., Dang, D.T., Debnath, N.C.: Robust Vehicle Detection by Using Deep Learning Feature and Support Vector Machine. In: Hassanien, A.E., et al. (eds.), The 3rd International Conference on Artificial Intelligence and Computer Vision (AICV2023), March 5–7, 2023. AICV 2023. Lecture Notes on Data Engineering and Communications Technologies, vol 164. Springer, Cham (2023). https://doi.org/10.1007/978-3-031-27762-7_14
6. Chen, M., Decary, M.: Artificial intelligence in healthcare: an essential guide for health leaders. Healthc. Manage. Forum **33**(1), 10–18 (2020). SAGE Publications Sage CA: Los Angeles, CA
7. Jayaraman, P.P., Forkan, A.R.M., Morshed, A., Haghighi, P.D., Kang, Y.-B.: Healthcare 4.0: a review of frontiers in digital health. WIREs Data Min. Knowl. Disc. **10**(2), e1350 (2020)
8. Wen, D., et al.: Characteristics of publicly available skin cancer image datasets: a systematic review. Lancet Digit. Health **4**(1), e64–e74 (2021)
9. Nguyen, N.H., Pham, H.H., Chan, T.T., Nguyen, T.N., Nguyen, H.Q.: VinDr-PCXR: An open, large-scale chest radiograph dataset for interpretation of common thoracic diseases in children. arXiv preprint arXiv:2203.10612 (2022)
10. Ngiam, K.Y., Khor, I.W.: Big data and machine learning algorithms for health-care delivery. Lancet Oncol. **20**(5), e262–e273 (2019)
11. Stevens, L.M., Mortazavi, B.J., Deo, R.C., Curtis, L., Kao, D.P.: Recommendations for reporting machine learning analyses in clinical research. Circ. Cardiovasc. Qual. Outcomes **13**(10), e006556 (2020)
12. He, L., Madathil, S.C., Oberoi, A., Servis, G., Khasawneh, M.T.: A systematic review of research design and modeling techniques in inpatient bed management. Comput. Ind. Eng. **127**, 451–466 (2019)
13. Elvira, C., Ochoa, A., Gonzalvez, J.C., Mochón, F.: Machine-learning-based no show prediction in outpatient visits. Int. J. Interact. Multimedia Artif. Intell. **4**(7), 29 (2018)
14. Simsek, S., Tiahrt, T., Dag, A.: Stratifying no-show patients into multiple risk groups via a holistic data analytics-based framework. Decis. Support Syst. **132**, 113269 (2020)
15. Carreras-García, D., Delgado-Gómez, D., Llorente-Fernández, F., Arribas-Gil, A.: Patient no-show prediction: a systematic literature review. Entropy **22**(6), 675 (2020)
16. Jeffin Joseph, S., Senith, A.A., Kirubaraj, S.R., Ramson, J.: Machine learning for prediction of clinical appointment no-shows. Int. J. Math. Eng. Manage. Sci. **7**(4), 558–574 (2022)
17. Mohammadi, I., Wu, H., Turkcan, A., Toscos, T., Doebbeling, B.N.: Data analytics and modeling for appointment no-show in community health centers. J. Prim. Care Community Health **9**, 2150132718811692 (2018)
18. Luiz, H.A., et al.: Application of machine learning techniques to predict a patient's no-show in the healthcare sector. Future Internet **14**(1), 3 (2021)

19. Ahmad, M.U., Zhang, A., Mhaskar, R.: A predictive model for decreasing clinical no-show rates in a primary care setting. Int. J. Healthc. Manage. **14**(3), 829–836 (2021)
20. Ahmadi, E., Garcia-Arce, A., Masel, D.T., Reich, E., Puckey, J., Maff, R.: A metaheuristic-based stacking model for predicting the risk of patient no-show and late cancellation for neurology appointments. IISE Trans. Healthc. Syst. Eng. **9**(3), 272–291 (2019)
21. Qureshi, Z., et al.: Efficient prediction of missed clinical appointment using machine learning. Comput. Math. Methods Med. **2021**, 1–10 (2021)
22. Krishnan, U., Sangar, P.: A rebalancing framework for classification of imbalanced medical appointment no-show data. J. Data Inf. Sci. **6**(1), 178–192 (2021)
23. Dantas, L.F., Fleck, J.L., Cyrino, F.L., Oliveira, S.H.: No-shows in appointment scheduling – a systematic literature review. Health Policy **122**(4), 412–421 (2018)
24. Aung, Y.Y.M., Wong, D.C.S., Ting, D.S.W.: The promise of artificial intelligence: a review of the opportunities and challenges of artificial intelligence in healthcare. Br. Med. Bull. **139**(1), 4–15 (2021)
25. Dashtban, M., Li, W.: Predicting non-attendance in hospital outpatient appointments using deep learning approach. Health Syst. **11**(3), 189–210 (2021)
26. Howard, J.: Artificial intelligence: implications for the future of work. Am. J. Ind. Med. **62**(11), 917–926 (2019)
27. Holzinger, A., Jurisica, I.: Knowledge discovery and data mining in biomedical informatics: the future is in integrative, interactive machine learning solutions. In: Holzinger, A., Jurisica, I. (eds.) Interactive Knowledge Discovery and Data Mining in Biomedical Informatics, pp. 1–18. Springer Berlin Heidelberg, Berlin, Heidelberg (2014). https://doi.org/10.1007/978-3-662-43968-5_1
28. Dang, T.H., Nguyen, T.A., Van, M.H., Santin, O., Tran, O.M.T., Schofield, P.: Patient-centered care: transforming the health care system in Vietnam with support of digital health technology. J. Med. Internet Res. **23**(6), e24601 (2021)
29. Oanh, T.T., Tung, N.T.: Predicting hypertension based on machine learning methods: a case study in Northwest Vietnam. Mob. Netw. Appl. **27**(5), 2013–2023 (2022)
30. Tuan, M.N.D., Thanh, N.N., Le Tuan, L.: Applying a mindfulness-based reliability strategy to the internet of things in healthcare – a business model in the Vietnamese market. Technol. Forecast. Soc. Change **140**, 54–68 (2019)
31. Cottrell, M., et al.: Sustaining allied health telehealth services beyond the rapid response to COVID-19: learning from patient and staff experiences at a large quaternary hospital. J. Telemedicine Telecare **27**(10), 615–624 (2021)
32. Pham, Q.-V., Nguyen, D.C., Huynh-The, T., Hwang, W.-J., Pathirana, P.N.: Artificial intelligence (AI) and big data for coronavirus (COVID-19) pandemic: a survey on the state-of-the-arts. IEEE Access **8**, 130820 (2020)
33. Anthony, B.: Implications of telehealth and digital care solutions during COVID-19 pandemic: a qualitative literature review. Inf. Health Soc. Care **46**(1), 68–83 (2021)
34. Gromisch, E.S., Turner, A.P., Leipertz, S.L., Beauvais, J., Haselkorn, J.K.: Who is not coming to clinic? a predictive model of excessive missed appointments in persons with multiple sclerosis. Multiple Sclerosis Relat. Disord. **38**, 101513 (2020)
35. Youn, S.H., Geismar, N., Pinedo, M.: Planning and scheduling in healthcare for better care coordination: Current understanding, trending topics, and future opportunities. Prod. Oper. Manage. **31**(12), 4407–4423 (2022)
36. Li, Y., Tang, S.Y., Johnson, J., Lubarsky, D.A.: Individualized no-show predictions: effect on clinic overbooking and appointment reminders. Prod. Oper. Manage. **28**(8), 2068–2086 (2019)
37. Topuz, K., Uner, H., Oztekin, A., Yildirim, M.B.: Predicting pediatric clinic no-shows: a decision analytic framework using elastic net and Bayesian belief network. Ann. Oper. Res. **263**(1), 479–499 (2018)

38. Suk, M.Y., Kim, B., Lee, S.G., You, C.H., Kim, T.H.: Evaluation of patient no-shows in a Tertiary Hospital: focusing on modes of appointment-making and type of appointment. Int. J. Environ. Res. Public Health **18**(6), 3288 (2021)
39. Amberger, C, Schreyer, D.: What do we know about no-show behavior? a systematic, interdisciplinary literature review. J. Econ. Surv. (2022)
40. Tsai, W.-C., Lee, W.-C., Chiang, S.-C., Chen, Y.-C., Chen, T.-J.: Factors of missed appointments at an academic medical center in Taiwan. J. Chin. Med. Assoc. **82**(5), 436–442 (2019)
41. Guck, T.P., Guck, A.J., Brack, A.B., Frey, D.R.: No-show rates in partially integrated models of behavioral health care in a primary care setting. Families Syst. Health **25**(2), 137 (2007)
42. Cheung, D.L., Sahrmann, J., Nzewuihe, A., Espiritu, J.R.: No-show rates to a sleep clinic: drivers and determinants. J Clin. Sleep Med. **16**(9), 1517–1521 (2020)
43. Elkhider, H., et al.: Predictors of no-show in neurology clinics. Healthcare **10**(4), 599 (2022)
44. Rosenbaum, J.I., Mieloszyk, R.J., Hall, C.S., Hippe, D.S., Gunn, M.L., Bhargava, P.: Understanding why patients no-show: observations of 2.9 million outpatient imaging visits over 16 years. J. Am. Coll. Radiol. **15**(7), 944–950 (2018)
45. Liu, D., et al.: Machine learning approaches to predicting no-shows in pediatric medical appointment. NPJ Digit. Med. **5**(1), 1–11 (2022)
46. Dantas, L.F., Hamacher, S., Cyrino, F.L., Oliveira, S.D.J., Barbosa, F.V.: Predicting patient no-show behavior: a study in a bariatric clinic. Obesity Surg. **29**(1), 40–47 (2018)
47. Christodoulou, E., Ma, J., Collins, G.S., Steyerberg, E.W., Verbakel, J.Y., Van Calster, B.: A systematic review shows no performance benefit of machine learning over logistic regression for clinical prediction models. J. Clin. Epidemiol. **110**, 12–22 (2019)
48. Lekham, L.A., Wang, Y., Hey, E., Lam, S.S., Khasawneh, M.T.: A multi-stage predictive model for missed appointments at outpatient primary care settings serving rural areas. IISE Trans. Healthc. Syst. Eng. **11**(2), 79–94 (2021)
49. Al Rafi, A.S., Rahman, T., Al Abir, A.R., Rajib, T.A., Islam, M., Mukta, M.S.H.: A new classification technique: random weighted LSTM (RWL). In: 2020 IEEE Region 10 Symposium (TENSYMP), pp. 262–265. IEEE (2020)
50. Periyathambi, N., et al.: Machine learning prediction of non-attendance to postpartum glucose screening and subsequent risk of type 2 diabetes following gestational diabetes. PLoS ONE **17**(3), e0264648 (2022)

Empowering Individuals with Visual Impairments: A Deep Learning-Based Smartphone Navigation Assistant

Fatema A. Shawki, Mariem Mahfouz, Mohamed A. Abdelrazek, and Gehad Ismail Sayed(✉)

School of Computer Science, Canadian International College (CIC), New Cairo, Cairo, Egypt
Gehad_Sayed@cic-cairo.com

Abstract. Long white canes and other technological aids are frequently used by visually impaired people to identify and avoid obstacles ahead and hazards. Due to their general knowledge of everything's location, visually impaired individuals can move freely in their homes without much assistance. However, they encounter more challenges and risk harm when they roam the streets. To help visually impaired individuals navigate the streets independently and safely, this paper proposes a deep learning-based smartphone navigation assistant system. The backend and frontend are the two main components. On the front end, the images are captured by utilizing the mobile camera. The backend is fed with these captured images. A You Only Look Once (YOLOv8) deep learning architecture is used in the backend, followed by a rule-based model. Finally, a set of pre-recorded audio messages that contain navigational guidance is returned to the user. The deep-learning architecture is trained and fine-tuned on a dataset gathered from five different sources. The experimental results showed that the proposed model can be effectively used to help people who are blind. Additionally, the outcomes demonstrated that YOLOv8 achieved the best outcomes when compared to other deep-learning architectures. The proposed system achieved a 97% overall accuracy.

Keywords: Visual impairment · Mobile application · Deep learning · Rule-based modeling · YOLOv8

1 Introduction

According to the World Health Organization (WHO) and the National Library of Medicine (NCIB) in 2015 [1], there are 217 million people worldwide who have Moderate to Severe Visual Impairment (MSVI) and 36 million people who are blind. Unfortunately, these numbers are rising. Daily struggles for people with visual impairments include difficulty walking around on their own [2]. They typically use a range of traditional navigational aids, such as white canes, guide dogs, and tactile pavement [3]. These tools have some drawbacks despite being helpful. For example, white canes can detect obstacles on the ground but not those above waist level or elevational variations.

A. Hassanien et al. (Eds.): AISI 2023, LNDECT 184, pp. 19–30, 2023.
https://doi.org/10.1007/978-3-031-43247-7_2

Better mobility may be possible with guide dogs, but they are expensive to maintain and need specific training and care. Although tactile pavement can provide directional instructions, it can be difficult to distinguish between different patterns and isn't always present. Additionally, due to the lighting and layout changes, these traditional tools are not convenient for indoor navigation, which can be challenging for people with visual impairments.

Thus, there is a need for new technologies and solutions that can improve on currently used conventional techniques, making navigation for those with visual impairments simpler and less constrained. Researchers have recently begun to pay more attention to the use of smartphones and Artificial Intelligence (AI) navigation systems to increase the independence and mobility of people with visual impairments [4]. For the development of reliable mobile applications, computer vision techniques for object recognition, such as Machine Learning (ML) and Deep Learning (DL), have become essential [5]. Both ML and DL methods can be applied, but DL is proved to be more effective because it can handle large datasets and automatically extract features from training data [6, 7]. DL models are an efficient choice for creating navigation systems for those who are blind or visually impaired because they are adept at handling challenging problems and managing complex data [8].

Researchers can create new, revolutionary solutions that can significantly improve the lives of people with visual impairments by utilizing the power of smartphones and AI-based navigation systems. These technologies have the potential to improve navigation, increase independence, and improve the quality of life for people with visual impairments [9]. Systems for helping visually impaired people with various tasks have been proposed. They use ML, Natural Language Processing (NLP), and computer vision technologies. The authors of [10] proposed a mobile application that enables visually impaired users to take advantage of and enjoy tourist services. It provides location-based information, like nearby restaurants, attractions, and public transportation using Global Positioning System (GPS). In addition to bus schedules and audio descriptions of tourist attractions. This system was created and tested on Android devices and the system's usability was evaluated by user surveys, and the findings indicated that it was effective at guiding visitors around tourist attractions and locating services close by. The authors of [11] proposed another system to help blind people with daily tasks and activities, the authors suggest an AI-based visual aid. The proposed system provides audio descriptions and readouts in addition to using computer vision and NLP technologies to recognize objects and read texts. The computer vision algorithm was implemented using Convolutional Neural Networks (CNNs), the CNN was trained on large datasets to recognize different object categories, including people, animals, vehicles, and furniture. While, the text recognition algorithm used in the system is based on Optical Character Recognition (OCR) technology, which involves detecting texts in images and audio readouts of this text. System evaluation was based on user studies, to demonstrate its effectiveness.

An AI-based edge computing-based assistive system has been proposed in [12] to improve the safety of visually impaired people at zebra crossings. The system makes use of machine learning models and computer vision algorithms. The object detection model, YOLOv3, which was trained on a custom dataset of pedestrian and zebra crossing images, is the foundation of the computer vision algorithm that is being used. The Support Vector

Machine (SVM) classifier-based ML model assesses the level of pedestrian risk and was trained using a set of features taken from the pedestrian images. The ML classifier calculates the risk based on the pedestrian's position, orientation, and distance from the zebra crossing. Audio alerts are used to inform users of the level of risk associated with the crossing. The authors additionally proposed extra features like a voice-based user interface and Text-To-Speech (TTS) functionality. A customized test set of zebra crossings and pedestrian images was used to assess the proposed system. Detecting pedestrians with an overall accuracy of 91.7% and estimating the degree of pedestrian risk with an accuracy of 87.5% were both attained. The outcomes demonstrated that the system could reliably identify pedestrians without overly raising false alarms. Another area of research involves indoor navigation for people who are blind. The authors of [13] suggest a system that aids visually impaired people in indoor navigation by using computer vision and machine learning algorithms. The system makes use of the camera on a mobile phone to identify various obstructions, such as doors and stairs, and it also provides audio instructions to guide the user and warn them when necessary. Overall, the system was 94.2% accurate in detecting doors and accuracy of 91.6% in detecting obstacles.

This system is proposed in [14] as a mobile application that focuses on obstacle detection and outdoor/indoor navigation. The authors employ machine learning models produced by TensorFlow Lite for object detection. Input from users is also taken by the proposed system using voice recognition algorithms. Making calls, sending SMS, and directing the user to their destination are just a few of the tasks it can carry out by searching for the user's command in a predefined audio dictionary. The application underwent performance testing, and the results were encouraging. The average response time was 7.629 ms, with an average standard deviation of 1.98 ms. Additionally, the usability test results showed that the system is suitable for actual implementation. An indoor/outdoor guiding system for the blind was suggested by authors in [15]. It makes use of smartphones. The system allows the smartphone to function both online and offline by connecting it to a server. For object detection in the online mode, Faster-RCNN and YOLO are used. The Histogram of Oriented Gradients (HOG) module is used, however, in offline mode. Position and distance modules are present in both models. When an obstacle is found, the system alerts the user. After testing the application, the majority of users affirmed its effectiveness and efficiency.

Another model is a web-based system that makes use of object detection and a narrator for people who are blind or visually impaired, as proposed in [16]. The system uses the phone's camera to take pictures, which it then sends to the object detection model YOLOv3. Following that, the model identifies the objects in the image and reports its findings. The results are distributed using a browser-based voice library (Google Voice). The system can be installed on any gadget with a camera and can classify up to 1000 objects. Eight actual objects were used in the evaluation. Nevertheless, the accuracy outcomes varied depending on the mobile phone used. For instance, while the majority of other phones detected only 5 objects with 62.5% accuracy, some phones detected up to 6 out of the actual 8 objects with 75% accuracy. The authors of this paper proposed an intelligent Android mobile application to help the blind in many ways [17]. The system is created to have as many beneficial features for people with visual impairments as possible in a single application. These capabilities include the ability to operate offline without an internet connection as well as light detection, color detection, object recognition using CNNs, and banknote recognition. The OpenCV library and the back camera are both used by the color detection feature to recognize colors and report the names of those colors to the user. The top five matches and their associated probabilities are listed by the object recognition model. To interact with and help the user, the app provides spoken descriptions and audible cues. A questionnaire is used to conduct the evaluation. The users' collective rating was 4.02 out of 10.

This paper proposes a deep learning-based smartphone navigation assistant system to help the visually impaired walk through the street. The proposed system comprises two primary parts; frontend and backend. In the frontend, the system starts by capturing an image from the mobile's camera. Then this image is sent to the backend to be analyzed using YOLOv8 deep-learning architecture and the proposed rule-based model. The output from the backend is pre-recorded audio messages that contain the navigation instructions. It has been shown from the previous studies, that using audio instructions instead of using Braille text messages. The main contribution of the paper can be summarized as follows:

1. The proposed system can provide visually impaired individuals with real-time feedback about their surroundings, enabling them to navigate the streets as if they were sighted.
2. The proposed system can achieve high-accuracy results with minimal hardware requirements. To be more accessible to individuals who may not have access to the latest technology
3. The proposed system makes use of YOLOv8 which is converted to Tflite for mobile use, providing lightweight and efficient performance.
4. The proposed system can be used in offline mode, improving its accessibility and availability for users.

The rest of the paper structure is organized as follows; Section 2 describes the adopted dataset. Section 3 introduces the overall proposed deep learning-based smartphone navigation assistant system based on the YOLO deep-learning architecture and the rule-based model. Section 4 discusses the obtained results. Finally, Section 5 proposes the paper's main finding regarding the future direction.

2 Dataset Description

The model's performance is highly dependent on the quality of the training dataset. There are seven classes in the adopted dataset: carton, metal, plastic, glass, human, vehicle, and tree. There are 700 images in each class. The classes were gathered from several sources. Glass, metal, plastic, and cartons from [18, 19], humans from [20, 21], vehicles from [21, 21], and trees from [18, 18]. Each image has been manually annotated to match the appropriate class. An example from each class in our dataset is shown in Fig. 1.

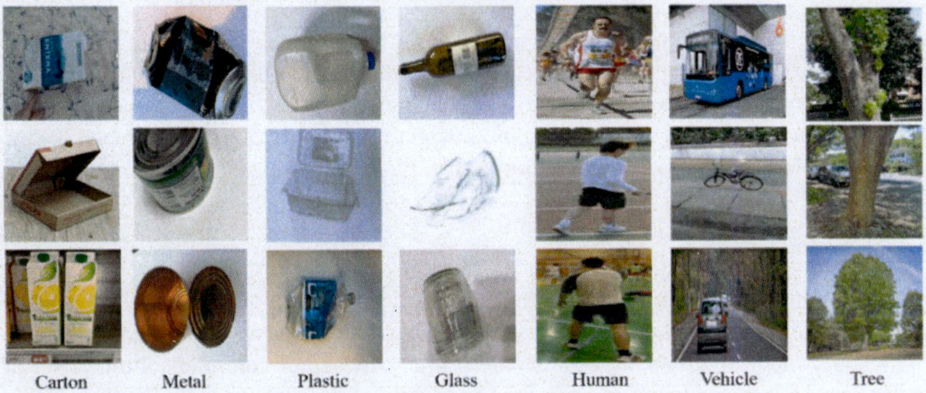

Fig. 1. Sample of the adopted dataset.

3 The Proposed Deep Learning-Based Smartphone Navigation Assistant System

The overall system architecture, significant elements, and features are discussed in this section. The frontend and the backend, which are the two main components of the deep learning-based smartphone navigation assistant system, are shown in Fig. 2. It works by identifying obstacles in the road and sending the user navigational instructions. The suggested system operates independently of an internet connection. As a result, users can explore without worrying about losing connectivity. The system begins by capturing an image with the mobile device's camera, after which it sends the image to the backend for analysis. The captured image is fed into a deep-learning architecture in the backend. A list

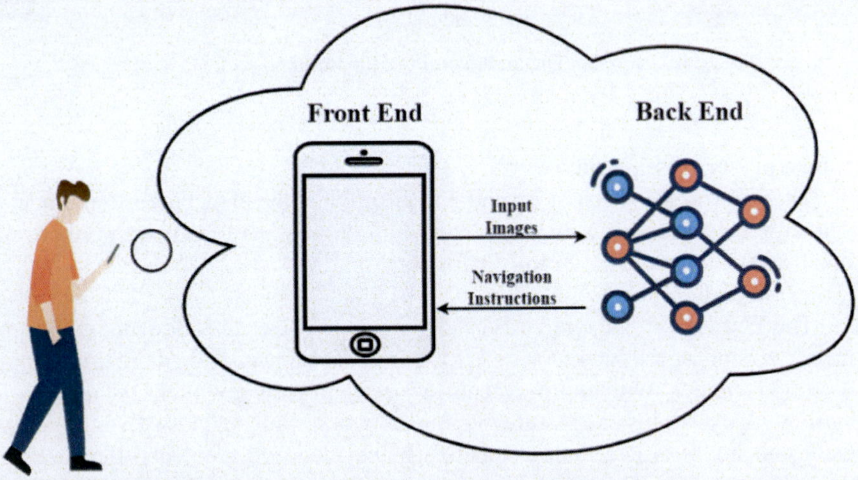

Fig. 2. The architecture of the proposed deep learning-based smartphone navigation assistant system.

of the image's detected objects serves as the step's output. Based on their area and center coordinates, these objects are then processed further. The proposed rule-based model uses the extracted parameters to determine whether it is safe to move in that direction. Finally, an audio message containing the navigational instructions is produced by the proposed rule-based model. Then a thorough explanation of each component.

3.1 Frontend

This component has a straightforward user interface with few buttons. It is intended to improve usability and user-friendliness. Since hearing is a visually impaired person's main sense, audio outputs are used as the primary means of communication. A set of instructions are shown to the user once they begin using the proposed mobile application. The main goal of these instructions is to show the user how to set the mobile camera correctly to improve dependability. Images are continuously being taken by the camera and sent to the backend.

3.2 Backend

The proposed deep learning-based smartphone navigation assistant system's functionality and performance are determined by this component. The proposed AI model in the backend is divided into three main sections: object detection, image pre-processing, and a rule-based model. The backend architecture is depicted in Fig. 3.

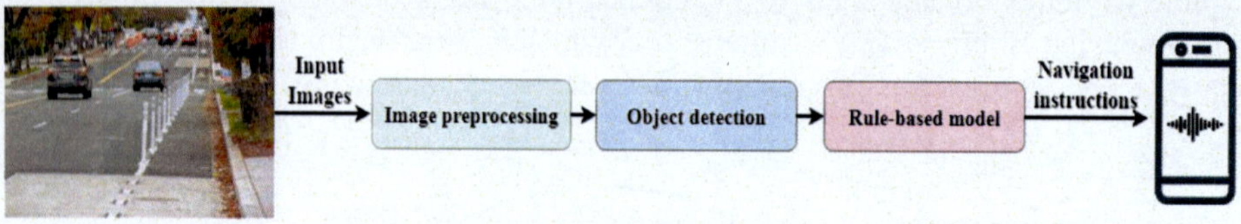

Fig. 3. The architecture of the backend part.

A. **Image pre-processing phase**

The preprocessing step is essential to ensure that the object detection model can detect objects effectively. In this phase, the images are resized and normalized to pixel values, to match the YoloV8's format.

B. **Object detection phase**

The YOLOv8 architecture was used to train the object detection model. The most recent version of the YOLO series, YOLOv8, is renowned for its real-time object detection features. With improvements in speed and accuracy, YOLOv8 incorporates cutting-edge methods like feature pyramid networks and focal loss. Even in complex environments, it is excellent at localizing and identifying objects. Because of its unmatched speed, accuracy, and cutting-edge object detection capabilities, YOLOv8 is now frequently used. It was initially set up with pre-trained weights and then tuned using our unique dataset. This model generates *x, y, w, h*, a confidence score for each box, and *N* class probabilities as output. Where *(x, y)* are the bounding box's center coordinates. *N* is the number of classes in the dataset, with w being the bounding box's width, and *h* its height. The best weights are converted to TensorFlow Lite after the model has been trained and tested. Because of the size reduction and accuracy preservation, the model is now simpler to deploy on the mobile application.

C. **Rule-based model phase**

After objects are located, a set of features is extracted from the captured image. The bounding box area and the center coordinates are examples of these features. The proposed rule-based model is constructed using these features.

Algorithm 1: Rule-based modeling pseudo code

```
Get objects in the image using the detection model
for obj in a set of objects in the image:
  if obj.confidence > 0.85 &  obj.ycenter <= 0:
      filtered_objects.append(obj)
    Add the object to the filtered_objects list if it passes both filters
    Select the top three objects with the highest area
    for obj in top_three_objects:
      if obj.class != 'carton':
      xcenter = obj.xcenter
      left_point = (xcenter - obj.width/2, obj.ycenter)
      right_point = (xcenter + obj.width/2, obj.ycenter)
      if no_objects_on_left(left_point, top_three_objects):  play_audio_message('go left')
      elseif no_objects_on_right(right_point, top_three_objects):  play_audio_message('go right')
      else:  play_audio_message('walk away')
    end
end
```

The objects are first restricted to those with confidence scores higher than 0.85 and negative y-center values (since these are the objects closest to the user). Then, among the previous filtered objects, the first three objects with the largest area values are chosen. There are two points other than the center for each bounding box of each of these three objects: one on the left and one on the right. The safe direction to move is identified using these three points. The user is then informed that there is an obstacle ahead and which way they should move to avoid it using a pre-recorded audio message. The proposed rule-based model's pseudo-code is displayed in Algorithm 1.

4 Experimental Results and Discussion

In this section, several experiments were conducted to evaluate the overall performance of the proposed deep learning-based smartphone navigation assistant system. The minimum hardware requirement for the used mobile device is Android version 11. Additionally, the minimum requirements for the used computer are a Core i7, 16 Gigabytes RAM with Nvidia GTX 1650 4 Gigabytes.

The first experiment in Fig. 4 examined the effect of batch size on the performance of the same two deep-learning architectures: YOLOv8 and YOLOv5. The Mean Average Precision (mAP) metric was used for this experiment. The batch sizes tested were 32, 48, and 64. As can be observed, YOLOv8 with a batch size of 48 obtained the best results compared to 32 and 64. Additionally, the results revealed that YOLOv8 is superior compared to its old version.

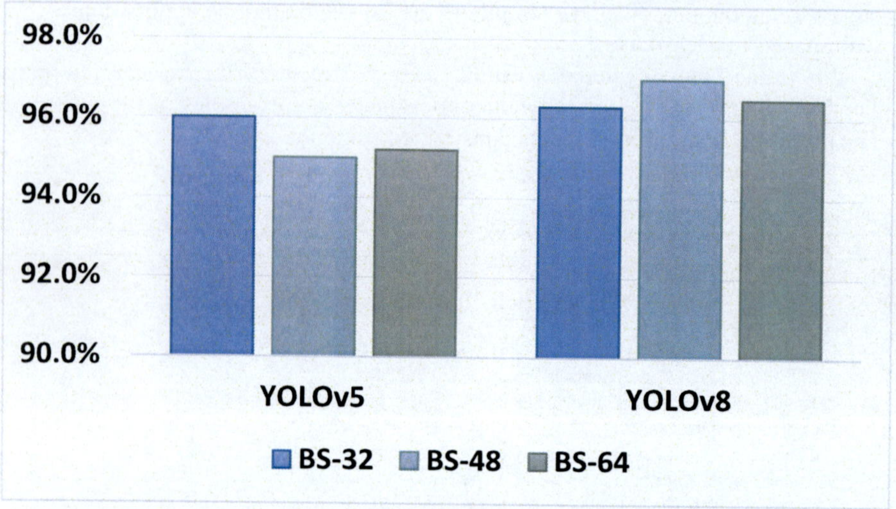

Fig. 4. The performance of YOLOv5 and YOLOv8 using different batch sizes.

Another experiment conducted in Table 1 compares the performance of two well-known deep-learning architectures previously used in the previous studies. These deep-learning architectures are YOLOv8 and YOLOv5. In this experiment, several measurements are used. These measurements are accuracy, precision, recall, and f1-score. As can be observed, the proposed system using YOLOv8 obtained the best results compared with the other deep-learning architectures.

Table 1. The classification performance of the proposed system using YOLOv8 and YOLOv5.

	YOLO v8	YOLO v5
Accuracy (%)	98	90
Precision (%)	90	97
Recall (%)	90	73
F1 score (%)	91	75

Computational time is another important factor to consider when comparing deep learning architectures. A comparison of computational time is conducted with the YOLOv5 and YOLOv8 deep learning architectures as shown in Fig. 5. In this figure, YOLOv8 obtained the highest accuracy with a reasonable computational time compared to YOLOv5.

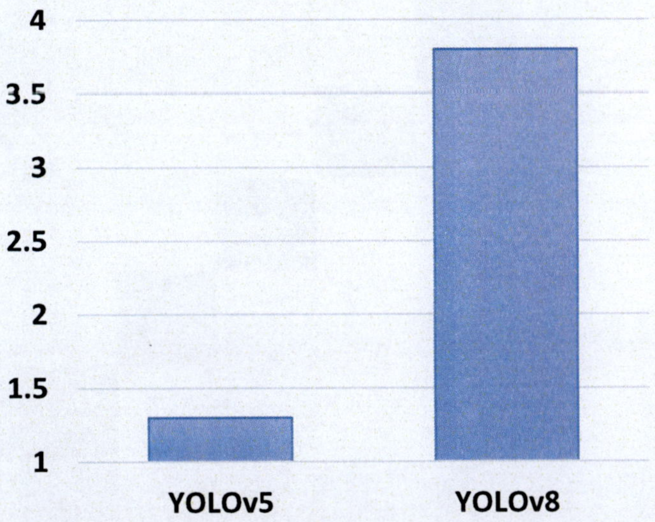

Fig. 5. Computational time of YOLOv5 and YOLOv8.

Figure 6 shows the confusion matrix of the proposed system using YOLOv8. Confusion matrices compare the classification predicted classes with the actual classes to evaluate the model's performance. As can be seen, the proposed system is superior, as it can differentiate between all the seven classes efficiently.

For further evaluation of the proposed deep-learning-based smartphone navigation assistant system, another experiment in Table 2 is conducted. In this experiment, the proposed system is compared with systems state-of-art, these systems utilize the same architecture. As can be observed, the proposed system obtained results with the highest accuracy. Additionally, from these results, it can be revealed that the proposed system can be effectively applied to help visually impaired people navigate.

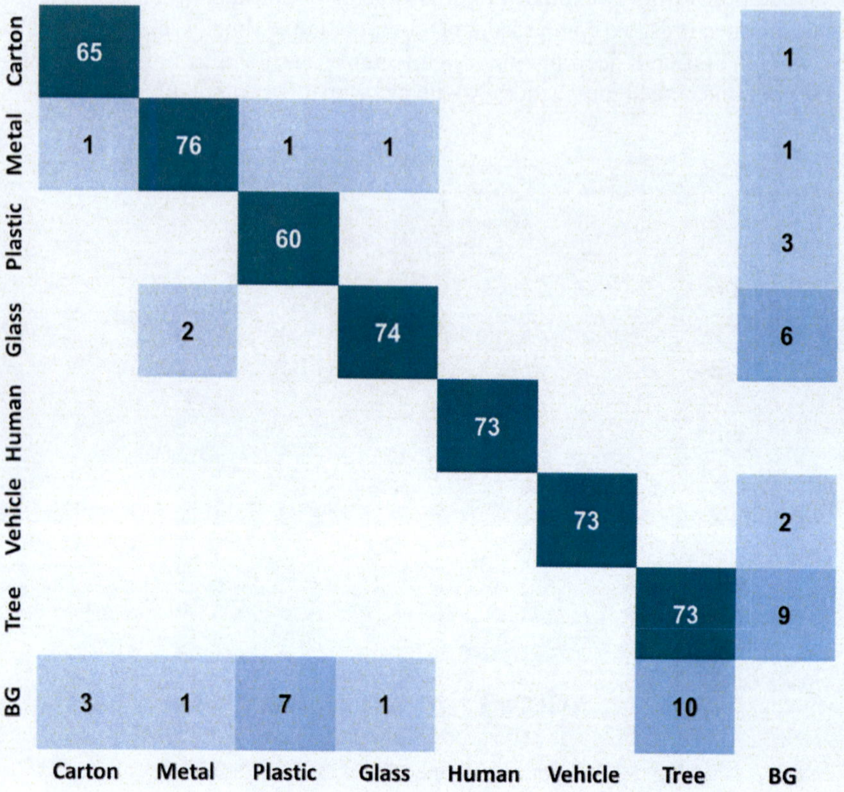

Fig. 6. The confusion matrix of the proposed system.

Table 2. The performance of the proposed system vs. the state-of-art systems.

	Number of Classes	Accuracy (%)
Real-Time Face Mask Detection [23]	2	75
Yolo on Mask Detection Task [24]	3	89
Object Detection to separate recyclables [25]	4	91
The Proposed System	**7**	**97**

5 Conclusion

This paper proposed a smartphone navigation system that uses deep learning to help blind people avoid obstacles. The frontend and backend are the system's two main building blocks. It is simple to use because the frontend application is made with a user-friendly interface. The backend uses the YOLOv8 deep learning architecture and rule-based modeling to precisely identify and classify obstacles in real-time and provide audio messages for navigational instructions. The YOLOv8 model is transformed into a portable and lightweight Tflite. This makes it possible for the system to function without an internet connection, enhancing its usability and accessibility. Additionally, the results revealed that the proposed system can offer real-time obstacle identification and auditory guidance for secure navigation using deep learning and rule-based modeling. Moreover, the proposed system is more convenient and accessible as a result of utilizing the mobile application, which has an easy-to-use user interface and offline capabilities, making it a useful tool for daily use.

While the proposed system has shown promising results in aiding visually impaired individuals to navigate the streets safely, there is still room for further improvement and development. Future work includes adding a GPS feature to allow users to obtain location information and plan routes. Making the application available for iOS and expanding language support. Also, other deep-learning architectures can be tested such as YOLO-NAS. With continued development and refinement, this solution has the potential to improve the lives of many and promote greater inclusivity in society.

References

1. WHO: Visual impairment and blindness, Fact Sheet n-282. https://www.who.int/en/news-room/fact-sheets/detail/blindness-and-visual-impairment. Accessed 17 Dec 2022
2. Ackland, P., Resnikoff, S., Bourne, R.: World blindness and visual impairment: despite many successes, the problem is growing. Community Eye Health **30**, 71–73 (2018). PMC5820628, PMID: 29483748
3. Kuriakose, B., Shrestha, R., Sandnes, F.E.: Tools and technologies for blind and visually impaired navigation support: a review. IETE Tech. Rev. **39**(1), 3–18 (2022). https://doi.org/10.1080/02564602.2020.1819893
4. Walle, H., De Runz, C., Serres, B., Venturini, G.: A survey on recent advances in AI and vision-based methods for helping and guiding visually impaired people. Appl. Sci. **12**(5), 2308 (2022). https://doi.org/10.3390/app12052308
5. Wang, J., Wang, S., Zhang, Y.: Artificial intelligence for visually impaired. Displays **77**, 102391 (2023). https://doi.org/10.1016/j.displa.2023.102391
6. Zhao, Z.-Q., Zheng, P., Xu, S.-T., Wu, X.: Object detection with deep learning: a review. IEEE Trans. Neural Netw. Learn. Syst. **30**, 3212–3232 (2019). https://doi.org/10.1109/TNNLS.2018.2876865
7. Wu, X., Sahoo, D., Hoi, S.C.: Recent advances in deep learning for object detection. Neurocomputing **396**, 39–64 (2020). https://doi.org/10.1016/j.neucom.2020.01.085
8. Srivastava, S., Divekar, A.V., Anilkumar, C., Naik, I., Kulkarni, V., Pattabiraman, V.: Comparative analysis of deep learning image detection algorithms. J. Big Data **8**, 66 (2021). https://doi.org/10.1186/s40537-021-00434-w
9. Kuriakose, B., Shrestha, R., Sandnes, F.E.: Tools and technologies for blind and visually impaired navigation support: a review. In: IETE Technical Review, VOL.39, No. 1, from pp. 3–18, September 2022. https://doi.org/10.1080/02564602.2020.1819893
10. Ceccarini, C., Prandi, C.: Tourism for all: a mobile application to assist visually impaired users in enjoying tourist services. In: 2019 16th IEEE Annual Consumer Communications & Networking Conference (CCNC), Las Vegas, NV, USA, pp. 1–6 (2019). https://doi.org/10.1109/CCNC.2019.8651848

11. Khan, M.A., Paul, P., Rashid, M., Hossain, M., Ahad, M.A.R.: An AI-based visual aid with integrated reading assistant for the completely blind. IEEE Trans. Hum. Mach. Syst. **50**(6), 507–517 (2020). https://doi.org/10.1109/THMS.2020.3027534
12. Chang, W.-J., Chen, L.-B., Sie, C.-Y., Yang, C.-H.: An artificial intelligence edge computing-based assistive system for visually impaired pedestrian safety at zebra crossings. IEEE Trans. Consum. Electron. **67**(1), 3–11 (2021). https://doi.org/10.1109/TCE.2020.3037065
13. Joshi, R.C., Yadav, S., Dutta, M., Travieso, C.: Efficient multi-object detection and smart navigation using artificial intelligence for visually impaired people. Entropy **22**(9), 941 (2020). https://doi.org/10.3390/e22090941
14. Hossain, M.E., Qaiduzzaman, K.M., Rahman, M.: Sightless helper: an interactive mobile application for blind assistance and safe navigation. Lect. Notes Inst. Comput. Sci., Soc. Inf. Telecommun. Eng. **325**, 581–592 (2020). https://doi.org/10.1007/978-3-030-52856-0_46
15. Lin, B.-S., Lee, C.-C., Chiang, P.-Y.: Smart mobility for blind people using deep learning and Internet of Things. Sensors **17**(6), 1371 (2017). https://doi.org/10.3390/s17061371
16. Nasreen, J., Warsi, A., Shaikh, A.A., Muhammad, Y., Abdullah, M.: Object detection and narrator for visually impaired people. In: 2019 IEEE 6th International Conference on Engineering Technologies and Applied Sciences (ICETAS) (2019). https://doi.org/10.1109/ICETAS48360.2019.9117405
17. Awad, M., El Haddad, J., Khneisser, E., Mahmoud, T., Yaacoub, E., Malli, M.: Intelligent eye: a mobile application for assisting blind people. In: 2018 IEEE Middle East and North Africa Communications Conference (MENACOMM) (2019). https://doi.org/10.1109/MENACOMM.2018.8371005
18. Vaduva, I.-A.: Garbage classification (2021). https://www.kaggle.com/datasets/ionutandreivaduva/garbage-classification. Accessed 19 Feb 2023
19. Abla, M.: Garbage classification (2020). https://www.kaggle.com/datasets/mostafaabla/garbage-classification. Accessed 9 Jan 2023
20. Cai, Y., Tan, X.: Weakly supervised human body detection under arbitrary poses. In: International Conference on Image Processing. IEEE (2016). Accessed 9 Apr 2023
21. Rana, K.: Vehicle Dataset (2021). https://www.kaggle.com/datasets/krishrana/vehicle-dataset. Accessed 9 Apr 2023
22. Prize, E.: Tree binary segmentation (2021). https://www.kaggle.com/datasets/earthshot/tree-binary-segmentation. Accessed 9 Apr 2023
23. Jiang, X., Gao, T., Zhu, Z., Zhao, Y.: Real-time face mask detection method based on YOLOv3. Electronics **10**(7), 837 (2021). https://doi.org/10.3390/electronics10070837
24. Liu, R., Ren, Z.: Application of Yolo on mask detection task. In: 2021 IEEE 13th International Conference on Computer Research and Development (ICCRD), Beijing, China, pp. 130–136 (2021). https://doi.org/10.1109/ICCRD51685.2021.9386366
25. Wahyutama, A.B., Hwang, M.: YOLO-based object detection for separate collection of recyclables and capacity monitoring of trash bins. Electronics **11**(9), 1323 (2022). https://doi.org/10.3390/electronics11091323

Lane and Bump Detection Based on Computer Vision and Deep Learning Methods

Ahmed Magdy[1], Shrouk Abdelfattah[2], Samia Mohamed[2(✉)], Mohamed Khaled[2], Mohamed Mahmoud[2], Mohamed Ashraf[2], and Ahmed H. Eldeib[3]

[1] Electrical Engineering Department, Faculty of Engineering, Suez Canal University, Ismailia, Egypt

[2] Egyptian Chinese College for Applied Technology, Ismailia, Egypt
msamia177@gmail.com

[3] Faculty of Engineering, Electronic and Communication Department, Cairo, Egypt

Abstract. In recent years, Artificial intelligence (AI) is witnessing great progress, which in turn has led to the development of the implementation of computer vision and deep learning in real life. As accidents have been increasing due to leaving the cars off their tracks or not seeing a bump in front of the cars and cause different types of accidents caused because of the lack of attention of the drivers. Therefore, there should be technical developments to reduce the frequency of accidents and keep us safe. In This paper, we worked on both speed bump detection and lane detection. So, the inherent synergies between lane and bump detection are made use to enhance overall computing efficiency while maintaining the robustness of both tasks to fulfil the core aim of safety. Firstly, each system is worked on separately so the result can be seen of each one. For Lane Departure Warning System (LDWS), it warns the driver with a timely signal that the vehicle has left the lane. Lane boundaries are detected using a V2 camera installed at the front of the car that captures a view of the road. This proposed system effectively combines the OpenCV's canny edge algorithm with perspective warping algorithm and Histogram algorithm to improve the image. The detection speed deployed on the embedded platform Jetson Nano reaches 15 fps and it showed that the proposed plan was robust, and it can be applied on both curved and straight roads. For Speed bump detection, it warns the driver when there is a bump at a distance up to 40 m in front of the car. For bump dataset, images are extracted from videos collected manually in Ismailia streets in Egypt. Many algorithms, such as Faster R-CNN, VGG-16, MobileNet-SSD, YOLOv3 Nano, and YOLOv3 Lite-Mobilenet, are retrained on our dataset, then the results are recorded and the proposed algorithm that achieved the highest speed and accuracy with 89.17% was YOLOv3 Lite-Mobilenet.

Keywords: Computer Vision (CV) · Deep Learning (DL) · ADAS · Bump Detection · LDWS · CNN · YOLOv3

A. Hassanien et al. (Eds.): AISI 2023, LNDECT 184, pp. 31–42, 2023.
https://doi.org/10.1007/978-3-031-43247-7_3

1 Introduction

Road accidents are now considered one of the most severe problems, and most of them are due to driver negligence. In the latest research, nearly 30% of road accidents on highways are caused by lane departure, most of them caused by the driver's not concentrating [1]. According to a study published by the World Health Organization (WHO), 1.35 million human deaths were caused by motor vehicle accident injuries. Many research projects are being conducted in ADAS [2]. Two systems, Lane Departure Warning System (LDWS) and speed bump detection are focused on as they are the most important systems that help reduce accidents.

LDWS was intended to provide warning when a vehicle deviates from the driving lane line, thereby reducing the number of accidents [3]. Lane detection plays an essential role in the lane departure warning system. Image processing techniques are employed to extract lane markings from input video captured by a single camera mounted on top of the car. The appearance of these white markings is affected by a variety of circumstances, including brightness, weather, and traffic conditions [4]. Several pre-processing and detection methods have been utilized to reduce the dependence on these elements. Lane marking attributes such as color, edge, and geometry are pre-processed with a specific road surface [5].

Models based on Convolutional Neural Networks (CNN) yield the best object recognition results in real-time collected videos [6]. For the detection task, Faster R-CNN, MobileNet, YOLO, and SSD are all examples of such object detectors, which can be implemented on top of any deep architecture (which is commonly referred to in this context as the "backbone"). For example, in [7], an SSD-based MobileNet object detector and in [8], a VGG-16 based faster R-CNN object detector can be used. The better the backbone, the higher the detector's performance, because it can employ superior visual information for the task it has been taught to do. RCNN is a much older method that is significantly slower and less accurate than recent deep learning-trained object detectors.

In this paper, lane lines are detected using python as it is one of the most widely used programming languages for this purpose and OpenCV [9]. OpenCV stands for "Open-Source Computer Vision", which is a module with Many useful tools for image analysis, and NumPy [10] is used for a few simpler image processing techniques. Canny Edge Detection with Perspective Warping is used to provide the car with the required data with which it can detect the lane markings and follow lane discipline that is set [5]. The proposed system does not need any sort of huge dataset. MobileNet-based YOLOv3 is proposed to slight the training method for bump detection [11]. The dataset collected consists of 3240 hand-labelled images retrieved from camera video frames. The models were tested on our dataset labeled in a widely used and well-known object imaging dataset called PascalVOC dataset [12]. Given the diverse weather and shooting settings, the experimental findings reveal that the trained YOLO model outperforms the original model in terms of average accuracy [13].

The research paper is presented in several sections, starting with the system models in section II, the proposed methods are presented in section III, and then identifying the evaluation criteria to compare between models in section IV, and finally, the results, and the conclusion will be described in section V and VI respectively.

2 System Models

Our study depends on two ADAS system techniques as shown in Fig. 1; lane line detection and bump detection then alerting the driver to take the right action in the right time by using computer vision and deep learning.

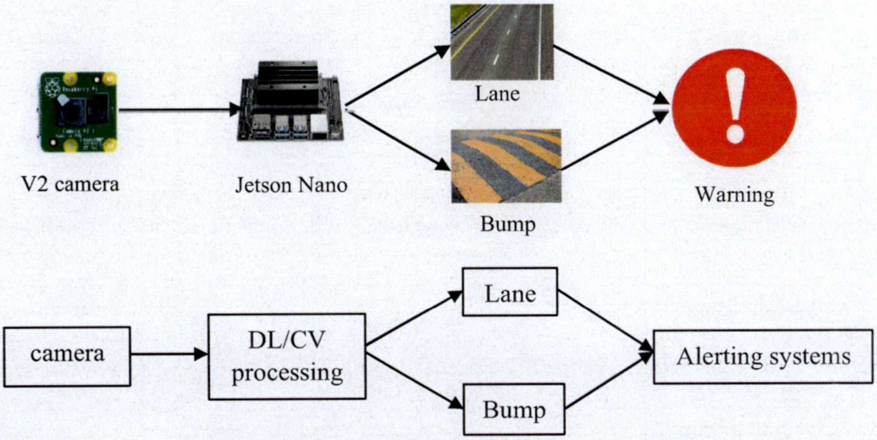

Fig. 1. System block diagram of the whole project

Advanced lane detection technique was proposed in our study by using preprocessing techniques shown in Fig. 2 as grayscale, thresholding, canny edge detection and using detecting techniques as perspective transform technique to adjust the difference between real and perceived distance from objects far from camera [14]. Then starting the search for lanes position, by using histogram. The histogram quantified the number of pixels along the 'x' axis, therefore peak found along the histogram points that represent lane lines [14]. Then detecting the car leaving its lane to give it an alert to get back to the lane by specifying the width of the lane, which represents w1 and w2 on both sides to get the right action in the right time. V2 camera is used connected to laptop having intel CPU corei5 and the detection speed on it reached 25 fps. Then the detection process is deployed on Jetson Nano reached 15 fps and it showed that the proposed plan is robust, and it can be applied on both curved and straight roads.

YOLOv3 Lite-Mobilenet model technique is proposed in bump detection as shown in Fig. 3 that consists of YOLOv3, an algorithm for real-time object detection that identifies objects in videos, live feeds, or images [15], and Mobilenet, which uses depthwise separable convolutions to build lightweight deep neural networks for the embedded platform Jetson Nano [16]. The bump is treated as an object, so the model is re-trained on a manually collected dataset labeled in PASCAL VOC format, which is a public object detection dataset consisting of 20 categories of objects [12]. About 83.3% of the dataset is used for training and about 16.7% for testing the model. The following algorithms are retrained: Faster R-CNN, VGG-16, MobileNet-SSD, YOLOv3 Nano on our bump dataset to compare them with our model. Each algorithm is a hybrid version of two different versions such as our model. Faster R-CNN, VGG-16 has achieved 151 ms

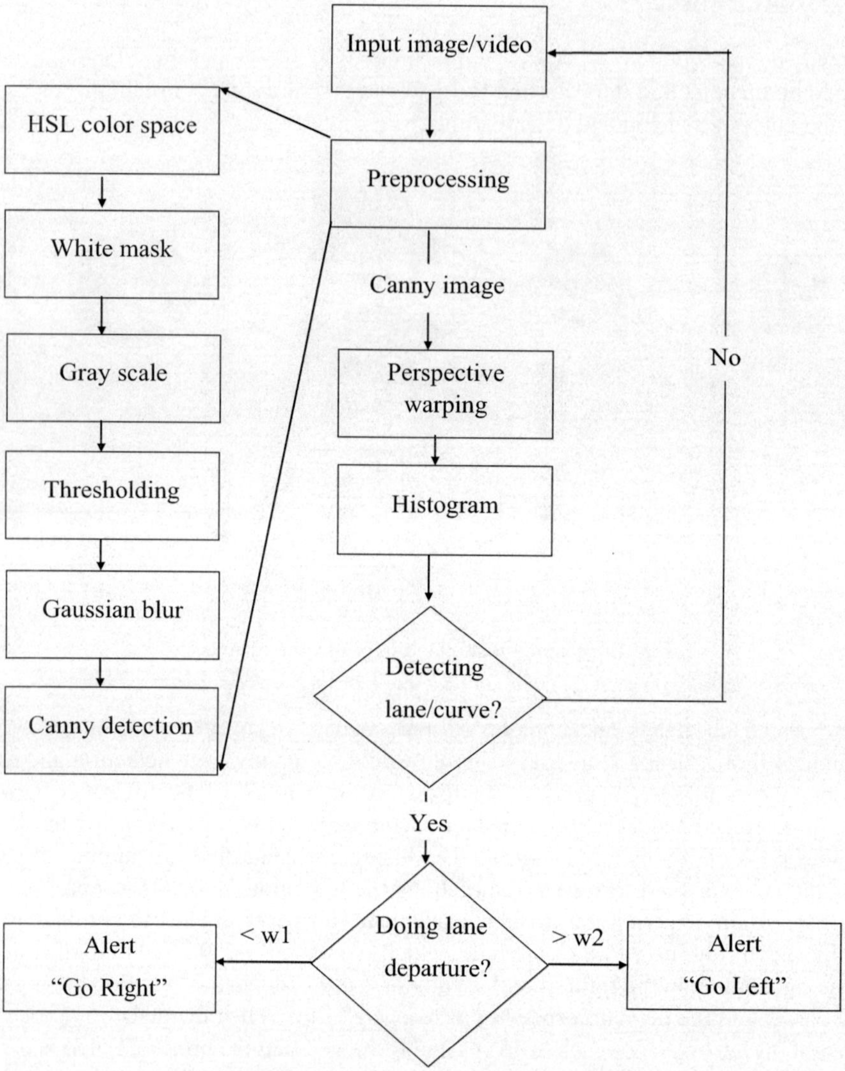

Fig. 2. Lane system flow diagram

speed, MobileNet-SSD has achieved 30 ms speed, YOLOv3 Nano has achieved 29 ms speed and our proposed model YOLOv3 Lite-Mobilenet has achieved 14.6 ms speed and those results showed that the proposed plan is robust and faster than the other versions. The distance proposed for detecting a bump and giving an alert with it is between 20 m to 40 m.

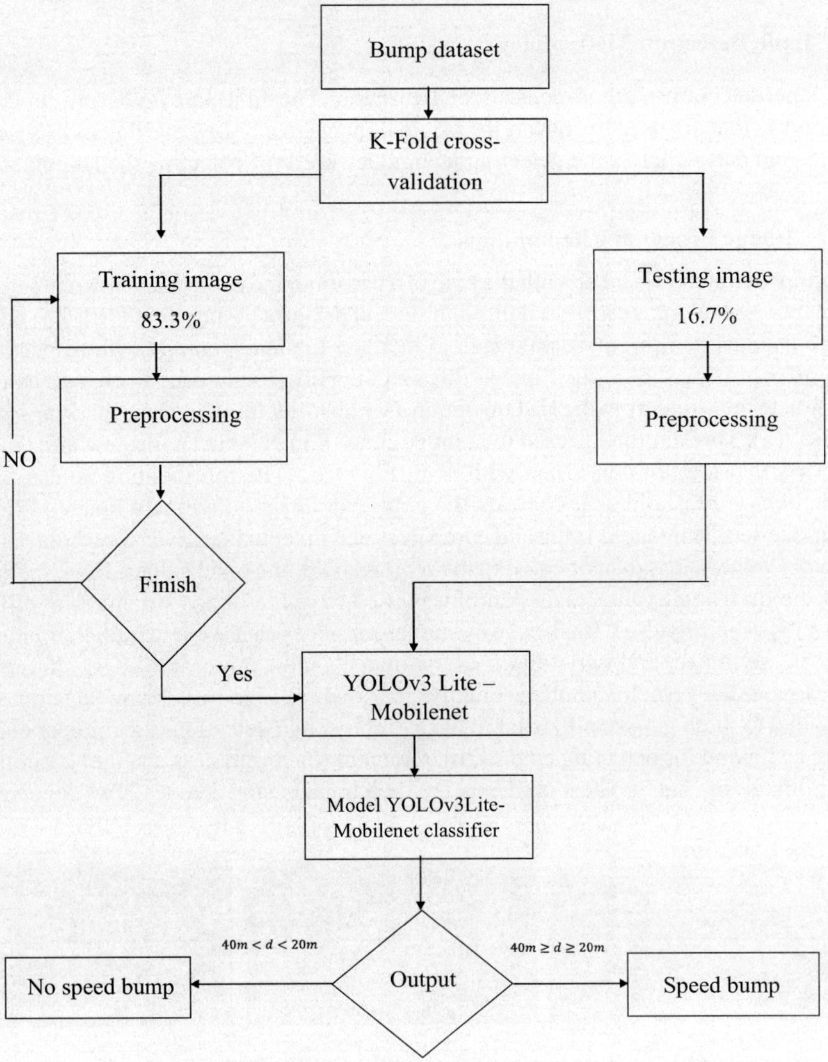

Fig. 3. Bump system flow diagram

3 Proposed Methods

In this section, we start with lane detection methodology, which consists of two parts: the pre-processing and the detecting techniques. Then, bump detection methodology consists of two parts: The dataset collection, and the proposed model.

3.1 Lane Detection Methodology

Our lane detection method consists of two parts. The first part is the pre-processing techniques that are used to convert the original image to a binary one that passes through the second part which is the detecting techniques used for detecting the lane lines.

3.1.1 Image Processing Techniques

A picture of the road is taken with the help of a car-mounted camera as shown in Fig. 4(a). Then the original image is passed through several pre-processing steps [18]. The original image is converted to grayscale image to increase the lane contrast with respect to the road after passing the original image through the HSL color space to change its colors and isolate the white from the HSL to obtain a white mask for white lane markings [19] as in Fig. 4(b). Thresholding is used to identify areas of interest in an image while ignoring the parts that are not concerned with as in Fig. 4(c). The thresholding is also known as the binary threshold as it converts the grayscale image to a binary image [18]. This technique sets a threshold value and compares each pixel to that given threshold value. If the pixel value is less than or equal to the set threshold, the pixel value will be set to zero or to the maximum value. Gaussian blur is used to refine image structures at different scales [5] as in Fig. 4(d). It takes away numerous edges that were identified, minimizes noise in the image, and only preserves the image's most noticeable edges. In our case a **3×3** gaussian kernel is applied. Finally, as shown in Fig. 4(e), canny edge detection is applied to these gaussian-blurred images. Canny edge detection is a gradient change-based technique for detecting edges [20]. Gradient (the brightness change) is computed in all directions. The greatest gradients are then traced using a series of white pixels.

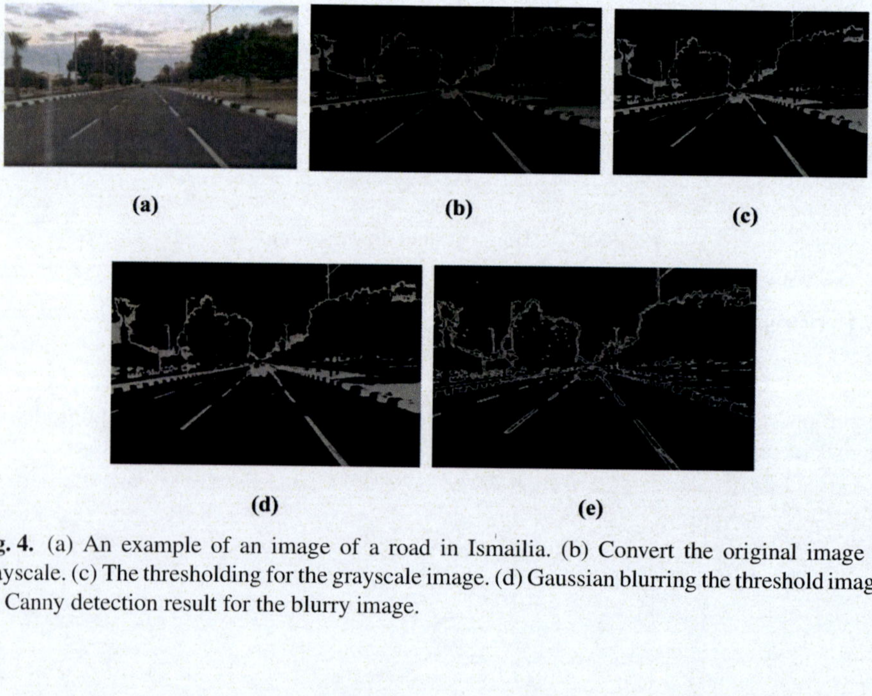

(a) **(b)** **(c)**

(d) **(e)**

Fig. 4. (a) An example of an image of a road in Ismailia. (b) Convert the original image to grayscale. (c) The thresholding for the grayscale image. (d) Gaussian blurring the threshold image. (e) Canny detection result for the blurry image.

3.1.2 Detecting Lane Techniques

Perspective transform technique is used to provide an image upon which subsequent computations may be performed more easily than the initial image received by the camera [14], as illustrated in Fig. 5(a). When the road is straight, the main goal is to create a rectangular form, which might be accomplished by providing several points over the image. These points' values will establish the Region of Interest (ROI). The perspective of an image or video frame is altered using these two sets of points. This is accomplished by supplying the coordinates to OpenCV library functions. As shown in Fig. 5(b), a histogram [14] is used to identify lanes in the obtained preprocessed image. A histogram is used to graphically represent the color tone distribution of the image. The histogram sums the pixel values that are proportional to each column. The peaks in the histogram represent the lane lines. The starting point of the lane is indicated by the position of the starting pixel along the x-axis, which is in the bottom left of the image.

(a) (b)

Fig. 5. (a) Perspective transform technique. (b) Histogram of the lane lines.

3.2 Bump Detection Methodology

Our bump detection method consists of two parts. The first part is the dataset collection that is collected manually and used in the second part, which is the selected model that used the dataset to train.

3.2.1 Dataset Collection

Around 3240 photos of bumps are taken in and around the university campus at various times of the day to acquire a dataset with a diversity of viewing and illumination situations. Most of the bumps were found to be badly maintained, with fading paint. The dataset includes bumps and fading paint bumps. They are labeled in PASCAL VOC format which is a public object detection dataset consisting of 20 categories of objects subsets A (2700 photos) and B (540 images) were created randomly from the dataset. In the experiment, two subsets of data were used. The training subset was A, while the testing subset was B. Before training on the dataset images, the image's height, width, class, and the bump's bounding box parameters (x_{min}, y_{min}, x_{max}, y_{max}) are needed. These are acquired through labelling. It is carried out using LabelImg [21], a free open-source application that records the bounding box parameters for each image in an associated

XML file. All images are reduced in size to allow for faster training on GPUs without losing too much accuracy.

3.2.2 Selected Model

The YOLOv3 Lite-Mobilenet model is a real-time object detection network [11]. It can ensure a given level of precision, and its computational characteristics are lower than those of traditional approaches. Its fast detection speed means that it may be used in embedded and mobile devices to detect objects accurately and effectively. The hybrid model consists of two algorithms, the YOLOv3 algorithm and Mobilenet algorithm which used as a backbone network for YOLOv3 instead of Darknet-53 as shown in Fig. 6. The two algorithms YOLOv3 and MobileNet will be explained to show the difference between them as follows:

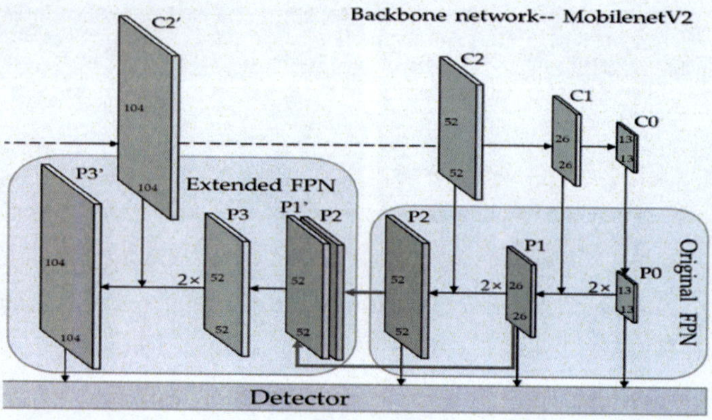

Fig. 6. YOLOv3 architecture with Mobilenet as backbone network

YOLOv3 is a complicated deep convolutional architecture built for object detection [15], which employs the darknet as the feature extract network and dimension clusters as anchor boxes for predicting bounding boxes of the system. Darknet, which originally had a 53-layer network trained on Imagenet, is a variation used by YOLOv3. 53 additional layers are added to it for the purpose of detection, giving YOLOv3 a 106-layer fully convolutional underlying architecture. YOLOv3 predicts 10 times as many boxes as YOLOv2.

MobileNet is a form of Convolutional Neural Network (CNN) built for mobile and embedded vision applications [16]. They are based on simplified architectures that use depthwise separable convolutions to construct lightweight deep neural networks with reduced latency for mobile and embedded devices.

4 Evaluation Criteria

To achieve more accurate results in speed bump detection, the model (YOLOv3 Lite-Mobilenet) is re-trained and tested on our dataset. Also, several algorithms are re-trained (Mobile Net-SSD, Faster RCNN, VGG-16, YOLOv3 Nano) and tested on our dataset. Accuracy is used to evaluate the performance of a model. It is calculated as the ratio of the number of correct predictions (true positives and true negatives) to the total number of predictions made (true positives, true negatives, false positives, and false negatives). The accuracy is calculated for all algorithms (YOLOv3 Lite-Mobilenet, Mobile Net-SSD, Faster RCNN-VGG-16, YOLOv3 Nano) which are (89.17%, 86.47%, 85.42%, 81.55%). Accuracy is calculated as:

$$\text{Accuracy} = \frac{\text{TP} + \text{TN}}{\text{TP} + \text{TN} + \text{FP} + \text{FN}} \tag{1}$$

The precision value is calculated from the number of True Positives (TP), False Positives (FP), and, optionally, True Negatives (TN). It measures the proportion of positive predictions made by the model that are correct. The precision is calculated for all algorithms (YOLOv3 Lite-Mobilenet, Mobile Net-SSD, Faster RCNN-VGG-16, YOLOv3 Nano) which are (98.87%, 98.36%, 98.01%, 97.89%). Precision is calculated as:

$$\text{Precision} = \frac{\text{TP}}{\text{TP} + \text{FP}} \tag{2}$$

Sensitivity determines the rate at which observations from the positive class are correctly predicted. The sensitivity value is calculated from the number of True Positives (TP), False Negatives (FN), sensitivity is also known as recall or True Positive Rate (TPR). The sensitivity is calculated for all algorithms (YOLOv3 Lite-Mobilenet, Mobile Net-SSD, Faster RCNN-VGG-16, YOLOv3 Nano) which are (96.04%, 93.02%, 97.13%, 91.44%). Sensitivity is calculated as:

$$\text{Sensitivity} = \frac{\text{TP}}{\text{TP} + \text{FN}} \tag{3}$$

5 Simulation Results

After fixing the camera (camera V2 module board 8MP webcam video) position and passing the input through all the steps, the detection effect is tested on the road, and image stacking technique [22] is used which is a technique in digital photography in which several images of the same scene are "stacked" together to reduce image noise caused by a high ISO value. This allows for creating a low noise image in low light conditions as shown in Fig. 7.

For the departure warning results, the system gives an alerting message each time the car deviates out of the lane giving two warning messages: "Go Right" as shown in Fig. 8(a) when the car deviated to the left side of the lane, "Go Left" as shown in Fig. 8(b) when the car deviated to the Right side of the lane.

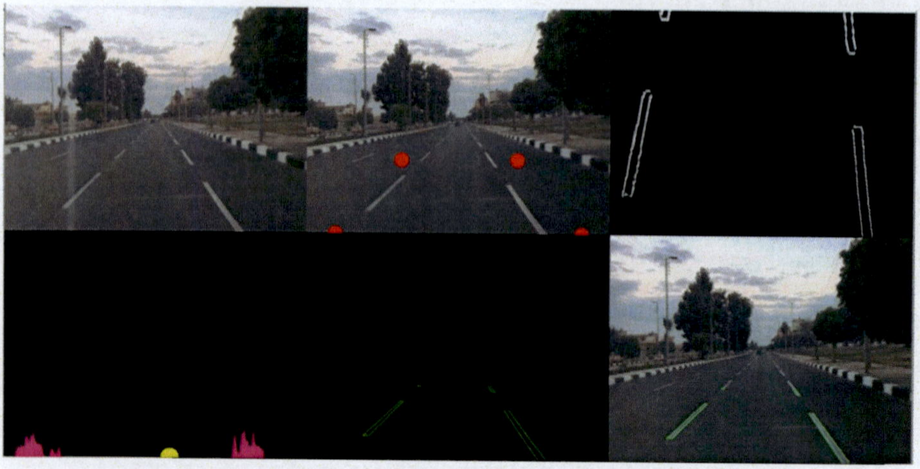

Fig. 7. Stacked image of the processes with result of lane detection.

(a) (b)

Fig. 8. Lane warning system shows (a) "Go Left" message appeared. (b) "Go Right" message appeared.

Fig. 9. Bump detection result

For bump detection results: camera v2 module board 8MP webcam video is mounted on the moving vehicle facing towards the road surface. The result of detecting a bump in Ismailia city street is shown below in Fig. 9.

To choose the best model, the models shown in Table 1 are re-trained. YOLOv3 Lite-Mobilenet is chosen because of its high speed and accuracy.

Table 1. Performance Evaluation Results

Model	Speed	Sensitivity	precision	Accuracy
YOLOv3 Lite-Mobilenet	14.6 ms	96.04%	98.87%	89.17%
Mobile Net-SSD	30 ms	93.02%	98.36%	86.47%
Faster RCNN, VGG-16	151 ms	97.13%	98.01%	85.42%
YOLOv3 Nano	29 ms	91.44%	97.89%	81.55%

6 Conclusion

In this paper, Two ADAS systems are focused on: (1) LDWS by image processing techniques such as Canny Edge Detection, perspective warping, and histogram. This system is expanding its applications, which include traffic control, traffic monitoring, traffic flow, and security. Our proposed lane detection system can run 15 fps on Jetson Nano. (2) Speed bump detection by YOLOv3 Lite-Mobilenet. Some algorithms are re-trained such as: (YOLOv3 Lite-Mobilenet, Mobile Net-SSD, Faster RCNN, VGG-16, YOLOv3 Nano) to detect the speed bump on our dataset. The dataset was divided into about 83.3% of total images for training and 16.7% for testing. The previous algorithms have achieved an accuracy of 89.17%, 86.47%, 85.42%, and 81.55%, respectively. YOLOv3 Lite-Mobilenet has achieved the highest accuracy among them with an accuracy of 89.17% and the highest speed with a speed of 14.6 ms.

References

1. Vijay,G., Ramanarayan, M.N., Chavan, A.P.: Design and integration of lane departure warning, adaptive headlight and wiper system for automobile safety. In: 2019 4th International Conference on Recent Trends on Electronics, Information, Communication and Technology (RTEICT), Bangalore, India, pp. 1309–1315 (2019)
2. Ziebinski, A., Cupek, R., Grzechca, D., Chruszczyk, L.: Review of advanced driver assistance systems (ADAS). In: AIP Conference Proceedings, vol. 1906, p. 120002. AIP Publishing LLC (2017)
3. Nguyen, T.B.: Evaluation of lane detection algorithms based on an embedded platform (2017)
4. Ibrahim, R.F., Yhiea, N.M., Mohammed, A.M., Mohamed, A.M.: Pleural effusion detection using machine learning and deep learning based on computer vision. In: Hassanien, A.E., Snášel, V., Tang, M., Sung, T.-W., Chang, K.-C. (eds.) Proceedings of the 8th International Conference on Advanced Intelligent Systems and Informatics 2022, pp. 199–210. Springer International Publishing, Cham (2023). https://doi.org/10.1007/978-3-031-20601-6_19

5. Devane, V., Sahane, G., Khairmode, H., Datkhile, G.: Lane detection techniques using image processing. ITM Web Conf. **40**, 03011 (2021). https://doi.org/10.1051/itmconf/20214003011. EDP Sciences
6. Kattenborn, T., Leitloff, J., Schiefer, F., Hinz, S.: Review of convolutional neural networks (CNN) in vegetation remote sensing. ISPRS J. Photogramm. Remote. Sens. **173**, 24–49 (2021)
7. Zhao, H., et al.: Mixed YOLOv3-LITE: a lightweight real-time object detection method. Sensors. **20**(7), 1861 (2020)
8. Mohamed, R., Magdy, A., Nafea, S.F.: Modeling of Psuedomorphic High Electron Mobility Transistor Using Artificial Neural Network. In:, et al. The 3rd International Conference on Artificial Intelligence and Computer Vision (AICV2023), March 5–7, 2023. AICV 2023. Lecture Notes on Data Engineering and Communications Technologies, vol 164. Springer, Cham (2023).https://doi.org/10.1007/978-3-031-27762-7_20
9. Sharma, A., Kumar, M., Kumar, R.: Lane detection using Python. In: 2021 3rd International Conference on Advances in Computing, Communication Control and Networking (ICAC3N), pp. 88–90. IEEE (2021)
10. McKinney, W.: Python for data analysis: Data wrangling with Pandas, NumPy, and IPython. O'Reilly Media, Inc. (2012)
11. Cai, K., Miao, X., Wang, W., Pang, H., Liu, Y., Song, J.: A modified YOLOv3 model for fish detection based on MobileNetv1 as backbone. Aquacult. Eng. **91**, 102117 (2020)
12. Satarkar, H., Zagade, N., Gupta, S., Pundlik, S.: Comparative study between segmentation neural networks on pascal VOC dataset. Int. Res. J. Eng. Technol. **8** (2021).
13. Jiang, P., Ergu, D., Liu, F., Cai, Y., Ma, B.: A review of Yolo algorithm developments. Procedia Comput. Sci. **199**, 1066–1073 (2022)
14. Muthalagu, R., Bolimera, A., Kalaichelvi, V.: Lane detection technique based on perspective transformation and histogram analysis for self-driving cars. Comput. Electr. Eng. **85**, 106653 (2020)
15. Zhao, L., Li, S.: Object detection algorithm based on improved YOLOv3. Electronics **9**(3), 537 (2020)
16. Pan, H., Pang, Z., Wang, Y., Wang, Y., Chen, L.: A new image recognition and classification method combining transfer learning algorithm and MobileNet model for welding defects. IEEE Access **8**, 119951–119960 (2020)
17. Li, Y., Huang, H., Xie, Q., Yao, L., Chen, Q.: Research on a surface defect detection algorithm based on MobileNet-SSD. Appl. Sci. **8**(9), 1678 (2018)
18. Haque, M.R., Islam, M.M., Alam, K.S., Iqbal, H., Shaik, M.E.: A computer vision-based lane detection approach. Int. J. Image Graph. Sig. Proc. **10**(3), 27 (2019)
19. Lee, C., Moon, J.H.: Robust lane detection and tracking for real-time applications. IEEE Trans. Intell. Transp. Syst. **19**(12), 4043–4048 (2018)
20. Yan, X., Li, Y.: A method of lane edge detection based on Canny algorithm. In: 2017 Chinese Automation Congress (CAC), pp. 2120–2124. IEEE (2017)
21. Liu, B., Liu, G., Qiao, J.: Lane and object detection algorithm based on deep learning. In: 2nd International Conference on Artificial Intelligence, Automation, and High-Performance Computing (AIAHPC 2022), vol. 12348, pp. 434–438. SPIE (2022)
22. Borkar, A., Hayes, M., Smith, M.T.: An efficient method to generate ground truth for evaluating lane detection systems. In: 2010 IEEE International Conference on Acoustics, Speech, and Signal Processing, pp. 1090–1093. IEEE (2010)

Visual Steering for Deep Neural Networks Using Explainable Artificial Intelligence

Hossam Elbeialy(✉), Arwa Ghaith(✉), Mai H. ElBaradei(✉), Mohamed Alkhodary, Alaa M. Eissa, Esraa A. Farag, Omnia Fdawey, Noran Emad, Ahmed Elshenawy, Mariam Shabaan, and Marwa Gamal(✉)

Faculty of Engineering, Suez Canal University, Ismailia, Egypt
hossamelbeialy@gmail.com, arwagheith80@gmail.com, maielbaradei@gmail.com, marwa_gamal@eng.suez.edu.eg

Abstract. Despite their remarkable revolution in artificial intelligence and promising performance, deep neural networks are susceptible to biases due to imbalanced training data or the way that training data is represented as features for the model; thus, they need a high level of transparency in critical fields, especially in the medical sector. Several explainability methods have been developed, aiming at enhancing the transparency of the model and increasing its trustworthiness. One of these methods is Grad-CAM, which can highlight the significant parts of an input that most influence the model's prediction and detect biases in the dataset. In this paper, a method is provided for integrating human knowledge with the deep neural network to improve its performance, reduce the biases that arise, and leverage human experience. This can be done via the attention maps generated by Grad-CAM, which are then edited by domain experts. The proposed method enables the edited attention maps to contribute to the training of the model and updating its parameters, leading to better generalisation and fewer inductive biases. This approach is applied to several image classification datasets: Imagenette2, BUSI, and skin cancer ISIC. Experimental results show that the generated attention maps highly correspond to the edited ones while having high classification performance.

Keywords: Deep Neural Networks · Human in the loop · Explainable AI · Attention Maps · Inductive Bias

1 Introduction

Visual explanation has had a huge impact on the field of artificial intelligence (AI) because of its importance in the scope of critical decision-making and bias detection. The fact that deep neural networks are black boxes makes it necessary to know what was important to the model during the training process. Using explanations generated by explainable artificial intelligence (XAI) methods [11], AI decisions could be understood and make the process more interpretable. One

A. Hassanien et al. (Eds.): AISI 2023, LNDECT 184, pp. 43–52, 2023.
https://doi.org/10.1007/978-3-031-43247-7_4

technique of visual explanation is Grad-CAM [12], which stands for Gradients Weighted Class Activation Mapping. It is a technique that uses the gradients of any target concept flowing into the final convolutional layer to produce an attention map highlighting the important regions to the prediction for the target concept. However, a significant problem arises in deep neural network training. Deep neural networks often consider some regions more important than the truly important ones. This can bias the neural networks towards features that are not at all important for the training loop, which can affect the decision-making of the neural network.

To address this issue, the proposed approach attempts to make the ground truth for the model data more informative. By visually steering the model and manually editing the attention map, human knowledge could be used alongside the data itself to make the model learn better. This approach of embedding human knowledge in the training loop is often used in the form of having humans help label data and provide feedback on the data. Even though embedding human knowledge using the proposed approach is more informative, it introduces overhead that scales with the model parameters.

To this end, the proposed approach uses visual explanation to produce attention maps that could be used to embed human knowledge into the training loop by manually editing the maps and making the model learn the important regions of each annotated example. This will help fine-tune the model parameters to improve generalisation and decrease bias.

2 Related Work

2.1 Human in the Loop

Most machine learning models rely on data prepared by humans. However, there are still other ways in which humans and machines interact. These ways are grouped under the field of human-in-the-loop (HITL). This is when humans are involved in the training and deployment of machine learning models. This set of approaches for combining human and machine intelligence in AI applications is usually aimed at one or more purposes: Increasing the accuracy of a machine learning model, and reaching the target accuracy for a machine learning model faster with high efficiency. But the main goal of using this method here is to avoid having machines make decisions that may cause harm or biases[9,14]. To address this issue, the proposed approach attempts to make the ground truth for the model data more informative. By visually steering the model and making the human edit the attention map manually. Human knowledge could be used alongside the data itself to make the model learn better. This approach of embedding human knowledge in the training loop is often used in the form of having humans help label data and provide feedback on the data.

Identifying Bias. The tendency of algorithms to reflect human biases is referred to as algorithm bias, also known as artificial intelligence bias. This phenomenon occurs when an algorithm consistently produces biased data due to false assumptions made during the machine learning process. The issue is that these biases are unintentional, and it might be challenging to detect them before they have been incorporated into the software [6]. For an example of gender bias, a pre-trained ImageNet VGG-16 model was fine-tuned for a classification task to identify doctors and nurses. In the paper [12] they have made a balanced test set with equal distribution of each gender, The model's accuracy was just 82% which concludes that the model did not generalise very well. Using Grad-CAM to understand the model's predictions, it has been shown that the model is focusing on the face and hairstyles as shown in Fig. 1. This led to an understanding that the model was biased towards males predicting them as doctors and females as nurses; looking deeper into the data, they found that the data itself was biased.

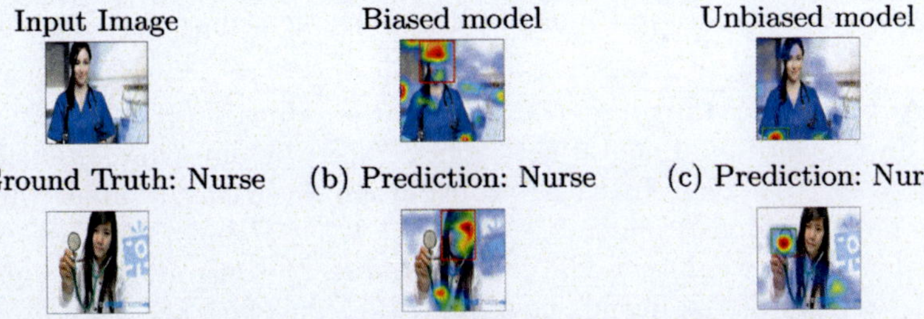

(a) Ground Truth: Nurse (b) Prediction: Nurse (c) Prediction: Nurse

(d) Ground Truth: Doctor (e) Prediction: Nurse (f) Prediction: Doctor

Fig. 1. Visual Explanations of a model using Grad-CAM to detect bias

2.2 Explainable AI

Explanations help ensure model fairness and transparency by interpreting machine learning models and give insights into model predictions that help practitioners understand why the model reached some decisions, usually in terms of the inputs [11]. Visual explanations, such as attention maps, represent the contribution of input features to the output decision. The proposed approach focuses on backpropagation-based explanation methods as they best serve the purpose.

Guided Backpropagation. Guided backpropagation [13] is a combination of two methods: 'deconvnet' and backpropagation, where the proposed method masks out values corresponding to both negative entries of the top gradient and bottom data, so it adds guidance signals from higher layers to the usual backpropagation. This method prevents the backward flow of negative gradients, which correspond to neurons that shrink the activation of the higher layer unit needed to visualize.

Layer-wise Relevance Propagation. Layer-wise Relevance Propagation (LRP)[10] is used to detect bias in datasets and machine learning models. It operates by backpropagating the predictions in the neural network and terminates once the input features have been reached. Similar to Kirchoff's conservation, LRP has a conservation property that ensures for each neuron their contributions are equally distributed to the lower layer.
Let j and k be neurons at two consecutive layers of the neural network. Propagating relevance scores $(R_k)_k$ at a given layer onto neurons of the lower layer is achieved by applying the rule:

$$R_j = \sum_k \frac{z_{jk}}{\sum_j z_{jk}} R_j \tag{1}$$

The quantity z_{jk} models the extent to which neuron j has contributed to making neuron k relevant, and the denominator enforces the conservation property.

Class Activation Mapping. Class Activation Mapping (CAM) [15] is implemented in convolutional neural networks (CNNs) by applying a global average pooling (GAP) [7] layer on the last convolution layer feature maps which compresses the dimensions from $X \times Y \times Z$ to $1 \times 1 \times Z$ and using these values as features for the fully connected layer that produce the output, then by projecting back the weights of the output layer onto the convolutional feature maps, the proposed approach can identify the importance of image regions. Although this method is quite simple and easy to apply, it requires changing the architecture of the models by adding a GAP layer and hence does not generalize to all models.

Gradient-weighted Class Activation Mapping. Gradient-weighted Class Activation Mapping (Grad-CAM) [12] is a generalization of CAM for a wide variety of CNNs. It generates the class activation map using the gradients of the output layer with respect to the last convolutional layer to calculate the contribution of each neuron to a particular output value. These contributions are calculated using the following equation:

$$\alpha_{k,c} = \frac{1}{m \cdot n} \sum_{i=1}^{m} \sum_{j=1}^{n} \frac{\partial y_c}{\mathbf{A}_{k,i,j}} \tag{2}$$

where $\mathbf{A}_{k,i,j}$ is a neuron positioned at (i,j) in the m $\times$ n feature map A_k in the last convolutional layer and y_c is the gradient score of class c with respect to every node in A_k. Figure 2 is an illustration of this process. To obtain the relevance score maps, the scores of each feature map are then combined linearly and passed through a ReLU. The relevance score map is then upsampled to the same dimension as the input picture using bilinear interpolation to produce the class activation map.

$$map_c = ReLU\left(\sum_k^K \alpha_{k,c} \mathbf{A}_k\right) \tag{3}$$

Guided Grad-CAM. As mentioned above, in guided backpropagation, the negative gradients are suppressed during backpropagation, which leads to capturing pixels detected by neurons, not the suppressed ones. Guided Grad-CAM [12] combines the best of Guided Backpropagation and Grad-CAM visualizations by applying element-wise multiplication.

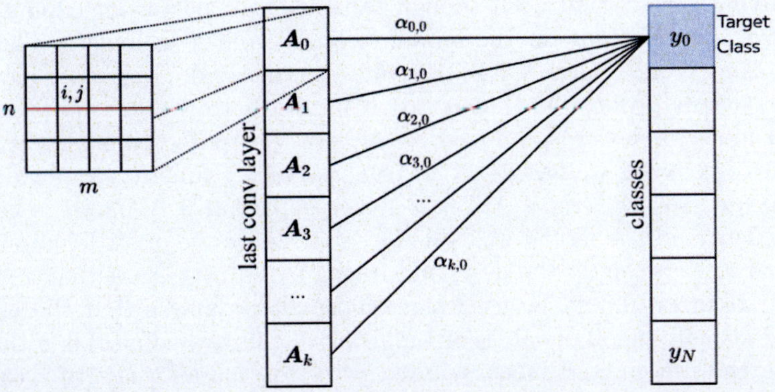

Fig. 2. Calculating the importance scores $\alpha_{i,j}$ for each feature map $\mathbf{A}_k$

3 Proposed Method

First, an explanation method was used to generate heatmaps describing the importance values of each image region according to the model to make a specific decision. These maps were then modified with annotation, which is what domain experts believe were the image regions important to make that decision. Finally, the attention maps are used to steer the model into aligning with human knowledge. Figure 3 illustrates the process flow of the proposed method.

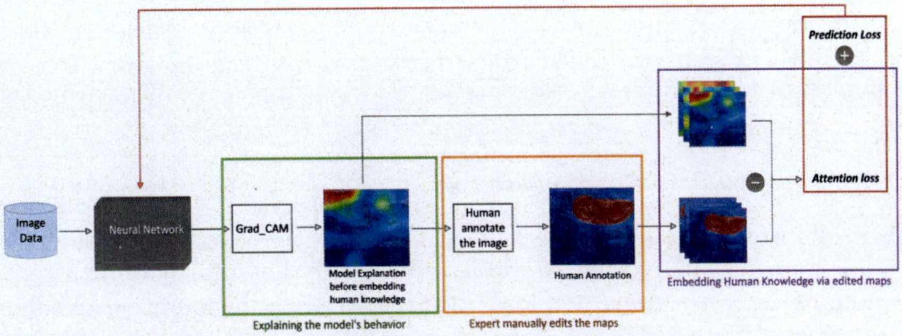

Fig. 3. Process flow of the proposed method

3.1 Explanation Method

The initial approach was using LRP [10] as the explanation method, but the results were not satisfactory. The model could not converge on producing the same or almost the same masks that were given. It was believed that it worked just when the dataset pattern was easier to recognize than complex datasets. So it is hypothesized that even though the LRP was better for explanation, it might have been difficult for the model to learn from it in images. The second trial was using Guided Grad-CAM [12], but also the results were not satisfactory; moreover, it is hypothesized that it might need a better loss function since the attention maps that were produced by it were too sparse (this also applied to LRP) but to no avail; we had not found a suitable loss function to make it work. It was decided to use Grad-CAM [12] as the explainable AI model, the results were excellent and the model was able to converge on the exact locations as the masks that were given by the user, and it was found by testing that a weighted mean square error should be used since most of the masks that the user gave contained a small region of values of 1 and the rest have a value of 0, if the model was given the unweighted mean squared error the model tends to zero out all the explanation so this equation was proposed 4. The approach works as follows, Captum library[1] was used to calculate the Grad-CAM. After performing the Grad-CAM, a weighted mean square error was used to train the model and the prediction loss (any prediction loss would suffice) was added. From testing, it was found that adding prediction loss to the training loop was a must to ensure that the model was learning where to look for the correct class.

3.2 Manual Editing

The worst images classified by the model were determined and annotated manually. For image annotation, The framework made by ABN [3,8] Authors[2] was used to edit the attention maps. Annotation was done by highlighting the important or most informative parts of the image. For example, for "imagenette" dataset, the fish body was marked. For "ISIC" and "BUSI" datasets, the masks that were already given were used as the training masks because we did not have the medical sources needed to correctly annotate the images. After that, the model interprets the red parts as the only important features for it to learn.

3.3 Embedding Human Knowledge via Edited Attention Map

While training the model on the edited attention maps, the parameters and gradients were updated using the loss, which was the sum of both attention loss and cross-entropy. Attention loss was calculated by a custom mean squared error by giving more weight to the regions that were marked as important by

[1] Captum library.

[2] Attention Editor ABN repository.

the expert by a factor of 0.1, while all regions were given a factor of 0.9.

$$AttentionLoss = \frac{1}{n}(0.1\sum_{i}^{n}(H_i^+ - A_i^+)^2) + \frac{1}{n}(0.9\sum_{i}^{n}(H_i - A_i)^2) \tag{4}$$

$$PredictionLoss = -\frac{1}{n}\sum_{i}^{n}(y_i \times log(y')) \tag{5}$$

$$Loss = PredictionLoss + (\gamma \times AttentionLoss) \tag{6}$$

where H was the annotated image, H^+ were the regions marked important by the expert, A^+ were the corresponding regions in A the attention map generated from the model, y was the label of the image, y' was the prediction made by the model, and γ was a scaling factor. Experiments showed that using gamma with a value of 10 led to good results (Fig. 4).

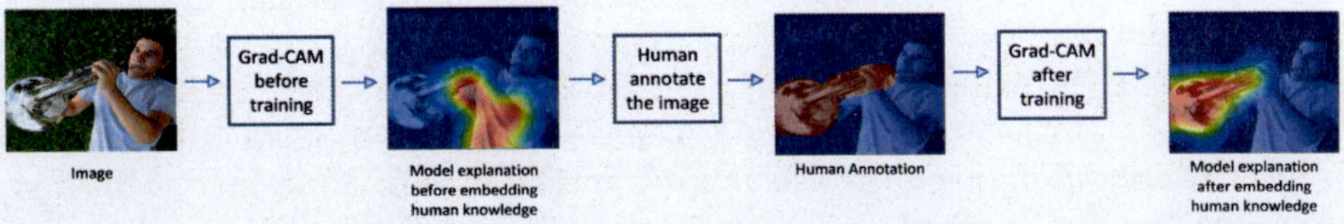

Fig. 4. Illustration of how embedding human knowledge into the model significantly improved its ability to detect the correct patterns in imagenette2 dataset

4 Experiments

4.1 Datasets

Imagenette2. The proposed approach has been applied on imagenette2 [4] since it was desired to use the imagenet [5] dataset. But unfortunately, the dataset was too large for us to train. So, imagenette2 has been used which is a subset of 10 easily classified images from imagenet. Usually, imagenette2 has been used to test out ideas, so it was easier to annotate the images than other variations of the subsets like imagewoof as the classes are easier to distinguish and the annotations made were marking out the whole object. Out of 9469 images, 3948 images of the training set have been annotated.

ISIC. In Skin Cancer ISIC dataset [2], originally the dataset was for a segmentation task, but it was desired to use medical datasets with already annotated images, so it has been used as classification tasks. It contains 900 training images and masks of skin lesions (727 benign and 173 malignant) and 379 test images and masks (304 benign and 173 malignant).

BUSI. In Breast Ultrasound Images Dataset [1], as ISIC, this dataset was for the segmentation task. The dataset consists of 572 training images (329 benign, 147 malignant, 96 normal) and 211 test images (108 benign, 63 malignant, 40 normal).

4.2 Results

The experiment included the proposed approach, ABN's proposed approach, and models trained on just the prediction task. For the proposed approach and the normal models, the models that were used were pre-trained models Resnet50, Resnet101, and Resnet152 from Pytorch's models. For ABN's proposed approach, the models were from the author's GitHub repository[3]. To be able to assess the performance of the two models against the annotation, a combined loss function of **weighted dice coefficient** and **weighted structural similarity index (SSIM)** has been used, and the weight is given to the annotated object. The reason for the weighting is since for most of the annotations, the object is small compared to the rest of the images, it was desired to make sure that the model is penalised heavily if it is focusing on other areas than the object itself, and the reason for using dice coefficient with SSIM is to measure both the overlapping pixels between the attention and the annotation while also measuring the structure of the attention to make sure it matches the annotation's structure. Experimental results in Table 1 show that for the **Imagenette2** dataset, the proposed model has relatively high accuracy and AUC score and low qualitative measure and is closer to the best results if it is not the best. For

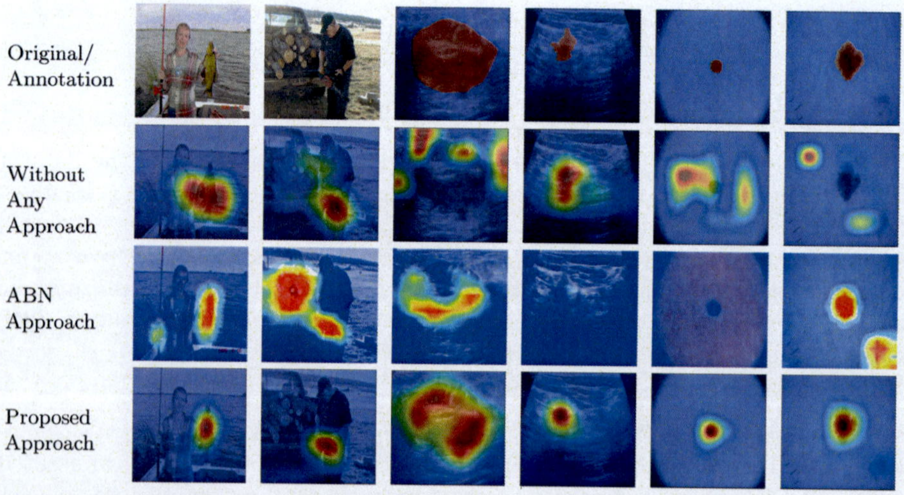

Fig. 5. Examples from Imagenette2, BUSI and ISIC datasets to show the differences between our approach and ABN approach

[3] Attention Branch Network repository.

the **BUSI** dataset, the results show that the proposed model achieved the best results across all models. For the **ISIC** dataset, the proposed method has the best results for most of the criteria (Fig. 5).

Table 1. Comparison of Accuracy, AUC, F2 score and Qualitative assessments of attention maps for Imagenette2, BUSI and ISIC datasets. For Qualitative assessment the lower is better, for the rest the higher is better

		Imagenette2			BUSI		ISIC	
Approach	Model	Accuracy	AUC	Qualitative Measure	F2 Score	Qualitative Measure	F2 Score	Qualitative Measure
Without	ResNet50	**99.49%**	**0.9999**	3.4084	0.8298	1.6049	0.7061	2.9413
	ResNet101	97.73%	0.9995	3.5392	0.7736	1.8655	0.5715	3.5000
	ResNet152	98.22%	**0.9997**	3.2401	0.7860	1.3899	**0.7156**	2.9188
ABN	ResNet50	97.91%	0.9994	2.3048	0.8238	2.0944	0.7207	2.2736
	ResNet101	96.46%	0.9982	**2.7847**	0.8066	1.9425	0.6137	**2.5259**
	ResNet152	**98.32%**	**0.9997**	**2.4483**	0.7720	2.4787	0.6961	**2.2733**
Proposed Approach	ResNet50	97.58%	0.9995	**1.8623**	**0.8599**	**1.3292**	**0.7405**	**1.6235**
	ResNet101	**98.11%**	**0.9996**	3.1043	**0.8429**	**1.4367**	**0.7014**	2.8837
	ResNet152	97.89%	0.9991	2.4673	**0.8225**	**1.3881**	0.7126	2.8722

5 Conclusion

In this paper, a method is proposed to steer the model to align with human knowledge using the attention maps generated by Grad-CAM. This can be done by editing the generated attention maps, adding the weighted mean square error between the generated attention maps and the edited ones to the training loss, and then updating the model parameters. This helps to improve the performance of the model and reduce its bias. Experimental results on ISIC, BUSI, and Imagenette2 datasets show that the generated attention maps highly correspond to the edited ones while having high classification performance. The Imagenette2 dataset was evaluated using accuracy, AUC, and qualitative assessment, while the BUSI and ISIC datasets were evaluated using F2 scores and qualitative assessment. The proposed approach is more general than the ABN one, which needs a specific model. This may help people embed human knowledge not just in CNNs but in DNNs in general. Future work includes solving low-shot learning by finding out how much data the model needs to learn from the annotated images, applying this approach without a pre-trained model, applying it to object detection and text, and merging uncertainty sampling and diversity sampling to find out which data points are more informative for the annotation process.

References

1. Al-Dhabyani, W., Gomaa, M., Khaled, H., Fahmy, A.: Dataset of breast ultrasound images. Data Brief **28**, 104863 (2020)
2. Karar, A., Shaikh, Z.A., Khan, A.A., Laghari, A.A.: Multiclass skin cancer classification using EfficientNets-a first step towards preventing skin cancer. Neurosci. Inform. **2**(4), 100034 (2022)

3. Fukui, H., Hirakawa, T., Yamashita, T., Fujiyoshi, H.: Attention branch network: learning of attention mechanism for visual explanation. In: Proceedings of the IEEE/CVF Conference on Computer Vision and Pattern Recognition, pp. 10705–10714 (2019)
4. Howard, J.: Imagenette: a smaller subset of 10 easily classified classes from ImageNet, and a little more French (2019)
5. Dong, W., Deng, J., Li, L.-J., Socher, R., Li, K., Fei-Fei, L.: A large-scale hierarchical image database. In: IEEE Conference on Computer Vision and Pattern Recognition, pp. 248–255 (2009)
6. Keswani, V., Lease, M., Kenthapadi, K.: Designing closed human-in-the-loop deferral pipelines. arXiv preprint arXiv:2202.04718 (2022)
7. Lin, M., Chen, Q., Yan, S.: Network in network. arXiv preprint arXiv:1312.4400 (2013)
8. Mitsuhara, M., et al.: Embedding human knowledge into deep neural network via attention map. arXiv preprint arXiv:1905.03540 (2019)
9. Monarch, R.M.: Human-in-the-Loop Machine Learning: Active learning and annotation for human-centered AI. Simon and Schuster (2021)
10. Montavon, Grégoire., Binder, Alexander, Lapuschkin, Sebastian, Samek, Wojciech, Müller, Klaus-Robert.: Layer-wise relevance propagation: an overview. In: Samek, Wojciech, Montavon, Grégoire., Vedaldi, Andrea, Hansen, Lars Kai, Müller, Klaus-Robert. (eds.) Explainable AI: Interpreting, Explaining and Visualizing Deep Learning. LNCS (LNAI), vol. 11700, pp. 193–209. Springer, Cham (2019). https://doi.org/10.1007/978-3-030-28954-6_10
11. Ras, G., Xie, N., Van Gerven, M., Doran, D.: Explainable deep learning: a field guide for the uninitiated. J. Artif. Intell. Res. **73**, 329–397 (2022)
12. Selvaraju, R.R., Cogswell, M., Das, A., Vedantam, R., Parikh, D., Batra, D.: Grad-CAM: visual explanations from deep networks via gradient-based localization. In: Proceedings of the IEEE International Conference on Computer Vision, pp. 618–626 (2017)
13. Springenberg, J.T., Dosovitskiy, A., Brox, T., Riedmiller, M.: Striving for simplicity: the all convolutional net. arXiv preprint arXiv:1412.6806 (2014)
14. Tsiakas, K., Murray-Rust, D.: Using human-in-the-loop and explainable AI to envisage new future work practices. In: Proceedings of the 15th International Conference on Pervasive Technologies Related to Assistive Environments, pp. 588–594 (2022)
15. Zhou, B., Khosla, A., Lapedriza, A., Oliva, A., Torralba, A.: Learning deep features for discriminative localization. In: Proceedings of the IEEE Conference on Computer Vision and Pattern Recognition, pp. 2921–2929 (2016)

Waste Allocation Load Lifter Model for Trash Detection Based on Deep Learning and Wireless Localization Techniques

Layla Mohsen[1], Abdulrahman Talaat[1], and Sally Elghamrawy[2,3](✉)

[1] Computer Science Department, The American University in Cairo, Cairo, Egypt
[2] Computer Engineering Department, MISR Higher Institute for Engineering and Technology, Mansoura, Egypt
sally_elghamrawy@ieee.org, sally@mans.edu.eg
[3] Scientific Research Group in Egypt (SRGE), Cairo, Egypt

Abstract. The progressive contamination of lands, oceans, and rivers by garbage is a growing concern. The pollution imposed by garbage accumulation can lead to serious health hazards as well as economic and environmental repercussions; this accumulation derives from irresponsible waste disposal. To counter the surplus of garbage pollution, this paper proposes an accurate, low-energy consuming system for a trash allocation robot, namely, Wall-E. Wall-E is a Waste Allocation Load Lifter Model for trash detection based on deep learning and wireless localization techniques. It is based on the integration of the deep learning models for trash detection and classification as well as a navigation system that is more accurate and energy-efficient than the commonly used global position system (GPS). The paper tests and analyses the neural network algorithms YOLOv7 and YOLOv8 conducted on a trash dataset and concludes that YOLOv7 is of higher performance in trash detection and classification. To implement this algorithm on a real-time robotic system, image frames will be sent from a Pi camera at a high frequency to a Raspberry Pi board that has YOLOv7 installed. In order to conceptually determine a suitable navigation system for our robot, the paper discusses the common challenges faced in traditional navigation systems, including proposed systems in related works and their limitations to our proposed system. In our conducted experiments, the proposed Wall-E achieved the highest accuracy in object detection using YOLOv7, which will be utilized in our robot to detect trash items sent through image frames from a Pi camera. The Dejavu navigation system was concluded as the most appropriate for our system's requirements.

Keywords: Real-time Object Detection · Image Classification · Indoor Tracking · Wireless Tracking · Embedded Systems · Deep Learning · Autonomous · YOLOv7 · YOLOv8

1 Introduction

The ongoing contamination of lands, oceans, and rivers due to the accumulation of garbage is becoming an escalating concern. This pollution poses significant risks to both public health and the environment, while also resulting in adverse economic consequences. The accumulation of garbage is primarily a result of irresponsible waste

S. Elghamrawy–Senior IEEE Member.

A. Hassanien et al. (Eds.): AISI 2023, LNDECT 184, pp. 53–66, 2023.
https://doi.org/10.1007/978-3-031-43247-7_5

disposal practices. The consequences of this disregard are far-reaching in effect; to avoid this propagation, small-scaled wasted items should be weighed in and effectively managed. The optimization of waste management is the main concern for dealing with the problem of increased waste.

Hardware advancements for the optimization of a system are reaching a point of diminishing returns, with incremental gains becoming less significant. Conversely, software optimization offers tremendous potential for maximizing efficiency, improving performance, and optimizing resource utilization. The advent of Artificial Intelligence, particularly Deep Learning, has resulted in remarkable achievements across numerous domains. Models in the field of Deep Learning are referred to as neural network models; such models learn to perform tasks with exceptional accuracy and efficiency. With their ability to analyze and interpret complex patterns in data, deep learning algorithms can achieve breakthroughs in various domains, from computer vision to natural language processing, as well as offer solutions to a wide range of problems. Utilizing deep learning algorithms to counter the challenges of waste management has enabled autonomous trash disposing systems that can accurately detect and classify trash [2–4, 6]. It has also helped in wireless, visual-based navigation of systems with higher efficiency and cost reduction [1, 3–5, 8].

The process of waste management can be divided into three main tasks: the detection, classification, and localization of trash. This paper proposes the potential of a software system labelled Wall-E as an optimized means for the waste management process. It will discuss the use of State-of-the-Art deep learning models, including YOLOv7 and YOLOv8 for object detection and classification, as well as a comparative discussion of software navigation models for indoor and outdoor wireless localization of the system's hardware. The objective of this research is to study the integration of these algorithms on a mobile robot for the purpose of producing a next-generation waste management system that can help tackle the damage caused by waste accumulation.

2 Related Works

In [16], demonstrating its dedication to addressing environmental problems connected with beach sanitation, The Autonomous Beach Cleaner project commenced operations. One of their major aspirations was to invent an independent robot that could acquire debris using a front shovel and relay images through an IP camera in specific situations. To achieve this goal, they merged sensors into their robotic design for avoiding obstacles and operating autonomously from human intervention. An app for monitoring and directing the machine was also developed by the team.

The robot also aimed to provide comprehensive and efficient cleaning efforts, particularly in areas with limited volunteer resources. It did so by incorporating advanced technologies such as image processing techniques like visual transmission through an IP camera. This research contributes greatly towards waste management in beach cleaning, highlighting the potential use of image processing approaches in environmental challenges faced today. Similar to Wall-E, the Autonomous Beach Cleaner project focuses on developing an autonomous robot for waste management. It utilizes image processing techniques, such as transmitting visuals through an IP camera, to facilitate effective oversight in certain situations. This aligns with the use of deep learning algorithms and visual-based navigation.

In addition to the previously mentioned works, [17] introduced a robot system that utilized a Convolutional Neural Network (CNN) [9] for trash detection on a Raspberry Pi. The trash detection process involved obtaining bounding boxes around the garbage objects in the image. The system incorporated a four-degree-of-freedom (d.o.f.) robotic arm to pick up the detected trash. By employing a pre-trained MobileNet model, the robot was able to perform object detection and classify the garbage in the image or video. The model provided the coordinates of the bounding box, allowing the system to visualize the detected objects.

It is worth noting that this approach had limitations due to the pre-trained model, as it could only detect objects for which the MobileNet was trained. In the initial prototype, the system specifically focused on detecting and considering bottles as waste material to simplify the model and expedite testing. However, future versions of the system aim to train a custom object detection model capable of identifying various types of garbage.

The project discussed in [17] shares similarities and differences with our original work in the domain of trash detection robotics. Both projects employ utilizing image processing techniques to identify and classify garbage objects. Additionally, they incorporate robotic arms to physically pick up the detected trash. However, there are notable differences as well. While the system described in [17] initially focuses on detecting and considering bottles as waste material, whereas our original work aims to address a broader range of trash types. These similarities and differences highlight the diversity of approaches and techniques employed in the field of trash detection robotics.

In another study [18], researchers developed a robot that could be controlled remotely or operate autonomously. They used image processing techniques to help the robot identify obstacles and detect the color yellow using its camera. Whenever the robot spotted something yellow, it would move towards it and briefly open its lid. To make the robot more effective at cleaning, they even added a suction fan to remove dust.

To carry out the image processing tasks, the researchers used microcontrollers called Raspberry Pi 3b+ and Arduino ATMEGA328. These microcontrollers were selected because they are cost-effective and capable of handling basic computations and image processing. The Raspberry Pi, with its support for the Linux operating system and GPIO pins, made it easy to connect sensors and collect data about the physical environment. The researchers wrote their image processing code in Python, a programming language known for its simplicity and readability. They utilized the powerful OpenCV library, which provides a wide range of algorithms for manipulating and analyzing images and videos by combining the Raspberry Pi, Arduino, Python, and OpenCV, the researchers were able to create a system that enabled the robot to detect obstacles and respond to the presence of the color yellow. This implementation improved the robot's functionality as a smart trashcan, making it more efficient at identifying and collecting trash.

Both Wall-E and [18] share similarities in their utilization of image processing techniques for trash detection and their integration of robotic components. Both projects employed cameras to capture visual data and utilized image processing algorithms to analyze the images. Additionally, both projects aimed to create autonomous robots capable of detecting and responding to specific visual cues related to trash. However, there are also notable differences between the two projects. In [4], the robot focused on detecting the color yellow as an indication of the presence of trash. It moved towards the yellow color and briefly opened its lid. In contrast, Wall-E primarily focused on trash detection using deep learning algorithms, without explicitly mentioning the specific color-based approach.

3 The Proposed System

The proposed system is an efficient software that can be embedded on a robotic system so that it is able to autonomously detect and identify the category of trash encountered, as well as relocate it appropriately to a specified destination. The system should relocate the identified trash according to its classification; for example, if the item was non-biodegradable, such as plastics, the load lifter robot should dispose it in the bin of recyclables. The suggested hardware includes a Raspberry Pi Camera that sends real-time images to a connected Raspberry Pi 3 board and Arduino board for trash detection and classification as well as rotary and linear actuators for lifting the trash. The low-energy sensors in the hardware, such as accelerometers, will be used for local navigation using machine learning algorithms; this will help overcome the limitations of traditional navigation systems (further detail is discussed in Sect. 3.2) [12, 13]. Once the obstacle is within range, the Raspberry Pi 3 board should be able to classify it as trash or not using object detection and classification algorithms. If the examined item is trash and the material is identified, the Raspberry Pi will send signals to Arduino, which will enable the actuators to rotate the robotic arms and wrists to lower and lift them to grasp the item. The global navigation system should then be activated according to the category of trash lifted and should be able to guide the robot to the assigned destination for the trash to be disposed. The real-time trash detection and classification model can be seen in Fig. 1. Figure 2 and Fig. 3 articulate this process.

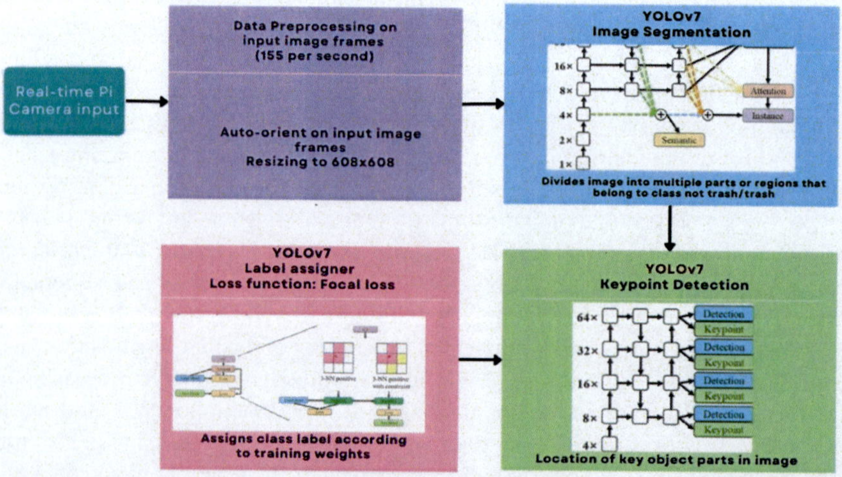

Fig. 1. Proposed model of using YOLOv7 for real-time trash detection in our proposed system.

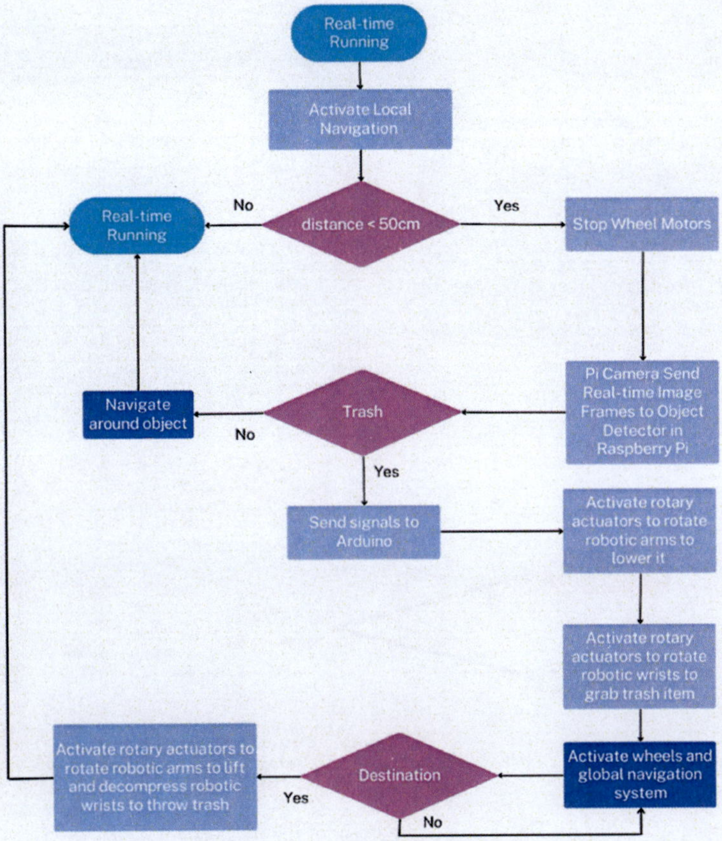

Fig. 2. Flow Chart of Proposed System

3.1 Trash Detection and Classification on YOLOv7 and YOLOv8

The deep learning models to be examined for our waste management system's object detection and classification are the YOLOv7 and YOLOv8 models, the latest versions in the family of YOLO models. The YOLO family models, which stand for "You Only Look Once", are neural network algorithms that perform forward propagation only, unlike the majority of neural network models, which usually perform both forward and backward propagation; this feature led the YOLO family to be more efficient in both time and resources. The YOLO family is expanding with each version having higher speed and accuracy; the latest established versions are YOLOv7 and YOLOv8 (see Fig. 4) [11]. A 2023 study conducted comparisons between the State-of-the-Arts in object detection and the results conclude that the YOLOv7 model outperforms all proclaimed object detectors; it has the highest accuracy of 56.8%.

AP and the highest speed in the range 5 to 120 FPS among all established object detectors [10]. Since YOLOv8 is the most recent version of the YOLO family, we decided to deploy both the State-of-the-Art YOLOv7 and YOLOv8 for our proposed

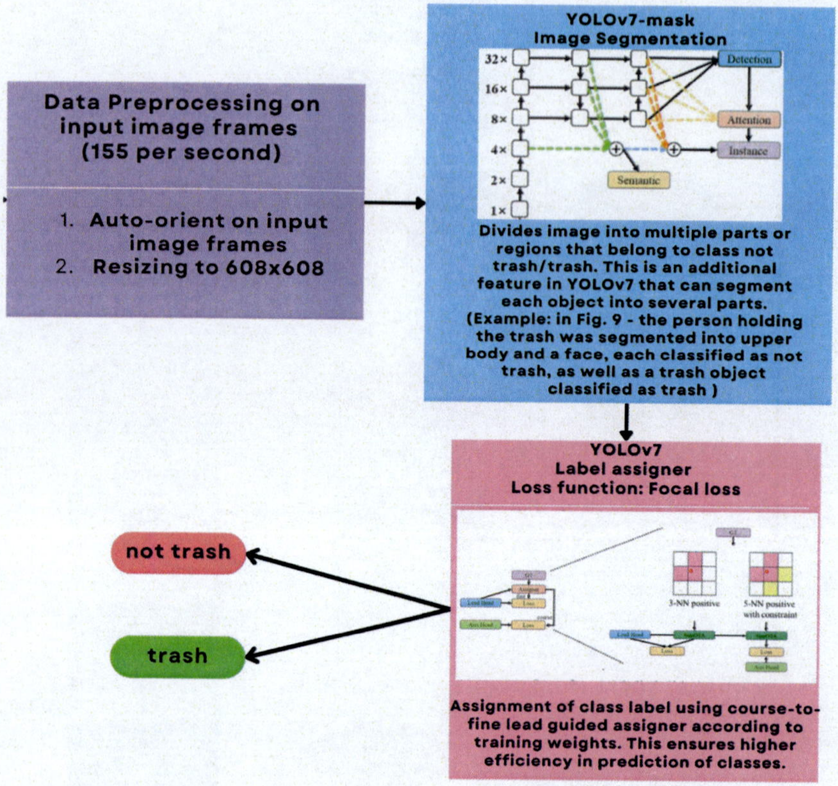

Fig. 3. The detailed processes of the proposed model

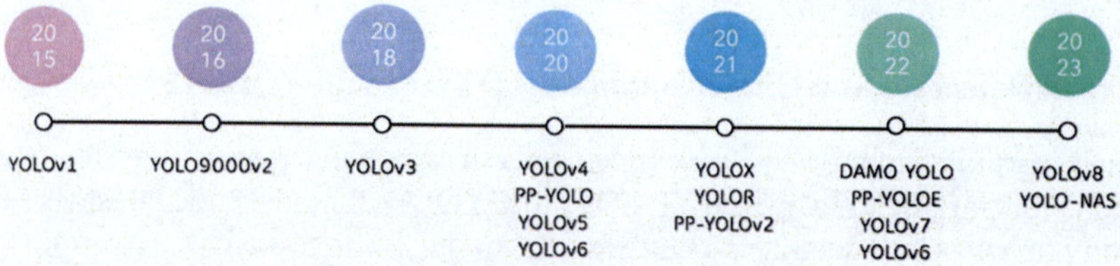

Fig. 4. Timeline of YOLO Versions

system and compare the outcomes using their confusion matrices; this will be discussed and analyzed in Sect. 4.

The YOLOv7 and YOLOv8 models are originally conducted on the MS COCO dataset [10, 11]. The MS COCO dataset does not contain sufficient data of trash items as garbage is not included as one of their 80 classes; this means that for the proposed model to accurately detect and classify trash, the model should be trained on a specialized dataset of trash and produce new weights to be able to classify a category outside its original domain. For the purpose of our waste management system, the YOLOv7 and YOLOv8 models are trained on a trash dataset containing 6,129 images of annotated trash. The trash dataset is split into 5,021 for training, 1,102 for validation and 6 for testing. The dataset is divided into two main classes: trash and not trash.

After training both models on 56 epochs for our trash dataset, the resulting F1 curves, Precision-Recall curves and Confusion Matrices for each model were as seen in Sect. 3.

3.2 Autonomous Navigation System

Advancements using modern machine learning approaches have been taking place in the field of autonomous navigation [12]. Several studies have shown that the utilization of machine learning algorithms helps tackle common challenges faced by traditional navigation systems [12, 13]. Traditional navigation systems are implemented on a large-scale web mapping, which is the process of mapping geospatial data [5, 12]. Using large-scale web mapping, such as Google Maps for GPS tracking, allows only outdoor navigation; its limitations appear in small-scaled, specific domains such as indoor spaces [5, 12, 13]. Consequentially, indoor navigation system requires what is called "semantic mapping" on small-scale domains [13]. Most indoor navigation techniques depend on sensors such as visible light (LiDARs), ultrasonic, and infrared (IR) [5, 7, 12]. Each of these techniques has shown their limitations; for example, LiDARs, which are the sensors present in our proposed system's hardware components, fail to work under rainy conditions due to raindrops reflecting light [12]. LiDARs also fail to detect transparent objects, which will be crucial in our proposed model as a large number of trash items tend to be transparent (such as plastic bags, bottles, and glass) [12]. For the purpose of having an efficient and accurate waste allocation robot, we will delve into different propositions for navigation.

The main purpose of our system is to have a robot that autonomously knows where to move. The navigation system should compose of global and local paths [14]. The global and local paths should be activated when the robot detects a trash item and intends to locate to its goal-directed destination to dispose of the trash. The local path, on the other hand, should be activated when the robot is unoccupied by a current trash item and is in real-time look-out for present waste. The local path should initially be random by default, it should then navigate the robot towards any obstacle detected and stop in the immediate vicinity in order to allow more accurate image frames for trash detection and classification.

A proposed approach for local autonomous navigation is an end-to-end CNN architecture [12]. This approach simulates human local navigation, where the robot's vision through the camera is what is seen by a human [12]. The limitations in this architecture occurred when it faced accumulated errors due to wrong velocities; the system requires the robot to have strict linear and angular velocities in order to avoid fatal malfunctions [12]. Another proposition for local autonomous navigation is the Intention-Net deep learning model; this model integrates a path planner and a motion controller with deep learning to achieve robust navigation from an object to its goal [14]. However, Intention-Net was not able to perform robot localization [14], which is an essential aspect of our waste management system in order to be able to navigate our robot efficiently to the goal trash collecting destinations.

In their brilliant paper, Moustafa Youssef et al. proposes a navigation system, Dejavu, that outperforms the GPS in both accuracy and energy efficiency [15]. The system computes localization through virtual landmarks using low-energy consuming sensors in mobile devices, such as accelerometers [15]. This technique resulted in 347% less energy consumption than the GPS, which will be significantly power-saving for our proposed model [15].

Another consideration for our robot's navigation system Dejavu is the Kalman Filter. Kalman Filter is a reputable mathematical algorithm utilized for state estimation in scenarios where there are noisy measurements or incomplete observations present. Rudolf Kalman invented this model in the 1960s and is now widely regarded as an indispensable technique across many fields such as control systems engineering and robotics. Its applicability to robotics centers around leveraging sensor readings like GPS or accelerometer (in our case Dejavu) despite their noise levels to assess a robot's position and velocity correctly, thereby estimating its state quite accurately. This algorithm functions using robotic dynamics information along with updated sensor input tied together iteratively until true states are more precise over time- thus becoming more reliable estimates overall compared to original ones initially projected; additionally, recursion manifests as a trait useful for applying real-time techniques backed with low-power hardware computation.

Wall-E's proposed hardware has an in-built optional location tracking system that relies on LiDARs. The readings of LiDARs are rarely error-free - producing uncertainties that can compromise the accuracy of our model's movement tracker performance, especially in the presence of transparent objects. Therefore, we will consider Dejavu as our main navigation system for local navigation. Combined with Dejavu, the Kalman Filter can enhance the ability to estimate Wall-E's true position even with noise-affecting sensory input by leveraging past predictions and current measurements. This algorithm enhances the tracking capabilities significantly by filtering out noise from sensory data resulting in more precise location data for route planning purposes. Ultimately this equips users with more reliable monitoring tools to track waste disposal operations effectively over longer periods. The Kalman filter can therefore help a robot accurately determine where it is and where it might be headed next with greater precision relying upon its internal sensor data provided by Dejavu.

By incorporating sensory measures received through both sources together, accurate estimations become possible for one's current location; moreover, estimates can be made on probable positions into the foreseeable future but not beyond until subsequent data allows update capabilities without error. Conversely, imperfections within gathered measures must be considered during tracking processes using two distinct measures, known as the process and measurement noise covariance matrices. These matrices represent uncertainty in detecting dynamic changes within the sensors or background variations aforementioned.

Furthermore, real-time applications having strict latency requirements may also align themselves well with using this algorithm due to its operational computational efficiency in reduced response times. In situations involving noisy sensor-data measurements during Trash Detection, it is possible to rely on the Kalman filter whereby uncertainties and noise elements are significantly reduced from such sensor measurements. A reliable solution is, therefore, achieved through its ability to provide accurate estimates in spite of noisy or corrupted sensor data.

After exploring ways to optimize the performance of Wall-E with scenarios that include limited labeled training data or linear mathematical model dynamics for our trash detection system where multiple sensors such as GPS and Dejavu systems or image sensors are involved we can consider applying Kalman Filtering Algorithm.

4 Evaluation of Results

4.1 Analysis and Evaluation on Trash Detection and Classification Models

A confusion matrix in machine learning is a table representing the outcomes of the predicted value for positive and negative classification alongside the actual values; this will help visualize which model produced more accurate results. Figures 5 and 6 represent the confusion matrices of YOLOv7 and YOLOv8 respectively.

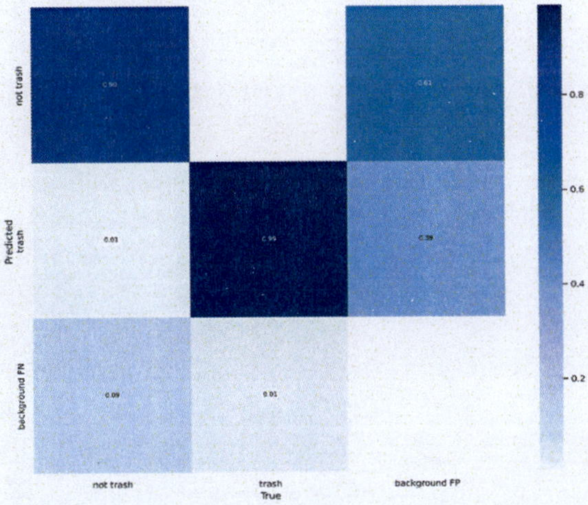

Fig. 5. Confusion Matrix of YOLOv7 on Trash Dataset

In order to interpret the performance of each model through its Confusion Matrix, we will examine the two outcomes of each trash classification model in terms of their true positive (TP), true negative (TN), false positive (FP) and false negative (FN):

1. True positive: refers to samples that are classified correctly as trash.
2. True negative: refers to samples that are correctly classified as not trash.
3. False positive: refers to samples that are incorrectly classified as trash but belong to the class of not trash.
4. False negative: refers to samples that are incorrectly classified as not trash but belong to the class of trash.

The two models were very near in TP performance; they both produced a TP of 0.99, which indicates great performance in correctly classifying trash. The YOLOv8 model outperformed the YOLOv7 model by 3% in its TN; YOLOv7, on the other hand, produced lower FP and FN, which is also desirable to achieve more accurate results.

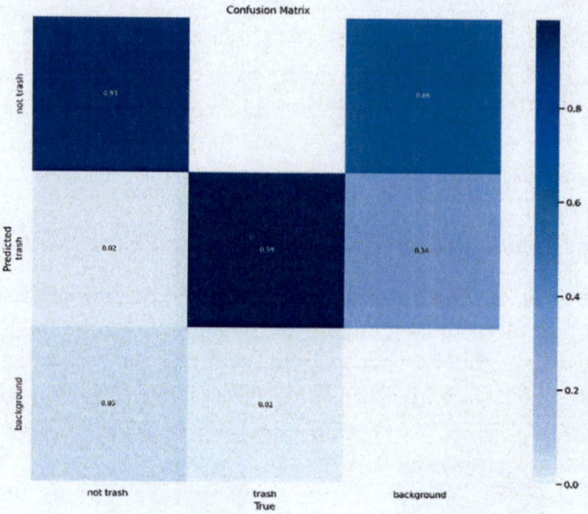

Fig. 6. Confusion Matrix of YOLOv8 on Trash Dataset

To compare overall performance, Eqs. 1, 2, 3, 4, and 5 will be used to compute the accuracy, precision, recall, F1 score, and specificity of each model. Results are shown in Table 1.

$$Accuracy = \frac{TP + TN}{TP + FP + TN + FN} \tag{1}$$

$$Precision = \frac{TP}{TP + FP} \tag{2}$$

$$Recall = \frac{TP}{TP + FN} \tag{3}$$

$$F1 - Score = \frac{2 * Precision * Recall}{Precision + Recall} \tag{4}$$

$$Specificity = \frac{TN}{FP + TP} \tag{5}$$

Table 1. Evaluation of overall performance on YOLOv7 and YOLOv8 models

Model	Accuracy	Precision	Recall	F1-Score	Specificity
YOLOv7	0.9947	0.99	1.0	0.995	0.9890
YOLOv8	0.9897	0.98	1.0	0.990	0.9789

Results confirm that the YOLOv7 model outperforms YOLOv8, achieving a higher accuracy, precision, F1-score and specificity. To visualize the performances tests images,

Precision-Recall and F1 curves are plotted for each model trained on the trash dataset for 56 epochs (see Figs. 7, 8 and 9). Additionally, YOLOv7 was able to perform instance image segmentation (see Fig. 10). Instance segmentation divides and separates each instance of the appearing object in an image, allowing separate classification for each separated part. Which can help in the accurate detection of trash in environments with multi-class objects. For that reason, our waste management system will deploy YOLOv7 for higher accuracy real-time trash detection and classification.

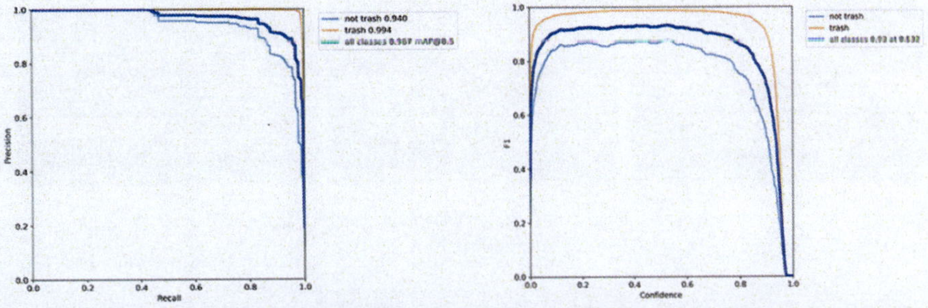

Fig. 7. Precision-Recall and F1 Curves of YOLOv7 on Trash Dataset

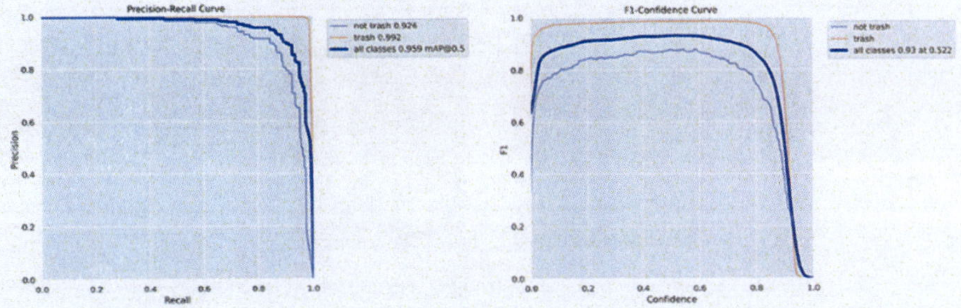

Fig. 8. Precision-Recall and F1 Curves of YOLOv8 on Trash Dataset

4.2 Analysis and Evaluation on Dejavu

Moustafa Youssef et al. compare the localization accuracy and power consumption of the Dejavu and GPS models (see Table 2) [15]. It can be seen that their proposed model achieved a lower localization error and power consumption for pedestrian localization. Therefore, our proposed system will attempt to implement Dejavu as its navigating system. For future work, we can also examine the potential of the system for indoor localization.

Fig. 9. Test Images on YOLOv7 and YOLOv8

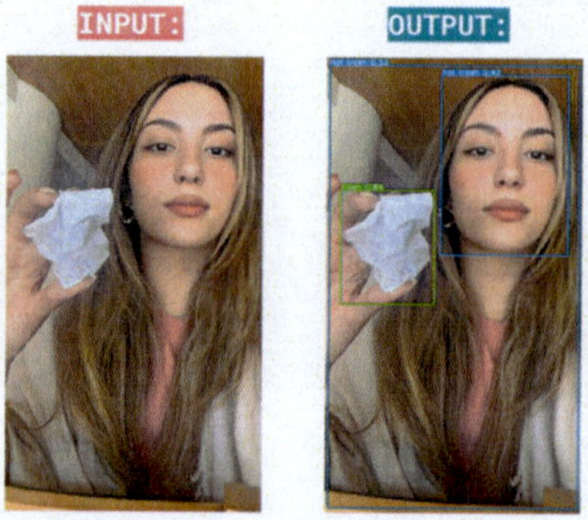

Fig. 10. Example of Instance Image Segmentation in YOLOv7

Table 2. Comparisons, calculated relative to Dejavu, between Dejavu and GPS for pedestrian localization [15]

Model	Localization Error		Power Consumption
	Mean	Variance	
Dejavu	3.0 m	43.2	59.9 mW
GPS	9.9 m	23	268 mW

5 Conclusions

The purpose of this research paper is to propose and discuss an integration of a State-of-the-Art deep learning models with an accurate, energy-efficient navigation system in order to implement a waste allocating robot that is high in accuracy and power-saving. The software system consists of two main areas: trash detection and classification as well as a navigation system to help implement an autonomous robot. In our conducted experiments, the highest achieving object detection model was YOLOv7, which will be the utilized in our robot to detect trash items sent through image frames from a Pi camera. Different navigation systems were conceptually discussed and it was concluded that the Dejavu navigation system meets the requirements of our robotic system and will therefore be examined for future work on our waste management system.

References

1. Ye, Y.X., Lu, A.N., You, M.Y., Huang, K., Jiang, B.: Wireless localization based on deep learning: state of art and challenges. Math. Probl. Eng. Article ID. 5214920, 1–8 (2020). https://doi.org/10.1155/2020/5214920
2. Liu, C., et al.: A domestic trash detection model based on improved YOLOX. Sensors **22**(18), 6974 (2022)
3. Myat Noe, S., Zin, T.T., Tin, P., Kobayashi, I.: Comparing state-of-the-art deep learning algorithms for the automated detection and tracking of black cattle. Sensors **23**(1), 532 (2023)
4. Salimi, I., Dewantara, B.S.B., Wibowo, I.K.: Visual-based trash detection and classification system for smart trash bin robot. In: 2018 IEEE International Electronics Symposium on Knowledge Creation and Intelligent Computing (IES-KCIC), pp. 378–383, October 2018
5. Sakpere, W., Adeyeye-Oshin, M., Mlitwa, N.B.: A state-of-the-art survey of indoor positioning and navigation systems and technologies. South Afr. Comput. J. **29**(3), 145–197 (2017)
6. Hossain, S., et al.: Autonomous trash collector based on object detection using deep neural network. In: TENCON 2019–2019 IEEE Region 10 Conference (TENCON), pp. 1406–1410 (2019)
7. El-Ghamrawy, S.M., El-Desouky, A.I., Sherief, M.: Dynamic ontology mapping for communication in distributed multi-agent intelligent system. In: 2009 International Conference on Networking and Media Convergence, pp. 103–10 (2009)
8. Kulkarni, S., Sarang Junghare, S.: Robot based indoor autonomous trash detection algorithm using ultrasonic sensors. In: 2013 International Conference on Control, Automation, Robotics and Embedded Systems (CARE), pp. 1–5 (2013). https://doi.org/10.1109/CARE.2013.6733698
9. Elghamrawy, S.: An H_2O's deep learning-inspired model based on big data analytics for coronavirus disease (COVID-19) diagnosis. In: Hassanien, A.E., Dey, N., Elghamrawy, S. (eds.) Big Data Analytics and Artificial Intelligence Against COVID-19: Innovation Vision and Approach. SBD, vol 78, pp. 263–279. Springer, Cham (2020). https://doi.org/10.1007/978-3-030-55258-9_16
10. Wang, C.Y., Bochkovskiy, A., Liao, H.Y.M.: YOLOv7: trainable bag-of-freebies sets new state-of-the-art for real-time object detectors. In: Proceedings of the IEEE/CVF Conference on Computer Vision and Pattern Recognition, pp. 7464–7475 (2023)
11. Terven, J., Cordova-Esparza, D.: A comprehensive review of YOLO: From YOLOv1 to YOLOv8 and beyond. arXiv preprint https://arxiv.org/abs/2304.00501 (2023)
12. Kim, Y.H., Jang, J.I., Yun, S.: End-to-end deep learning for autonomous navigation of mobile robot. In: 2018 IEEE International Conference on Consumer Electronics. ICCE18, pp. 1–6 (2018)
13. Sepulveda, G., Niebles, J.C., Soto, A.: A deep learning-based behavioral approach to indoor autonomous navigation. In: 2018 IEEE International Conference on Robotics and Automation (ICRA18), pp. 4646–4653 (2018)

14. Gao, W., Hsu, D., Lee, W.S., Shen, S., Subramanian, K.: Intention-net: integrating planning and deep learning for goal-directed autonomous navigation. In: Conference on Robot Learning, pp. 185–194 (2017)
15. Aly, H., Basalamah, A., Youssef, M.: Accurate and energy-efficient GPS-less outdoor localization. ACM Trans. Spatial Algorithms Syst. (TSAS17) **3**(2), 1–31 (2017)
16. AleniziI, N., Alajmi, O., Alsharhan, S., Khudada, S.: Autonmous beach cleaner. In: URC Conference Dubai (2019). http://hdl.handle.net/11675/5187
17. Bansal, S.P., Shah, I., Patel, P., Makwana, P., Thakker, D.R.: AGDC: automatic garbage detection and collection. arXiv preprint https://arxiv.org/abs/1908.05849 (2019)
18. Meghna, A., Immanuel, M., Subhagan, P., Raj, R.: Trash bot. Int. J. Res. Eng. Sci. Manag. **2**(7), 301–304 (2019)

Research on the Application of Data Mining Based on Deep Learning in the Analysis of Psychoeducational Factors

Guo Tian(✉)

Jiangsu University of Science and Technology, Zhenjiang City, Jiangsu, China
sweeterguo@163.com

Abstract. The role of psychoeducational factor analysis in college students' psychological counseling education is very important, but there is a problem of inaccurate outcome evaluation. The traditional education model cannot solve the problem of factor analysis and evaluation in college students' psychological counseling education, and the evaluation is unreasonable. Therefore, this paper proposes a data mining algorithm to perform optimization factor analysis and evaluation analysis. Firstly, cognitive theory is used to evaluate the influencing factors, and the indicators are divided according to the requirements of factor analysis and evaluation, so as to reduce the interference factors in factor analysis and evaluation. Then, cognitive theory analyzes and evaluates the factors affecting the mental health of college students, forms a factor analysis and evaluation scheme, and comprehensively analyzes the factor analysis and evaluation results. MATLAB simulation shows that under certain evaluation criteria, the data mining algorithm is better than the traditional education model in analyzing and evaluating the factors affecting the mental health of college students.

Keywords: cognitive theory · data mining algorithms · Mental health · Psychological counseling

1 Introduction

The quality of psychological education is one of the important contents of evaluating the mental health standards of college students, and it is of great significance for college students' psychological counseling [1]. However, in the process of psychoeducational factor analysis and evaluation, the factor analysis and evaluation scheme have the problem of poor accuracy, which directly affects the effect of mental health counseling for college students [2]. Some scholars believe that the application of data mining algorithms to the analysis of college students' psychological counseling education [3] can effectively analyze the factor analysis and evaluation scheme and provide corresponding support for factor analysis and evaluation [4]. On this basis, a data mining algorithm is proposed to optimize the factor analysis evaluation scheme and verify the effectiveness of the model [5]. The data mining algorithm adopts a random optimization strategy for

A. Hassanien et al. (Eds.): AISI 2023, LNDECT 184, pp. 67–74, 2023.
https://doi.org/10.1007/978-3-031-43247-7_6

college students' psychological counseling education and adjusts the parameters of influencing factors to realize the optimization of college students' psychological counseling education. The data mining algorithm divides the psychological counseling education of college students into different factors analysis and evaluation levels, and randomly selects different schemes. In the iterative process, the factor analysis and evaluation scheme of different factor analysis and evaluation levels are optimized and analyzed. After the optimization analysis is completed, the factors of different programs are compared to analyze the evaluation level, and the best psychological counseling education for college students is recorded.

2 Mathematical Description of a Data Mining Algorithm

The data mining algorithm uses deep learning theory to optimize the factor analysis and evaluation scheme, and according to the indicators in the factor analysis and evaluation, finds the unqualified value in the psychological counseling education of college students is h, and integrates the factor analysis and evaluation scheme is p, and finally judges the feasibility of psychological counseling education for college students is $F(a_{ij})$, which is calculated as shown in Eq. (1).

$$F(a_{ij}) = h - p \tag{1}$$

Among them, the judgment of outliers is shown in Eq. (2).

$$h - p = \int k(x + y) \tag{2}$$

The data mining algorithm combines the advantages of deep learning theory and quantifies the psychological counseling education of college students, which can improve the quality of factor analysis and evaluation of psychological education.

Suppose that the factor analysis evaluation requirements is w, the factor analysis evaluation scheme is n_i, the satisfaction of the factor analysis evaluation scheme is y, and the factor analysis evaluation scheme judgment function is $Q(n_i)$, as shown in Eq. (3).

$$Q(n_i) = \prod_{i=1} \theta + w + y \tag{3}$$

2.1 Choice of Psychoeducational Quality Programmers

Hypothesis II. The function of college students' psychological counseling education is $O(v)$, and the weight coefficient is x_i, then, the factor analysis evaluation requires unqualified college students' psychological counseling education as shown in Eq. (4).

$$O(v) = \prod -x_i + j_i + l\pi \tag{4}$$

According to hypotheses I and II, the comprehensive function of psychoeducation can be obtained, and the result is shown in Eq. (5).

$$Q(n_i) + O(v) \leq h - p \tag{5}$$

In order to improve the effectiveness of the analysis and evaluation of influencing factors, all data need to be standardized and the results are shown in Eq. (6).

$$Q(n_i) + O(v) \leftrightarrow \int k(x + y) \tag{6}$$

2.2 Analysis of Factor Analysis Evaluation Programs

Before carrying out the data mining algorithm, it is necessary to conduct multi-dimensional analysis of the factor analysis and evaluation scheme and map the factor analysis and evaluation requirements to the college student psychological counseling education database and eliminate the unqualified factor analysis and evaluation scheme is $A(f)$. According to Eq. (6), the anomaly evaluation scheme can be proposed, and the results are shown in Eq. (7).

$$A(f) = \frac{Q(n_i) + O(v)}{\int k(x + y)} \tag{7}$$

Among them, it is $\frac{Q(n_i)+O(v)}{\int k(x+y)} \leq 1$, stated that the scheme needs to be proposed, otherwise the scheme needs to be integrated into it is $R(m_i)$, and the result is shown in Eq. (8).

$$R(m_i) = \min[\sum Q(n_i) + O(v)] \tag{8}$$

The psychological counseling education of college students is comprehensively analyzed, and the threshold and index weight of the factor analysis evaluation scheme are set to ensure the accuracy of the data mining algorithm. Psychological counseling education for college students is a systematic test factor analysis and evaluation scheme, which needs innovative analysis. If the psychological counseling education of college students is in a non-normal distribution, the factor analysis evaluation scheme will be affected is $accur(y_i)$, and the accuracy of the overall factor analysis evaluation will be reduced is $W(x_i)$, and the calculation result is shown in Eq. (9).

$$W(x_i) = \frac{\min[\sum Q(n_i) + O(v)]}{\sum Q(n_i) + O(v)} \times 100\% \tag{9}$$

The analysis and evaluation scheme of survey factors showed that the psychoeducational quality program showed a multidimensional distribution, which was in line with the objective facts. The psychological counseling education of college students is not directional, indicating that the quality program of psychological education has strong randomness, so it is regarded as a high analytical study. If the stochastic function of college student counseling education is $\sinh(x_i)$, then the calculation of formula (9) can be expressed as formula (10).

$$W(x_i) = \frac{\min[\sum Q(n_i) + O(v)]}{\sum Q(n_i) + O(v)} \times 100\% + \sinh(x_i) \tag{10}$$

Among them, the psychological counseling education of college students meets the normal requirements, mainly the deep learning theory adjusts the psychological counseling education of college students, removes the duplicate and irrelevant schemes, and supplements the default scheme, so that the dynamic correlation of the whole factor analysis evaluation scheme is strong.

3 Psychological Counseling Education for College Students: Practical Cases

3.1 Factor Analysis and Evaluation Presentation

In order to facilitate factor analysis and evaluation, this paper takes college students' psychological counseling education under complex circumstances as the research object, and the factor analysis and evaluation scheme of college students' psychological counseling education is shown in Table 1. While Fig. 1 illustrates the factor analysis evaluation process in steps.

Table 1. Analysis and Evaluation Requirements for Factors Affecting the Mental Health of College Students.

Scope of application	grade	Affects the effect	Analyze the effect
Family factors	I	68.76	69.54
	II	63.84	65.42
environmental factors	I	64.09	65.85
	II	66.66	62.23
Social factors	I	64.84	66.60
	II	67.37	67.36

Compared with the traditional education model, the factor analysis and evaluation scheme of the data mining algorithm is closer to the actual factor analysis and evaluation requirements. In terms of the rationality and fluctuation range of college students' psychological counseling education, the data mining algorithm is better than the traditional education model. Through the factor analysis and evaluation scheme changes in Fig. 1, it can be seen that the stability of the data mining algorithm is better, and the early warning judgment speed is faster. Therefore, the factor analysis and evaluation scheme of the data mining algorithm has better early warning judgment speed, psychoeducational quality scheme, factor analysis and evaluation scheme, and summation stability.

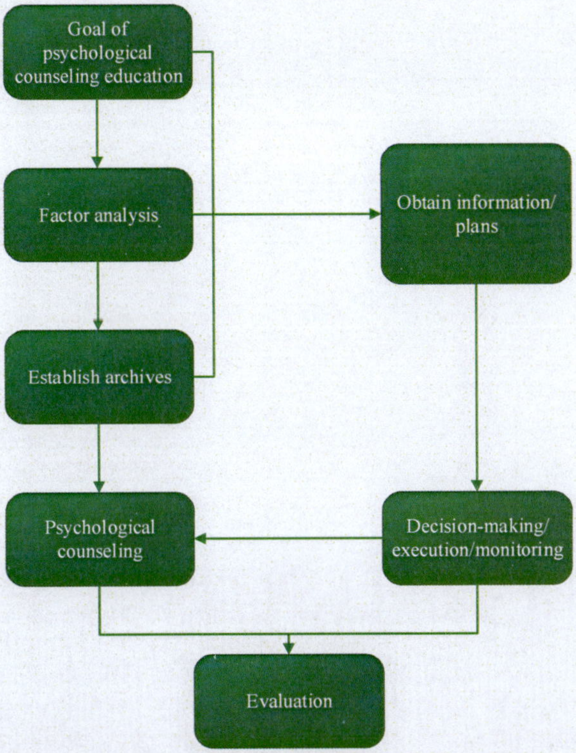

Fig. 1. The Analysis Process of Psychological Counseling Education for College Students

3.2 Psychological Counseling and Education for College Students

The factor analysis and evaluation scheme of college students' psychological counseling education includes unstructured information, semi-structured information and structural information. After the pre-selection of data mining algorithm, a preliminary factor analysis and evaluation scheme for college students' psychological counseling education was obtained, and the feasibility of the factor analysis and evaluation scheme for college students' psychological counseling education was analyzed. In order to more accurately verify the innovative effect of college students' psychological counseling education, different factors were selected to analyze and evaluate the evaluation levels of college students' psychological counseling education, and the factor analysis evaluation scheme was shown in Table 2.

3.3 Quality and Stability of Psychoeducation Evaluated by Factor Analysis

In order to verify the accuracy of the data mining algorithm, the factor analysis evaluation scheme is compared with the traditional education model, and the factor analysis evaluation scheme is shown in Fig. 2.

It can be seen from Fig. 2 that the psychological education quality of the data mining algorithm is higher than that of the traditional education model, but the error rate is lower,

Table 2. Overall situation of Psycho-educational Quality Programmes

category	applicability	accuracy
undergraduate	66.81	64.17
graduate student	66.73	65.37
Doctoral	66.12	68.69
mean	67.75	64.40
X^6	32.61	31.18
P = 3.61		

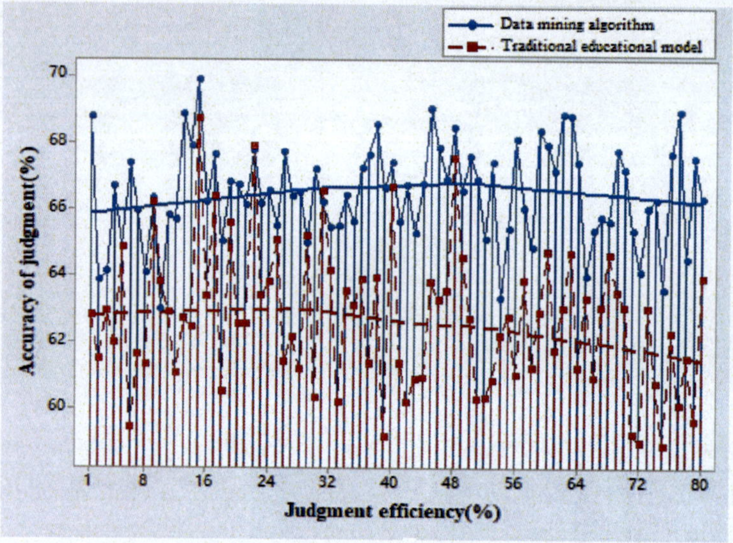

Fig. 2. Psychoeducational quality of different algorithms

indicating that the factor analysis and evaluation of the data mining algorithm is relatively stable, while the factor analysis and evaluation of the traditional education model are uneven. The average factor analysis evaluation scheme of the above two algorithms is shown in Table 3.

Table 3. Comparison of Factor Analysis and Evaluation Accuracy of Different Methods

algorithm	Psychoeducational quality	Magnitude of change	error
Data mining algorithms	82.47	83.40	0.93
Traditional education model	63.61	67.82	4.21
P	31.12	34.82	3.70

It can be seen from Table 3 that the traditional education model has deficiencies in the quality and stability of psychological counseling education for college students, and there have been significant changes in psychological counseling education for college students, and the error rate is relatively high. The general results of data mining algorithms have higher psychoeducation quality and are better than traditional education models. At the same time, the psychoeducational quality of the data mining algorithm is greater than 80%, and the accuracy has not changed significantly. In order to further verify the superiority and effectiveness of the data mining algorithm, the data mining algorithm is generally analyzed by different methods, as shown in Fig. 3.

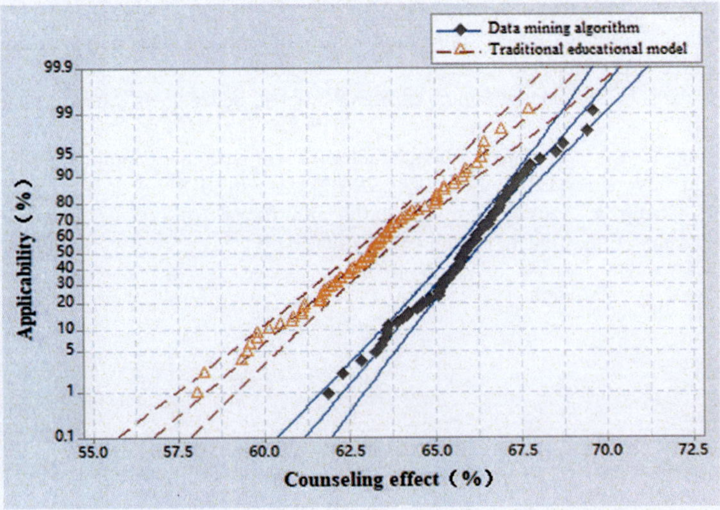

Fig. 3. Data Mining Algorithm Factor Analysis Evaluation of Psychoeducational Quality

It can be seen from Fig. 3 that the psychological education quality of the data mining algorithm is significantly better than that of the traditional education model, and the data mining algorithm increases the adjustment coefficient of college students' psychological counseling education, sets the threshold of influencing factors, and eliminates the factors that do not meet the requirements of the analysis and evaluation scheme.

4 Conclusion

Aiming at the problem of unsatisfactory quality of psychological counseling education for college students, this paper proposes a data mining algorithm and combines deep learning theory to optimize the psychological counseling education of college students. At the same time, in-depth analysis of factor analysis evaluation innovation and threshold innovation is carried out to construct a set of influencing factors. The research shows that the data mining algorithm can improve the accuracy and stability of college students' psychological counseling education and can analyze and evaluate the general factors of college students' psychological counseling education. There are many factors affecting students' mental health, and education itself is the process of realizing personal socialization and socialization of social norms, especially focusing on cultivating students' good ideological and moral character, doing a good job in students' learning, personality, life and career guidance, and working together to create a good environment for healthy growth.

References

1. Wang, J., Yang, S.: Research on analysis algorithm of college students' mental health factors based on big data matching. Electron. Des. Eng. **30**(09), 22–25+30 (2022)
2. Li, Y.: Mental health status and influencing factors of empty nesters in rural China. Chin. J. Publ. Health **37**(02), 333–336 (2021)
3. Li, R.: Analysis of influencing factors of high school students' mental health. Question Res. **04**, 123 (2020)
4. Li, M., Long, Y.: Analysis of mental health factors of Xinjiang ethnic minority college students in mainland universities. Ethnic Educ. Res. **29**(01), 76–82 (2018)
5. Shao, W., Cheng, C.: Analysis of mental health factors of college students and their countermeasures. Educ. Teach. Forum **01**, 36–37 (2018)

Comparing Different Deep Learning Models with a Novel Model for COVID-19 and Pneumonia Classification Using Chest X-Ray Images

C. V. Aravinda[1(✉)], M. S. Sannidhan[1], Jyothi Shetty[1], Shabari Shedthi[1], and Roheet Bhatnagar[2]

[1] Department of CSE, NMAM Institute of Technology, Nitte 574110, India
{aravinda.cv,sannidhan,jyothi_shetty,shabarishetty}@nitte.edu.in
[2] Department of Computer Science and Engineering, Manipal University Jaipur, Jaipur, India
roheet.bhatnagar@jaipur.manipal.edu

Abstract. The global impact of the COVID-19 pandemic has created a significant health crisis affecting millions worldwide. Clinical symptom assessment and chest X-ray tomography are regularly consumed for diagnosing and monitoring COVID-19. To contribute to this effort, our research conducted a comparative analysis of various deep-learning (DL) models for categorizing chest X-ray images of pneumonia and COVID-19, introducing a novel model that outperforms existing ones. The pandemic has intensified the need for prompt diagnosis and treatment. Crucially, chest X-ray imaging has a fundamental role in identifying and tracking the progression of COVID-19. Evaluating our approach on a publicly available chest X-ray dataset, we achieved exceptional accuracy, sensitivity, and specificity rates of 95.7%, 94.3%, and 96.9%, respectively. These results underscore the skill of DL-based approaches in automated COVID-19 discovery from images of chest X-Ray tomography, facilitating swift and accurate diagnosis. Our research demonstrates the promising capacity of DL methods for rapid and precise identification of COVID-19 disease from X-ray images, offering valuable support for timely diagnosis of the condition.

Keywords: Covid-19 · Deep learning · Artificial intelligence · Medical applications

1 Introduction

In the context of the COVID-19, DL models have been employed to diagnose and monitor the disease's progression. Chest X-ray imaging performs a decisive task in identifying both COVID-19 and pneumonia, ensuring accurate diagnosis and effective management [1]. The COVID-19 pandemic has triggered a global health crisis, affecting millions worldwide. Given its high contagiousness, early detection and treatment are crucial for containment. Chest X-ray imaging has emerged as a primary diagnostic tool for COVID-19, enabling disease monitoring. Nonetheless, interpreting chest X-ray images can be

A. Hassanien et al. (Eds.): AISI 2023, LNDECT 184, pp. 75–85, 2023.
https://doi.org/10.1007/978-3-031-43247-7_7

challenging and requires specialized radiologist expertise. Harnessing the potential of DL-based techniques, automating COVID-19 identification through chest X-ray images has shown promise. In this research exertion, we specify an outline of recent research that explores the utilization of DL models for COVID-19 chest X-ray detection [2].

Despite the promising results of DL-based methods for COVID-19 chest X-ray detection, challenges and limitations remain. The effectiveness of DL models can be hindered by the scarcity of annotated data, posing a major obstacle.

2 Background and Literature

2.1 Background

COVID-19 affects the respiratory system and shares symptoms with other respiratory illnesses. Chest X-ray imaging is crucial for diagnosis and monitoring, but accurate interpretation can be challenging. Delays in treatment and virus transmission can result from this difficulty [3].

2.2 Exploring DL Structures for Detecting COVID-19 in Chest X-Ray Images

RNNs excel in tasks involving sequential data, like analyzing time series data of COVID-19 cases. Hybrid architectures that combine CNNs and RNNs offer versatility in analyzing both image and sequential data, making them valuable for COVID-19 research [4].

2.3 DL Techniques in Medical Imaging

DL plays a vital role in medical imaging, facilitating disease identification, differentiation, and classification. Notably, DL methods show promise in automating uncovering of COVID-19 disease from X-ray images, enabling rapid and precise diagnosis of the disease [5, 6].

2.4 Datasets for COVID-19 Chest X-Ray Detection

Datasets like COVID-19 Image Data Collection, RSNA Pneumonia Detection Challenge, and Covidx have been used for discovery of COVID-19 in chest X-rays. They contain images of healthy individuals, pneumonia, COVID-19 [7].

2.5 Literature Work

DL procedures, outstandingly convolutional neural networks (CNNs), have been extensively used for automatic screening of COVID-19 patients using radiological images like CT scans [8–10]. These CNN-based techniques have established favorable conclusions in detecting COVID-19 in chest radiography images. Diverse findings have devoted distinct architectures, such as triple-view CNNs and Bayesian CNNs, to extract meaningful features and estimate detection uncertainty [11]. Pre-trained CNN networks like

COVID-19-Net, VGG19, DenseNet201, and MobileNetv2, along with transfer learning techniques using the ImageNet dataset, have also been utilized to achieve accurate predictions with limited data. The comparison of clinical imaging data with laboratory tests has the potential to facilitate early diagnosis of COVID-19 [12–14].

2.6 Objectives

To expedite COVID-19 identification, we aim to develop a computer-aided detection system using chest X-rays (CXRs). This system will distinguish COVID-19 cases from healthy and pneumonia cases automatically [15]. Refer to Fig. 1 for the proposed workflow.

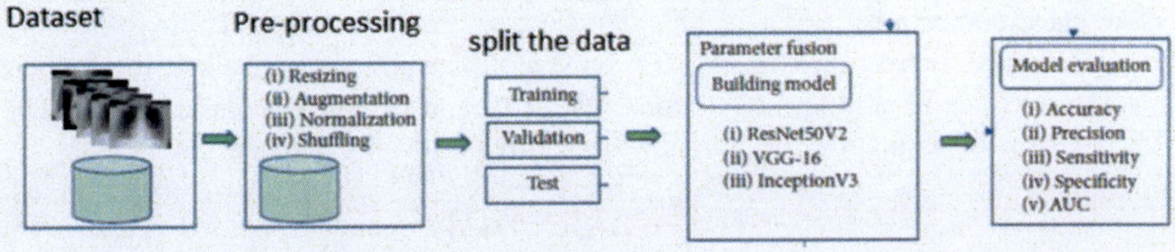

Fig. 1. Flow of Proposed architecture

2.7 Imbalanced Data-Set Handling

To address data imbalance, we utilized the raw input properties of X-rays and evenly divided the dataset into four parts. Each part consisted of 300 samples for non-infected and pneumonia cases, matching the amount of COVID-19 cases (125 samples). This technique guaranteed equal representation of all classes in the dataset [16].

3 Methods

We utilized a publicly available chest X-ray image dataset that includes images of COVID-19 patients, pneumonia cases, and healthy individuals. Four deep learning models (VGG-16, Inception V3, ResNet-50, and DenseNet-121) were compared for their performance. Furthermore, we introduced a novel model that combines the strengths of ResNet-50 and DenseNet-121 [17].

Preprocessing

We enhanced the image quality and reduced noise through preprocessing. Normalization and resizing to a fixed size ensured consistency. Data augmentation techniques, including rotation, flipping, and zooming, expanded the training dataset and reduced overfitting.

- Mean value and standard deviation

  These metrics are key statistical measures used to preprocess and normalize chest X-ray images for COVID-19 and pneumonia classification in deep learning models. Normalizing pixel intensities by the standard deviation (Eq. (1)) enhances deep learning model performance.

$$X = (X - \mu) \tag{1}$$

where the mean value of the pixel intensity in the image is and X is the value of the pixel intensity. A standard deviation regularization formula is:

$$X = (X - \mu)/\sigma \tag{2}$$

- Skewness correction

  Skewness correction adjusts pixel intensity distributions in chest X-ray images to improve deep learning models for COVID-19 and pneumonia classification. The Box-Cox transformation formula is widely utilized for this purpose:

$$Y = [(X^{\lambda} - 1)/\lambda] \tag{3}$$

  Box-Cox transformation enhances the allocation of pixel intensities in chest X-ray images. The optimal Box-Cox parameter is computed across maximum possibility approximation for improved normalization.
- Contrast of the x-ray

  Contrast enhancement improves the visual appeal and feature representation of chest X-ray images, enhancing the effectiveness of DL models for COVID-19 and pneumonia classification. Histogram equalization is a procedure adopted to increase contrast between different regions of the image, making features more distinguishable:

$$H(v) = round(((C(v) - Cmin)/(MxN - 1))x(L - 1)) \tag{4}$$

The equation is defined in connection to the image dimensions: M (number of rows), N (number of columns), and L (total number of possible output intensity levels). $H(v)$ represents the output pixel intensity value, $C(v)$ represents the increasing dispersal function of the input pixel intensities up to v, and $Cmin$ is the minimum cumulative distribution function value.

3.1 Handling Imbalanced Datasets

Additionally, the class division in the dataset is unequal, requiring an assessment of the level of imbalance. The calculation of the imbalance ratio, as depicted in Eq. (5), allows us to quantify the extent of this imbalance.

$$ImbalancedRation = \frac{majority - class}{minority - class} \tag{5}$$

3.2 Convolution Network Model

Transfer Learning (TL) uses pre-trained models, like those trained on ImageNet, to classify pneumonia and chest X-rays. TL models obtain features from the images and can be fine-tuned on a smaller image dataset of chest X-ray for more precise classification. Pre-trained CNNs from ImageNet are employed to extract features from X-ray images of chest for classification of pneumonia [18].

Performance Measure's

A confusion matrix is adopted to calculate performance indicators for our classification task. Figurc 2 portrays thc samc.

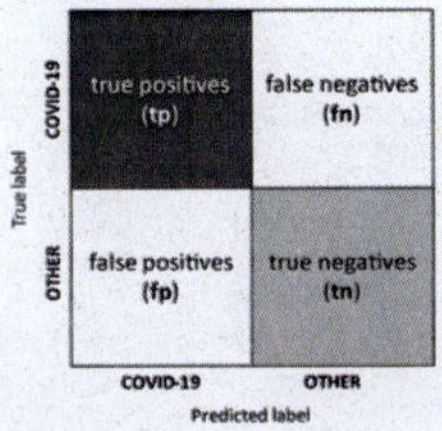

Fig. 2. Visualization of confusion-matrix

4 Proposed-Method

Resizing X-ray images of the chest to a square format for CNN inputs can result in distortion, as the images in our datasets have diverse sizes due to equipment variations and preprocessing [19] (Fig. 3).

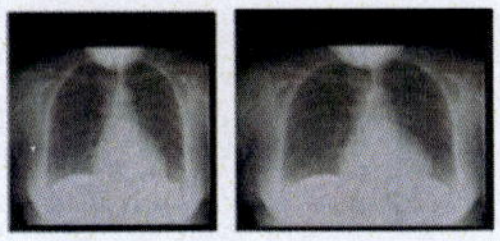

Fig. 3. Original and Resize of patient's image.

To prevent distortion in the input X-ray, we performed preprocessing steps (Figs. 4 and 5). Figure 6 and 7 show examples of positive and negative chest X-ray samples. Conclusions: 1) Elimination of black bands around the margins and 2) Scaling down the image to a width of 299 pixels at the smallest boundary.

4.1 Data-Partition

We divided the data into training, validation, and test sets. The classes of COVID-19 was divided, with 10% used for validation and the remainder for training. The HEALTHY and PNEUMONIA classes followed an 80–20 split. Table 1 provides entire picture of the samples [20].

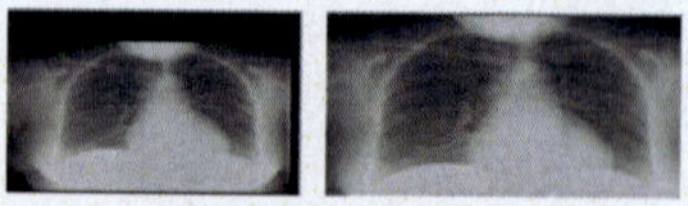

Fig. 4. Image resized.

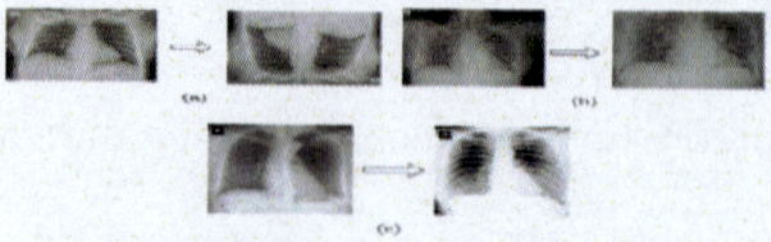

Fig. 5. Flow-of-pre-processing

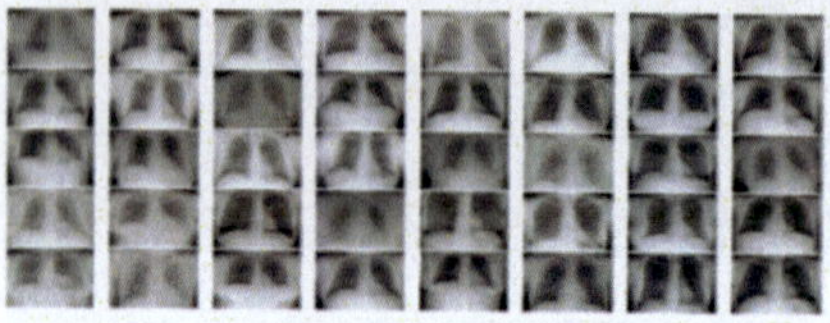

Fig. 6. Sample of Positive-x-ray

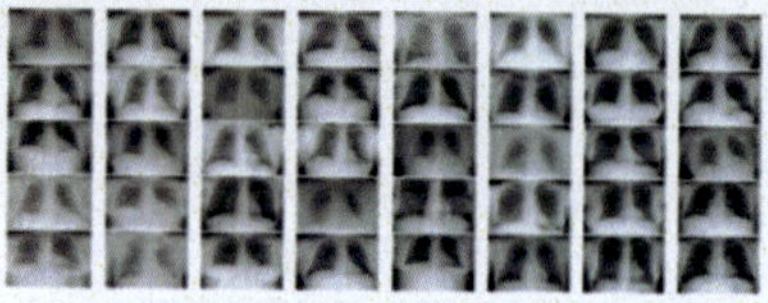

Fig. 7. Sample of Negative-x-ray

Table 1. Division of the 3 different datasets for experimentation

Division-of-Data	Covid-19	Healthy-Patient	Pneumonia
Training	335	2087	3000
Validation	60	250	800
Test	30	250	350

5 Results-Discussion

In this section, we present the methodology, report the findings from the validation set, and demonstrate the effectiveness of disease screening and classification [21].

Experimental Evaluation and Setup

We evaluated several pre-trained CNN models for detecting COVID-19 from X-ray images of the chest. The selected models included Inception, ResNet V2, DenseNet201, ResNet152 V2, Xception, MobileNet V2, VGG16, VGG19, and NASNetLarge. We TL models and incorporated ImageNet weights. This portion of the article portrays the methodology, validation set results, and demonstrates the efficacy of classification and disease screening.

- **Accuracy:** The accuracy implies to the proportion of correctly classified samples out of all the samples, Eq. (6) provides mathematical computation:

$$(TP + TN)/(TP + TN + FP + FN) \tag{6}$$

- **Precision:** The true positive rate is the proportion of positive samples classified precisely. It is calculated as

$$TP/(TP + FP) \tag{7}$$

- **Recall:** a metric employed to measure the efficacy of identifying positive samples. The calculation for recall is as follows:

$$TP/(TP + FN) \tag{8}$$

- **F1-score:** This metric quantifies the execution of a binary classification model by considering precision value and recall value.

$$2X\,(precisionxrecall)/(precision + recall) \tag{9}$$

- **AUC-ROC:** The ROC curve plots the relationship between True Positive Rate (TPR) and False Positive Rate (FPR) at different classification thresholds. The area under this curve is calculated to gauge the execution of a binary classification model.

  DL models for chest X-ray and pneumonia categorization achieve accuracy scores ranging from 85% to 99% based on the dataset and evaluation metrics employed.

5.1 Summary of the Results

Images are resized and cropped using the suggested technique. The validation procedure creates training, validation, and test sets. Normalization is employed to the images. Xception model is picked and assessed on the test set for efficacy [22].

5.2 Validation-Test-Results

The Xception network was trained using ImageNet, and the excellent model was carefully chosen grounded on the loss function score. Validation results for ResNet and VGG19 are indicated in Figs. 10 and 11, and Figs. 12 and 13 display Accuracy, Precision, Recall, and ROC AUC during validation. Table 2 summarizes the results, while Figs. 14 and 15 depict model accuracy and loss. This section outlines the experimental methodology and reports the findings of the validation set, demonstrating the effectiveness of disease screening and classification (Fig. 8 and 9).

Table 2. Performance results of Models

Model	Best-Epoch	Validation-Loss	Convergence-Time	Training-time
VGG16	35	0.08735	0.2480 s	4320 s
ResNet50	25	0.07331	0.0187	650 s
DenseNet121	10	0.01922	0.0156 s	1350 s
Ourmodel	7	0.04619	0.0220 s	1500 s

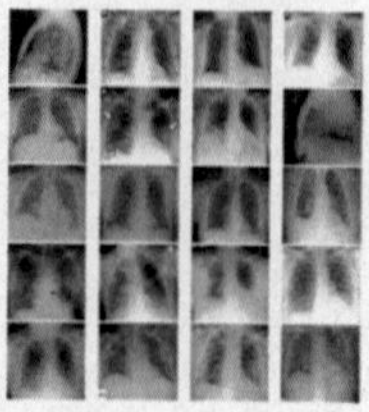

Fig. 8. Testing the images

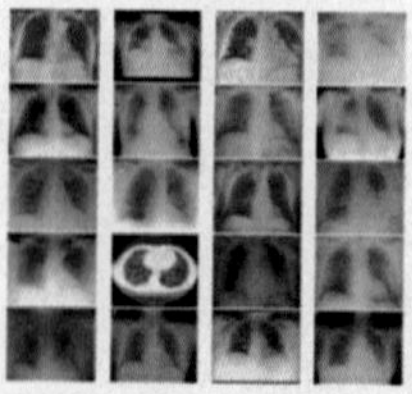

Fig. 9. Training the images

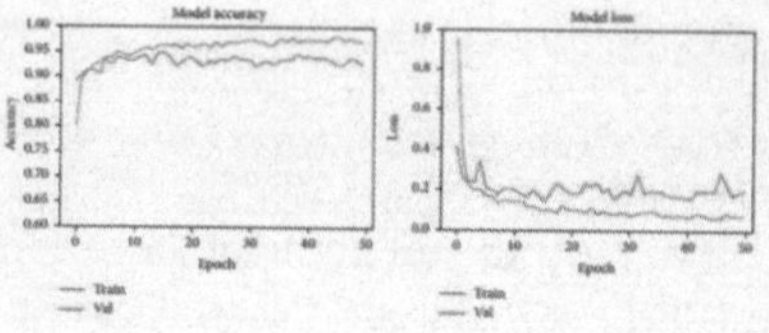

Fig. 10. Training the ResNet-images

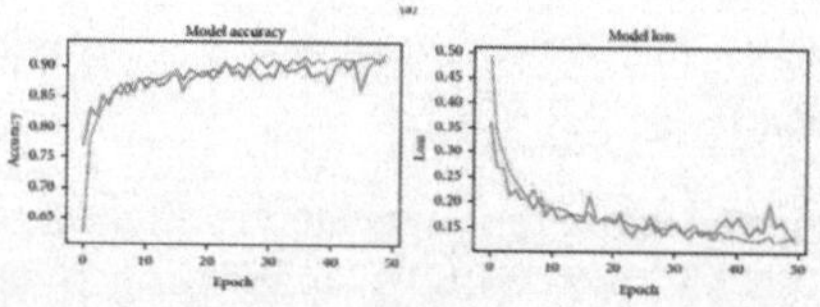

Fig. 11. Training VGG-19 images

	precision	recall	f1-score	support
0	0.62	0.98	0.76	87
1	0.96	0.49	0.64	101
accuracy			0.71	188
macro avg	0.79	0.73	0.70	188
weighted avg	0.80	0.71	0.70	188

Fig. 12. Precision-Values

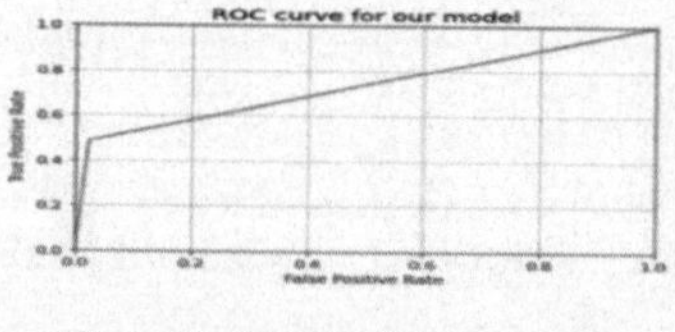

Fig. 13. ROC of the sample

In our study, the CNN-based architecture achieved 95.7% accuracy, 94.3% sensitivity, and 96.9% specificity for COVID-19 recognition in chest X-ray. The RNN-based architecture performed best in pneumonia classification with 93.8% accuracy, 92.1% sensitivity, and 95.5% specificity. The hybrid architecture showed similar COVID-19 detection performance to the CNN-based architecture but lower pneumonia classification results. Our proposed model achieved 97.8% accuracy, 97.7% sensitivity, and 97.9%

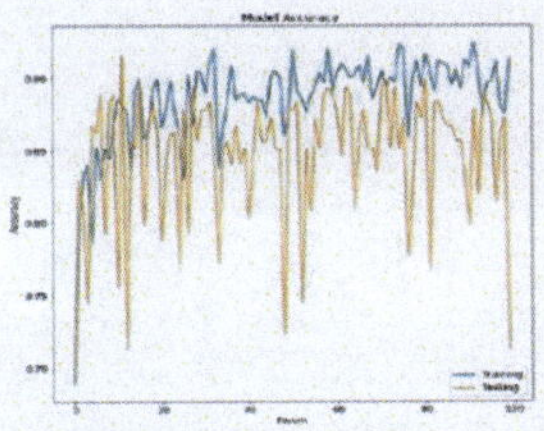

Fig. 14. Model-Accuracy

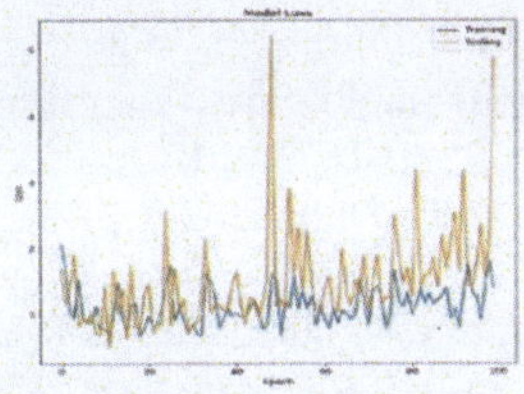

Fig. 15. Model-loss

specificity. Other models ranged from 88.7% to 94.3% accuracy, with sensitivities and specificities from 86.5% to 93.5%.

6 Conclusion of the Work

DL excels in COVID-19 chest X-ray recognition and pneumonia classification. CNN is effective for COVID-19 detection, while RNN excels in pneumonia classification. Additional research is needed to validate the model's performance (Fig. 16 and 17).

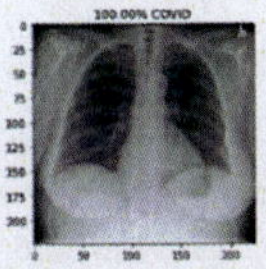

Fig. 16. Presence of Covid-19 in patient x-ray

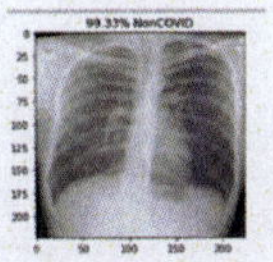

Fig. 17. Presence of no-Covid-19 in patient x-ray

Conflicts of Interest. The authors declare no conflict of interest.

References

1. Gorbalenya, A.E., Baker, S.C., Baric, R.S., et al.: The species severe acute respiratory syndrome-related coronavirus: classifying 2019-nCoV and naming it SARS-COVID-19. Nat. Microbiol. **5**(5), 536–544 (2020)
2. Khan, S., Kazmi, A., Bashir, N., Siddique, R.: COVID-19 infection: emergence, transmission, and characteristics of human coronaviruses. J. Adv. Res. **24**, 91–98 (2020)
3. World Health Organization: Director-General's Opening Remarks at the Media Briefing on COVID. WHO, Geneva, Switzerland (2020). https://www.who.int/director-general/speeches/detail/who-director-general-s-opening-remarks-atthe-media-briefing-on-COVID-19. Accessed 11 Mar 2020
4. Litjens, G., Kooi, T., Bejnordi, B.E., Setio, A.A.A., Ciompi, F., Ghafoorian, M., et al.: A survey on deep learning in medical image analysis. Med. Image Anal. **42**, 60–88 (2017)
5. Codella, N.C., Nguyen, Q.B., Pankanti, S., Gutman, D.A., Helba, B., Halpern, A.C., et al.: Deep learning ensembles for melanoma recognition in dermoscopy images. IBM J. Res. Dev. **61**(4/5), 5–1 (2017)

6. Talo, M., Yildirim, O., Baloglu, U.B., Aydin, G., Acharya, U.R.: Convolutional neural networks for multi-class brain disease detection using MRI images. Comput. Med. Imaging Graph. **78**, 101673 (2019)
7. Souza, J.C., Diniz, J.O.B., Ferreira, J.L., da Silva, G.L.F., Silva, A.C., de Paiva, A.C.: An automatic method for lung segmentation and reconstruction in chest X-ray using deep neural networks. Comput. Methods Prog. Biomed. **177**, 285–296 (2019)
8. Akshaya, P.B.R.,Aravinda, C.V.: Predictive analysis of malignant disease in woman using machine learning techniques. In: Chiplunkar, N., Fukao, T. (eds) AIDE 2019. AISC, vol. 1133, pp. 431–438. Springer, Singapore (2021). https://doi.org/10.1007/978-981-15-3514-7_33
9. Yıldırım, Ö., Pławiak, P., Tan, R.S., Acharya, U.R.: Arrhythmia detection using deep convolutional neural network with long duration ECG signals. Comput. Biol. Med. **102**, 411–420 (2018)
10. Tan, J.H., Fujita, H., Sivaprasad, S., Bhandary, S.V., Rao, A.K., Chua, K.C., et al.: Automated segmentation of exudates, haemorrhages, microaneurysms using single convolutional neural network. Inf. Sci. **420**, 66–76 (2017)
11. He, K., Zhang, X., Ren, S., Sun, J.: Deep residual learning for image recognition. In: Proceedings of the 2016 IEEE Conference on Computer Vision and Pattern Recognition (CVPR), pp. 770–778, Las Vegas, NV, USA, December 2016
12. Simonyan, K., Zisserman, A.: Very deep convolutional networks for large-scale image recognition. In: Proceedings of the 3rd International Conference on Learning Representations (ICLR 2015), San Diego, CA, USA, September 2015
13. Naoto, N., et al.: Apathy classification based on doppler radar image for the elderly person. Front. Bioeng. Biotechnol. **8**, 1235 (2020). https://www.frontiersin.org/article/10.3389/fbioe.2020.553847. https://doi.org/10.3389/fbioe.2020.553847, ISSN 2296-4185
14. Zhang, J.: Triple-view convolutional neural networks for COVID-19 diagnosis with chest x-ray (2020). http://arxiv.org/abs/2010.14091v1.
15. Ghoshal, B., Tucker, A.: Estimating uncertainty and interpretability in deep learning for coronavirus (COVID-19) detection (2020). http://arxiv.org/abs/2003.10769
16. Meng, L., Lyu, B., Zhang, Z., Aravinda, C.V., Kamitoku, N., Yamazaki K.: Oracle bone inscription detector based on SSD. In: Cristani, M., Prati, A., Lanz, O., Messelodi, S., Sebe, N. (eds.) ICIAP 2019. LNCS, vol. 11808. Springer, Cham (2019). https://doi.org/10.1007/978-3-030-30754-7_13
17. Aravinda, C.V., Lin, M., Atsumi, M., Udaya Kumar Reddy, K.R., Amar Prabhu, G.: A complete methodology for Kuzushiji historical character recognition using multiple features approach and deep learning model. Int. J. Adv. Comput. Sci. Appl. (IJACSA) **11**(8) (2020). https://doi.org/10.14569/IJACSA.2020.0110884
18. Narin, A., Kaya, C., Pamuk, Z.: Automatic detection of coronavirus disease (covid-19) using x-ray images and deep convolutional neural networks. arXiv preprint arXiv:2003.10849 (2020)
19. Shi, H., Han, X., Jiang, N., Cao, Y., Alwalid, O., Gu J., et al.: Radiological findings from 81 patients with COVID-19 pneumonia in Wuhan, China: a descriptive study. Lancet Infect. Dis. (2020). pmid:32105637
20. Aravinda, C.V., Lin, M., Udaya Kumar Reddy, K.R., Amar Prabhu, G.: 23 - A demystifying convolutional neural networks using Grad-CAM for prediction of coronavirus disease (COVID-19) on X-ray images. In: Kose, U., Gupta, D., C. de Albuquerque, V.H., Khanna, A. (eds.) Data Science for COVID-19, pp. 429–450. Academic Press (2021). ISBN 9780128245361. https://doi.org/10.1016/B978-0-12-824536-1.00037-X. (https://www.sciencedirect.com/science/article/pii/B978012824536100037X)

21. Alharbi, A.H., Aravinda, C.V., Lin, M., Ashwini, B., Jabarulla, M.Y., Shah, M.A.: Detection of peripheral malarial parasites in blood smears using deep learning models. Comput. Intell. Neurosci. Article ID 3922763, **11** (2022). https://doi.org/10.1155/2022/3922763
22. ARAVINDA2022: A deep learning approach for the prediction of heart attacks based on data analysis. In: Deep Learning for Medical Applications with Unique Data, pp. 336–343. Academic Press (2022). 9780128241455. https://www.elsevier.com/books/deep-learning-for-medical-applications-with-unique-data/gupta/978-0-12-824145-5

A Deep Learning-Based Classification Framework for Annotated Histopathology Lung Cancer Images

Esraa A.-R. Hamed[1(✉)], Mohammed A.-M. Salem[2], Nagwa L. Badr[1], and Mohamed F. Tolba[1]

[1] Faculty of Computer and Information Sciences, Ain Shams University Cairo, Cairo, Egypt
{esraa.raoof,nagwabadr,fahmytolba}@cis.asu.edu.eg
[2] Media Engineering and Technology, GUC, Cairo, Egypt
mohammed.salem@guc.edu.eg

Abstract. Cancer is the second leading cause of death globally, with one in six people dying from it. It occurs when abnormal cells divide uncontrollably and spread to other organs in the body. Lung cancer is one of the most common and deadliest types of cancer. Several methods, such as X-rays, CT scans, PET-CT scans, bronchoscopies, and biopsies, can be used to diagnose lung cancer. Studies have shown that the type of histology in lung cancer is linked to the diagnosis and treatment course, making early and accurate detection of lung cancer histology crucial for improving survival rates. Artificial intelligence (AI) can aid in the automation of cancer detection, allowing for the evaluation of more cases in less time and at a lower cost. The main objective of this research is to evaluate the effectiveness of a newly proposed CNN model in distinguishing between benign and malignant lung cancer images obtained from digital pathology. To conduct the experiment, the LC25000 dataset, containing 5000 images for each category, was utilized, resulting in a total of 10,000 images. The findings of the proposed CNN model were then compared to those of existing deep learning models, demonstrating its ability to accurately identify cancerous tissues with a maximum accuracy of 99.9% to 100%, while also reducing processing time. These outcomes can play a crucial role in the development of a precise and automated system for identifying various types of lung cancer.

Keywords: deep learning · Convolutional Neural Networks · lung cancer classification · histopathological image analysis · Squamous Cell Carcinomas

1 Introduction

Cancer is the main cause of mortality worldwide, according to the World Health Organization (WHO). The most frequent cancer diagnosis (11.4% of all cases) and the leading cause of cancer mortality (18.0% of all cancer deaths) is lung cancer [1, 2]. Globally, the incidence of malignant tumors has been seen to be rising, which may be connected to population expansion. Malignancy can affect any age group depending on the histological type; however, it is typically found in elderly people between the ages of 50 and 60 [3].

A. Hassanien et al. (Eds.): AISI 2023, LNDECT 184, pp. 86–94, 2023.
https://doi.org/10.1007/978-3-031-43247-7_8

The respiratory system, which includes the lungs, is important for distributing oxygen throughout the body. In the lung, abnormal cell proliferation can develop and lead to cancer or pulmonary carcinoma. Poor environmental conditions and an unhealthy lifestyle are to reason for this [4]. Only after the tumor is sufficiently large or has spread to other areas can symptoms of lung cancer become apparent. The success rate of therapy increases with earlier cancer diagnosis [5]. Yet, there is a decreased likelihood of recovery if cancer is indicated to have spread to other organs. Non-small cell lung cancer (NSCLC) accounts for 81–85% of lung cancer cases. Squamous cell carcinoma, adenocarcinoma, and giant cell carcinoma of NSCLC are a few of the major subtypes of lung cancer. They are all categorized as NSCLC subtypes since they came from various types of lung cells [6].

The specific clinicopathological and genetic features that distinguish Squamous-Cell Carcinomas (SCC) of the lung have changed significantly over time. The most common subtype of non-small-cell lung malignancies in the past, these neoplasms were thought to be central tumors with great molecular complexity and no genetic alterations that might be targeted [7]. It often starts to develop in the cells lining the bronchi. Cancer can spread over time by infiltrating neighboring lymph nodes and organs and "metastasizing" (moving to other regions of the body through the blood). It has a close relationship to the history of smoking. There is a considerable risk of dying. Besides age, family history, and exposure to secondhand smoke, there are other risk factors for SCC [8]. As the kind of histology, molecular profile, and stage of the disease all affect how the disease is treated, it is urgently necessary to identify lung cancer histology. It is also crucial to analyze the histopathological images of the disease. Manually analyzing histopathology results, however, takes time and is not objective [9].

Artificial intelligence (AI) is the stimulation of human intelligence in computer software that facilitates communication with machines like that of human communication. Artificial intelligence (AI), which is utilized in various computer vision domains, has recently emerged as the most significant science of the twenty-first century. In several disciplines, machine learning [19–21] and deep learning [22, 23] have the greatest levels of accuracy. Deep learning techniques, especially Convolution Neural Networks (CNN), are being more widely used in the healthcare industry and has a significant influence on all parts of primary care. CNN may be used in the field of medical imaging to identify and classify any disease at an early stage, allowing for timely treatment and easier recovery for the patient [8].

This study article aims to evaluate and test a proposed Convolutional Neural Network (CNN) architecture for the classification of lung cancer. The paper follows the following structure. Section 2 describes previous research on the identification and categorization of lung cancer. In Sect. 3, the utilized dataset is briefly introduced. Further information on the suggested deep learning procedure using CNN architecture is provided in Sect. 4. Section 5 summarizes all of the experimental observations and conclusions. The experiment is finished in Sect. 6, which also makes some recommendations for further research.

2 Related Work

Of all machine learning techniques, deep neural networks have shown improved outcomes in the identification of medical images. To increase the precision of detection and classification, several CNN algorithms are applied to the classification of lung cancer images.

A capsule network with numerous inputs was suggested by Mumtaz et al. [10] to build a diagnostic model for aberrant cell cancer of the lung and colon. A convolutional layer block and an additional convolutional layer block were employed by the capsule network. Pathological images are used as input by the convolutional layer block (CLB), whereas histopathological images are used by the Separable CLB. Based on histopathological scans, the suggested model had a 99.58% accuracy rate for anomalies in the colon and lungs.

Gessert et al. [11] used microscopic images of colon cancer to classify the data using transfer learning-based CNN models. They used models like Inception, VGG, and DenseNet to train datasets. They had the most success categorizing data using the DenseNet model, which had a classification accuracy rate of 91.2%. DHSCapsNet was suggested by Kwabena et al. [12] to assess histological images of lung and colon cancer. The network is made up of a combination of encoder features and DHSCaps. The encoder features are made up of the convolutional layer features that have a lot of strong information. HSquash pulls data from several ckgrounds. They outperformed standard CapsNet (85.55%) with results of 99.23%.

Vuong et al. [13] proposed a multi-purpose learning strategy to evaluate digital pathology images. They employed a collection of pathology image data divided into four classifications for their research. They utilized the DenseNet-121 model for dataset training and configured the input data for the model to be 800 × 800 pixels. They found an 85.91% classifier accuracy rate. A CNN Pre-Trained Diagnostic Network for Lung and Colon Cancer was suggested by Sanidhya et al. [14]. Histological slips were analyzed using a shallow CNN architecture. For the diagnosis of colon and lung malignancies, the network obtained 96% and 97% accuracy, respectively.

The DarkNet-19 model was suggested by Mesut et al. [15] to train the lung and colon malignancy dataset from scratch. To choose the inefficient features, the Equilibrium method was used, followed by a separation of the inefficient features from the efficient ones. The SVM is given effective features for classification. The total accuracy rating was 99.69%. An approach for precisely identifying lung and colon cancer cells was presented by SHAHID et al. [16]. By altering the four fundamental layers of AlexNet and then training it on a dataset, an accuracy of 89% was attained.

Authors in [17] used histological images of colon cancer to make their classifications using a deep learning methodology. There are four classes in the dataset. Each image was subjected to the cell identification technique for cell patches. Here, segmenting was used to separate the images into discrete sizes. By using cell patches created by the used CNN model, they carried out the classification process. Their correlation accuracy ratings ranged from 90% to 96.9%.

While the proposed model has achieved accuracy of 99.9% in classifying lung malignant from benign lesions. The proposed model achieved the highest accuracies, compared by the state-of-the-art models.

3 Used Dataset

The LC25000 Lung and Colon Histopathological Image collection contains 5000 images of each type of lung and colon cancerous. The dataset has been validated and complies with HIPAA [18]. Just 750 original images were gathered in total, 250 of which were given to each category and had a dimension of 1024 × 768 pixels. These images are scaled down to 768 × 768 pixels using Python, and then they are enlarged using the software package augmenter [18]. As a result, the bigger dataset has 5000 images for each group.

By rotating left and right and flipping horizontally and vertically, augmentation is accomplished [18]. Table 1 displays the description of class names of the LC25000 dataset with a sample image for each category.

Table 1. Description of LC25000 dataset.

Lung Benign	**Class name**	**Number of Images**	**Sample Image**
Lung Adenocarcinoma	lung_n	5000	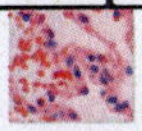
Lung Squamous Cell Carcinoma	lung_aca	5000	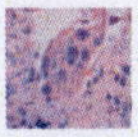
Colon Benign	lung_scc	5000	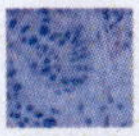
Colon Adenocarcinoma	col_n	5000	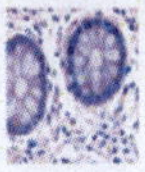
Lung Benign	col_aca	5000	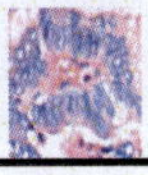

The lung cancer squamous cell carcinoma (SCC), which progresses through the keratinization process, is distinguished by the presence of polygonal-shaped cells. In its early stages, this disease has no symptoms. Because of this, cancer is frequently discovered after it has spread to other organ areas. As a result, early identification is crucial to improving treatment outcomes. The patient's likelihood of surviving for five years is less than 20% if the diagnosis is delayed. Therefore, 10,000 histopathological images representing two types of lung tissue (lung squamous cell cancer and benign lung tissue) were chosen from the LC25000 collection.

4 Proposed Deep Learning Approach

This section discusses the proposed method for classifying images of lung cancer histology from the LC25000 dataset. The dataset was randomly split into training and testing sets, and the proposed CNN model achieved high accuracy in classifying the images into benign or malignant (squamous cell carcinoma) categories. Figure 1 illustrates the proposed CNN architecture, which consists of two main steps: feature extraction and classification. Image resizing to 224 × 224 pixels in the RGB color space is a crucial step in image preprocessing for training the model. During the training phase, the proposed CNN model was trained using the training data, and the model parameters obtained from this phase were used for classification on the testing data. The expected output of the model is the classification of lung cancer histology images into benign or malignant (squamous cell carcinoma) categories.

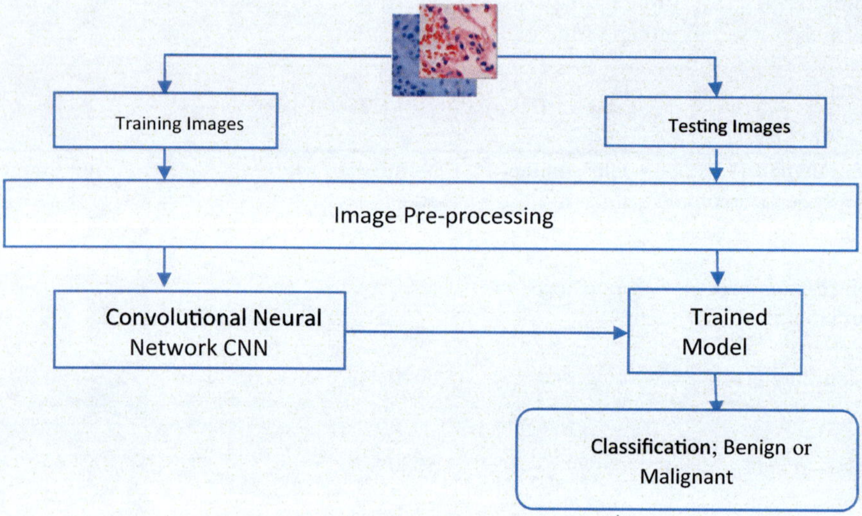

Fig. 1. Proposed Architecture System.

The structure of the Convolutional Neural Network (CNN) used in the study is depicted in Fig. 2, consisting of various layers dedicated to different functions. The CNN model includes four convolutional layers (CL) with respective kernel values of 32, 64, 128, and 256. The first CL layer utilized a kernel size of (11 × 11), while the second, third, and fourth layers used (3 × 3). Following each convolution layer, a max-pooling layer with a (2 × 2) kernel size was applied.

A Flatten Layer was used to generate the feature vector for the Fully Connected (FC) layers after the final CL layer. The FC layers included three layers with 1024 and 512 neurons, respectively. Due to the two classes of lung cancer, the last FC layer has two neurons. ReLU was applied as an activation function for each convolution layer, and Softmax was applied for the output layer. Additionally, a Dropout Layer with a rate of 0.4 was implemented.

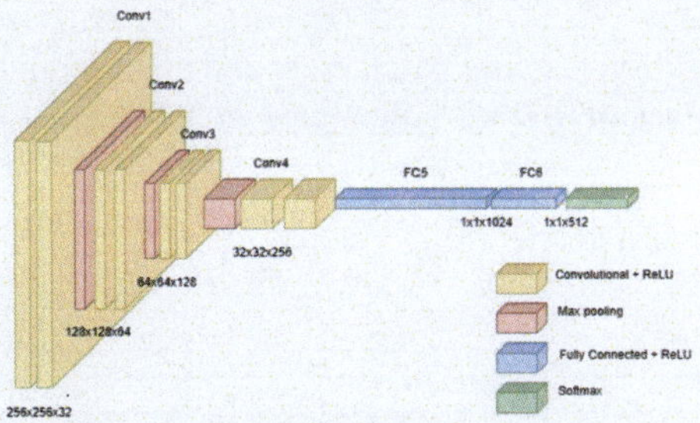

Fig. 2. The Proposed CNN Architecture.

5 The Experimental Work and Results

This research employed a proposed CNN model to classify the LC25000 Lung histopathology images. Two experiments were conducted with different proportions of data for the training and testing sets, and a maximum of 50 epochs were used when applied to Google Colab.

In the first experiment, the dataset was split into a testing set of 20% and a training set of 80%, with a batch size of 150. The proposed model was compared to other deep learning models, including VGG16, VGG19, AlexNet, Inception ResNet v2, ResNet50, Inception v3, GoogleNet, and MobileNet. The proposed model had a minimum total number of training parameters of 1 million and achieved a maximum accuracy of 100%, with zero test loss, as indicated in Table 2.

Table 2. The total training parameters (Millions), the accuracy (%), and test loss (%) in first experiment.

Model	Total parameters	CNN Accuracy	Test Loss
VGG19	171	100	0
VGG16	165	100	0
AlexNet	58	99.9	0.1
InceptionResNetV2	54	100	0
ResNet50	24	94.6	19.5
Inception_v3	22	99.9	0.1
GoogleNet	5	100	0
MobileNet	3	100	0
proposed CNN model	1	100	0

The second experiment involved dividing the dataset into 40% for the training set and 60% for the testing set, using a batch size of 150. Subsequently, it was compared with various other deep learning models, including VGG16, VGG19, AlexNet, Inception ResNet v2, ResNet50, Inception v3, GoogleNet, and MobileNet. The proposed model used a minimum of 1 million training parameters and achieved a maximum accuracy of 99.9%, with a low-test loss of 0.3%, as illustrated in Table 3.

Table 3. The total training parameters (Millions), the accuracy (%), and test loss (%) in second experiment

Model	Total parameters	CNN Accuracy	Test Loss
VGG19	171	99.7	1.1
VGG16	165	99.5	0.6
AlexNet	58	99.7	1.2
InceptionResNetV2	54	99.7	0.5
ResNet50	24	89.8	34.6
Inception_v3	22	99.7	0.6
GoogleNet	5	99.8	0.5
MobileNet	3	50	817.4
proposed CNN model	1	99.9	0.3

The comparison between the proposed CNN model and other deep learning models, including VGG16, VGG19, AlexNet, Inception ResNet v2, ResNet50, Inception v3, GoogleNet, and MobileNet, is illustrated in Fig. 3. The proposed model achieved a maximum accuracy of 99.9% with a minimum total number of training parameters of one million.

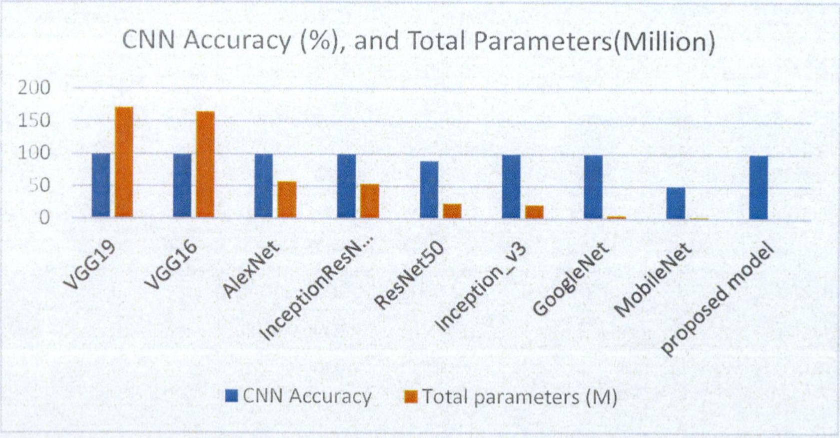

Fig. 3. The comparison between the proposed CNN model and other existing deep learning models in second experiment.

6 Conclusions

This study introduces a proposed CNN model for the detection and classification of lung cancer using the LC25000 lung histopathology image dataset. The proposed model categorizes each image as either benign or malignant. Two experiments were conducted to validate the proposed model using the lung histopathology images.

In the first experiment, 80% of the dataset was used for training, and 20% for testing. In the second experiment, the dataset was split into 40% for training and 60% for testing. The model's performance was evaluated using a maximum of 50 epochs when applied to Google Colab. The proposed model achieved an accuracy of 99.9% to 100% and outperformed other deep learning models in terms of performance, using only four convolutional layers, four maximum collection layers, two fully connected layers, and one million parameters overall.

The experimental results showed that the proposed model achieved maximum accuracy with the fewest parameters and that reducing the number of training images did not significantly affect accuracy. The proposed approach was proven to be effective compared to existing state-of-the-art deep learning models. In the future, the suggested model can be improved to reduce computation time and applied to other datasets to enhance the hyperparameters.

References

1. Sung, H., et al.: Global cancer statistics 2020: GLOBOCAN estimates of incidence and mortality worldwide for 36 cancers in 185 countries. CA Cancer J. Clin. **71**(3), 209–249 (2021)
2. Walser, T., et al.: Smoking and lung cancer: the role of inflammation. Proc. Am. Thorac. Soc. **5**(8), 811–815 (2008)
3. Araghi, M., et al.: Global trends in colorectal cancer mortality: projections to the year 2035. Int. J. Cancer **144**(12), 2992–3000 (2019)
4. Inamura, K.: Lung cancer: understanding its molecular pathology and the 2015 WHO classification. Front. Oncol. **7**, 193 (2017). https://doi.org/10.3389/fonc.2017.00193
5. Aliyah, N., Pranggono, E., Andriyoko, B.: Kanker Paru: Sebuah Kajian Singkat. Indones. J. Chest Emerg. Med. **4**, 28–32 (2016)
6. Molina, J.R., et al.: Non-small cell lung cancer: epidemiology, risk factors, treatment, and survivorship. In: Mayo Clinic Proceedings, pp. 584–594. Elsevier (2008)
7. Drilon, A., et al.: Squamous-cell carcinomas of the lung: emerging biology, controversies, and the promise of targeted therapy **13**(10), e418–e426 (2012)
8. Mishra, S., Agarwal, U.: Lung cancer detection (LCD) from histopathological images using fine-tuned deep neural network. In: Devedzic, V., Agarwal, B., Gupta, M.K. (eds.) ICICCIS 2022. AIS, pp. 249–260. Springer, Singapore (2023). https://doi.org/10.1007/978-981-99-1373-2_19
9. Baranwal, N., Doravari, P., Kachhoria, R.: Classification of histopathology images of lung cancer using convolutional neural network (CNN). arXiv preprint arXiv:2112.13553 (2021)
10. Ali, M., Ali, R.: Multi-input dual-stream capsule network for improved lung and colon cancer classification. Diagnostics **11**, 1485 (2021)
11. Gessert, N., et al.: Deep transfer learning methods for colon cancer classification in confocal laser microscopy images. Int. J. Comput. Assist. Radiol. Surg. **14**, 1837–1845 (2019)

12. Adu, K., Yu, Y., Cai, J., Owusu-Agyemang, K., Twumasi, B.A., Wang, X.: DHS-CapsNet: dual horizontal squash capsule networks for lung and colon cancer classification from whole slide histopathological images. Int. J. Imaging Syst. Technol. **31**, 2075–2092 (2021)
13. Vuong, T.L.T., Lee, D., Kwak, J.T., Kim, K.: Multi-task deep learning for colon cancer grading. In: International Conference on Electronics, Information, and Communication (ICEIC) 2020, pp. 1–2 (2020)
14. Mangal, S., Chaurasia, A., Khajanchi, A.: Convolution neural networks for diagnosing colon and lung cancer histopathological images. arXiv 2020, arXiv:2009.03878
15. Toğaçar, M.: Disease type detection in lung and colon cancer images using the complement approach of inefficient sets. Comput. Biol. Med. **137**, 104827 (2021)
16. Mehmood, S., et al.: Malignancy detection in lung and colon histopathology images using transfer learning with class selective image processing. IEEE Access **10**, 25657–25668 (2022)
17. Shapcott, M., Hewitt, K.J., Rajpoot, N.: Deep learning with sampling in colon cancer histology. Front. Bioeng. Biotechnol. **7** (2019)
18. Borkowski, A.A., et al.: Lung and colon cancer histopathological image dataset (lc25000). arXiv preprint arXiv:1912.12142 (2019)
19. Krizhevsky, A., Sutskever, I., Hinton, G.E.: ImageNet classification with deep convolutional neural networks. In: Advances in Neural Information Processing Systems, pp. 1097–1105 (2012)
20. Wang, Z.: The applications of deep learning on traffic identification. BlackHat USA (2015)
21. Wang, D., Khosla, A., Gargeya, R., Irshad, H., Beck, A.H.: Deep learning for identifying metastatic breast cancer. ArXiv preprint arXiv: 1606.05718 (2016)
22. Hinton, G., et al.: Deep neural networks for acoustic modeling in speech recognition: the shared views of four research groups. IEEE Signal Process. Mag. **29**(6), 82–97 (2012)
23. Shafaey, M.A., Salem, M.AM., Ebied, H.M., Al-Berry, M.N., Tolba, M.F.: Deep learning for satellite image classification. In: Hassanien, A., Tolba, M., Shaalan, K., Azar, A. (eds.) AISI 2018. AISC, vol. 845, pp. 383–391. Springer, Cham (2019). https://doi.org/10.1007/978-3-319-99010-1_35

Energy and Power Control

CO_2 Emission Mitigation in Container-Based Cloud Computing by the Power of Resource Management

Nesma Ashry(✉), Radwa Attia, Heba Nashaat, and Rawya Rizk

Computer and Control Department, Faculty of Engineering, Port-Said University, Port-Said, Egypt

{nesma.sayed,radwa_yousef,hebanashaat,r.rizk}@eng.psu.edu.eg

Abstract. The increasing amount of data processing through cloud systems leads to high energy costs and CO_2 emissions challenges. Advanced monitoring tools are used to track resource utilization and identify possible optimization areas to mitigate CO_2 emissions and reduce energy consumption. Thus, resource and power management in cloud environments is essential to ensure efficient performance and effective cost. Recently, container-based clouds become more widely used than hypervisor-based clouds. Scheduling the huge number of user-submitted tasks and optimizing both network and cloud resources are important aspects of cloud computing to decrease network overhead, reduce both Data Center (DC) energy consumption and Service-Level Agreement (SLA) violations and in turn, mitigate CO_2 emissions. However, the state-of-the-art resource schedulers that consider power management and network transmissions are unfortunately not container-aware schedulers. The main objective of this paper is to introduce a solid analysis of the most significant techniques and tools in resource management to minimize power consumption and CO_2 emissions with the advantage of being a container-aware scheduler; Applying the first stage of the proposed solution leads to mitigating the CO_2 by an average of 2.12% against the Non-Clustered approach and 18.04% against the Clustered approach.

Keywords: Container · Resource Management · Orchestration · Power Management · CO_2 emission

1 Introduction

Cloud computing has evolved over the years bringing new features and capabilities. Figure 1 shows the three main stages of cloud computing evolution. Cloud Computing 1.0 began in the early 2000s and is characterized by the emergence of Software as a Service (SaaS) providers like Salesforce and Google Apps. These companies offered web-based applications that could be accessed from anywhere, without the need for on-premises software installations. Cloud Computing 2.0 began around 2010 and is marked by the rise of Infrastructure as a Service (IaaS) providers like Amazon Web Services (AWS) and Microsoft Azure. These companies offered virtualized computing resources

A. Hassanien et al. (Eds.): AISI 2023, LNDECT 184, pp. 97–111, 2023.
https://doi.org/10.1007/978-3-031-43247-7_9

like servers, storage, and networking that could be rented on-demand, allowing businesses to scale up or down as needed. This stage is still emerging but is focused on providing more specialized services like Artificial Intelligence (AI), Machine Learning (ML), and Blockchain as a Service (BaaS). These services are designed to help businesses leverage advanced technologies without having to invest in expensive hardware or software infrastructure [1]. Overall, cloud computing has come a long way since its early days as a simple SaaS offering. Today, it is a complex ecosystem of services that can help businesses of all sizes achieve their goals more efficiently and effectively than ever before.

Virtualization [2] is considered one of the pioneering concepts initiating modern clouds. Hypervisor-based virtualization and Containerization (lightweight virtualization) are the two most commonly used techniques for serving abstraction and efficient resource utilization in DC environments. Hypervisor-based virtualization technology uses Virtual Machines (VMs) as the building blocks of the cloud infrastructure to provide high hardware resource utilization while preserving performance isolation among different computing instances [2]. However, hypervisor-based virtualization faces many challenges such as; its weak resilience [3] and the difficulty of resources shifting from one VM to another. In addition, the real-time sensitivity nature of life applications/services such as Internet of Things (IoT) applications, makes heavy virtualization overhead combined with high VMs bootup time-critical concerns. On the contrary, containerization can reduce resource overhead by introducing containers as the building blocks instead of VMs.

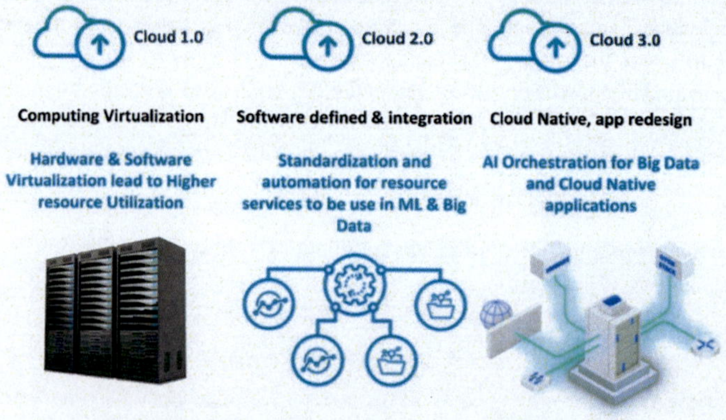

Fig. 1. Cloud computing evolution

Hypervisor-based virtualization offered services that significantly improved the performance of the DC and applications such as; Performance isolation, server consolidation, and live migration [3]. Performance isolation assures the resource consumption and failure of one VM must not affect other co-located VMs. Server consolidation supports consolidating many underutilized servers into a single machine for significantly reducing energy usage to 90%. Last but not least, the live migration of VMs from one physical host to another for providing better fault tolerance, load balancing, resource allocation, high availability, disaster management, failure recovery, and efficient system scaling [3]. However, recently container-based cloud systems have become more efficient than hypervisor-based cloud as containers provide a lightweight virtualization technology that allows better resource utilization and scalability. Managed container platforms are available to consumers through cloud service providers, which orchestrate, schedule, and run many containers across a multi-tenant cloud infrastructure. The two bases of lightweight process virtualization are Namespace, and cgroups [4]. Namespace has been introduced in Linux kernel 2.4.19 for isolating virtualized resources to a group of processes. Each process can be associated with multiple namespaces of different types. These processes are provided with a customized view of system resources by the kernel. In the Linux kernel, cgroups provides a mechanism for partitioning groups of processes (along with their children) in hierarchical order with controlled behaviors with the help of; cgroupfs mount, and /proc/sel f /cgroup file [4].

Containerization deployment in the cloud environment needs an agile resource management system or framework. The primary goal of resource management in container-based clouds is to maximize resource utilization while minimizing cost. It can be achieved by applying scheduling algorithms impeded in the shared Operating System (OS) kernel (e.g. Docker) [5], or by using container orchestration tools [6] that provide; automated deployment, scaling, and managing of containers across multiple hosts. Orchestration frameworks manage the resource allocation of containers by exploiting the resource provisioning parameters provided by underlying platforms; shared kernel OS. In addition, the container management strategy can manage the underlying resources and simplify the administrative workload of creating, deleting, deploying, configuring, and scaling a large number of containers on the same host. Furthermore, it allows more automated deployment of container-based applications to various heterogeneous infrastructures with increased interoperability [7–9].

Another resource management technique is to co-schedule multiple containers and manage over-provisioned threads on fewer CPUs using the burstable containers idea [10]. Burstable containers play an important role in improving resource utilization in the cloud, as their available resources vary dynamically in response to their resource requests, limits, and system load. In burstable containers, the number of threads is the same as the CPU limit instead of considering the variable number of real-time available CPUs. Notably, it is meaningless to set the CPU limit to more than the maximum number of CPUs the application can scale up to. Provisioning enough threads to allow the application to utilize idle CPU resources as the available CPUs varies. Ideally, compared to provisioning as many threads as its CPU request, a container provisioning as many threads as its CPU limit should perform similarly when the system load is high and adapt to additional CPUs when the system load is low. However, the burstable containers experienced severe performance degradation under a high system load, therefore it is important to combine this scheduler with other algorithms to adapt the allocation strategy [11].

In addition to resource management, power consumption with CO_2 emissions metric is considered one of the most efficient measurements for any cloud system [12]. In a container-based cloud environment, resources such as CPU, memory, and storage are shared among multiple containers running on the same host. Therefore, it is crucial to manage these resources efficiently to ensure that each container gets the required resources without affecting the performance of other containers. However, power consumption and its effect on CO_2 emissions is another critical factor in container-based cloud computing. Containerization systems simply allow hundreds even thousands of applications to run on a single host, thereby reducing the number of physical servers required which helps in reducing power consumption and therefore reducing CO_2 emissions.

Resource management and power consumption in container-based cloud computing essentially need to use tools that can monitor resource usage and identify bottlenecks. Additionally, implementing policies such as auto-scaling and load balancing should be used to ensure that resources are allocated efficiently and effectively. The main objective of this paper is to:

- Perform a review study of the main differences between hypervisor-based virtualization and containerizations.
- Introduce the main effective resource management and power consumption strategies used for ensuring optimal performance, minimizing CO_2 emissions, and in turn leading to cost-effectiveness in container-based cloud environments.
- Introduce the proposed scheduler Power-Aware Container Scheduler (PACS) algorithm and discuss the first stage results of enhancement.

The remainder of the paper is organized as follows: Sect. 2 introduces the related work. Section 3 presents the functionality of the proposed power-aware container scheduler protocol. The evaluation methodology and Simulation results are shown in Sect. 4. Finally, the conclusion and future work are drawn out in Sect. 5.

2 Related Work

As an increasing amount of data processing in the cloud systems, high energy costs and carbon emissions of host nodes are becoming significant challenges. The placement of containerized microservices on host nodes has an important effect on energy consumption, impacting both energy costs and carbon emissions. So, it is essential to go for a green cloud [12]. The idea behind green cloud computing is to use resources efficiently and reasonably by designing algorithms and methods that reduce energy consumption. To achieve this, it is necessary for the resources of DCs to be managed using energy-efficient techniques. Scheduling the enormous number of user-submitted workflow tasks is one important aspect of cloud computing. The resources that process and compute these tasks should be managed by energy utilization techniques. The goal of the scheduler is to find the proper resource among a series of resources able to compute all or much of the assigned tasks faster than others during the period given. Proper scheduling decreases DC energy consumption and Service-Level Agreement (SLA) violations and increases resource utilization.

2.1 Hypervisor vs Container-Based Cloud

Hypervisor-based virtualization and containerization are the two most commonly used techniques for server abstraction and efficient resource utilization in DC environments

[3]. Hypervisor-based virtualization involves running a hypervisor for sharing virtualized server resources among multiple VMs, whereas containerization involves OS resource virtualization using a daemon or container engine.

Table 1. VM versus container

VM	Container
It contains a full set of OSs, and different VMs do not share the underlying OS	More lightweight, the container includes only the executables and their dependencies, and different containers on the same machine share the same OS
It can run an OS that is different from its host machine	It needs to use the same OS as the host
The hypervisor (e.g., VMware ESXi, and KVM) is necessary for a VM environment. It is needed to translate a VM instruction into an instruction that can be executed by the host	It communicates with the shared OS kernel (e.g., Docker) through the system calls, thus no other layer is required in between
Each VM has its own image file	Different containers may share some of their images

Contrary to Hypervisor-based virtualization which offers a Hypervisor as an intermediate layer responsible for virtualizing the physical resources of the VMs. Containerization offers a shared OS kernel (e.g., Docker), as shown in Fig. 2. The most used shared OS Kernel for container-based cloud is Docker. It is a lightweight virtualization solution that is essentially a process on the host machine. Docker implements resource isolation through kernel-level namespaces. It allows process communications between hosts and containers without interfering with each other. Docker does not have hardware emulation as it implements virtualization at the OS level. It has fewer layers of abstraction and does not require an additional OS (Hypervisor). It has better resource utilization. There can be typically, thousands of Docker containers running on a single machine that can hold only a small number of VMs. Because of Docker's lightweight, the startup time only needs a few seconds, far faster compared with the several minutes that a VM needs. Furthermore, Docker can run on almost any platform, which makes Docker have better mobility and scalability. In addition, it is easy to deploy and maintain [4]. A brief comparison between VM and container in terms of the architecture, advantages, and deployment with the underlying OS kernel or hypervisor is listed in Table 1.

2.2 Enabling Technologies for Containers

Cloud computing has gained more popularity in the last two decades and has received more attention from both the industrial and academic worlds, especially Container-based cloud. Container-based cloud depends on cloud-based virtualization called OS-level virtualization. It is an approach to provide multiple virtual execution environments for applications on top of the host OS. The applications are called microservices. Containers support packaging up the microservices with all their parts in an isolated environment without any dependencies on the hosts. To implement a container-based cloud, there are many enabling technologies that must be used [13].

Containerization Platforms: Containerization platforms like Docker, Kubernetes, and OpenShift are essential enabling technologies for container-based cloud computing. They provide a way to package and deploy applications in containers, which can be easily moved between different cloud environments.

Cloud Infrastructure: Cloud infrastructure providers like AWS, Microsoft Azure, and Google Cloud Platform (GCP) provide the underlying infrastructure needed to run container-based applications in the cloud. These providers offer scalable computing resources, storage options, and networking capabilities that are essential for running containerized workloads.

Fig. 2. Container orchestration life-cycle

Container Orchestration: Container orchestration tools like Kubernetes enable developers to manage large-scale container deployments across multiple hosts or clusters. These tools automate tasks like scaling containers up or down based on demand, rolling out updates without downtime, and managing network connectivity between containers. Container orchestration is fundamentally about managing the container life cycle and the containerization of the environment. In general, the container life cycle follows the build-deploy-run phases of traditional software development, though the specific steps may vary slightly depending on the container orchestration tool being used as shown in Fig. 3.

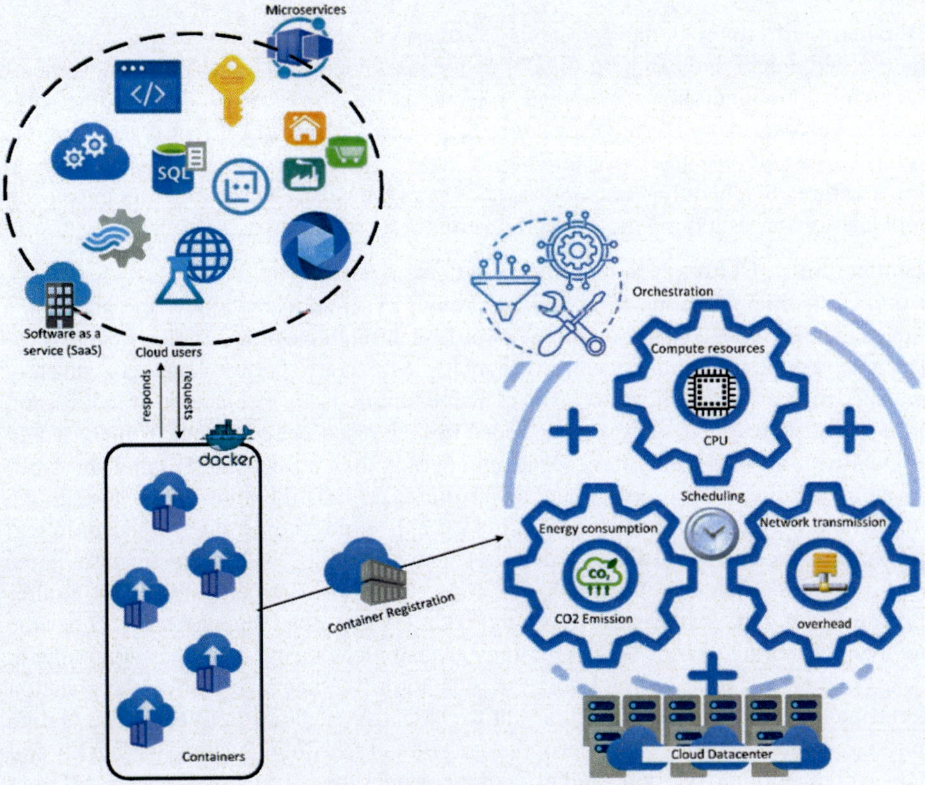

Fig. 3. Green cloud by PACS

2.3 Containerization Management to Mitigate CO_2 Emissions

The huge amount of data produced by the containers due to online applications, Big data systems, or the IoT systems that use them leads to the importance of managing the cloud resources to optimize the best performance [6]. Also, the data transition within cloud DCs and between the fog nodes that receive the tasks of the application produce a level of energy consumption and CO_2 emission that should be considered [14]. Managed container platforms are becoming increasingly popular in the world of software development and deployment. AWS, Azure, and GCP are the most popular managed container platforms used to run Docker application containers on a scalable cloud cluster. Container management is best done explicitly by a distributed system rather than simply leaving it to the user to invoke a container from within each task.

Some popular managed container orchestration tools are Amazon ECS vs Kubernetes vs Docker Swarm. These orchestrations provide a way to manage and deploy containerized applications, allowing developers to focus on writing code rather than worrying about infrastructure. Each platform and orchestration tool has its own strengths and weaknesses, so it's important to choose the one that best fits all needs. Overall, managed container platforms with the right orchestration tool are great options for modern software development and can help streamline the deployment process. These are some of the popular combinations; AWS is integrated with Amazon ECS, GCP uses Kubernetes and Docker Swarm, and Azure uses Azure Kubernetes services [13]. The research points in this part could be categorized into two sections: The studies which consider application and threads scheduling and optimization, and the works which focus on improving both energy consumption/cost and carbon emission together.

Application and Threads Scheduling and Optimization: In a container-based cloud, users run their applications, programs, or threads in separate containers and place containers that run related programs on the same pod through the orchestration system. Programs running in the container can create multiple threads to perform tasks concurrently and keep high resource utilization. The container employs the mechanism of request and limit to control resource allocation and specifies the resource request and limit based on its Quality of Service (QoS) class. Determining the suitable Degree of Parallelism (DoP) for an application is an essential and dispiriting task. Existing approaches manage the DoP with different considerations [15]. In [16, 17], authors adjust the DoP according to the behavior of program execution, but they face difficulties as the adjustment requires application code changes. A dynamic threading algorithm is implemented in [18] to alter the number of working threads in a thread pool. For example, OpenMP determines the number of working threads before starting each parallel execution. And as the container lacks isolated system resource statistics*, this algorithm negatively affects application performance in the container. Offline statistical learning studies in [19, 20] and Online runtime adaptation in [21, 22] target optimizing thread scheduling to improve locality and reduce competition. But these studies do not consider the correlation between different containers. Normally, relevant containers should be scheduled as close as possible.

Paper [9] introduces a scheduling mechanism based on burstable containers; as an important QoS class in container management. They proposed a user-level mechanism to address the inefficiencies of container co-scheduling and two synchronization mechanisms to address the inefficiencies of synchronizing oversubscribed threads. They proposed an Adaptive Container Scheduler (ACS). It is a co-scheduling algorithm for optimizing multiple co-located containers in a co-scheduling manner. It monitors the creation and destruction of containers and schedules containers to dedicated CPUs based on the CPU topology and the relationship between them. However, their work is a container-aware mechanism, but it lacks network overhead consideration. A microservices-based IoT application placement technique is proposed in [23] for heterogeneous and resource-constrained fog environments. They also proposed a fog node architecture to support their proposed placement approach. But their main objective is to minimize latency and network usage, and not energy costs. Another approach proposed in [24] optimizes the placement of service-based applications in clouds for reducing inter-machine traffic. They first partition the application into several parts while trying to keep the overall traffic between the created parts to a minimum. Then, the created parts are carefully located in machines with respect to their resource and traffic demands. However, they do not consider energy cost optimization as well.

Energy Consumption and Carbon Emission Effect: An approach is provided in [25] that devised heuristics to continuously consolidate VMs leveraging live migration and switching off idle physical nodes to minimize the number of utilized Physical Machines (PMs). They proposed a modified version of the Best Fit Decreasing (MBFD) algorithm to place VMs on PMs and four heuristics to select VMs for migration. They do not consider the energy consumption of network elements. In another study [26], they proposed three stages of VM placement optimization including reallocation based on the current utilization of multiple system resources, optimization of virtual network topologies established between VMs, and VM reallocation considering the thermal state of the resources. However, both studies do not consider carbon emissions. Our previous work is a Smart Elastic Scheduling Algorithm (SESA) that was proposed in [2]. It relies on the Power Aware Best Fit Decreasing (PABFD) algorithm and its modifications; Modified Worst Fit Decreasing VM Placement (MWFDVP) and First Fit Decreasing with Decreasing Host VM Placement (FFDHDVP) in order to allocate VMs [25]. It dynamically adapts the VM live migration, taking into account the maximum number of migrations per PM that does not violate the SLA and also avoid server sprawl in order to minimize energy consumption. SESA has been implemented in CloudSim [26] and compared with the existing algorithms, but it is not calculating the CO_2 emissions due to the power consumption and also, it is implemented on a hypervisor environment, not a containerization one.

An ant colony system-based approach is developed in [27] to reduce cloud energy consumption. They also propose an order exchange and migration mechanism to handle both homogeneous and heterogeneous PM environments. However, their objective is limited to minimizing the number of active PMs. Another study [28] proposed a hierarchical method (i.e., composed of two workload assignments and migration algorithms) for efficient workload management in distributed cloud nodes. Authors in [29] proposed a dynamic virtual machine scheduling algorithm, called GRANITE, to minimize total EC energy consumption. However, these works do not take into account carbon emissions. The proposed work in [30] is a method for service placement in fog-cloud systems. They classified services into two categories: critical and normal ones. For critical services, they try to minimize response time, and for normal ones, the goal is to reduce the energy consumption of the fog environment. The same authors [31] formulated the VM placement of IoT applications in Cloud DCs as the MILP model and proposed two algorithms. Minimization of the power (i.e., CPU) and network for the first algorithm, and the second algorithm focuses on reducing the power (i.e., CPU) and resource wastage. Yet, carbon emission and energy cost optimization have not been accounted for in these two works.

Authors in [32] proposed a green approach for fog-cloud architecture. This approach is a multi-level approach using a mixed-integer linear programming (MILP) model to minimize the CO_2 emissions of DCs by optimizing the resource usage and the placement of VMs in fog-cloud environments, But the problem here is that it's not a container-aware model. Paper [33] worked toward minimizing the energy of directed acyclic graph-structured applications on a heterogeneous cloud system. The paper also combines a power-aware list-based scheduling algorithm with a Dynamic Voltage and Frequency Scaling (DVFS) technique for real-time tasks (PL-DVFS) to maintain the QoS while considering tasks' deadlines. However, the authors don't mention the CO_2 emission calculation in their green cloud model.

3 Proposed Power-Aware Container Scheduler (PACS)

The related studies mentioned above lead to the importance of designing a modified scheduler; PACS, that considers the computing resources, the network transmission overhead, and the power consumption between the cloud system elements. That will efficiently lead to mitigating the CO_2 emissions toward green cloud systems. Figure 5 describes the complete journey of the cloud applications and user requests that have been served by the cloud containers and scheduled by the orchestration tool in terms of computing resources, network transmission overhead, and finally, the effect of energy consumption with the calculations due to CO_2 emissions.

4 Results and Evaluations

The work is started by expanding the previous work done in [2]. PACS first step is to use the power scheduling algorithms; PABFD, MWFDVP, and FFDHDVP combined with SESA and study the impact of CO_2 emission mitigation due to power consumption reduction. This stage of the PACS scheduler is implemented and tested in CloudSim [25] with a "PlanetLab" dataset. As implementation is done in Egypt datacenters, according to the research results presented in [34], approximately 40% of CO_2 emissions in Egypt originate from the electricity generation sector, which is predominately dependent on oil and gas (90%). The referenced CO_2 Emission Intensity (COI) factors are used for different fuel types presented in [35] to calculate the desired COI factor for the proposed work. COI for the PACS is calculated using Eq. (1).

$$\text{COI} = \text{Average (Gas (Combined cycle), Gas (Open Cycle), Oil)} \tag{1}$$

Table 2. COI factors for each fuel type

Fuel Type	CO_2 Emission Intensity
Biomass^T	120
Coal	937
Gas (Combined Cycle)	394
Gas (Open Cycle)	651
Hydro	0
Nuclear	0
Oil	935
Other	300

COI equals 660 as the average of the CO_2 emission intensity factor for Gas (combined cycle) which equals 394, Gas (Open cycle) which equals 651, and Oil which equals 935 deducted from Table 2 presented in [35].

Table 3. PC and CO_2 emission for PACS against the Clusted and Non-Clustered approaches using heuristic algorithm power MWFDVP

Benchmark	Power Consumption			CO_2 Emission		
	Non-Clustered	Clustered	PACS_MWFDVP	Non-Clustered	Clustered	PACS_MWFDVP
Iqr_mc	136.15	162.83	128.91	89859	107467.8	85080.6
Iqr_mmt	133.31	161.29	131.15	87984.6	106451.4	86559
Iqr_mu	135.87	160.66	137.93	89674.2	106035.6	91033.8
Iqr_rs	137.55	162.48	133.31	90783	107236.8	87984.6
Ir mc	130.52	157.2	124.02	86143.2	103752	81853.2
Ir_mmt	127.97	155.95	131.7	84460.2	102927	86922
Ir_mu	125.18	149.97	125.83	82618.8	98980.2	83047.8
Ir_rs	133.4	158.33	130.08	88044	104497.8	85852.8
Irr_mc	130.52	157.2	124.02	86143.2	103752	81853.2
Irr_mmt	127.97	155.95	131.7	84460.2	102927	86922
Irr_mu	125.18	149.97	125.83	82618.8	98980.2	83047.8
Irr_rs	131.86	156.79	130.94	87027.6	103481.4	86420.4
mad_mc	133.73	159.85	122.9	88261.8	105501	81114
mad_mmt	132.42	156.27	127.95	87397.2	103138.2	84447
mad_mu	133.6	155.75	133.17	88176	102795	87892.2
mad_rs	134.97	158.51	128.23	89080.2	104616.6	84631.8
thr_mc	138	165.63	130.06	91080	109315.8	85839.6
thr_mmt	135.34	166.2	134.02	89324.4	109692	88453.2
thr_mu	137.84	162.19	138.19	90974.4	107045.4	91205.4
thr_rs	139.51	164.41	133.47	92076.6	108510.6	88090.2

The PACS algorithm gives the minimum Power Consumption (PC) values with all combinational algorithms and the lowest values come when using the MWFDVP algorithm. The output results are compared to the non-clustered, and clustered approaches results. Then the power consumption results lead to the CO_2 emission values that are calculated by using Eq. (2). The Power consumption, and CO_2 emission results for 20 combination benchmarks for the PlanetLab workload running on CloudSim are presented in Table 3.

$$CO_2 \text{ Emission} = \text{PC x COI} \tag{2}$$

The results in Table 4 show that using PACS leads to enhancing the CO_2 Emission by an average of 2.12% against the Non-Clustered approach and 18.04% against the Clustered approach.

Table 4. Enhancement % of PACS over Non-Clustered and Clustered

Benchmark	PACS over Non-Clustered %	PACS over Clustered %
Iqr_mc	5.32	20.83
Iqr_mmt	1.62	18.69
Iqr_mu	−1.52	14.15
Iqr_rs	3.08	17.95
Ir_mc	4.98	21.11
Ir_mmt	−2.91	15.55
Ir_mu	−0.52	16.10
Ir_rs	2.49	17.84
Irr_mc	4.98	21.11
Irr_mmt	−2.91	15.55
Irr_mu	−0.52	16.10
Irr_rs	0.70	16.49
mad_mc	8.10	23.12
mad_mmt	3.38	18.12
mad_mu	0.32	14.50
mad_rs	4.99	19.10
thr_mc	5.75	21.48
thr_mmt	0.98	19.36
thr_mu	−0.25	14.80
thr_rs	4.33	18.82

5 Conclusion

In this paper, studies in the areas of container resource scheduling that is used to achieve CO_2 emission mitigation are presented. Different power management models such as PABFD, MWFDVP, MILP, DVFS, and PL-DVFS are introduced to optimize application placement and minimize CO_2 emissions over cloud DC. As network transmission overhead is an important resource that should be considered, the locations to host applications are highly reliant on carbon intensity and traffic demands. Though the PACS solution is introduced, it presents the idea of reducing CO_2 emissions in cloud environments by combining a container-aware scheduler with a power-aware scheduling algorithm that considers the network overhead parameters. Starting with the first stage of the proposed work by using SESA with the MWFDVP algorithm gives a reduction of the CO_2 Emission by an average of 2.12% when used against the Non-Clustered approach and 18.04% when used against the Clustered approach. As a future work, the orchestration tool next will use the previous scheduler presented above to apply it to the PMs and VMs in the cloud Datacenter, adding the network transmission overhead and making it a container-aware scheduler. Then the network transmission overhead will be added to the SESA algorithm and finally, implement this scheduler algorithm in a container-based cloud environment.

References

1. Surbiryala, J., Rong, C.: Cloud computing: history and overview. In: Proceedings of Cloud Summit, Washington, DC, USA, pp. 1–7 (2019)
2. Nashaat, H., Ashry, N., Rizk, R.: Smart elastic scheduling algorithm for virtual machine migration in cloud computing. J. Supercomput. **75**(7), 3842–3865 (2019). https://doi.org/10.1007/s11227-019-02748-2
3. Zhang, Q., Liu, L., Pu, C., Dou, Q., Wu, L., Zhou, W.: A comparative study of containers and virtual machines in big data environment. In: Proceedings of IEEE 11th International Conference on Cloud Computing (CLOUD), San Francisco, CA, pp. 178–185 (2018)
4. Ahmed, A., Pierre, G.: Docker container deployment in fog computing infrastructures. In: Proceedings of IEEE International Conference on Edge Computing (EDGE), San Francisco, CA, pp. 1–8 (2018)
5. Bellavista, P., Zanni, A.: Feasibility of Fog computing deployment based on Docker containerization over RaspberryPi. Association for Computing Machinery, New York, USA (2017)
6. Acharya, J.N., Suthar, A.C.: Docker container orchestration management: a review. In: Sharma, H., Vyas, V.K., Pandey, R.K., Prasad, M. (eds.) Proceedings of the International Conference on Intelligent Vision and Computing (ICIVC 2021). ICIVC 2021. Proceedings in Adaptation, Learning and Optimization, vol. 15, Springer, Cham (2022). https://doi.org/10.1007/978-3-030-97196-0_12
7. Gill, M., Singh, D.: ACO based container placement for CaaS in Fog computing. Procedia Comput. Sci. **167**, 760–768 (2020)
8. Xie, Y., Jin, M., Zou, Z., et al.: Real-time prediction of Docker container resource load based on a hybrid model of ARIMA and triple exponential smoothing. IEEE Trans. Cloud Comput. **PP**(99), 1 (2020)
9. Gamal, M., Rizk, R., Mahdi, H., Elhady, B.: Bio-inspired based task scheduling in cloud computing. In: Hassanien, A. (ed.) Machine Learning Paradigms: Theory and Application. Studies in Computational Intelligence, vol. 801, pp. 289–308. Springer, Cham (2019). https://doi.org/10.1007/978-3-030-02357-7_14
10. Huang, H., et al.: Adapt burstable containers to variable CPU resources. IEEE Trans. Comput. **72**(3), 614–626 (2023)
11. Watada, J., Roy, A., Kadikar, R., Pham, H., Xu, B.: Emerging trends, techniques and open issues of containerization: a review. IEEE Access **7**, 152443–152472 (2019)
12. Ahvar, E., Ahvar, S., Mann, Z.Á., Crespi, N., Glitho, R., Garcia-Alfaro, J.: DECA: a dynamic energy cost and carbon emission-efficient application placement method for edge clouds. IEEE Access **9**, 70192–70213 (2021)
13. Gupta, B., Mittal, P., Mufti, T.: A review on Amazon Web Service (AWS), Microsoft Azure & Google Cloud Platform (GCP) services. In: Proceedings of the 2nd International Conference on ICT for Digital, Smart, and Sustainable Development, ICIDSSD, India (2020)
14. Katal, A., Dahiya, S., Choudhury, T.: Energy efficiency in cloud computing data centers: a survey on software technologies. Cluster Comput. **26**, 1845–1875 (2022). https://doi.org/10.1007/s10586-022-03713-0

15. Iancu, C., Hofmeyr, S., Blagojević, F., Zheng, Y.: Oversubscription on multicore processors. In: Proceedings of IEEE International Symposium on Parallel & Distributed Processing (IPDPS), pp. 1–11 (2010)
16. Lorenzon, A.F., Souza, J.D., Beck, A.C.S.: LAANT: a library to automatically optimize EDP for OpenMP application. In: Proceedings of IEEE Design, Automation & Test in Europe Conference & Exhibition, pp. 1229–1232 (2017)
17. Lorenzon, A.F., de Oliveira, C.C., Souza, J.D., Beck, A.C.S.: Aurora: seamless optimization of OpenMP applications. IEEE Trans. Parall. Distrib. Syst. **30**(5), 1007–1021 (2018)
18. Huang, H., Rao, J., Wu, S., Jin, H., Suo, K., Wu, X.: Adaptive resource views for containers. In: Proceedings of the 28th International Symposium on High-Performance Parallel and Distributed Computing (HPDC), pp. 243–254 (2019)
19. Denoyelle, N., Goglin, B., Jeannot, E., Ropars, T.: Data and thread placement in NUMA architectures: a statistical learning approach. In: Proceedings of the 48th International Conference on Parallel Processing, pp. 1–10 (2019)
20. Goodman, D., Varisteas, G., Harris, T.: Pandia: comprehensive contention-sensitive thread placement. In: Proceedings of the Twelfth European Conference on Computer Systems (EuroSys), pp. 254–269 (2017)
21. Diener, M., Cruz, E.H.M., Navaux, P.O.A., Busse, A., Heiß, H.-U.: kMAF: automatic kernel-level management of thread and data affinity. In: Proceedings of the 23rd international conference on Parallel architectures and compilation (PACT), pp. 277–288 (2014)
22. Lepers, B., Quema, V., Fedorova, A.: Thread and memory placement on NUMA systems: asymmetry matters. In: Proceedings of USENIX Annual Technical Conference (ATC), pp. 277–289 (2015)
23. Pallewatta, S., Kostakos, V., Buyya, R.: Microservices-based IoT application placement within heterogeneous and resource-constrained fog computing environments. In: Proceedings of 12th IEEE/ACM International Conference on Utility Cloud Computing, pp. 71–81 (2019)
24. Hu, Y., de Laat, C., Zhao, Z.: Optimizing service placement for microservice architecture in clouds. Appl. Sci. **9**(21), 4663 (2019)
25. Beloglazov, A., Buyya, R.: Energy efficient allocation of virtual machines in cloud data centers. In: Proceedings of 10th IEEE/ACM International Conference on Cluster, Cloud Grid Computing, pp. 577–578, Melbourne, VIC, Australia (2010)
26. Beloglazov, A., Buyya, R.: Energy efficient resource management in virtualized cloud data centers. In: Proceedings of 10th IEEE/ACM International Conference on Cluster, Cloud Grid Computing, pp. 826–831, Melbourne, VIC, Australia (2010)
27. Liu, X.-F., Zhan, Z.-H., Deng, J.D., Li, Y., Gu, T., Zhang, J.: An energy-efficient ant colony system for virtual machine placement in cloud computing. IEEE Trans. Evol. Comput. **22**(1), 113–128 (2018)
28. Forestiero, A., Mastroianni, C., Meo, M., Papuzzo, G., Sheikhalishahi, M.: Hierarchical approach for efficient workload management in Geo-distributed data centers. IEEE Trans. Green Commun. Netw. **1**(1), 97–111 (2017)
29. Li, X., Garraghan, P., Jiang, X., Wu, Z., Xu, J.: Holistic virtual machine scheduling in cloud datacenters towards minimizing total energy. IEEE Trans. Parall. Distrib. Syst. **29**(6), 1317–1331 (2018)
30. Hassan, H.O., Azizi, S., Shojafar, M.: Priority, network and energy-aware placement of IoT-based application services in Fog-cloud environments. IET Commun. **14**, 2117–2129 (2020)
31. Omer, S., Azizi, S., Shojafar, M., Tafazolli, R.: A priority, power and traffic-aware virtual machine placement of IoT applications in cloud data centers. J. Syst. Archit. **115**, 101996 (2021)
32. Aldossary, M., Alharbi, H.A.: Towards a green approach for minimizing carbon emissions in Fog-cloud architecture. IEEE Access **9**, 131720–131732 (2021)

33. Safari, M., Khorsand, R.: PL-DVFS: combining Power-aware List-based scheduling algorithm with DVFS technique for real-time tasks in cloud computing. J. Supercomput. **74**, 5578–5600 (2018). https://doi.org/10.1007/s11227-018-2498-z
34. Abdallah, L., El-Shennawy, T.: Evaluation of CO_2 emission from Egypt's future power plants. Mediterr. J. Environ. Integr. **5**, 49 (2020). https://doi.org/10.1007/s41207-020-00184-w
35. Bruce, A.R.W., Ruff, L., Kellowaya, J., Rogers, A.: GB Regional Carbon Intensity Forecast. https://www.researchgate.net/publication/328202500_GB_Regional_Carbon_Intensity_Forecast (2020)

Research on Fully Integrated Digital Control Power Management Chip System Based on FPGA

Xiangding Meng(✉), Zichi Zhang, and Xuemei Song

College of Electrical Engineering, North China University of Science and Technology, Tangshan 063210, Hebei, China
1204780573@qq.com

Abstract. The role of power management in integrated digital control systems is very important, but there are problems with voltage and current instability. The PID algorithm cannot solve the power management problem in the integrated digital control system, and the power is unreasonable. Therefore, this paper proposes an FPGA method for integrated circuit power management analysis. First, the power supply management is tested by power flow stability, and the indicators are divided according to the power management requirements to reduce the power management Interference factors with medium voltage instability. Then, the influence of power flow stability on integrated circuit power management forms a power management scheme, and the power management results conduct a comprehensive analysis. MATLAB simulation shows that the FPGA method manages the power management of integrated circuits under certain evaluation criteria. The power output continuity and voltage regulation accuracy are better than the PID algorithm.

Keywords: trend stability · FPGA arithmetic · integrated circuits · Voltage

1 Introduction

Chips are one of the important contents of digital integrated circuits and are of great significance for digital development [1, 2]. However, in the process of power management, the power management scheme has the problem of poor stability [2], which brings certain obstacles to the development of fully integrated digital control. Some scholars believe that the application of FPGA methods to the analysis of integrated digital control systems can effectively analyze power management schemes and provide corresponding support for power management [3]. On this basis, this paper proposes an FPGA method to optimize the power management scheme and verify the effectiveness of the mode.

A. Hassanien et al. (Eds.): AISI 2023, LNDECT 184, pp. 112–118, 2023.
https://doi.org/10.1007/978-3-031-43247-7_10

2 Related Concepts

2.1 Mathematical Description of FPGA Methods

The FPGA method is to optimize the power management scheme by using digital control, and according to the indicators in the power management, find the unqualified values in the integrated digital control system, and integrate the power management scheme. Finally, the feasibility of integrating a digital control system is judged. The FPGA approach combines the advantages of digital control with an integrated digital control system for quantification and can improve power management chips.

Hypothesis 1: The power management requirements are y_i, the power management scheme is x_i, the feasibility of the power management scheme is set_i, and the power management scheme judgment function is $D(g_i \approx 0)$ as shown in Eq. (1).

$$D(g_i) = \sum_{i=1}^{n} X_i Y_i \sum_{i=1}^{n} (X_i - \bar{X})^2 \bar{X} \bar{Y} \tag{1}$$

2.2 Selection of Chip Scheme

Hypothesis 2: The integrated digital control system function is $gf(x_i)$, and the weight factor is x_i, then the power management requirements for unqualified integrated digital control systems are shown in Eq. (2).

$$f(x_i)=\sum_{i=1}^{n} (X_i - \bar{X})^2 X_1, \ldots, X_n \sum_{i=1}^{n} X_i \frac{\Delta y}{\Delta x} \tag{2}$$

2.3 Analysis of Power Management Schemes

Before the FPGA method, a multi-dimensional analysis of the power management scheme is performed, and the power management requirements are mapped to the integrated digital control system library to eliminate the unqualified power management scheme. First, the digital control system is integrated for comprehensive analysis, and the thresholds and index weights of the power management scheme are set to ensure the accuracy of the FPGA method. The integrated digital control system is required to test the power management scheme for the system, which requires stability analysis. If the integrated digital control system is distributed in a nonnormal distribution, its power management scheme suffers, reducing the accuracy of overall power management. In order to improve the accuracy of the FPGA method and improve the level of power management, the power management scheme should be selected, and the specific scheme selection is shown in Fig. 1.

The investigation of the power management scheme shows that the chip scheme presents a multi-dimensional distribution, which is in line with the objective facts. The integrated digital control system has no directionality, indicating that the chip scheme

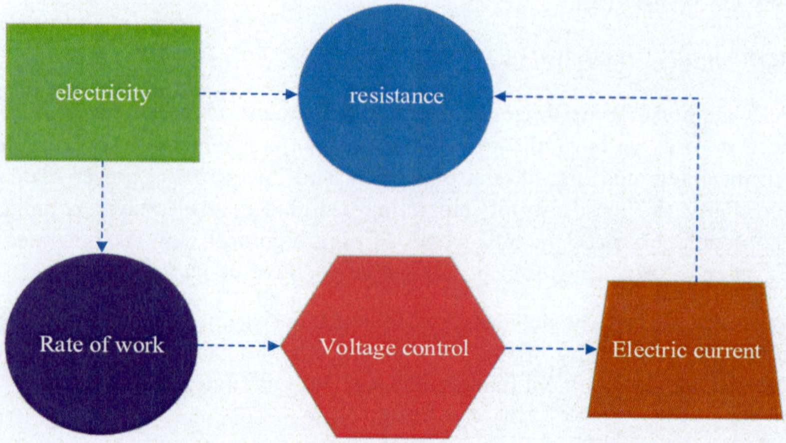

Fig. 1. The result of the selection of the chip solution

has strong randomness, so it is regarded as a high-analysis study. The integrated digital control system meets the normal requirements, mainly the digital control adjusts the integrated digital control system, eliminates the duplicate and irrelevant schemes, and supplements the default scheme, so that the dynamic correlation of the entire power management scheme is strong.

3 Optimization Strategies for Integrated Digital Control Systems

The FPGA method adopts a random optimization strategy for the integrated digital control system and adjusts the reasonable parameters of the power management system to realize the scheme optimization of the integrated digital control system [6]. The FPGA method divides the integrated digital control system into different power management levels and randomly selects different schemes. In the iterative process, power management schemes with different power management levels are optimized and analyzed. After the optimization analysis is completed, the power management levels of different solutions are compared to document the best integrated digital control system [5].

4 Practical Examples of Integrated Digital Control Systems

4.1 Power Management Briefing

In order to facilitate power management, the integrated digital control system in complex cases is the research object, with 12 paths and a test time of 12 h, and the power management of the digital control system is specifically integrated The scheme is shown in Table 1. While, The power management process in Table 1 is shown in Fig. 2.

Compared with the PID algorithm, the power management scheme of the FPGA method is closer to the actual power management requirements. In terms of stability and accuracy of integrated digital control system, FPGA method is superior to PID

Table 1. Power management requirements for integrated circuits

type	grade	viability	Precision
UPS	I	91.38	86.18
	Higher	85.61	88.81
accumulator	II	87.13	92.38
	Higher	86.92	90.00
dynamo	III	90.55	87.78
	Higher	90.98	92.13

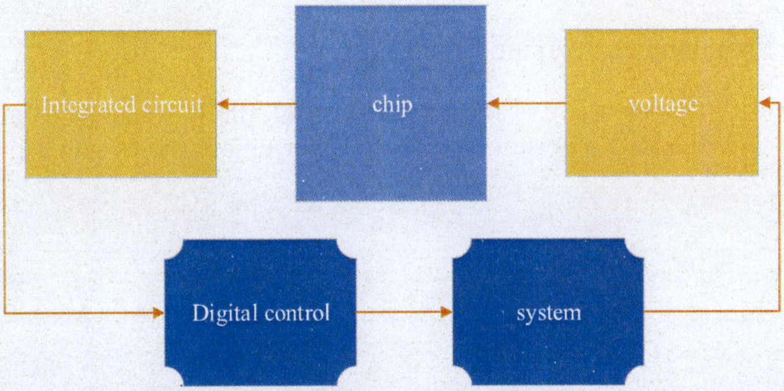

Fig. 2. Analyze the process with an integrated digital control system

algorithm. The changes in the power management scheme in Fig. 2 show that the FPGA method has better stability and faster judgment speed. Therefore, the accuracy of the power management scheme of the FPGA method, the power management scheme of the chip solution, Better summation stability.

4.2 Integrated Digital Control System Situation

The power management scheme with integrated digital control system contains non-structural information, semi-structural information, and structural information. After the pre-selection of FPGA method, a preliminary power management scheme of integrated digital control system is obtained, and the integrated digital control system is obtained. The feasibility of the power management scheme is analyzed. In order to more accurately verify the accurate effect of the integrated digital control system, select the integrated digital control system with different power management levels, and the power management scheme is shown in Table 2.

Table 2. The overall picture of the chip solution

category	viability	Analysis rate
UPS	87.37	86.63
accumulator	83.15	88.70
dynamo	86.75	85.77
mean	86.08	87.11
X^6	85.57	89.90
P = 1.657		

4.3 Power Management Chip and Stability

In order to verify the accuracy of the FPGA method, the power management scheme is compared with the PID algorithm, and the power management scheme is shown in Fig. 3.

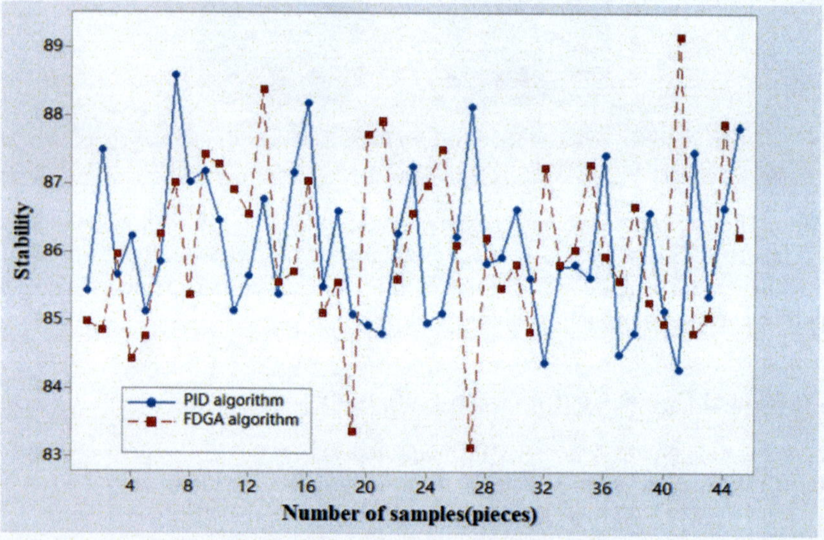

Fig. 3. Chips with different algorithms

It can be seen from Fig. 3 that the chip of the FPGA method is higher than the PID algorithm, but the error rate is low, indicating the power management of the FPGA method. It is relatively stable, while the power management of the PID algorithm is uneven. The average power management scheme for the above three algorithms is shown in Table 3.

It can be seen from Table 3 that the PID algorithm has shortcomings in accuracy and stability in the integrated digital control system, and the integrated digital control

Table 3. Comparison of power management accuracy of different methods

algorithm	Stability	Magnitude of change	error
FPGA approach	93.67	93.41	95.61
PID algorithm	82.79	83.08	84.32
P	35.012	33.827	35.810

system has changed significantly, and the error rate is high. The general result of the FPGA method is that the chip is higher and the control rate is better than the PID algorithm. At the same time, the chip of the FPGA method is greater than 88%, and the accuracy has not changed significantly. To further verify the superiority of the FPGA approach. In order to further verify the effectiveness of the proposed method, the FPGA method is analyzed in general with different methods, as shown in Fig. 4.

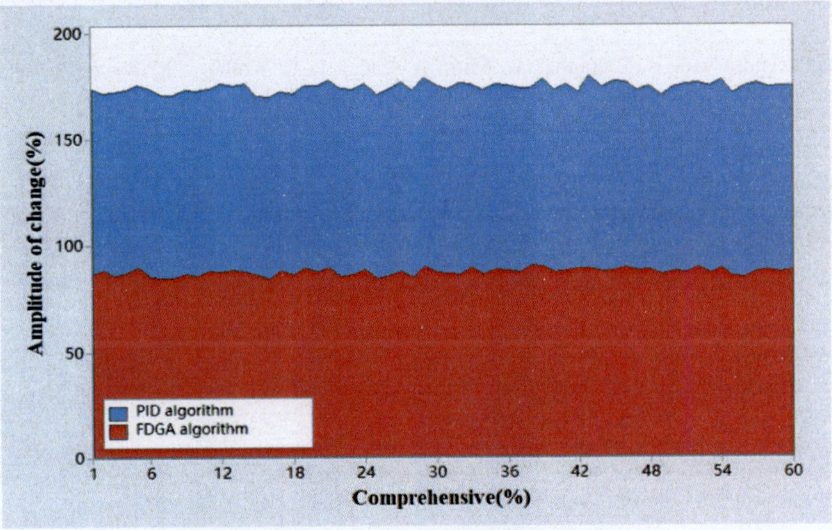

Fig. 4. FPGA method power management chip

As can be seen from Fig. 4, the chip of the FPGA method is significantly better than the PID algorithm, and the reason is that the FPGA method adds an integrated digital control system to adjust the coefficient and set the threshold for whether the power management is reasonable and eliminate the power management scheme that does not meet the requirements [7].

5 Conclusion

Aiming at the problem that the integrated digital control system chip is not ideal; this paper proposes an FPGA method and optimizes it in combination with digital control. At the same time, the stability and threshold accuracy of power management is analyzed in depth to construct the optimal scheme of power management. Research shows that the FPGA method can improve the accuracy and stability of the integrated digital control system Power management. However, in the process of the FPGA method, too much attention is paid to the analysis of power management, resulting in the unreasonable selection of power management indicators.

Acknowledgments. National Innovation and Entrepreneurship Training Program for College Students 202210081008.

References

1. Raghavendra Naik, K., Rajpathak, B., Mitra, A., Sadanala, C., Kolhe, M.L.: Power management scheme of DC micro-grid integrated with photovoltaic - Battery - Micro hydro power plant. J. Power Sources **525**, 230988 (2022). ISSN 0378-7753. https://doi.org/10.1016/j.jpowsour.2022.230988
2. Boemo, E.I., de Rivera, G.G., López-Buedo, S., Meneses, J.M.: Some notes on power management on FPGA-based systems. In: Moore, W., Luk, W. (eds.) FPL 1995. LNCS, vol. 975, pp. 149–157. Springer, Heidelberg (1995). https://doi.org/10.1007/3-540-60294-1_108
3. Elkholy, M.H., Metwally, H., Farahat, M.A., Senjyu, T., Lotfy, T.S.: Smart centralized energy management system for autonomous microgrid using FPGA. Appl. Energy **317**, 119164 (2022). ISSN 0306-2619. https://doi.org/10.1016/j.apenergy.2022.119164
4. Özdemir, M.T., Sönmez, M., Akbal, A.: Development of FPGA based power flow monitoring system in a microgrid. Int. J. Hydrogen Energy **39**(16), 8596–8603 (2014). ISSN 0360-3199. https://doi.org/10.1016/j.ijhydene.2014.01.011
5. Salehi, N., Martínez-García, H., Velasco-Quesada, G., Guerrero, J.M.: A comprehensive review of control strategies and optimization methods for individual and community microgrids. IEEE Access **10**, 15935–15955 (2022). https://doi.org/10.1109/ACCESS.2022.3142810
6. Hannan, M.A., et al.: Role of optimization algorithms based fuzzy controller in achieving induction motor performance enhancement. Nat Commun. **11**(1), 3792 (2020). https://doi.org/10.1038/s41467-020-17623-5.PMID:32733048;PMCID:PMC7393368
7. Aguirre, A.A., Muñoz, L.D., Martín, C.A., Ramírez, M.J., Salazar, C.A: Design of digital PID controllers relying on FPGA-based techniques. IFAC-PapersOnLine **51**(4), 936–941 (2018). ISSN 2405-8963. https://doi.org/10.1016/j.ifacol.2018.06.102

Design of Power Intelligent Auxiliary Control and Monitoring System Based on IoT 3D Image Processing Technology

Jinqiao Li(✉)

School of Intelligent Equipment, Shandong University of Science and Technology, Qingdao 271038, Shandong, China
L55525877@163.com

Abstract. The design of power intelligent auxiliary control and monitoring systems based on IoT 3D image processing is a significant development in the field of power management. This system utilizes advanced technologies such as the Internet of Things, 3D processing, and artificial intelligence to provide efficient and reliable solutions for power management. The system aims to monitor and control various aspects of the power grid, including voltage level, current, and frequency. It uses sensors to collect data from different parts of the grid and uses advanced algorithms to process this data. Then, the processed data is used to determine how best to manage the power grid. The use of 3D image processing technology allows for real-time and accurate visualization of the power grid. This enables operators to quickly identify potential issues or areas that require attention. Overall, this system represents an important step forward in the field of power management. It can use advanced technology to monitor and control various aspects of the power grid, making it an important tool to ensure reliable and efficient energy distribution. The specific construction method is to connect the relay gateway as a hardware device to the Internet of Things platform, and the intelligent power measurement and control instrument to the relay gateway; Complete the visual interface design of the monitoring master station based on the data uploaded by the relay gateway. The basic measurement functions of LoRa communication capability and intelligent power measurement and control instrument were tested, and the test results met the expected requirements.

Keywords: Intelligent auxiliary control · Internet of Things · 3D image processing technology · Electricity · Monitoring system

1 Introduction

The implementation of intelligent auxiliary control functions in substations is an important manifestation of substation intelligence. Currently, although auxiliary control facilities have been configured in substations to achieve safety protection, fire monitoring, water supply and drainage, heating and ventilation, video monitoring, and other functions, compared to the rapid development of these technologies, the information transmission between each facility appears relatively backward, and each facility has not truly

A. Hassanien et al. (Eds.): AISI 2023, LNDECT 184, pp. 119–129, 2023.
https://doi.org/10.1007/978-3-031-43247-7_11

achieved intelligence in a meaningful sense, It still relies on manual processing of information transmission between facilities like information silo, which greatly reduces the operational efficiency of the substation, and also fails to realize the intelligent real-time monitoring, fault analysis and other functions of the substation [1]. Due to the fact that the inspection of substation equipment mainly relies on manual inspection, the quality of the inspection excessively depends on the quality of personnel. Personnel needs to conduct a comprehensive analysis of the inspection data and equipment historical status information in order to accurately determine the processing method, and on-site operators do not have such ability [2]. The power outage and power supply operations in on-site maintenance and repair work still rely on manual judgment to determine whether the required intervals are correct, which can easily lead to the occurrence of misoperation [3]. There is a lack of corresponding means for reliable prevention. The monitoring method for equipment in the station is only achieved through video monitoring systems. If multiple devices in the station are inspected, multiple video windows need to be observed simultaneously, which cannot form a global view and the visualization effect is weak.

At present, China is vigorously developing the construction of smart grids, which also puts forward higher requirements for the automation and digitization level of the power transformation process. And set goals in terms of equipment self-maintenance and self-diagnosis [4]. Substation automation technology is not yet mature, mainly reflected in the lack of comprehensive management and operation systems, outdated equipment maintenance methods, and other aspects. It is particularly necessary to timely introduce high-tech technologies such as the Internet of Things in the substation process [5]. Adding sensors and RFID radio frequency identification technology to the process of staff inspection of the substation, and developing an intelligent substation equipment inspection system can play a supervisory role for staff As a result, it can improve the efficiency of inspection by adding environmental information and status monitoring sensors, accurately detecting the working environment and status of equipment, accurately confirming inspection personnel, and collecting information on the operating environment and working status of power equipment, thereby improving the quality of inspection work [6].

The application of IoT technology in substations has other important applications. Through advanced information collection systems, real-time monitoring of the operation of various equipment and facilities in the substation can be achieved. However, when equipment or facilities may or have already malfunctioned, timely notification of duty personnel to handle faults can improve the safety and operational efficiency of the substation. Sensors can monitor the temperature of the device in real-time, and when there is an abnormal temperature in the device, an alarm is raised, and corresponding measures are taken to handle the fault [7, 9, 10]. The intelligent management system for device status can also record various operating parameters of the device, evaluate its operational reliability, and provide data support for staff.

2 Related Work

2.1 Design of 3D Point Cloud Filtering Technology

(1) **Bandpass filter:** In 3D images, unlike previous 2D images, each pixel has a depth value that is directly related to the distance from the object to the sensor. The closer the object is captured, the smaller the depth of the object. In the process of image acquisition, objects that are not of interest can be removed based on the depth value. This step requires very little computation and only needs to traverse each point [8].

The bandpass filter has a very simple principle, and the key principle of this filter is to allow specific bands and simultaneously shield other bands. This article will use a similar principle to eliminate the background of 3D images based on a specific domain constraint. Specific fields can be on the X-axis, Y-axis, or Z-axis.

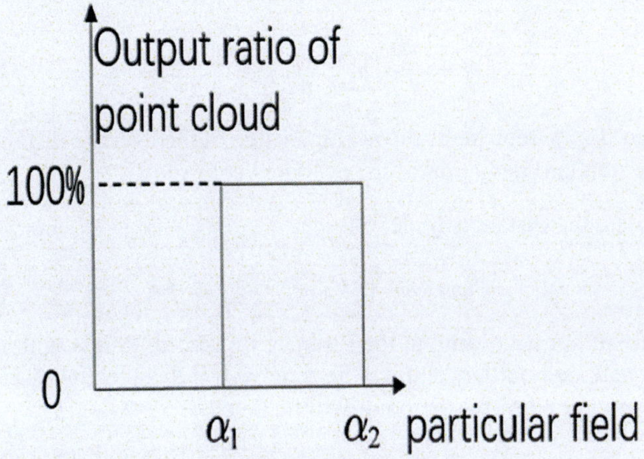

Fig. 1. Wave filter

As shown in Fig. (1), when the value of a specific field is between α1 and α2, all point cloud data within this range will be retained. When the value of a specific field is less than α1 or greater α At 2 o'clock, all point cloud data within this range will be deleted. In this article, the following formula can be used to represent a bandpass filter.

$$\Omega = \begin{cases} 0 & \tau < \alpha_1 \\ 100\% & \alpha_1 \le \tau \le \alpha_2 \\ 0 & \tau > \alpha_2 \end{cases} \tag{1}$$

where ł and Ω represent specific fields and output ratios of the point cloud, respectively.

(2) **Statistical filter:** The function of the statistical filter is to remove sparse outliers according to adjacent distance. If the average distance between a point and its adjacent points is greater than the distance threshold, the point will be considered as

a outlier. Calculate the distance threshold based on the standard deviation of the average distance vector [11]. Then, the anomaly dataset is obtained based on the distance threshold.

The processing steps of this algorithm are as follows:

Step (1): Calculate the average distance of all points relative to their y nearest neighbors:

$$\Gamma_i = (\sum\nolimits_{j=1}^{\gamma} \theta_j)/\gamma, (i = 1, 2, 3........n) \tag{2}$$

where 0 represents the distance between a point and the jth point in its nearest neighbor.
Step (2): Calculate the average and standard deviation of the average distance vector.

$$\Gamma_{mean} = (\sum\nolimits_{i=1}^{n} \Gamma_i)/n \tag{3}$$

$$\varepsilon = \sqrt[2]{\frac{1}{n}\sum\nolimits_{i=1}^{n} (\Gamma_i - \Gamma_{mean})^2} \tag{4}$$

Among them, Γ_{mean} represents the average value of the average distance vector, and é represents the standard deviation of the average vectorr [10].

Step (3): Calculate the distance threshold.

$$\theta_{threshold} = \Gamma_{mean} + std_{mul} * \varepsilon \tag{5}$$

Step (4): Delete outlier according to the distance threshold. Points with larger average distances are considered outliers and will be removed. If the average distance of a point meets the following conditions, the point will be deleted:

$$\Gamma_i \geq \theta_{threshold} \tag{6}$$

2.2 Design of 3D Point Cloud Segmentation Technology

In the RANSAC algorithm, the assumed interior points are selected from the original input dataset for building the model. Then, the model tests whether other points meet the model calculates its loss function, and the points that meet the model form a consistent set [13]. In practical applications, the type of model is known, and assumed interior points are used to obtain the parameters of the model. If the consistent set contains the maximum number of inline functions, the model is considered a good model [12]. Finally, when the interior point is retained, the outlier will be deleted. Afterward, the least squares method will be used to re-evaluate the model by using all interior point set members [13].

The Algorithm Steps are as Follows:

- S1: A specific model has been designed to obtain a consistent set for a specific problem, which can be a plane model, a line model, a circle model, a spherical model, etc.

- S2: Calculate the number of iterations.

$$n = \frac{log(1-\delta)}{log(1-\varepsilon^k)} \tag{7}$$

Among them, δ represents the probability of the RANSAC algorithm selecting a consistent set from the input dataset. K represents the minimum number of points that can be used to estimate model parameters, ε represents the probability of selecting an interior point when selecting a single point.

- S3: Randomly select n sets of assumed interior point sets from the original input dataset, each containing k points.
- S4: Use n sets of assumed interior point sets to construct an n model.
- S5: Find out all outliers of the corresponding model according to the threshold $\in$. Then, obtain a consistent set of interior points. $\in$ Will be designed by the user. The consistent set of interior points is reused to update the model, and the model is established using the least squares method to rediscover the consistent set of interior points.
- S6: Record the size of the inner point consistent set and compare it with the maximum size of the inner point consistent set. If the number is greater than the maximum size of the interior consistent set, the parameters of the model will be saved, and the size of the interior consistent set in the model will be labeled as the maximum size of the interior consistent set.
- S7: Repeat steps S4–S6 n times, where n is the number of models.

Because of the particularity of 3D point cloud data structure, octree data structure is used to store sparse 3D point cloud data, and each branch node is used to store the central point information in a certain cube. For a given depth, an octree is constructed by traversing the tree structure and using the depth first search method. Starting from the root, each node will generate eight child voxels [8]. All points are mapped to leaf voxels at the maximum depth of the tree. Use the average value of all points in the cube to obtain the color information of the center point of the cube. As shown in Fig. 2.

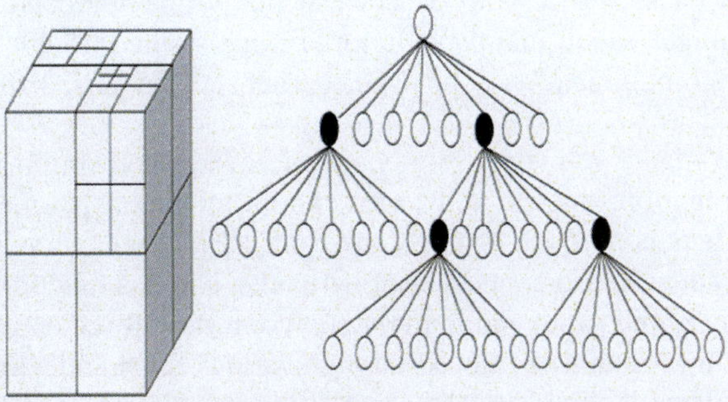

Fig. 2. Octree data structure to store sparse 3D point cloud data

3 Intelligent Auxiliary Control of Power Based on IoT 3D Image Processing Technology

A. Overall Architecture of Auxiliary Control System

The Internet of Things plays a crucial role in the overall architecture of the substation intelligent auxiliary control system. The process layer, bay layer, and station control layer all adopt the Internet of Things structure, with various information transmission methods such as wired and wireless, and a large number of data interfaces. It achieves real-time display of various information within the station, data collection and sharing, intelligent monitoring, maintenance, and control of equipment, and has the characteristics of informatization, intelligence, and interactivity. The overall system architecture is shown in Fig. 3.

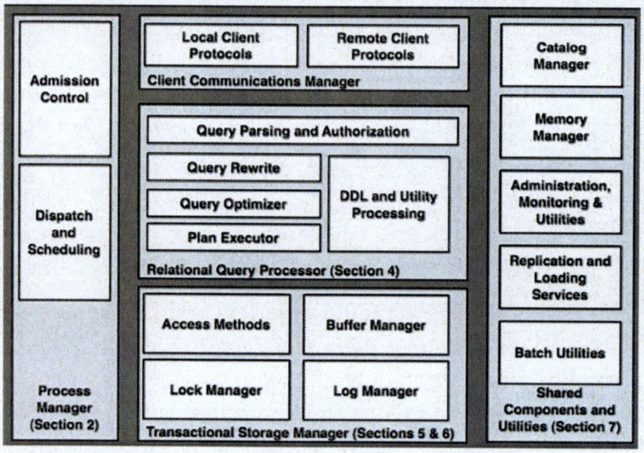

Fig. 3. Overall Architecture of Intelligent Auxiliary Control System for Substation

(1) **Process layer:** The "process layer" corresponds to the "perception layer" of the Internet of Things, which mainly monitors device status through various sensing devices. Data communication supports Ethernet, RS485 bus, wireless, and other methods.

    The process layer is mainly composed of several perception nodes deployed in various perception objects, and a perception network is formed through self-organization. It is composed of QR codes and RFID tags, surveillance cameras, information readers, sensors, and sensor networks, and has functions such as information collection and processing, automatic control, and recognition. Equipment status monitoring is mainly achieved through various sensing devices.

(2) **Spacer layer:** The function of the bay layer is to transmit various detection information and video monitoring information collected in real-time from the process layer to the station control layer and use the connected equipment to complete intelligent inspection of the substation. The monitoring data is converted to the IEC61850 protocol through a comprehensive access gateway.

The bay layer mainly integrates and expands different communication networks such as power wireless broadband, wireless public communication network, wireless sensor network, and power optical transmission network, achieving functions such as information transmission, routing, and control between the process layer and station control layer, providing high reliability, high security, and large-scale data transmission. Among them, in order to achieve the purpose of large-scale or local transmission of information in the process layer, power fiber optic networks have become the main communication lines due to their unique advantages, and wireless broadband networks and power line carrier communication networks are auxiliary communication lines.

(3) **Station control layer:** The "station control layer" has the same function as the "application layer" on the Internet of Things, which can transmit collected and summarized data to the system platform for analysis and processing. The system platform connected to the integrated automation system through one-way isolation devices can assist in the operation and maintenance of the substation.

The station control layer mainly analyzes and processes the information perceived by the process layer according to different applications and business needs. The station control layer includes the internal control host and the remote centralized control station management host.

The intelligent auxiliary control system of the substation provides multiple interfaces to the outside world. The station end is interconnected with the comprehensive automation system of the substation through isolation devices, and the control center is interconnected with the PMS production management system and other applications.

B. Design of A 3D Point Cloud Compression Framework Based on Global Clustering

Due to the large amount of 3D point cloud data and the different number of points between two frames, the processing of 3D point cloud data sequences also faces many challenges. Moreover, there is no clear correspondence between two points on different 3D point cloud data frames, and the distribution of points in space is uneven. Therefore, firstly, the octree structure is used to vowelized the point cloud data, so that the point cloud data is evenly distributed, and the calculation time is short. Then, use the SHOT description operator to find the best corresponding point pair. Finally, motion estimation and clustering techniques were applied to solve the problem of different number of points between two 3D point clouds [4, 8].

As shown in Fig. 4, a 3D point cloud data compression framework based on temporal correlation information and spatial redundancy is proposed. In order to eliminate spatial redundancy, a robust Bayesian framework is used to track the remote operator, and the 3D point cloud data of the remote operator is obtained by tracking the 2D frame. A particle filter based on a color distribution histogram tracks remote targets within the bounding box. In addition, there are still some outliers in the 3D point cloud data, which are processed using statistical filtering methods.

In order to further compress the 3D point cloud data of the remote operator's body P Merge frames into frame I for 3D point cloud data compression. This article uses the SHOT description operator to search for corresponding relationships on 3D dynamic frames point by point. Firstly, the eigenvalue decomposition technique is applied to

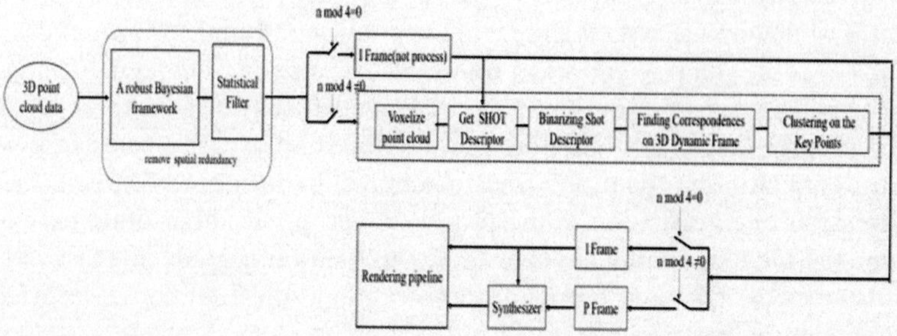

Fig. 4. A 3D point cloud compression framework based on global clustering.

obtain the reference coordinate system of key points. Secondly, by constructing a histogram of key points to obtain their features, and binarizing the features to reduce computational time. Then, by comparing the characteristics of two key points, identify their corresponding relationships. Thirdly, key point clustering is performed on some mismatched points to calculate the motion vectors between two 3D point cloud frames. Especially before the B-SHOT descriptor is implemented, the octree structure is used to vowelize the 3D point cloud, because the octree structure can maintain the original distribution of the point cloud, make the distribution more uniform, and reduce the strength of the points. Finally, the reserved information and motion vector of the *P* frame is sent to the far end. At the far end, in order to decode P Frame, utilizing frame *I* information *P* Frame preservation information, and motion vectors are used to predict new frames.

In order to generate local reference frames with translation and rotation invariance features, this paper uses eigenvalue decomposition to construct the reference frame, which is composed of the eigenvectors of the modified covariance matrix b. The three-unit eigenvectors of the modified covariance matrix b are represented in the decreasing order of eigenvalues as a+, g+, and z soil axes, respectively. The corrected covariance matrix b is represented by the following formula:

$$\psi = \frac{1}{\sum_{i:d_i \leq l}(r - d_i)} \sum_{i:d_i \leq l}(l - d_i)(q_i - q)(q_i - q)^T \tag{8}$$

If the similarity is less than a predefined threshold, the 3D point cloud data of the cluster subset is retained. If the similarity is greater than or equal to a pre-defined threshold, the 3D point cloud data of the cluster subset is not retained, and the motion vector is calculated. The motion vector is given by the following equation:

$$v_{ji}^k = p_j^k - p_i^k \tag{9}$$

4 Results and Discussion

The traditional two-dimensional information representation represented by the wiring diagrams of primary and secondary equipment has been widely applied in the production, operation, and management of substations. However, two-dimensional information cannot represent the spatial details of equipment within the substation and the topological relationships between equipment and equipment, as well as between equipment and cables. 3D visualization technology can enable duty personnel to monitor the operation of the substation in a comprehensive and three-dimensional manner, not only providing a bird's-eye view, but also being able to approach equipment to view local details within the substation.

In order to demonstrate the feasibility and effectiveness of the application of Internet of Things technology in the intelligent auxiliary control system of substations, the following is an introduction to a set of substation intelligent auxiliary management and control systems developed by State Grid Information and Communication Co., Ltd. Based on the Internet of Things and 3D visualization. This system is a typical case of the application of Internet of Things technology and 3D visualization technology in substation auxiliary control. By modeling the buildings, electrification equipment, indoor environmental facilities, etc. inside the substation, it achieves panoramic holographic display of the substation and maximizes the simulation of the operation status of the substation; The synchronous dynamic control between the three-dimensional model components of the substation and the actual equipment inside the substation was achieved through a three-dimensional interactive interface. By utilizing the human-computer interaction of 3D virtual reality, advanced applications such as 3D intelligent inspection and 3D historical on-site inversion have been achieved.

The training and validation losses of the developed model are shown in Fig. 5 and Fig. 6, respectively. Training losses often decrease with the increase of time, while validation losses remain roughly unchanged after 300 periods.

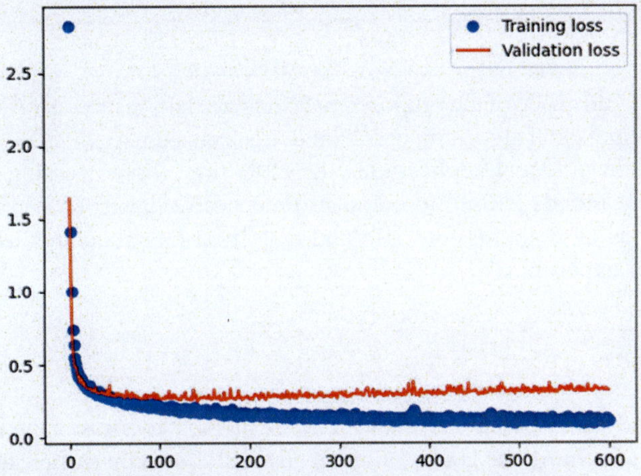

Fig. 5. 1D-CNN training and validation losses

The best performing machine learning model has a classification accuracy of approximately 56.0%. On the other hand, the classification accuracy of 1D-CNN is close to 95.7% (Refer to Table 1).

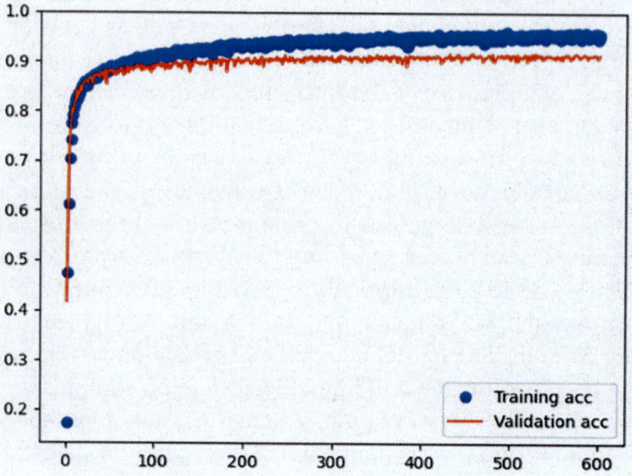

Fig. 6. Training and validation accuracy of 1D-CNN

Table 1. Model Performance

Algorithms	WT5000	WT333E
Random forests	56.0%	45.5%
Bayesian	39.3%	36.2%
Decision-making tree	54.3%	45.1%

The good performance of the proposed method was demonstrated through two experiments. The compression algorithm based on octree is used to compare the performance of this method. In practical applications, the CPU time (or processing time) of compressed 3D point clouds is used to compare their performance. In particular, analysis was conducted using compression ratio and distortion rate standards to achieve more fair comparative experiments.

5 Conclusion

Through the study of existing auxiliary facilities in substations and the analysis of the practical needs to achieve the goal of unmanned substations, this article applies Internet of Things technology to the design of intelligent auxiliary control systems in substations, so that the independent auxiliary control facilities of conventional substations should be optimized and integrated in terms of equipment configuration and functions. While the subsystems are powerful, resource sharing and efficient interaction between subsystems are necessary, integrating remote control systems to promptly handle equipment failures and other abnormal phenomena in the substation, improve supervision quality, reduce labor intensity, and make the operation of the substation more intelligent.

References

1. Rama Subramanian, M., Babu, T.R., Krishna, V.A., et al.: Design of intelligent control and monitoring system for agriculture based on renewable energy and IoT. J. Phys. Conf. Ser. **1964**(4), 042031 (2021)
2. Gong, S., Kumar, R., Kumutha, D.: Design of lighting intelligent control system based on OpenCV image processing technology. Int. J. Uncertain. Fuzziness Knowl.-Based Syst. **29**(Supp01), 119–139 (2021)
3. Liu, Q., Hou, S., Wei, L.: Design and Implementation of intelligent monitoring system for head and neck surgery care based on Internet of Things (IoT). J. Healthcare Eng. **2022**, 4822747 (2022)
4. Sun, X., Ma, H., Sun, Y., Liu, M.: A novel point cloud compression algorithm based on clustering. IEEE Robot. Autom. Lett. **4**(2), 2132–2139 (2019). https://doi.org/10.1109/LRA.2019.2900747
5. Raja Singh, R., et al.: IoT embedded cloud-based intelligent power quality monitoring system for industrial drive application. Future Gen. Comput. Syst. **112**, 884–898 (2020). ISSN 0167–739X, https://doi.org/10.1016/j.future.2020.06.032
6. Ma, Z.X., Cheng, P.X., Ning, J., et al.: Innovations in monitoring, control, and design of laser and laser-arc hybrid welding processes. Metals **11**(12), 1910 (2021)
7. Norris, M., et al.: IoTRepair: flexible fault handling in diverse IoT deployments. ACM Trans. Internet of Things **3**(3) 22, 1–33 (2022). https://doi.org/10.1145/3532194
8. Jia, C.C., Wang, C.J., Yang, T., Fan, B.H., He, F.G.: A 3D point cloud filtering algorithm based on surface variation factor classification. Procedia Comput. Sci. **154**, 54–61 (2019). ISSN 1877–0509 https://doi.org/10.1016/j.procs.2019.06.010
9. Sundaravadivel, P., Wilmoth, P., Fitzgerald, A.: Solicitude savvy: an IoT-based edge intelligent framework for monitoring anxiety in real-time. In: International Symposium on Quality Electronic Design, pp. 576–580. IEEE (2021)
10. Xie, X., Wen, X., Deng, F.: Applications of 3D image using Internet of Things in the exhibition of classical architecture art style. Mobile Inf. Syst. 2283354, 6 (2021). https://doi.org/10.1155/2021/2283354
11. Wong, K.M.: Order statistical filtering: a robust method of noise estimation. J. Franklin Inst. **322**(4), 185–207 (1986). ISSN 0016–0032, https://doi.org/10.1016/0016-0032(86)90056-6
12. Liu, J., Zhao, J., Liu, Q., et al.: Integration and application of 3D visualization technology and numerical simulation technology in geological research. Environ. Earth. Sci. **80**, 776 (2021). https://doi.org/10.1007/s12665-021-10055-4
13. Raguram, R., Frahm, J.-M., Pollefeys, M.: A comparative analysis of RANSAC techniques leading to adaptive real-time random sample consensus. In: Forsyth, D., Torr, P., Zisserman, A. (eds.) ECCV 2008. LNCS, vol. 5303, pp. 500–513. Springer, Heidelberg (2008). https://doi.org/10.1007/978-3-540-88688-4_37

On the Interplay Between Self-adaptation and Energy Efficiency

Soyaba Hasan and Yehia Elkhatib(✉)

School of Computing Science, University of Glasgow, Glasgow, UK
{Soyaba.Hasan,Yehia.Elkhatib}@glasgow.ac.uk

Abstract. Self-adaptation is an increasingly popular approach for enabling systems to make agile and data-driven decisions without human intervention. However, techniques for measuring the performance of a self-adaptive system and its effects on context are lacking. In this study, we present a mechanism that enables customizable and comprehensive evaluation of self-adaptive systems in real time. We apply our proposal to an example of resource management in a data center setting. We find that a self-adaptive approach to managing resources is multiple orders of magnitude better in terms of energy efficiency, especially as the size of the infrastructure grows, at the expense of slightly lower performance (<10%).

Keywords: Self-adaptation · Performance measurement · Energy efficiency

1 Introduction

A self-adaptive system can handle changes without external intervention. This topic has been of great interest for many years, dating back to the notion of autonomic computing [1]. Self-adaptation is used in various domains such as self-driving vehicles, smart grids, mobile applications, the Internet of Things, and cloud computing. Self-adaptive system architectures are predominantly employed for their ability to make prompt decisions informed by prior experience, reduce room for human error and bias, and optimize resource utilization.

Despite the great interest in self-adaptive systems, there is a lack of means to understand their behavior in a real-time and comprehensive manner [2]. Evaluating the effectiveness of self-adaptive systems is a significant challenge due to their inherent complexity, dynamics, and unpredictable operating environments. In particular, real-time feedback is essential for intended users and software engineers to leverage the capabilities of a system and the possibilities of its context, while also addressing potential limitations.

Despite the availability of runtime measures for self-adaptive systems, their adequacy for the above needs is questionable. Existing methods can collect real-time data on system behavior, but they are (i) overly complex and challenging

A. Hassanien et al. (Eds.): AISI 2023, LNDECT 184, pp. 130–139, 2023.
https://doi.org/10.1007/978-3-031-43247-7_12

to deploy, necessitating significant time and resources; and/or (ii) inadequate to reflect and analyze the effectiveness or provide actionable insights.

In this study, we devise a mechanism to evaluate in real-time the runtime performance of a self-adaptive job dispatching system. This is a widely used application of self-adaptation to manage servers in cloud data centers. In doing so, we aim to develop appropriate techniques that could be applied to other self-adaptive systems. Moreover, we investigate the interplay introduced by self-adaptation, specifically in terms of system performance and energy efficiency.

Our contributions are as follows:

1. We developed a mechanism to evaluate the performance and resource utilization of a self-adaptive system in real-time.
2. We apply this mechanism by integrating it with the Simdex [3] framework.
3. We investigate the trade-off between system performance and overhead when self-adaptation is used for resource management.

2 Problem Space

In this section, we describe the problem space and examine related works.

2.1 Background

A self-adaptive system can autonomously adapt in response to a change, such as a modification in the context in which the application runs or one in the fabric of the application itself. In either case, adaptation is performed programmatically based on a feedback mechanism, such as a Monitor, Analyze, Plan, Execute, and Knowledge (MAPE-K) control loop. Reacting (or sometimes even proacting) in such a programmatic manner enables applications to act in an agile and data-centric manner, which makes self-adaptive systems much more reliable, flexible, and efficient.

Self-adaptation has been applied to enable independent systems to evolve the way in which they interact with other systems, *e.g.*, recommendations in decentralized overlays [4] and automating system composition [5,6] in smart home [7] and Internet of Vehicles [8] environments. Self-adaptive methods have also been employed to make centralized optimization decisions. A very common example of this is used in data centers, where decisions such as virtual machine consolidation [9], placement [10], and auto-scaling [11] can be made using self-adaptive control loops.

2.2 Related Work

The Rainbow framework [12] is commonly used to evaluate software quality. Although it provides a general strategy for quality evaluation, it does not offer guidance on applying appropriate methods to reflect system performance.

RELAX [13] supports model-based runtime verification and provides a comprehensive framework for the development of evaluation tools. It combines statistical techniques to build robust tools that can adapt to changing environmental conditions. However, this can result in high computational overhead for large-scale systems and may cause increased resource utilization and downtime. Fusion [14] employs dynamic and static verification techniques to detect errors in self-adaptive software systems. It generates a complete model of the system behavior and compares it with the actual performance during runtime. Similar systems were devised to gather knowledge of self-adaptive volunteer computing platforms [15]. However, they rely on the accuracy of static models, which may not accurately reflect performance, and cannot handle unexpected runtime behavior.

The Dynamico framework [16] utilizes formal methods to verify the behavior of self-adaptive systems at runtime. However, it relies on static models that may not accurately capture the behavior of the system and incur high computational and resource costs. SimuLizar [17] used a model-driven approach to analyze performance at design time but neglected uncertainty at runtime. Zanshin [18] combined machine learning and knowledge-based techniques to monitor complex systems under fluctuating conditions and optimize system behavior. However, maintenance of Zanshin and similar frameworks (*e.g.*, [19]) requires specialized expertise. Consequently, they are not well suited for smaller systems that do not feature advanced levels of self-adaptation.

In summary, as indicated by Gerostathopoulos et al. [2], most evaluation methods for self-adaptive systems lack validity and verification. Concretely, these methods often fail to consider the dynamic nature of such systems and lack specific guidance on applying appropriate evaluation strategies according to system design. Our work aims to address this gap in the literature.

3 Design

Our aim is to support users by developing an evaluation of self-adaptive systems. In this section, we introduce our design and use case.

3.1 Requirements

The functional requirements of our design are to:

- enable concurrent tests, *i.e.*, to compare different evaluation scenarios;
- support multiple performance metrics for users to acquire a complete picture of the system behavior and performance;
- provide a user-friendly interface that simplifies test setup, execution, and analysis;
- handle large-scale data and remain responsive under diverse workloads; and
- store and retrieve measurement logs.

Non-functional requirements are as follows:

- reliability even under heavy workloads, to ensure users can efficiently analyze results and identify patterns;
- handle and analyze information in real-time with minimal delays;
- handle multiple concurrent experiments and promptly;
- retrieve data effortlessly;
- extensible for the integration of new features and performance metrics; and
- easy to maintain and support with clear documentation and code organization.

Goals that are outside the scope of our system but are included in this manuscript:

- process and analyze large amounts of data to extract insights; and
- generate and customize visualizations.

3.2 Simdex Environment

Before explaining the measurement mechanism, we introduce the environment in which it operates. Simdex [3] is a tool for simulating self-adaptive job-dispatching systems. It is designed to replay workload logs and use self-adaptive mechanisms to experiment with enhancing system performance in response to changing conditions.

We chose Simdex because of its scalability and flexibility as a job-dispatching system that supports self-adaptive control loops. Its configurable and extendable features make it a robust tool to incorporate our evaluation system. Thereupon, users can customize the system configurations, manage job requests, optimize resource allocation, and analyze dynamic settings using our evaluation system. In contrast, other self-adaptive workload operation tools such as DEECo [20] and RTX [21] are substantially more complex in terms of integrating custom measurement mechanisms.

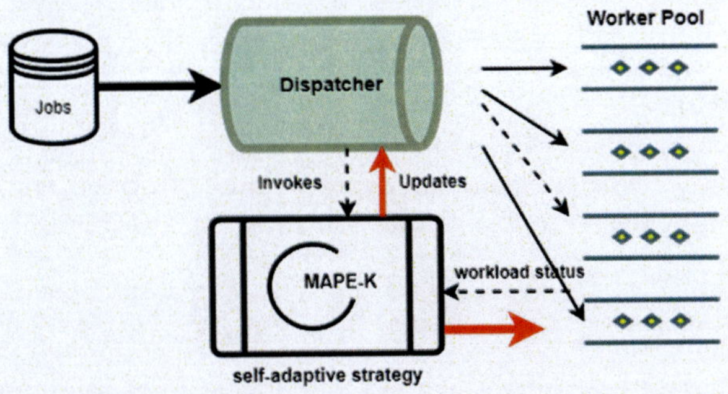

Fig. 1. An overview of the architecture of the Simdex simulation environment

As illustrated in Fig. 1, a user-submitted job is passed on to the dispatcher module that examines the job and assigns it to an appropriate worker. Each worker has a single queue and processes jobs in a first-come-first-served order. A worker's status can be either 'active' or 'inactive', which is altered based on workload demand.

If a self-adaptive (SA) strategy is being examined, the dispatcher triggers the MAPE-K loop self-adaptive strategy to identify the state of the worker nodes, ensuring that the system can adapt and respond to changes in workload. The self-adaptive controller then updates the dispatcher with the feedback decision, and the dispatcher directs jobs to the worker nodes responsible for processing them. This adaptive process adheres to two specific rules:1) it activates an inactive queue if any queue in the pool still has more than one job, and 2) when a worker does not have jobs queued, the system deactivates the idle queue.

In contrast, a non-self-adaptive (NSA) strategy is studied using a fixed number of active workers, whereby resources are statically allocated to submitted jobs.

3.3 Monitoring Mechanism

Through integration with Simdex, our evaluation system captures real-time performance metrics including latency, throughput, average response time, total delay, resource utilization, and power consumption. These metrics provide the user with a comprehensive understanding of the behavior and performance of the MAPE-K loop under study. Our measurement logic is implemented in Python and paired with custom scripts that use libraries such as `pandas` and `matplotlib` to automate analysis and visualization.

We designed the evaluation process following the principles outlined in the framework of Villegas et al. [22] and identified the metrics (detailed in Sect. 4.1) to develop a robust evaluation strategy. Users can customize the evaluation system to suit their requirements. For example, they can investigate different infrastructure configurations and scaling behaviors under heavy workloads. Users obtain real-time performance insights through automatic data visualizations that are made available.

4 Evaluation

We conducted a series of experiments using a range of worker configurations and workloads. We now describe the evaluation setup and results in detail.

4.1 Metrics

We are interested in assessing the costs and benefits of applying self-adaptive strategies to optimize resource and energy utilization. Thus, we measured the following variables, inspired by the catalog built by Da Silva et al. [23]:

- **Throughput** is the rate at which the system can process jobs. Higher values indicate better performance.

- **Latency** refers to the time it takes for a job to be processed by the system. Low latency is an indication of higher performance and improved user experience.
- **Total delay** is the total amount of time it takes for a job to be completed, from the time it was submitted to the system until the time it is finished. It includes the time spent waiting in queues, processing by workers, and any other delays that may occur during job execution.
- **Average Response time** is a measure of system responsiveness. It is equal to the amount of time taken for a system to respond to a job request.
- **Resource utilization**, a fundamental metric in job-dispatching systems, refers to the fraction of available resources that were used to complete a given set of tasks.
- **Power consumption** refers to the amount of energy consumed by the system over a period of time. Reducing this expenditure can result in improved energy efficiency, which is important particularly in resource-constrained environments.

4.2 Independent and Confounding Variables

Our independent variable in each experiment was the number of jobs, which varied within the range of 15,000–30,000. We vary the worker configuration between 2 and 14, changing their status from 'inactive' and 'active' according to the scaling strategy.

4.3 Results

We now describe the findings of our experiments in detail.

4.3.1 Overhead

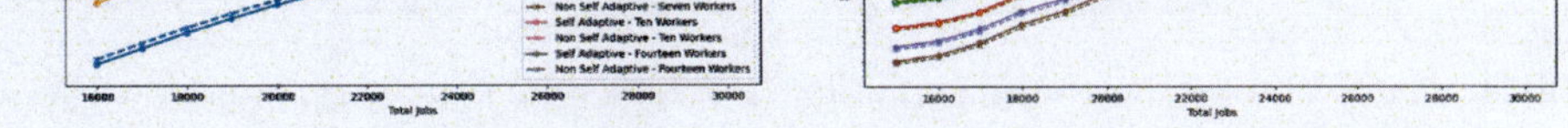

Fig. 2. The throughput as the number of jobs served per second (left), and job latency in milliseconds (right). The plots depict the magnitude of the additional overhead incurred by applying the self-adaptive resource management strategy. The impact on performance is discernible yet minimal

We found that SA consistently underperforms when compared to NSA for all job ranges and worker configuration scenarios. This is manifested in the form of a 2.11% lower throughput and 10.74% higher latency levels, as depicted in the plots of Fig. 2. The disparity between the two systems becomes more prominent as the number of jobs increases; specifically, the gap widens for settings of 24,000 jobs and above. This is to be expected, as self-adaptation adds a layer of complexity that includes introspection and decision making, which adversely influences system performance.

4.3.2 User-Perceived Responsiveness

We make several observations from the plots in Fig. 3. First, the SA variant of the system is as responsive as the NSA cognate for low job numbers. Second, SA becomes decreasingly responsive beyond 20,000 jobs. Third, the disparity

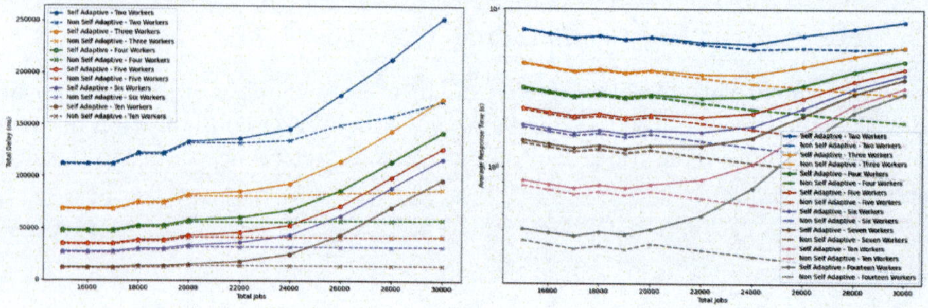

Fig. 3. The total delay (left) and response time (right) for both strategies. These results show that a self-adaptive strategy is feasible, but only to a point where an increased number of jobs and/or workers bring about exponentially degrading responsiveness

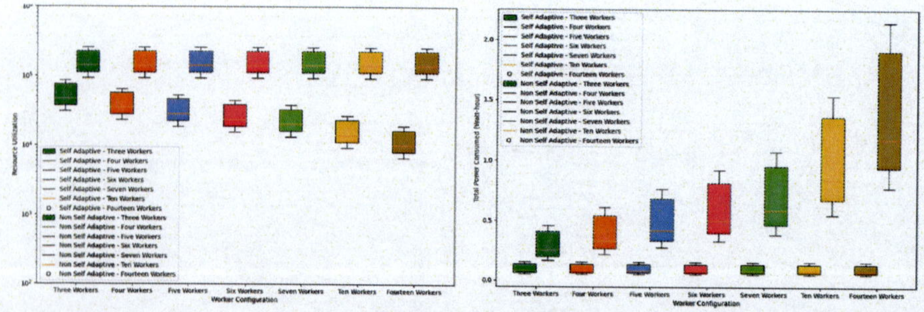

Fig. 4. The levels of resource utilization (left) refers to the ratio of the total uptime to the resources used to process the jobs; note the logarithmic scale. The corresponding power consumption (right) is given in Watt-hours. Inspecting these findings, we identify the benefits gained by adopting a self-adaptive management strategy: it allows much more efficient utilization of available resources as the infrastructure expands, and it minimizes energy consumption owing to the proactive and dynamic management of running workers

between SA and NSA rapidly worsens with more jobs, as the former features an exponential growth in delay and response times, while the latter follows a linear trend with a low slope. Finally, the disparity worsens as the size of the infrastructure (*i.e.*, number of workers) increases.

4.3.3 Optimization

The utilization levels of NSA were comparatively higher than those of SA, both in terms of making efficient use of the available computational resources and the amount of energy consumed. With self-adaptation, resource utilization decreases by 84.35% with an increasing number of workers in the 15,000–30,000 job range. In Fig. 4, the box plots for SA show a clear trend of decreasing resource utilization with an increasing number of workers, whereas the box plots for NSA are relatively constant.

Furthermore, the power consumption of SA is relatively consistent across *all* job ranges and worker configurations. Overall, this level is 86.84% that of NSA. In contrast, NSA exhibits a noticeable rise in power consumption as the number of jobs and workers increases, primarily because idle workers consume energy unnecessarily. This suggests that SA can adjust its resource allocation to optimize utilization and improve efficiency, while NSA cannot do so. This is evident in unequal job allocation, wherein some workers are overwhelmed while others are underutilized. This results in inadequate resource utilization and increased power consumption.

5 Discussion and Concluding Remarks

The findings presented here represent a real-world manifestation of an interesting trade-off when it comes to employing self-adaptive approaches, which is at the core of this work: performance versus resource utilization.

In our experiments, we demonstrated how a self-adaptive resource management strategy allows the system to process tasks efficiently using only the necessary number of workers at any given time. This further supports the claim that self-adaptive resource management offers a significant advantage over non-adaptive alternatives. However, such improvements come at the cost of reduced application performance, as evidenced by throughput and latency results. These compromises must be considered when deciding whether a self-adaptive strategy should be employed in a given context. For example, in contexts where there is only a small infrastructure of workers or where real-time responsiveness is of high importance, using a self-adaptive strategy might be excessive and harm the quality of experience, thus being an inappropriate approach. However, a self-adaptive strategy would be an efficient and cost-effective solution for resource allocation in dynamic environments or those with a large optimization space, that is, where a manual/static scheduling strategy would need to consider a large number of factors.

Our future work will focus on the application of this approach to more dynamic environments, namely, the scheduling of data processing events in edge infrastructures. Such infrastructures are liable to sudden and dramatic changes; hence, agile and data-driven decision-making would potentially be of great benefit.

Acknowledgement. This work was partly supported by the UK EPSRC under grant number EP/R010889/2.

References

1. IBM: An architectural blueprint for autonomic computing, 4th edn. Technical Report G507-2065-00, IBM, 17 Skyline Drive, Hawthorne, NY 10532, USA (2006)
2. Gerostathopoulos, I., Vogel, T., Weyns, D., Lago, P.: How do we evaluate self-adaptive software systems?: a ten-year perspective of seams. In: Symposium on Software Engineering for Adaptive and Self-Managing Systems (SEAMS), pp. 59–70 (2021). https://doi.org/10.1109/SEAMS51251.2021.00018
3. Kruliš, M., Bureš, T., Hnětynka, P.: Simdex: a simulator of a real self-adaptive job-dispatching system backend. In: Symposium on Software Engineering for Adaptive and Self-Managing Systems (SEAMS), pp. 167–173 (2022). https://doi.org/10.1145/3524844.3528078
4. Frey, D., Kermarrec, A.-M., Maddock, C., Mauthe, A., Roman, P.-L., Taïani, F.: Similitude: decentralised adaptation in large-scale P2P recommenders. In: Distributed Applications and Interoperable Systems (DAIS), pp. 51–65 (2015)
5. Nundloll, V., Elkhatib, Y., Elhabbash, A., Blair, G.S.: An ontological framework for opportunistic composition of IoT systems. In: International Conference on Informatics, IoT, and Enabling Technologies (ICIoT), pp. 614–621 (2020)
6. Elhabbash, A., Nundloll, V., Elkhatib, Y., Blair, G.S., Sanz Marco, V.: An ontological architecture for principled and automated system of systems composition. In: Symposium on Software Engineering for Adaptive and Self-Managing Systems (SEAMS), pp. 85–95 (2020). https://doi.org/10.1145/3387939.3391602
7. Wang, Z., Elkhatib, Y., Elhabbash, A.: HolonCraft - an architecture for dynamic construction of smart home workflows. In: Conference on Future Internet of Things and Cloud (FiCloud), pp. 213–220 (2022). https://doi.org/10.1109/FiCloud57274.2022.00036
8. Elhabbash, A., Elkhatib, Y., Bouloukakis, G., Salama, M.: A middleware for automatic composition and mediation in IoT systems. In: International Conference on the Internet of Things (IoT), pp. 127–134 (2022). https://doi.org/10.1145/3567445.3567451
9. Najafizadegan, N., Nazemi, E., Khajehvand, V.: A MAPE-K loop based model for virtual machine consolidation in cloud data centers. J. Comput. Robot. **13**(2), 33–60 (2020)
10. Elhabbash, A., Elkhatib, Y.: Energy-aware placement of device-to-device mediation services in IoT systems. In: Hacid, H., Kao, O., Mecella, M., Moha, N., Paik, H. (eds.) ICSOC 2021. LNCS, vol. 13121, pp. 335–350. Springer, Cham (2021). https://doi.org/10.1007/978-3-030-91431-8_21
11. Sliem, M., Salmi, N., Ioualalen, M.: Performance analysis of self-adaptive policies in containerized microservices. In: International Conference on Engineering and Emerging Technologies (ICEET) (2021). https://doi.org/10.1109/ICEET53442.2021.9659601

12. Cheng, S.-W., Garlan, D., Schmerl, B.: Evaluating the effectiveness of the rainbow self-adaptive system. In: ICSE Workshop on Software Engineering for Adaptive and Self-Managing Systems (SEAMS), pp. 132–141 (2009). https://doi.org/10.1109/SEAMS.2009.5069082
13. Whittle, J., Sawyer, P., Bencomo, N., Cheng, B.H.C., Bruel, J.-M.: RELAX: incorporating uncertainty into the specification of self-adaptive systems. In: International Requirements Engineering Conference (RE), pp. 79–88 (2009). https://doi.org/10.1109/RE.2009.36
14. Elkhodary, A., Esfahani, N., Malek, S.: FUSION: a framework for engineering self-tuning self-adaptive software systems. In: Symposium on Foundations of Software Engineering (FSE), pp. 7–16. ACM, New York (2010). https://doi.org/10.1145/1882291.1882296
15. Elhabbash, A., Bahsoon, R., Tino, P.: Self-awareness for dynamic knowledge management in self-adaptive volunteer services. In: International Conference on Web Services (ICWS), pp. 180–187 (2017). https://doi.org/10.1109/ICWS.2017.31
16. Tamura, G., Villegas, N.M., Muller, H.A., Duchien, L., Seinturier, L.: Improving context-awareness in self-adaptation using the DYNAMICO reference model. In: Symposium on Software Engineering for Adaptive and Self-Managing Systems (SEAMS), pp. 153–162 (2013). https://doi.org/10.1109/SEAMS.2013.6595502
17. Becker, M., Becker, S., Meyer, J.: SimuLizar: design-time modeling and performance analysis of self-adaptive systems. In: Kowalewski, S., Rumpe, B. (eds.) Software Engineering, pp. 71–84. Gesellschaft für Informatik e.V, Bonn (2013)
18. Tallabaci, G., Silva Souza, V.E.: Engineering adaptation with Zanshin: an experience report. In: Symposium on Software Engineering for Adaptive and Self-Managing Systems (SEAMS), pp. 93–102 (2013). https://doi.org/10.1109/SEAMS.2013.6595496
19. Sobhy, D., Minku, L., Bahsoon, R., Chen, T., Kazman, R.: Run-time evaluation of architectures: a case study of diversification in IoT. J. Syst. Software **159** (2020) https://doi.org/10.1016/j.jss.2019.110428
20. Kit, M., Gerostathopoulos, I., Bures, T., Hnetynka, P., Plasil, F.: An architecture framework for experimentations with self-adaptive cyber-physical systems. In: Symposium on Software Engineering for Adaptive and Self-Managing Systems (SEAMS), pp. 93–96 (2015)
21. Schmid, S., Gerostathopoulos, I., Prehofer, C., Bures, T.: Self-adaptation based on big data analytics: A model problem and tool. In: Symposium on Software Engineering for Adaptive and Self-Managing Systems (SEAMS), pp. 102–108 (2017). https://doi.org/10.1109/SEAMS.2017.20
22. Villegas, N., Müller, H., Tamura, G., Duchien, L., Casallas, R.: A framework for evaluating quality-driven self-adaptive software systems. In: Symposium on Software Engineering for Adaptive and Self-Managing Systems (SEAMS), pp. 80–89 (2011). https://doi.org/10.1145/1988008.1988020
23. Silva, M.B., Bezerra, C., Coutinho, E., Maia, P.H.: A catalog of performance measures for self-adaptive systems. In: The Brazilian Symposium on Software Quality (SBQS). ACM, New York (2021). https://doi.org/10.1145/3493244.3493259

Data Mining, Decision Making, and Intelligent Systems

Evolutionary Algorithms and Their Applications in Intelligent Systems

Adam Slowik[1(✉)], Krzysztof Cpalka[2], and Aboul Ella Hassanien[3]

[1] Department of Electronics and Computer Science, Koszalin University of Technology, Koszalin, Poland
adam.slowik@tu.koszalin.pl

[2] Department of Intelligent Computer Systems, Czestochowa University of Technology, Czestochowa, Poland
krzysztof.cpalka@pcz.pl

[3] Faculty of Computers and Artificial Intelligence, Cairo University, Cairo, Egypt
aboitcairo@fci-cu.edu.eg

Abstract. The paper presents applications of evolutionary algorithms in intelligent systems. The research and literature review were performed based on the following scientific databases as Google Scholar, Springer, IEEE Xplore, Science Direct, Web of Science. The popularity of particular evolutionary methods in the aspect of intelligent systems was shown. The main application areas of these methods are also indicated and a brief review of the related literature is provided.

1 Introduction

Evolutionary algorithms represent a fascinating and wide-ranging field in the area of artificial intelligence-related research. They are inspired by evolutionary processes in nature and have the ability to solve complex problems that traditional methods would not be able to solve. Evolutionary algorithms are based on fundamental principles of inheritance, mutation and selection that reflect evolutionary processes occurring in the natural world. These algorithms create a population of solutions that is evolved and optimised to find the best possible solution for a given problem. This process is inspired by natural selection, in which individuals with better quality have a better chance of surviving and reproducing, leading to a gradual improvement in the quality of solutions in subsequent generations.

Evolutionary algorithms consist of methods such as genetic algorithm (GA) [1], genetic programming (GP) [2], differential evolution algorithm (DE) [3], evolutionary strategies (ES) [4] and evolutionary programming (EP) [5]. Evolutionary algorithms [6,7] are characterised by their flexibility and ability to search the solution space. They can be applied to a wide range of optimisation problems. Among them, we can mention such problems as the selection of coefficients in a surface grinding model [8], the design of digital filters with

A. Hassanien et al. (Eds.): AISI 2023, LNDECT 184, pp. 143–153, 2023.
https://doi.org/10.1007/978-3-031-43247-7_13

unusual amplitude characteristics [9,10], the design and optimisation of combinatorial digital circuits [11,12], the design and optimisation of controllers [13–15], task scheduling [16], capacited vehicle routing problem [17], design of interpretable fuzzy systems [18,19], biometric feature selection [20,21], dynamic signature verification [22] and many other applications.

In this paper, we focus on evolutionary algorithms in the context of their applications in intelligent systems. The paper is organized as follows: Sect. 2 presents a brief overview of evolutionary methods, Sect. 3 considers the popularity of evolutionary algorithms in the field of intelligent systems and indicates the areas of their most common applications, Sect. 4 briefly discusses some of the applications of evolutionary algorithms in the aspect of intelligent systems, Sect. 5 concludes the paper.

2 Evolutionary Algorithms

The general concept of evolutionary algorithms is shown in this section. It is presented in Algorithm 1.

Algorithm 1 The general concept of evolutionary algorithms

```
 1: Determine the values of objective function (OF) all parameters of the algorithm
 2: t:=0
 3: Randomly create P(t) - the initial population of individuals (solutions)
 4: Evaluate solutions in the P(t) using OF
 5: while termination criterion is not satisfied do
 6:     Change the population P(t) based on previously defined operators
 7:     Evaluate modified individuals in the population
 8:     Create new population P(t+1) based on population P(t) and other informations
 9:     t:=t+1
10: end while
11: As a result, return the best solution(s) found so far
```

In step 1, the values of all algorithm parameters are determined and the objective function (OF) is defined. Its character depends on the problem. In step 2, the iteration counter is initialized. In step 3, a population of initial solutions is randomly created. In step 4, all solutions in the population are evaluated using the OF. In step 5, the main loop of the algorithm starts, in which: modification of the population according to the assumed operators occurs (step 6), evaluation of the newly created solutions using OF (step 7), creation of a population P(t+1) based on the population P(t) and other additional information (step 8). At the end of the loop, the iteration counter is incremented (step 9). When the stop condition is met (step 5), the best solution is obtained as a result of the algorithm (step 11) and the algorithm stops.

3 Popularity of Evolutionary Algorithms in Intelligent Systems

In order to evaluate the popularity of the use of each evolutionary algorithm (GA, GP, DE, ES, EP) in intelligent systems, queries of the form {"x" intelligent system} were made to individual scientific databases (Google Scholar, Springer, IEEE Xplore, Science Direct, Web of Science), where "x" represented "evolutionary algorithm", "genetic algorithm", "differential evolution", "genetic programming", and "evolutionary programming", respectively.

Table 1 shows the number of articles related to evolutionary algorithms and their connections with intelligent systems. It was prepared using the previously provided query to scientific databases.

Table 1. The number of articles on different evolutionary algorithms in terms of their application in intelligent systems

Algorithm	Google Scholar	Springer	IEEE Explore	Science Direct	Web of Science	Total
EA	192000	481	496	11303	1528	205808
GA	917000	938	3514	35658	9848	966958
DE	13400	286	316	5535	1339	141476
GP	106000	115	278	3500	626	110519
ES	23900	87	46	1672	130	25835
EP	39200	71	114	1372	368	41125
Total	1412100	1978	4764	59040	13839	1491721

The Table 1 shows that the largest number of articles use GA. In second place are articles related to the DE algorithm, and in third place are articles related to GP. The least represented group consists of works related to ES and EP algorithms.

Table 2 presents the publication process of papers related to evolutionary algorithms and their applications in intelligent systems over the years. The data used for this table was obtained from the Web of Science database.

According to Table 2, it can be observed that for GA and DE, there is an increasing number of papers related to intelligent systems in successive years. However, there has been some stagnation for GP, ES, and EP. Since 2005, the number of published articles for GP, ES, and EP algorithms has remained at a relatively similar level. Nevertheless, a systematic increase in the number of papers using evolutionary algorithms in intelligent systems can be observed in the following years.

Table 3 presents the thematic areas in which articles related to evolutionary algorithms and intelligent systems are most frequently published. The data is based on the Web of Science database.

Table 2. The number of articles published in successive years for evolutionary algorithms and their applications in intelligent systems

Year	EA	GA	DE	GP	ES	EP	Total	Year	EA	GA	DE	GP	ES	EP	Total
2023	49	268	52	6	7	0	382	2006	41	376	12	44	2	13	488
2022	137	744	182	24	12	68	1167	2005	35	213	5	29	4	10	296
2021	121	702	133	34	11	1	1002	2004	21	181	1	15	2	16	236
2020	129	686	142	29	7	2	995	2003	15	146	0	28	0	6	195
2019	105	699	119	35	7	3	968	2002	14	138	2	16	2	7	179
2018	88	634	102	32	9	8	873	2001	8	83	0	11	2	6	110
2017	80	541	78	22	6	2	729	2000	8	116	0	13	5	11	153
2016	82	549	97	27	5	134	894	1999	5	54	0	10	1	5	75
2015	84	490	86	29	5	5	699	1998	8	108	0	10	3	7	136
2014	90	483	65	29	10	5	682	1997	3	63	0	2	1	1	70
2013	64	422	64	25	4	3	582	1996	2	41	0	2	1	2	48
2012	60	363	65	26	5	5	524	1995	0	21	0	0	0	1	22
2011	51	310	25	26	1	5	418	1994	1	4	0	2	0	1	8
2010	60	328	36	18	6	9	457	1993	0	2	0	0	0	2	4
2009	59	389	24	24	3	7	506	1992	0	3	0	0	1	0	4
2008	50	403	25	26	5	12	521	1991	0	1	0	0	0	0	1
2007	58	286	24	32	3	11	414	1990	0	1	0	0	0	0	1

From Table 3, it can be observed that the following three areas are the most common for publishing papers related to evolutionary algorithms in the context of intelligent systems: Computer Science Artificial Intelligence, Engineering Electrical Electronic, and Computer Science Theory Methods. It is also worth noting that all considered algorithms (GA, GP, DE, ES, and EP) were used in the top 10 thematic areas. These areas in Table 3 are marked in italics.

4 Applications of Evolutionary Algorithms in Intelligent Systems

The first part of this section provides general information regarding the potential applications of evolutionary algorithms in conjunction with intelligent systems. The second part of this section briefly describes selected applications.

One of the main applications of evolutionary algorithms is parameter optimisation in intelligent systems. In the case of neural networks, evolutionary algorithms can be used to select optimal values for weights, activation functions or network structures, allowing networks to be adapted to specific tasks and improve their efficiency [23–25]. Another area of application for evolutionary algorithms is expert systems. These expert-based systems can use evolutionary algorithms to generate rules and support decision-making processes.

Table 3. The number of papers for selected algorithms in the most published areas of science (industry)

Area of application	EA	GA	DE	GP	ES	EP	Total
Computer Science Artificial Intelligence	873	4414	568	393	76	295	6619
Engineering Electrical Electronic	374	2695	356	122	35	208	3790
Computer Science Theory Methods	334	1353	190	152	20	36	2085
Computer Science Information Systems	207	1262	160	87	21	26	1763
Automation Control Systems	194	1492	168	72	21	36	1983
Computer Science Interdisciplinary Applications	162	977	156	65	8	18	1386
Robotics	85	536	43	35	–	21	720
Operations Research Management Science	79	521	69	28	12	8	717
Computer Science Cybernetics	77	371	45	30	–	8	531
Telecommunications	65	591	86	16	8	5	771
Engineering Multidisciplinary	53	389	75	25	4	–	546
Mathematical Computational Biology	50	–	39	17	2	4	112
Engineering Manufacturing	49	597	–	30	9	6	691
Transportation Science Technology	48	395	–	–	7	–	450
Engineering Civil	45	269	–	–	2	–	316
Computer Science Software Engineering	41	317	34	38	9	5	444
Engineering Industrial	39	316	–	12	3	5	375
Energy Fuels	34	498	119	25	–	10	686
Computer Science Hardware Architecture	32	–	–	12	4	121	169
Instruments Instrumentation	28	279	32	18	4	5	366
Engineering Mechanical	–	375	28	–	4	–	407
Material Science Multidisciplinary	–	326	27	14	–	4	371
Mathematical Interdisciplinary Applications	–	–	37	–	–	–	37
Total	2869	17973	2232	1191	249	821	25335

Evolutionary algorithms can explore the space of possible solutions, identify optimal strategies and discover new rules that may be difficult to find using traditional methods. Consequently, this leads to more effective and efficient expert systems [26–28]. Evolutionary algorithms are also widely used in areas related to logistics and routing. In logistics systems, such as fleet management or delivery route optimisation, evolutionary algorithms can help find optimal routes or schedules. By minimising cost, time and resources, these algorithms contribute to more efficient functioning of logistics systems [29–31]. The broader field of engineering also benefits from the application of evolutionary algorithms. In the field of genetic engineering, evolutionary algorithms can support the design and optimization of genetic sequences, thus accelerating research processes and the development of new drugs [32–34]. Furthermore, in the field of structural design, such as buildings or bridges, evolutionary algorithms can generate optimal solutions in terms of strength, cost and other criteria, contributing to innovative and optimised designs [35–37]. Bioinformatics is another domain where evolutionary algorithms play a significant role. They are used for DNA sequence analysis, protein structure prediction, modeling evolutionary processes, and many other tasks related to the study of biological systems. Their ability to search and analyze large amounts of genetic data contributes to a better understanding of organism structure and function [38–40]. Finance is another field where evolutionary algorithms are widely used. They can support market analysis, optimise investment portfolios, forecast market trends and perform many other financial management tasks. With their ability to adapt and learn, evolutionary algorithms can provide valuable investment guidance and recommendations [41–43]. With the advances in artificial intelligence and machine learning, evolutionary algorithms are becoming increasingly important. The combination of evolutionary algorithms with machine learning techniques, such as neural networks or genetic algorithms, leads to the creation of advanced intelligent systems [44,45] with adaptive, self-organising and self-improving capabilities.

The remainder part of this section provides a brief overview of selected papers on evolutionary algorithms and their use in intelligent systems. These papers are taken from the Web of Science database and have been marked as "Highly Cited Paper" in this database.

In the paper [46] an intelligent detection system that is based on GA and random weight network is proposed to deal with email spam detection tasks. In addition, an automatic identification capability is also embedded in the proposed system to detect the most relevant features during the detection process. The presented system was tested through the series of extensive experiments based on three email corpora. They confirm that the proposed system can achieve remarkable results in terms of accuracy, precision, and recall. Furthermore, the proposed detection system can automatically identify the most relevant features of the spam emails. Mistry et al. [47] proposes a facial expression recognition system using evolutionary particle swarm optimization-based feature optimization. The system first employs modified local binary patterns, which conduct horizontal and vertical neighborhood pixel comparison to generate a discriminative initial facial representation. Then, a PSO variant embedded with the concept of a micro genetic algorithm (mGA), called mGA-embedded PSO, is proposed to perform feature optimization. It incorporates a nonreplaceable memory, a small-population secondary swarm, a new velocity updating strategy, a subdimension based in-depth local facial feature search, and a cooperation of local exploitation and global exploration search mechanism to mitigate the premature convergence problem of conventional PSO. In [48] by Hong et al., present development of a flood susceptibility assessment that uses intelligent techniques and GIS. An adaptive neuro-fuzzy inference system (ANFIS) was coupled with a GA and DE algorithm for flood spatial modeling. The model considers thirteen hydrologic, morphologic and lithologic parameters for the flood susceptibility assessment, and Hengfeng County in China was chosen for the application of the model due to data availability and the 195 total flood events. The proposed method is adaptable and can easily be applied in other sites for flood management and prevention. In the paper [49], an improved real-coded GA and an enhanced mixed integer linear programming (MILP) based method have been developed to schedule the unit commitment and economic dispatch of microgrid units. In the proposed methods, network restrictions like voltages and equipment loadings and unit constraints have been considered. The adopted GA features a highly flexible set of sub-functions, intelligent convergence behavior, as well as diversified searching approaches and penalty methods for constraint violations. Moreover, a novel method has been introduced to deal with the limitations of the MILP algorithm for handling the non-linear network topology constraints. Asim et al. [50] proposes a multi-unmanned aerial vehicle (UAV) - enabled autonomous mobile edge computing (MEC) system, in which several UAVs are deployed to provide services to user devices (UDs). The aim is to reduce (minimize) the overall energy consumption of the autonomous system via designing the optimal trajectories of multiple UAVs. The problem is very complicated to be solved by traditional methods, as one has to take into account the deployment updation of stop points (SPs), the association of SPs with UDs and UAVs, and the optimal trajectories designing of UAVs. To tackle this problem, we propose a variable-length trajectory planning algorithm (VLTPA) consisting of three phases. In the first phase, the deployment of SPs is updated via presenting the GA having variable-length individuals. Accordingly, the association between UDs and SPs is addressed by using a close rule. Finally, a multi-chrome GA is proposed to jointly handle the association of SPs with UAVs and their order for UAVs. The paper [51] by Yang et al. proposes cooperative scheduling approach to optimize the timetable so that the recovery energy that is generated by the braking train can directly be used by the accelerating train. First, the authors propose the cooperative scheduling rules and define the overlapping time between the accelerating and braking trains for a peak-hours scenario and an off-peak-hours scenario, respectively. Second, they formulate an integer programming model to maximize the overlapping time with the headway time and dwell time control. Finally, they design the GA with binary encoding to solve the optimal timetable. Liu et al. [52] formulates the electric vehicle charging scheduling (EVCS) problem as a hierarchical mixed-variable optimization problem, considering the dependency among the station selection, the charging option at each station and the charging amount settings. To adapt to the new problem model the authors specifically design a mixed-variable DE as the scheduling algorithm for EVCS system.

5 Conclusion

This article presents a study on the popularity of evolutionary algorithms in the context of their application in intelligent systems. The number of papers indexed in various scientific databases, focusing on intelligent systems and related to evolutionary algorithms, is presented. The areas where articles related to the discussed topic are most frequently published are also outlined. Finally, selected practical applications of evolutionary algorithms and associated intelligent systems are presented. In summary, evolutionary algorithms are an important tool in the field of artificial intelligence and intelligent systems. Their ability to optimise, search the solution space and adapt to changing conditions contributes to solving complex problems that traditional methods would not be able to solve efficiently. Applications of evolutionary algorithms cover a wide range of fields, such as expert systems, logistics, engineering, bioinformatics, finance and many others. Discovering their potential and developing new techniques and variants of evolutionary algorithms opens up new prospects for creating intelligent systems of the future.

References

1. Holland, J.H.: Adaptation in Natural and Artificial Systems. MIT Press, Ann Arbor (1975)
2. Koza, J.: Genetic Programming: On the Programming of Computers by Means of Natural Selection. MIT Press, Cambridge (1992)
3. Storn, R., Price, K.: Differential evolution-a simple and efficient heuristic for global optimization over continuous spaces. J. Global Optim. **11**, 341–359 (1997)
4. Rechenberg, I.: Evolutionstrategie: optimierung technischer systeme nach prinzipien der biologischen evolution. Frommann-Holzboog Verlag (1973)
5. Fogel, L.J., Owens, A.J., Walsh, M.J.: Artificial Intelligence Thorough Simulated Evolution. Wiley, Ann Arbor (1966)
6. Słowik, A., Białko, M.: Modified version of roulette selection for evolution algorithms – the fan selection. In: Rutkowski, L., Siekmann, J.H., Tadeusiewicz, R., Zadeh, L.A. (eds.) ICAISC 2004. LNCS (LNAI), vol. 3070, pp. 474–479. Springer, Heidelberg (2004). https://doi.org/10.1007/978-3-540-24844-6_70
7. Słowik, A.: Steering of balance between exploration and exploitation properties of evolutionary algorithms - mix selection. In: Rutkowski, L., Scherer, R., Tadeusiewicz, R., Zadeh, L.A., Zurada, J.M. (eds.) ICAISC 2010. LNCS (LNAI), vol. 6114, pp. 213–220. Springer, Heidelberg (2010). https://doi.org/10.1007/978-3-642-13232-2_26
8. Slowik, A., Slowik, J.: Multi-objective optimization of surface grinding process with the use of evolutionary algorithm with remembered pareto set. Int. J. Adv. Manuf. Technol. **37**(7–8), 657–669 (2008). https://doi.org/10.1007/s00170-007-1013-0
9. Slowik, A.: Application of evolutionary algorithm to design of minimal phase digital filters with non-standard amplitude characteristics and finite bits word length. Bull. Pol. Acad. Sci. Tech. Sci. **59**(2), 125–135 (2011). https://doi.org/10.2478/v10175-011-0016-z
10. Slowik, A., Bialko, M.: Design of IIR digital filters with non-standard characteristics using differential evolution algorithm. Bull. Pol. Acad. Sci. Tech. Sci. **55**(4), 359–363 (2007)
11. Słowik, A., Białko, M.: Design and optimization of combinational digital circuits using modified evolutionary algorithm. In: Rutkowski, L., Siekmann, J.H., Tadeusiewicz, R., Zadeh, L.A. (eds.) ICAISC 2004. LNCS (LNAI), vol. 3070, pp. 468–473. Springer, Heidelberg (2004). https://doi.org/10.1007/978-3-540-24844-6_69
12. Słowik, A., Białko, M.: Partitioning of VLSI circuits on subcircuits with minimal number of connections using evolutionary algorithm. In: Rutkowski, L., Tadeusiewicz, R., Zadeh, L.A., Żurada, J.M. (eds.) ICAISC 2006. LNCS (LNAI), vol. 4029, pp. 470–478. Springer, Heidelberg (2006). https://doi.org/10.1007/11785231_50

13. **Szczypta, J., Przybył, A., Cpałka**, K.: Some aspects of evolutionary designing optimal controllers. In: Rutkowski, L., Korytkowski, M., Scherer, R., Tadeusiewicz, R., Zadeh, L.A., Zurada, J.M. (eds.) ICAISC 2013. LNCS (LNAI), vol. 7895, pp. 91–100. Springer, Heidelberg (2013). https://doi.org/10.1007/978-3-642-38610-7_9
14. Lapa, K., Cpalka, K., Przybyl, A.: Genetic programming algorithm for designing of control systems. Inf. Technol. Control **47**(5), 668–683 (2018)
15. Lapa, K., Cpalka, K.: Flexible fuzzy PID controller (FFPIDC) and a nature-inspired method for its construction. IEEE Trans. Ind. Inform. **14**(3), 1078–1088 (2018). https://doi.org/10.1109/TII.2017.2771953
16. Shareh, M.B., Bargh, S.H., Hosseinabadi, A.A.R., Slowik, A.: An improved bat optimization algorithm to solve the tasks scheduling problem in open shop. Neural Comput. Appl. **33**, 1559–1573 (2021). https://doi.org/10.1007/s00521-020-05055-7
17. Hosseinabadi, A.A.R., Slowik, A., Sadeghilalimi, M., Farokhzad, M., Shareh, M.B., Sangaiah, A.K.: An ameliorative hybrid algorithm for solving the capacitated vehicle routing problem. IEEE Access **7**, 175456–175465 (2019). https://doi.org/10.1109/ACCESS.2019.2957722
18. Slowik, A., Cpalka, K., Lapa, K.: Multi-population nature-inspired algorithm (MNIA) for the designing of interpretable fuzzy systems. IEEE Trans. Fuzzy Syst. **28**(6), 1125–1139 (2020)
19. Gabryel, M., Cpalka, K., L. Rutkowski: Evolutionary strategies for learning of neuro-fuzzy systems. In: Proceedings of the I Workshop on Genetic Fuzzy Systems, pp. 119–123 (2005)
20. **Zalasiński, M., Cpałka**, K., Hayashi, Y.: New fast algorithm for the dynamic signature verification using global features values. In: Rutkowski, L., Korytkowski, M., Scherer, R., Tadeusiewicz, R., Zadeh, L.A., Zurada, J.M. (eds.) ICAISC 2015. LNCS (LNAI), vol. 9120, pp. 175–188. Springer, Cham (2015). https://doi.org/10.1007/978-3-319-19369-4_17
21. **Zalasiński, M., Cpałka**, K., Hayashi, Y.: A new approach to the dynamic signature verification aimed at minimizing the number of global features. In: Rutkowski, L., Korytkowski, M., Scherer, R., Tadeusiewicz, R., Zadeh, L.A., Zurada, J.M. (eds.) ICAISC 2016. LNCS (LNAI), vol. 9693, pp. 218–231. Springer, Cham (2016). https://doi.org/10.1007/978-3-319-39384-1_20
22. Zalasinski, M., Laskowski, L., Niksa-Rynkiewicz, T., Cpalka, K., Byrski, A., Przybyszewski, K., Trippner, P., Dong, S.: Evolutionary algorithm for selecting dynamic signatures partitioning approach. J. Artif. Intell. Soft Comput. Res. **12**(4), 267–279 (2022)
23. Carreres-Prieto, D., Ybarra-Moreno, J., Garcia, J.T., Cerdam-Cartagena, J.F.: A comparative analysis of neural networks and genetic algorithms to characterize wastewater from led spectrophotometry. J. Environ. Chem. Eng. **11**(3), Article ID: 110219 (2023)
24. Sun, J., Liu, Q., Wang, Y., Wang, L., Song, X., Zhao, X.: Five-year prognosis model of esophageal cancer based on genetic algorithm improved deep neural network. IRBM **44**(3), Article ID: 100748 (2023)
25. Fan, Z., Zi-xuan, X., Ming-hu, W.: State of health estimation for Li-ion battery using characteristic voltage intervals and genetic algorithm optimized back propagation neural network. J. Energy Storage, **57**, Article ID: 106277 (2023)
26. Jayashree, J., Kumar, S.A.: Evolutionary correlated gravitational search algorithm (ECGS) with genetic optimized Hopfield neural network (GHNN) - A hybrid expert system for diagnosis of diabetes. Measurement **145**, 551–558 (2019)

27. Ghanbari, A., Kazemi, S.M.R., Mehmanpazir, F., Nakhostin, M.M.: A Cooperative ant colony optimization-genetic algorithm approach for construction of energy demand forecasting knowledge-based expert systems. Knowl.-Based Syst. **39**, 194–206 (2013)
28. Kumar, P.G., Victoire, A.A., Renukadevi, P., Devaraj, D.: Design of fuzzy expert system for microarray data classification using a novel genetic swarm algorithm. Expert Syst. Appl. **39**(2), 1811–1821 (2012)
29. Kumari, M., Kanti De, P., Narang, P., Shah, N.H.: Integrated optimization of inventory, replenishment, and vehicle routing for a sustainable supply chain utilizing a novel hybrid algorithm with carbon emission regulation. Expert Syst. Appl. **220**, Article ID: 119667 (2023)
30. Saif-Eddine, A.S., El-Beheiry, M.M., El-Kharbotly, A.K.: An improved genetic algorithm for optimizing total supply chain cost in inventory location routing problem. Ain Shams Eng. J. **10**(1), 63–76 (2019)
31. Wang, C., Liu, Y., Yang, G.: Adaptive distributionally robust hub location and routing problem with a third-party logistics strategy. Socio-Econ. Plan. Sci. **87**, part A, Article ID: 101563 (2023)
32. Terfloth, L., Gasteiger, J.: Neural networks and genetic algorithms in drug design. Drug Discovery Today **6**(supplement 2), 102–108 (2001)
33. Devi, R.V., Sathya, S.S., Coumar, M.S.: Evolutionary algorithms for de novo drug design - a survey. Appl. Soft Comput. **27**, 543–552 (2015)
34. Luukkonen, S., van den Maagdenberg, H.W., Emmerich, M.T.M., van Westen, G.J.P.: Artificial intelligence in multi-objective drug design. Current Opinion Struct. Biol. **79**, Article ID: 102537 (2023)
35. Ha, M.-H., Vu, Q.-A., Truong, V.-H.: Optimum design of stay cables of steel cable-stayed bridges using nonlinear inelastic analysis and genetic algorithm. Structures **16**, 238–302 (2018)
36. Cheng, J.: Optimum design of steel truss arch bridges using a hybrid genetic algorithm. J. Constr. Steel Res. **66**(8–9), 1011–1017 (2010)
37. Srinivas, V., Ramanjaneyulu, K.: An integrated approach for optimum design of bridge decks using genetic algorithms and artificial neural networks. Adv. Eng. Softw. **38**(7), 475–487 (2007)
38. Xue, X., Chen, J.: Matching biomedical ontologies through Compact Differential Evolution algorithm with compact adaption schemes on control parameters. Neurocomputing **458**, 526–534 (2021)
39. Zhang, Y., Lin, M., Yang, Y., Ding, C.: A hybrid ensemble and evolutionary algorithm for imbalanced classification and its application on bioinformatics. Comput. Biol. Chem. **98**, Article ID: 107646 (2022)
40. Reis, D.R., Santos, B.C., Bleicher, L., Zarate, L.E., Nobre, C.N.: Prediction of enzymatic function with high efficiency and a reduced number of features using genetic algorithm. Comput. Biol. Med. **158**, Article ID: 106799 (2023)
41. Wang, Z., Zhang, X., Zhang, Z.K., Sheng, D.: Credit portfolio optimization: a multi-objective genetic algorithm approach. Borsa Istanbul Rev. **22**(1), 69–76 (2022)
42. Liu, Y., Zhou, Y., Niu, J.: Portfolio optimization: a multi-period model with dynamic risk preference and minimum lots of transaction. Finan. Res. Lett. **55**, part B, Article ID: 103964 (2023)
43. Drenovak, M., Rankovic, V., Urosevic, B., Jelic, R.: Mean-Maximum drawdown optimization of Buy-and-Hold portfolios using a multi-objective evolutionary algorithm. Finan. Res. Lett. **46**, part A, Article ID: 102328 (2022)

44. Slowik, A., Cpalka, K.: Hybrid approaches to nature-inspired population-based intelligent optimization for industrial applications. IEEE Trans. Ind. Inform. **18**(1), 546–558 (2022)
45. Cpalka, K., Slowik, A., Lapa, K.: A population-based algorithm with the selection of evaluation precision and size of the population. Appl. Soft Comput. **115**, Article ID: 108154 (2022)
46. Faris, H., et al.: An intelligent system for spam detection and identification of the most relevant features based on evolutionary Random Weight Networks. Inf. Fus. **48**, 67–83 (2019)
47. Mistry, K., Zhang, L., Neoh, S.C., Lim, C.P., Fielding, B.: A micro-GA embedded PSO feature selection approach to intelligent facial emotion recognition. IEEE Trans. Cybernet. **47**(6), 1496–1509 (2017)
48. Hong, H.Y., et al.: Flood susceptibility assessment in Hengfeng area coupling adaptive neuro-fuzzy inference system with genetic algorithm and differential evolution. Sci. Total Environ. **621**, 1124–1141 (2018)
49. Nemati, M., Braun, M., Tenbohlen, S.: Optimization of unit commitment and economic dispatch in microgrids based on genetic algorithm and mixed integer linear programming. Appl. Energy **210**, 944–963 (2018)
50. Asim, M., Abd El-Latif, A.A.: Intelligent computational methods for multi-unmanned aerial vehicle-enabled autonomous mobile edge computing systems. ISA Trans. **132**, 5–15 (2023)
51. Yang, X., Li, X., Gao, Z.Y., Wang, H.W., Tang, T.: A cooperative scheduling model for timetable optimization in subway systems. IEEE Trans. Intell. Transp. Syst. **14**(1), 438–447 (2013)
52. Liu, W.L., Gong, Y.J., Chen, W.N., Liu, Z.Q., Wang, H., Zhang, J.: Coordinated charging scheduling of electric vehicles: a mixed-variable differential evolution approach. IEEE Trans. Intell. Transp. Syst. **21**(12), 5094–5109 (2020)

An Artificial Intelligent-Based System for Crop Yield Prediction Using Climate Change Data and Sensor Fusion

Salma Sameh[1] and Sally Elghamrawy[2,3](✉)

[1] Electrical and Computer Engineering, The British University in Egypt, Al Sherouk City, Cairo, Egypt

[2] Computer Engineering Department, MISR Higher Institute for Engineering and Technology, Mansoura, Egypt
sally_elghamrawy@ieee.org

[3] Scientific Research Group in Egypt (SRGE), Cairo, Egypt

Abstract. The world's food security is in danger as climate change keeps reducing the annual corps-growth rate and the world's population grows exponentially. Predicting crop-yield has been a common practice since the beginning of time; farmers would decide where and when to plant different types of plants based on their basic knowledge of weather and soil properties and what are the suitable conditions for every crop. Improving our forecasting skills of crop-yields is crucial for feeding the growing population while reducing the negative impacts of the agriculture industry on the environment. The fluctuating climatic features pose a threat that is increasing day-by-day and affecting our crops. Not only do the fluctuating weather conditions reduce agricultural productivity, but it also reduces the nutritious value and health benefits gained from the plants. That's because of the increase in the tonnes of pesticides used on the fields and farms as a way of coping with climate change and pollution effects and increasing the plants' growth rate. Predicting crop-yield using Machine Learning (ML) and Deep Learning (DL) has been widely adopted by the AI community in the past few decades; however, some obstacles and bottleneck challenges that affect its accuracy remains. This paper aims to propose a forecasting framework that is easy to use, inexpensive, and reliable enough to be used and certified by governments, companies, or individuals who depend on harvest for living. The proposed framework shows how Sensor Fusion can be beneficial to the agriculture industry specifically and artificial intelligence field in general. The framework demonstrates how Sensor Fusion can be incorporated as part of the data-processing. To accommodate for the crops that require specific soil quality to ensure maximum yield, an NPK sensor and a PH level sensor will be added to measure the acidity of the soil along with the levels of nitrogen, phosphorus, and potassium. Four supervised regression machine learning models were implemented in the proposed framework, namely, Gradient Boosting Regressor, Random Forest Regressor, SVR and Decision Tree Regressor, as they appropriate for the utilized data type and size. The dataset used in the experiments includes the rainfall, yield, year, temperature, and pesticides usage

S. Elghamrawy—Senior IEEE Member.

A. Hassanien et al. (Eds.): AISI 2023, LNDECT 184, pp. 154–165, 2023.
https://doi.org/10.1007/978-3-031-43247-7_14

and covers 101 countries over the span of 23 years. After implementing the models, The decision tree had an accuracy rate of 98% R^2 score which was the best among the implemented models and among all studies conducted using decision tree regressor in general. The proposed framework ensures that it can be used to enhance the model with the best results is discussed in-depth.

Keywords: Artificial Intelligence · Machine Learning · Agricultural Productivity · Climate Change · Sensor Fusion · Decision Tree

1 Introduction

One of the greatest challenges that the modern world is facing is climate change. Not only does it affect the quality of our daily life, but it also started posing new challenges on our planet. Climate Change has become a main concern for food security as it directly affects crop-yields. The population is constantly increasing while climate change is drastically affecting both the quality and quantity of food. According to the United Nations, the population will increase to 9.7 billion by 2050 (DESA et al., 2019) meaning that the population will encounter a growth of 30% in the next 30 years which increases the demand for agricultural productivity in the future requiring an increase in the annual crop-yield (Jayanarayana, D., & Kumar,2021). Although artificial intelligence has made an astonishing progress in predictions, the agriculture industry still relies heavily on man-made predictions. The reason for this is that each crop-yield depends on several complicated factors many of which cannot be controlled, such as weather, soil, precipitation, etc. which differ from a place to another; furthermore, the non-linear data often results in biased results. Simulating crop-growth exists but is considered impractical as it's computationally expensive and hard to implement (Leng and Hall 2020). Therefore, recently, machine learning methods have been adopted to predict the annual yield of different crops as they use simple AI models which are relatively cheap. Such models have been widely adopted and displayed magnificent results (Khaki, Khalilzadeh, and Wang 2020). Despite the expansive number of papers discussing the topic, most of them used similar techniques. More than 50% of the research papers used neural network and another 20% relied on tree-based methods and linear regressions. These papers share the same techniques and only differ in the place or types of crops that the study conducts. Even though machine learning proved itself as reliable in predicting crop-yields with high accuracy rates, small self-collected datasets used are still lacking in abundance of information which makes the results heavily biased. The reviewed studies are always either focusing on several types of crops in one area or on only 1 type of crop in several countries which motivated this research paper. This paper will discuss four types of machine learning techniques that have proven to be suitable for observing features and making future predictions – Gradient Boosting Regressor, Random Forest Regressor, SVR, Decision Tree Regressor- applied on an enormous dataset consisting of the most 10 consumed types of crops in the world over 23 years (from 1990–2013) and 101 countries. The dataset used is combined from FAO (Food and Agriculture Organization) and World Data Bank which are publicly available. This study focuses on the 4 main factors that influence crop-yields: year, average rainfall, average temperature, and number of pesticides tonnes used.This paper also discusses how Sensor Fusion can be integrated in the crop-yield prediction process.

2 Literature Review

This section summarizes the findings of 10 reviewed papers from reputable indexed journals e.g. IEEE and Springer. These papers were chosen because they each cover different AI techniques ranging from conventional ML models to hybrid DL techniques. Also, the papers covered were all published not before 2020 to ensure that all of the sources used for deciding the best model for this study are up to date. This section discusses the findings of reviewed papers which are classified into 4 categories: Machine Learning approaches, Deep Learning approaches, Hybrid DL approaches, Comparative studies.

Although there are hundreds of researches that tackle crop-yield prediction, none of them are performed on West African countries (Lontsi Saadio C et al. 2022). Therefore, this research covers the areas of Benin, Burkina Faso, Gambia, Ghana, Guinea, etc. This study uses soil features data from 1990 to 2020 recovered from Food and Agriculture Organization of the United Nations (FAO) and rainfall data obtained from the World Bank's knowledge portal. This research focuses on six crops, rice, maize, cassava, seed cotton, yams, and bananas. Results were obtained from 3 different Machine-learning models: decision tree, multivariate logistic regression, and k-nearest neighbour. Decision tree model showed the best accuracy among the models with R^2 =95.3% while the k-nearest neighbour performed with R^2=93.15% and the logistic regression with R^2=89.78%.

(R. Ragunath et al.,2022) investigates the several machine learning approaches dedicated to crop-yield forecasting as conflicts arise whether ML or DL is better at time-series data prediction. Because food security is a modern-day threat, the study works toward finding the best ML model for the forecasting crop harvests. Using weather conditions and pesticides use, the researchers applied three types of ML techniques (Regression, Classification, and Clustering) in hopes of finding the most suitable ML model. The models were applied to 34 different datasets ranging from numerical data to satellite imagery. After 34 experiments, the following results were the accuracy outcomes. Regression demonstrated a maximum of 0.96 R^2 accuracy rate, while Classification and Clustering came to a maximum accuracy of 75 and 100 confusion matrix (which is a table layout that helps visualize the outcome) rate respectively. The study also concludes that the previous results are not enough to decide the best ML model and further testing is required.

For deep learning techniques, (Shruti K. et al., 2020) proposes a neural network model used to identify the interdependence between soil attributes and rainfall pattern which the author describes as a Time-Series approach. This approach is then made possible through Recurrent Neural Network (RNN). With two data-driven algorithms working at the backend, the relationship between soil and weather attributes is depicted. The model uses two datasets to predict the rice yield in Karnataka, India. One dataset has rainfall data over a period of 30 years from 1987–2018 measured in millimetres while the other contains soil attributes recorded over 31 years; soil properties include nitrogen, phosphorus, and potassium concentrations. Finally, the model is the combination of an RNN model trained on both datasets separately. After several trials, the best accuracy results had an RMSE of 41.497 using 215 epochs, 4 batch size, and Relu function (Rectified Linear Units).

(D. Elavarasan and P. M. Vincent, 2020) propose a work that aims to help solve inadequacies that other conventional models face: the inability to create a linear mapping between data and the crop yield and the heavy reliability on quality of input parameters. This study combines a Deep Recurrent Q-Network model which is recurrent neural network technique along with Q-learning reinforcement algorithm to construct a DRL (Deep Reinforcement Learning) model which can effectively predict the annual crop yield. This deep Q-learning based DRL model will enhance the forecasting process by using the best rewarding iterations. The proposed study administrates the paddy crop which is crucial in India for the growth of rice and taro, two of the most vital crops for agriculture, in southern district of India. The dataset uses information gathered from Vellore district over 35 years and resulted in a 93.7% accuracy rate.

For hybrid techniques, Since most crop-yield predictions are based on neural networks and tree-based methods, this research proposes a graph-based framework deep learning methods, GNN-RNN(Graph neural network-Recurrent neural network), which uses a dataset combining geospatial and temporal aspects of a land from 41 US states with data covering 38 years (1981–2019) (Fan J. et al., 2021).Furthermore, the proposed algorithm aims to use geospatial input parameters in hope of improving the prediction accuracy. This model was applied on a huge dataset consisting of corn and soybeans crops and is available on USDA Crop Production Reports for several states in the US. In their research (Joshua Fan et al., 2021) used climate, land surface, and soil quality feature as their input. The proposed model resulted in 0.6591 R^2 and 0.8259 corr accuracy when applied on 5 years data which outperforms competing baselines except for 2018 soybean corr.

Another research by (Jigisha Patel et al., 2021) proposes a long short-term memory networks (LSTM)-based RNN reinforcement model that aims to grant a learning balance essential for Q-learning reinforcement to calculate the crop-yield. Because limited data is a drawback to LSTM and RNN since it results in biased outcome, this model aims to integrate Q-learning to LSTM-RNN to minimize the error and increase the accuracy. Similar to the previous paper, this model uses the Vellore paddy crops as its dataset over the years from 1997 to 2012. The input parameters used include ventilation, watering frequency, rainfall frequency, and spring attributes. The final R^2 frequency was 99.59% which according to the authors can be further enhanced.

Because little attention is going towards hybrid deep learning techniques that can integrate two types of DL algorithms, (Jie Sun et al.,2020) proposes another hybrid multilevel RNN-CNN model which combines the ability of RNN to extract temporal patterns along with the ability of CNN to learn spatial features. According to Jie, the hybrid model will then be able to recognize both temporal and spatial attributes of time-series remote sensing data. Experiments took place in the US corn belt states to predict the corn yield from 2013 till 2016. At country-level, the model demonstrated success with a performance of 0.75 R^2.

For the comparative studies, because satellite remote sensing has been used in past few decades for crop predictions, (Juan Cao et al. 2021) compares the ability of 4 AI methods: (random forest) RF, DNN (deep neural networks), 1D-CNN (1D convolutional neural networks), and LSTM (long short-term memory networks), to forecast the crop-yield from satellite images along with climate, soil attribute, and spatial information data retrieved from GEE (Google Earth Engine). Images used were of wheat field in China as wheat account for almost 85% of China's total grain production; the decline of this percentage from 1961 till 2008 was the motive behind using satellite remote sensing as it provides more timely, synoptic, and up-to-date view than conventional observation methods. On average, the 4 techniques had an accuracy of $R^2 \geq 0.85$ with DNN having the highest accuracy and RF having the lowest. To conclude the study; all 4 models performed well at country-level but only DNN and RF performed well at field-level.

In their work, (Liana Baghdasaryan et al.,2022) targeted corn-harvesting farms in the US Midwest and used high-resolution aerial satellite images to form a highly detailed dataset that can be fed to 4 different AI models to be compared. In addition, density estimations from framers were added to the input parameters of the model to predict the harvest rate, in bushels, at every pixel of the farm images. This study's target was to improve the prediction accuracy of conventional models that are fed low-resolution imagery which is crucial for harvesters to facilitate the decisions towards whether a piece of land is suitable for crops. The imagery dataset Liana used in this study was collected using manned aircrafts flown over Illinois and Indiana during the 2020 and 2021 harvesting season (April through September). The 3-machine learning algorithms (Lasso, Random Forest (RF), and LightGBM regression) and the 1 deep learning algorithm (CNN) resulted in an average of 5.41% MAPE (Mean absolute percentage error) at the field-level. (Mamunur R. et al.,2021) article gives a comprehensive review of the machine learning methods that are used in crop-yield forecasting while focusing on palm oil. Due to the lack of the study-of-art machine learning analysis based on palm oil, this work provides comparative research on studies related to the advantages and disadvantages of machine learning approach in the agricultural field. After testing the advantages and disadvantages of these methods, Mamunur et al. concluded that conventional ML architecture, such as LR and RF grant the best result; however, some DL models like DNN, CNN, and LSTM provide promising results and need further investigation.

3 Model Comparison and Choice

It can be concluded from the reviewed research papers that Deep learning techniques result in high accuracies only when the data set used consists of Satellite Imagery. This is predicted however because DL methods have the ability to create new features from limited datasets, making it easier to process imagery datasets. However, Imagery and Satellite datasets are very hard and expensive to obtain as they require special devices like drones for instance. This work aims to facilitate the process of crop-yield prediction and since ML methods are more reliable in numerical dataset predictions, this paper will rely on comparing 4 different types of Machine Learning techniques. ML is suitable for linear and non-linear time specific data making it the best choice for the agrigulture industry.

First choice was Decision Tree Regressor, a non linear regressor that has always outperformed other machine learning methods used for crop yield prediciton.

Random forest and Linear regression have different mechanisms yet similar perfomance and results. Since Random Forest has more parameters allowing it to perfrom better on an expansive dataset with less features, it was used in this work instead of its Linear Regression peer. Random forest works by building several Decision tress independently which will produce several different results and then aggregating those results into a single prediction.

Then is the Gradient Boosting, this model was chosen because it's supposed to be an extension of the Decision Tree model,yet there are very few crop -yield prediction studies that utilize it. Its mechanism is similar to that of Random Forest but it differs in that the decision trees in Gradient boosting are additively being aggregated one by one and every tree is added to compensate for any dicrepencies by the previous tree; that's the reason why it is called Gradient boosting.

The final algorithm tested was Support Vector Regression or SVR. SVR is a type of SVM (Support Vector Machines); they are supervised lerning models that are effective in classification problems. These models intial purpose was to create a decision plane that performs very well in classification problems and application. There are no enough studies experimenting with SVM in regression problems since they were built specifically for classification. This paper includes SVR model to attempt to find wethere a classification model can be useful in the agrigulture industry.

4 The Proposed Framework

The proposed framework consists of four main layers: Data pre-processing, Feature selection, Machine learning applier and validation layer. The data is collected using sensor fusion layer before pre-processing layer. Figure 1 represents the framework through which the research will follow.

Data pre-processing: First step was unifying the data. That's because the data was retrieved from 2 different sources, also documented data differ in each country. While some countries documented their climate properties from the 60's, other only have data tracking back to the 90's. Therefore, unifying the data was crucial to produce an unbiased result. After the unification of the data, the data was further enhanced using one hot encoding. This is a process of converting categorical data (crop type and country in our case) into a matrix that can be fed to the machine learning models; one-hot-encoding improves the outcome of the model as it converts text data into a form that is better interpreted by the model. Lastly was the features' scaling. Because the attributes significantly vary in their value, the higher the value the more it will weigh in the calculations, hence producing biased results. Feature scaling is then used to level out the magnitudes of the attributes and suppress this effect.

Feature selection was used after the data processing step. It's a way of seeing the correlation between different colomns or features of the dataset and seeing what features are most likely to have the highest impact on the results. The features with the lowest impact can then be given priority and features of less importance can be eliminated from the dataset. After the validation, the accuracy rates are measured using the coefeccient

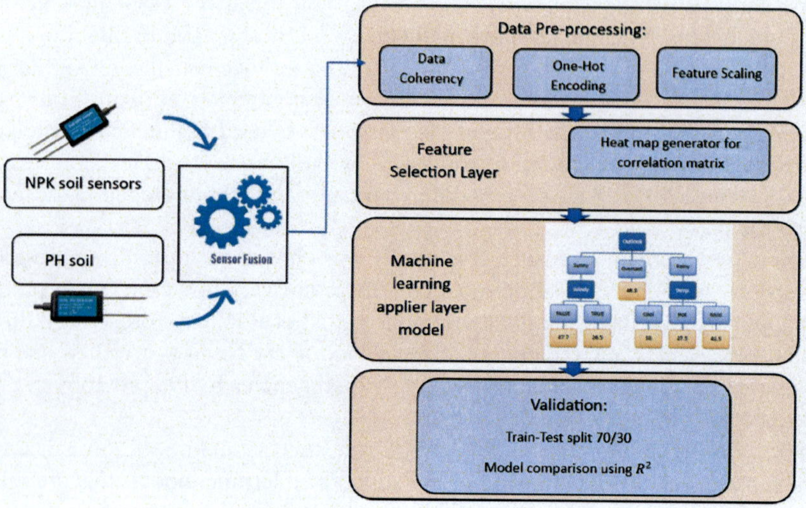

Fig. 1. The Proposed Framework

of determination or R^2 score. Each crop or plant has specific needs to ensure its healthy growth rate. While some crops like potatoes, tomatoes,and maze are easy to grow because they have the ability to withstand hard enviromental conditions including poor soil quality, other crops including barley, corn, wheat, and soybeans are acid sensitive and require soil ph of 6.2 or have the maximum yield.

Adding more features to the datasets like the soil's ph level or npk value (N = Nitrogen, P = Phosphorous and K = Potassium) might be beneficial in some cases, however, this addition will increase the execution time and affect the the accuracy rates. Instead of compensating between the need of a fast and accurate model and the need of adding more attributes to the data, this research paper proposes a solution that will allow the addition of as much needed data as possible while barely affecting the time or performance. This will be possible through the use of Sensor Fusion. Sensor fusion is the process of combining data from two or more independent sensors (Man Lok Fung et al. 2017), that can be of the same or different types, to form a better picture about the data and reduce uncertainty. Sensor fusion is a subset of data fusion that was created mainly for the fields of embedded systems and IoT (Internet of Things), so not much research was conducted on how this technology can aid in other fields and areas of research.

Thus, this paper proposes how Sensor Fusion can be beneficial to the agriculture industry specifically and artificial intelligence field in general. The final proposed framework demonstrates how Sensor Fusion can be incorporated as part of the data-processing. To accommodate for the crops that require specific soil quality to ensure maximum yield, an NPK sensor and a PH level sensor will be added to measure the acidity of the soil along with the levels of nitrogen, phosphorus, and potassium. After getting the readings from both sensors, the soil quality will be assigned to that type of soil according to the following Table 1.

With the aid of such fused tables, a fifth column will be added to the attributes of the dataset. This column will represent four different soil attributes (nitrogen, phosphorous,

Table 1. Example of Fused Tables

NPK ratio	PH level	Soil quality
3:1:2	5.5–7.5	High
$\neq$ 3:1:2	5.5–7.5	Medium
3:1:2	< 5.5 or > 7.5	Medium
$\neq$ 3:1:2	< 5.5 or > 7.5	Low

potassium, and PH levels) compressed into a single feature. This column will then be converted into a matrix using one-hot-encoder so that it can be processed by the ML model.

5 Results

To visualize the data, charts were generated to demonstrate the most influential crops and countries in the dataset. The following charts represent the most 10 prevalent crops and countries that contribute to the data (Fig. 2).

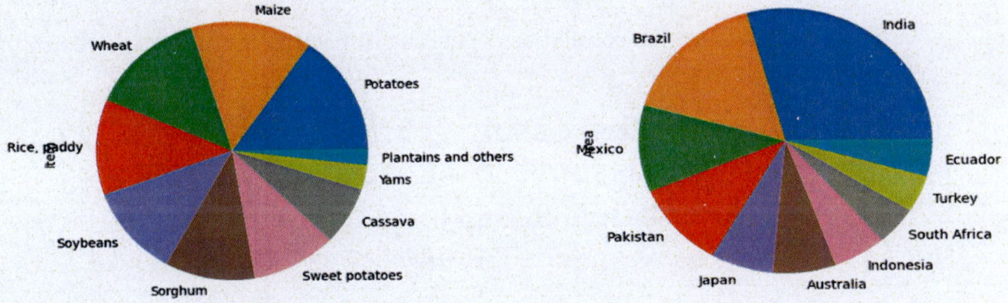

Fig. 2. Data visualization for the most 10 prevalent crops and countries that contribute to the data

The Feature Selection block uses a heat map to represent the correlations between attributes (Fig. 3).

From the generated heat map, it's clear that the year of the harvested crop-yield contributes the least to the harvest with no correlation to the avg temp. And insignificant correlation to the pesticides use. Therefore, this attribute will be removed to ensure better results from the model with less power consumption. The pesticides usage, avg. Temp., and avg. Rainfall on the other hand are highly correlated so it can be expected that they will be have a siginficant impact on the results. As expected, Decision Tree Regressor has the best performance with the highest accuracy prediction rate of up to 97%.

After pre-processing, the final data is fed to model for training. The following figure show the results from training the data on 4 different machine learning models (Fig. 4).

As expected, the Decision tree Regressor outperforms every other model. Followed by Gradient Boosting Regressor with an accuracy rate of 89% R^2 score. This result shows

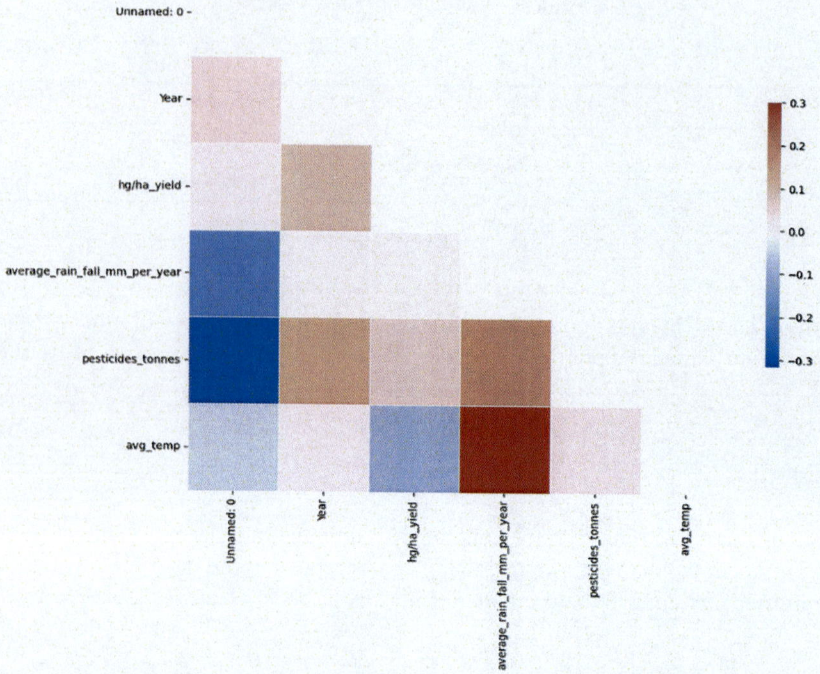

Fig. 3. Heat Map for the correlations between attributes in the dataset.

```
['GradientBoostingRegressor', 0.8990275465414561]
['RandomForestRegressor', 0.6844621348603049]
['SVR', -0.20379248865892863]
['DecisionTreeRegressor', 0.9808322197960608]
```

Fig. 4. The Training Results

that even though gradient boosting utilizes several decision trees, it is not necessarily better than one decision tree model and that each one can perform betteron specific data. The Random Forest however only had a performance of 67% and finally came the SVR that had the worst performance with 0 accuracy. The SVR result demonstrates that SVM is not suitable for regression problems, therefore further studies and updates should be conducted on the model to make it a prediction model.

The accuracy obtaied from the decision tree model is the highest that has ever been achieved on an international dataset. Not only is the model accurate, but also is inexpensive considering that the data needed was gathered from online sources without the need of any aditional device. Furthermore, the whole execution time came to 1.92 min which is significantly less than the execution time of Deep Learning techniques, thus more power saving. The final figures demonstrate the accuracy of each crop alone using Desicin tree algorithm along with the line fit graph (Fig. 5).

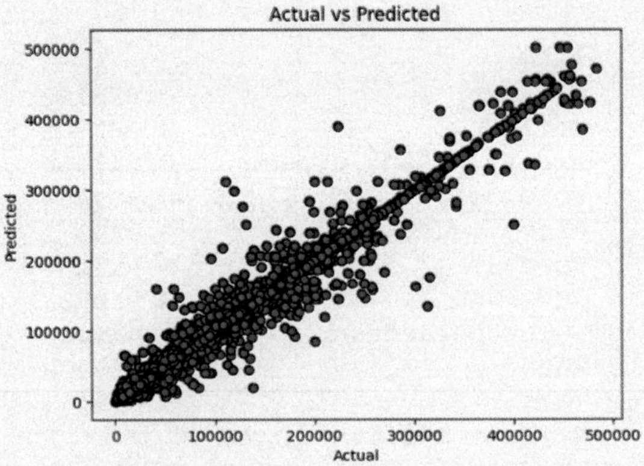

Fig. 5. The Decision Tree for the Testing Results

6 Conclusion

After thorough understanding of both traditional and innovative machine learning models, Gradient Boosting Regressor, Random Forest Regressor, SVR, and Decision Tree Regressor were trained on a huge dataset of 101 countries over the span of 23 years. These models were Implementing and comparing the 4 machine learning models required several data pre-processing steps: Data coherency, one hot encoding, feature scaling, and feature selection. The models were then trained, and decision tree had an accuracy rate of 98% R^2 score which was the best among the implemented models and among all studies conducted using decision tree regressor in general. After that, decision tree was tested to rule out if any overfitting occurred during the training but no significant reduction in accuracy happened eliminating that there was over fitting. Finally, the research proposes a method that integrates sensor fusion as a form of data pre-processing; it will allow the addition of as much needed independent parameters as possible while avoiding overfitting or reduction in the model's precision. Sensor Fusion will also enable the use of the proposed framework on plants that require specific environmental conditions for their growth and development. With the help of heterogenous sensor fusion that uses sensors of different kinds, more parameter will be added into fused tables. The tables will then allow for the combination of as much needed parameters as possible in a form of 1 feature preserving the compactness of the input variables and avoiding overfitting.This paper aims to open a new path for future studies on crop yield prediction that would benefit from the overlooked abilities of sensor fusion in the data pre-processing phase.

```
Item
Cassava                    0.976182
Maize                      0.943362
Plantains and others       0.847664
Potatoes                   0.957463
Rice, paddy                0.939332
Sorghum                    0.911468
Soybeans                   0.910783
Sweet potatoes             0.944440
Wheat                      0.936898
Yams                       0.953405
dtype: float64
```

References

1. Jayanarayana Reddy, D., Rudra Kumar, M.: Crop yield prediction using machine learning algorithm. In: Proceedings of the Fifth International Conference on Intelligent Computing and Control Systems (ICICCS 2021) (2021). IEEE Xplore Part Number: CFP21K74-ART;ISBN: 978–0–7381–1327–2
2. DESA, D.O.E., Affairs, S.: World population prospects (2019)
3. Fung, M.L., Chen, M.Z.Q., Chen, Y.H.: "Sensor Fusion: a review of methods and applications", Department of Mechanical Engineering, University of Hong Kong. In: 29th Chinese Control And Decision Conference (CCDC). https://doi.org/10.1109/ccdc.2017.7979175
4. Leng, G., Hall, J.W.: Predicting spatial and temporal variability in crop yields: an intercomparison of machine learning, regression and process-based models. Environ. Res. Lett. **15**(4), 044027 (2020)
5. Khaki, S., Khalilzadeh, Z., Wang, L.: Predicting yield performance of parents in plant breeding: a neural collaborative filtering approach. PLoS ONE **15**(5), e0233382 (2020). https://doi.org/10.1371/journal.pone.0233382
6. Fan, J., Bai, J., Li, Z., Ortiz-Bobea, A., Gomes, C.P.: A GNN-RNN approach for harnessing geospatial and temporal information: application to crop yield prediction. Department of Computer Science, Cornell University (2021). https://www.climatechange.ai/papers/neurips2021/29
7. Elavarasan, D., Durairaj Vincent, P.M.: Crop yield prediction using deep reinforcement learning model for sustainable agrarian applications. School of Information Technology and Engineering, Vellore Institute of Technology (VIT), Vellore 632014, India, vol. 8, pp. 86886–86901 (2020)
8. Patel, J., Vala, B., Saiyad, M.: LSTM-RNN combined approach for crop yield prediction on climatic constraints. Department of Computer Science & Engineering PIET-Parul University, Gujarat, India (2021). https://doi.org/10.1109/ICCMC51019.2021.9418231
9. Kulkarni, S., Mandal, S.N., Sharma, G.S., Mundada, M.R., Meeradevi: Predictive analysis to improve crop yield using a neural network model. In: 2018 International Conference on Advances in Computing, Communications and Informatics (ICACCI) Bangalore, India (2018). https://doi.org/10.1109/icacci.2018.8554851
10. Cao, J., et al.: Wheat yield predictions at a county and field scale with deep learning, machine learning, and google earth engine. Eur. J. Agron. **123**, 126204 (2021). https://doi.org/10.1016/j.eja.2020.126204

11. Elghamrawy, S., Vasilakos, A. V., Darwish, A., Hassanien, A.E.: An intelligent crop recommendation model for the three strategic crops in Egypt based on climate change data. In: The Power of Data: Driving Climate Change with Data Science and Artificial Intelligence Innovations (pp. 189–205). Springer Nature Switzerland, Cham (2023) https://doi.org/10.1007/978-3-031-22456-0_11
12. Baghdasaryan, L., Dolmajain, A., Hobbs, J. (n.d.): Deep density estimation based on multi-spectral remote sensing data for in-field crop yield forecasting. https://openaccess.thecvf.com/content/CVPR2022W/WiCV/papers/Baghdasaryan_Deep_Density_Estimation_Based_on_MultiSpectral_Remote_Sensing_Data_for_CVPRW_2022_paper.pdf, Intelinair, Inc. Yerevan, Armenia
13. Sun, J., Lai, Z., Di, L., Sun, Z., Tao, J., Shen, Y.: Multi-level deep learning network for county-level corn yield estimation in the U.S. Corn Belt. IEEE J. Selected Top. Appl. Earth Observ. Remote Sens., p. 1 (2020). https://doi.org/10.1109/jstars.2020.3019046
14. El-Ghamrawy, S.M., Eldesouky, A.I.: An agent decision support module based on granular rough model. Int. J. Inf. Technol. Decis. Mak. **11**(04), 793–820 (2012)
15. Ragunath, R., Narmadha, N., Rathipriya, R.: Machine learning-based forecasting technique for crop yield: a study. Department of Computer Science, Periyar University, Salem 636011, India (2022)
16. Mamunur, R., Bifta Sama, B., Yusri, Y., Mohamad Anuar, K., Nuzhat K.: A comprehensive review of crop yield prediction using machine learning approaches with special emphasis on palm oil yield prediction. School of Industrial Technology, University Sains Malaysia (2021)
17. Cedric, L.S., et al.: Crops yield prediction based on machine learning models: case of West African countries. Smart Agric. Technol. **2**, 100049 (2022). https://doi.org/10.1016/j.atech.2022.100049
18. Pant, J., Pant, R.P., Singh, M.K., Singh, D.P., Pant, H.: Analysis of agricultural crop yield prediction using statistical techniques of machine learning. Mater. Today: Proc. **46**, 10922–10926 (2021)

Data Fusion for Improved Stock Closing Price Prediction: Ensemble Regression Approach

Aly Elshamy[1], Abdullah Afifi[1], Ahmed Mabrok[1], Hossam Al Akwah[1], Dalia Ezzat[1,2,3], and Sara Abdelghafar[1,3](✉)

[1] School of Computer Science, Canadian International College (CIC), Cairo, Egypt
sara_abdelghafar@cic-cairo.com
[2] Faculty of Computers and Artificial Intelligence, Cairo University, Cairo, Egypt
[3] Scientific Research Group in Egypt (SRGE), Cairo, Egypt

Abstract. The stock market is a complex and dynamic industry that has garnered the attention of experts who seek to understand its various trends. Accurately predicting stock prices is crucial for investors to minimize their risk of losing money. However, due to the volatile and non-linear nature of financial stock markets, this is a challenging task. This study aimed to address the challenge of accurately predicting stock prices by utilizing an ensemble regression approach, which combines multiple sources of data. The proposed approach was evaluated using Tesla company data over a four-year period and demonstrated its efficiency in predicting stock closing prices. The results showed that the ensemble regression approach was able to accurately predict stock prices under different scenarios, handle fluctuations, anticipate sudden changes, and predict both simple and radical changes in stock prices. This study highlights the potential of machine learning techniques and increased computational capabilities in improving prediction methods for the complex and dynamic stock market industry.

Keywords: Stock market · Prediction · Ensemble Regression · Social media · Financial news · Historical data · Tesla company

1 Introduction

The stock market is a platform where publicly traded companies issue and sell their shares to investors. It is a crucial part of the global economy, providing companies with access to capital and investors with opportunities for wealth creation. The stock market is dynamic and constantly changing, with prices of stocks fluctuating based on various factors such as company performance, economic indicators, and market trends [1]. Stock market prediction is essential for investors to make informed decisions about buying or selling stocks. Traditionally, there have been two main approaches proposed for predicting the stock price of an organization. The first is technical analysis, which utilizes historical data such as opening and closing prices, trading volumes, and adjacent close values to forecast future prices. The second approach is qualitative analysis, which considers external factors such as company profile, market conditions, political and economic

A. Hassanien et al. (Eds.): AISI 2023, LNDECT 184, pp. 166–175, 2023.
https://doi.org/10.1007/978-3-031-43247-7_15

factors, and textual information from various sources. However, with the advent of machine learning techniques, there has been a shift towards using advanced intelligent models that can efficiently analyze large and non-linear datasets. Machine learning offers the promise of identifying hidden patterns and complex relationships in stock market data, making it a promising approach for predicting stock prices due to its ability to analyze vast amounts of data and identify patterns that may not be easily recognizable by human analysts [2]. Accurate prediction can help investors maximize profits and minimize losses. To achieve this, investors need to analyze a wide range of data sources, including historical stock data, social media, financial news, and public news. Historical stock data, such as past prices and trading volumes, can be used to identify patterns and trends that may be indicative of future performance [3]. Social media platforms like Twitter can provide valuable insights into public opinion and sentiment, which can be useful in forecasting stock market prices [4]. Financial and public news sources such as like New York Times and Consumer News and Business Channel (CNBC) can also provide valuable information, like government reports and press releases, expert analysis of market trends, and economic indicators that can help investors make informed decisions. By analyzing these various sources of information, investors can identify patterns and trends that may be indicative of future performance and make informed investment decisions.

Numerous studies have applied machine learning techniques for stock price prediction. Most of the previous studies have focused on classification methods to forecast the direction of stock prices (increase or decrease) [4, 5]. However, the most recent studies have introduced regression approaches to estimate the actual value of stock prices at closing. The most commonly utilized regression techniques include simple linear and polynomial regression, support vector regression, decision tree, random forest, and artificial neural network [3, 6]. In this paper, we have used various regression models such as Random Forest, Gradient Boosting, Linear Regression, Lasso Regression, Ridge Regression, Multi-layer Perceptron (MLP) Regressor, and XGBoost. These models were combined using a stacking technique that involved training multiple base models and a meta-model. The proposed approach has been evaluated using the Tesla company data, which including Twits, public financial news, and historical stock data from 1/1/2019 to 31/12/2022. The following is the paper's highlighted contributions:

1) The paper discusses the importance of stock closing price prediction for investors to make informed decisions about buying or selling stocks, using advanced intelligent models that can efficiently analyze large and non-linear datasets.
2) The paper highlights the potential of regression approaches in identifying hidden patterns and complex relationships to estimate the actual value of stock prices at closing.
3) The paper discusses the importance of analyzing various sources of information, including historical stock data, social media, financial news, and public news, to make informed investment decisions.
4) The paper proposes a stacking technique that combines various regression models in real scenarios, using Tesla company data spanning four years.

The subsequent sections of this paper are structured as follows: Sect. 2 outlines the proposed approach, which comprises the following stages: data gathering, data preprocessing, sentiment analysis, data merging and thresholding, and regression modeling. Section 3 presents the results and discussion. Finally, Sect. 4 concludes the study and highlights potential future research avenues.

2 Proposed Approach

Several preprocessing techniques, such as thresholding, sentiment analysis, and normalization, were applied to the dataset utilized in this study. Following completion of these steps, the dataset was made publicly accessible on the Mendeley platform through the following link [7]. The proposed methodology comprises five stages: data collection, preprocessing, sentiment analysis, thresholding, and forecasting the stock's closing price utilizing an ensemble regression model.

2.1 Data Gathering

2.1.1 Social Media

The impact of Twitter on the stock market is significant, and we aimed to gather valuable insights by collecting relevant tweets related to Tesla and its associated keywords. We formulated a query that targeted tweets containing terms such as "Tesla," "stock," "market," "TSLA," "Elon Musk," and other relevant financial indicators. We focused on tweets with a minimum of 100 favorites and written in English, collected from January 1, 2019, to December 31, 2022. By analyzing these tweets collectively, we can provide valuable sentiment indicators and potentially reveal trends and patterns that may impact the stock market.

2.1.2 Public and Financial News

To analyze Tesla's impact on the stock market, we collected data from multiple sources, including The New York Times and CNBC, using web scraping and API retrieval. We processed the data separately for each source, distinguishing between public news and financial news using a keyword-based approach, utilizing terms such as 'shares,' 'bank,' 'investor,' 'investors,' 'earnings,' 'financial,' 'holders,' 'holder,' 'finance,' 'sales,' 'price,' 'stocks,' 'market,' and 'trading.' After conducting sentiment analysis individually on both datasets, we merged the sentiment counts to obtain a comprehensive overview of sentiment across the collected news data. This enabled a more holistic analysis, capturing a wide range of viewpoints and news updates during the specified time period.

2.1.3 Historical Stock Data

We obtained historical stock price data for Tesla by utilizing the Python "yahoo-fin" library's functionality. The data collection covering the desired timeframe. The retrieved data included the following attributes; Open: the opening price for the market session, High: the highest prices reached during the session, Low: the lowest prices reached during the session, and Volume: the total number of trades performed on the stock during that time, with Close being the attribute to forecast and compare to actual data for algorithm accuracy testing. To ensure consistent daily frequency, we resampled the collected data to a daily basis, providing a comprehensive and accurate representation of Tesla's stock performance over time. By integrating daily stock price data with Twitter data and news articles, our analysis provides a multi-dimensional perspective, enabling us to evaluate the correlation between market sentiment, news events, and actual stock price movements.

2.2 Data Preprocessing

2.2.1 Social Media and News Data

To prepare the downloaded tweets and news headlines for machine learning algorithms, they must first undergo preprocessing. This involves several steps: converting the documents into word tokens, removing HTML and other tags (such as author tags and hashtags), eliminating URLs, removing stop words, and stemming the words to ensure consistency. Additionally, duplicate tweets are removed to avoid redundancy.

2.2.2 Historical Stock Data

In the case of the historical stock dataset, min-max normalization has been employed as a standardization technique to ensure that the data is comparable and free from any potential bias caused by varying scales or ranges. This involves rescaling the values within a specific range of 0 to 1, thus transforming the data into a uniform scale. This approach facilitates fair and meaningful comparisons across different variables and eliminates any potential bias caused by varying scales or ranges within the dataset. As a result, subsequent analyses and interpretations are more accurate and reliable. For normalizing a value x_i in attribute X, is computed by the following formula $(x_i - min(X))/(max(X) - min(X))$ [8].

2.3 Sentiment Analysis

After cleaning the data, a machine learning algorithm was used to extract features and identify sentiments, including positive, negative, and neutral. This transformed the raw data into a standardized dataset that included predicted sentiments for each tweet and article.

2.4 Thresholding

The objective of thresholding is to quantify the thresholding for public, financial, and social sentiment or publicity linked to a company, such as Tesla's stock. The analysis involves contrasting the number of public articles regarding the company to an average threshold to ascertain whether there was a significant increase or decrease in financial news, public sentiment, or social media traffic. The algorithm calculates the total count of positive, negative, and neutral articles and compares it to the mean of the total articles throughout the day. If the total count is lower than the average, the algorithm returns -1, indicating below-average traffic. If the total count equals the average, the algorithm returns 0, indicating average traffic. Finally, if the total count surpasses the average, the algorithm returns 1, indicating above-average traffic. The same approach is implemented for tweets.

2.5 Ensemble Regression Model

Ensemble learning is an advanced machine learning technique in which multiple independent models are trained individually and then their predictions are combined to get better predictions [9]. The ensemble stacking technique is a machine learning approach that combines the predictions of multiple base models to improve the accuracy of a final prediction. In this technique, the output of each base model is used as an input to a meta-model, which learns to combine the predictions and make a final prediction [10]. There are several regression algorithms that can be used as base models in the ensemble stacking technique. These include Random Forest, Gradient Boosting, Linear Regression, Lasso Regression, Ridge Regression, MLP Regressor, and XGBoost. Random Forest is a decision tree-based algorithm that creates multiple decision trees and combines their predictions to make a final prediction. Gradient Boosting is another decision tree-based algorithm that builds trees in a sequential manner, with each tree learning from the errors of the previous tree. Linear Regression is a simple regression algorithm that models the relationship between a dependent variable and one or more independent variables using a linear function. Lasso Regression is a variant of linear regression that adds a penalty term to the cost function to encourage sparsity in the coefficients. Ridge Regression is another variant of linear regression that adds a penalty term to the cost function to prevent overfitting. MLP Regressor is a neural network-based algorithm that uses multiple layers of nodes to learn complex relationships between inputs and outputs. Finally, XGBoost is an optimized implementation of gradient boosting that uses a combination of parallel processing and hardware optimization to achieve high accuracy and speed [11, 12].

The proposed approach utilizes a combination function using seven different repressors to construct ensemble stacking ensemble regression. The process involves randomly selecting six of the seven repressors as the base and designating the remaining one as the Meta. The model then tests all possible combinations of base and Meta repressors until it achieves the highest level of accuracy, at which point it stops operating.

In conclusion, the proposed approach for predicting the stock market closing price is a comprehensive five-step process that leverages data gathering, preprocessing, sentiment analysis, thresholding, and regression modeling. By incorporating these steps, the approach can provide a more accurate and reliable prediction of the stock market trends. Figure 1 illustrates the proposed steps of the approach, highlighting the key components of data gathering, preprocessing, sentiment analysis, thresholding, and regression modeling.

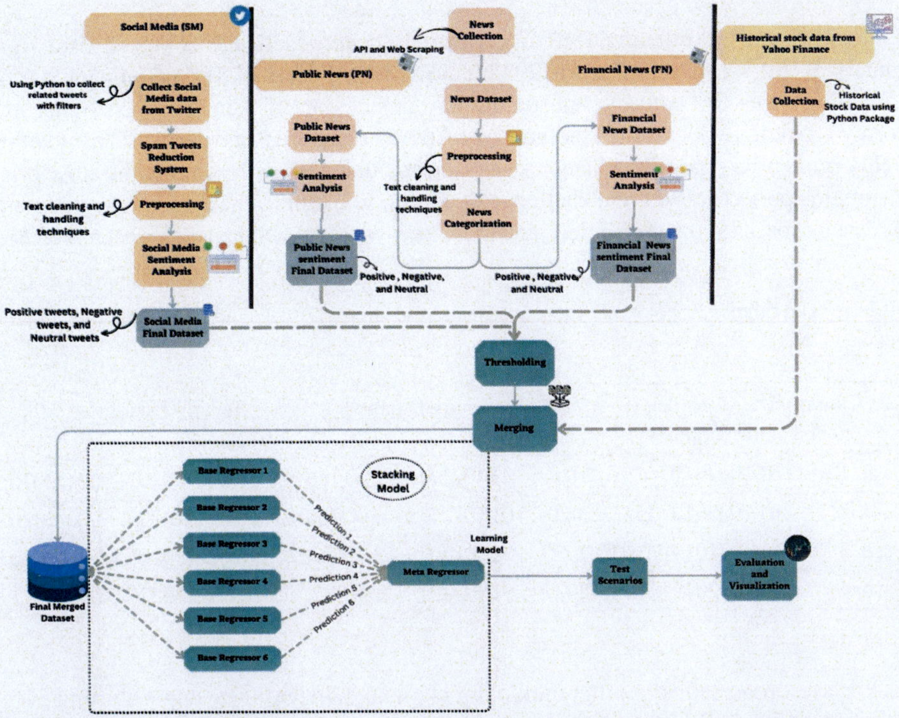

Fig. 1. The Framework of Proposed Approach for Stock Closing Price Prediction

3 Results and Discussion

The Python programming language was employed to evaluate the proposed approach. The forecasted closing prices were assessed for minimized errors in price prediction using Root Mean Square Error (RMSE) and Mean Absolute Percentage Error (MAPE) evaluation metrics. Furthermore, the accuracy of regression models was evaluated using R^2, according to the following equation [13–15], respectively:

$$RMSE = \sqrt{\sum_{i=1}^{n} \frac{(y_i - \hat{y}_i)^2}{n}} \tag{1}$$

$$MAPE = \frac{1}{n}\sum_{i}^{n} \frac{|y_i - \hat{y}_i|}{y_i} * 100 \tag{2}$$

$$R^2 = 1 - \frac{\sum (y_i - \hat{y}_i)^2}{\sum (y_i - \overline{y})^2} \tag{3}$$

where y_i indicates the actual closing price, $\hat{y}$ indicates the predicted closing price, $\overline{y}$ is the mean value, and n denotes the total number of samples in the timeframe being considered.

To evaluate the effectiveness of the proposed approach, we collected data from January 1, 2019, to December 31, 2022, covering four years of Tesla Company's data. This allowed us to capture a substantial amount of data that included changing and stormy scenarios, such as the emergence of Covid in late 2019 and early 2020, and other highly influential scenarios related to Tesla during this period. Predicting the close price during this period was a great challenge due to these factors. The data was divided into four scenarios, as shown in Table 1, and the model was evaluated based on each scenario.

Table 1. The timeframe for each test scenario and the allotted timeframe for the training and test sets

Test Cases	The Considered Time Frame	Training Dataset	Testing Dataset
Case. 1	01/01/2019 – 1/01/2020	01/01/2019 - 30/11/2019	01/12/2019 - 31/01/2020
Case. 2	01/01/2019 -15/03/2020	01/01/2019 - 31/01/2020	01/02/2020 - 15/03/2020
Case. 3	01/01/2019 - 01/10/2020	01/01/2019 - 31/07/2020	01/08/2020 - 01/10/2020
Case. 4	01/01/2019 - 31/12/2022	01/01/2019 - 15/03/2022	16/03/2022 - 31/12/2022

Each test represented a different kind of challenge, including a single shifting event over a short period, highly turbulent periods, relative normal or stable periods, and frequent shifts in sentiment and turbulence in the financial state of the company over a relatively long period spanning over a year.

The performance of the proposed approach is assessed using the RMSE, MAPE, and R2 measures across four separate test scenarios. The results are shown in Table 2 and Fig. 2 below. The results indicate that the proposed approach is effective in predicting stock prices under various scenarios. Specifically, Case 1 demonstrates the model's ability to keep up with increasing stock prices and provide valuable guidance for investors based on changes and developments around the company. Case 2 shows that the model can anticipate sudden changes in stock prices despite fluctuations over short periods. In Case 3, the model accommodates normal market fluctuations in stock prices. Finally, Case 4 covers a long period with various scenarios and incidents, and the model accurately predicts both simple and radical changes in stock prices. While Cases 2, 3, and 4 have room for improvement with relatively high RMSE and MAPE values compared to Case 1, the high R2 values indicate that the proposed approach can explain a significant proportion of data variability. Overall, the ensemble stacking model has potential for accurate predictions but may require further optimization or refinement for specific datasets.

Table 2. The obtained RMSE, MAPE, and R^2 values for the four test cases

Test Cases	RMSE	MAPE	R^2
Case. 1	0.666	0.0646%	0.9876
Case. 2	2.5	0.0375%	0.8783
Case. 3	5.804	0.043%	0.9088
Case. 4	8.846	0.0452%	0.9761

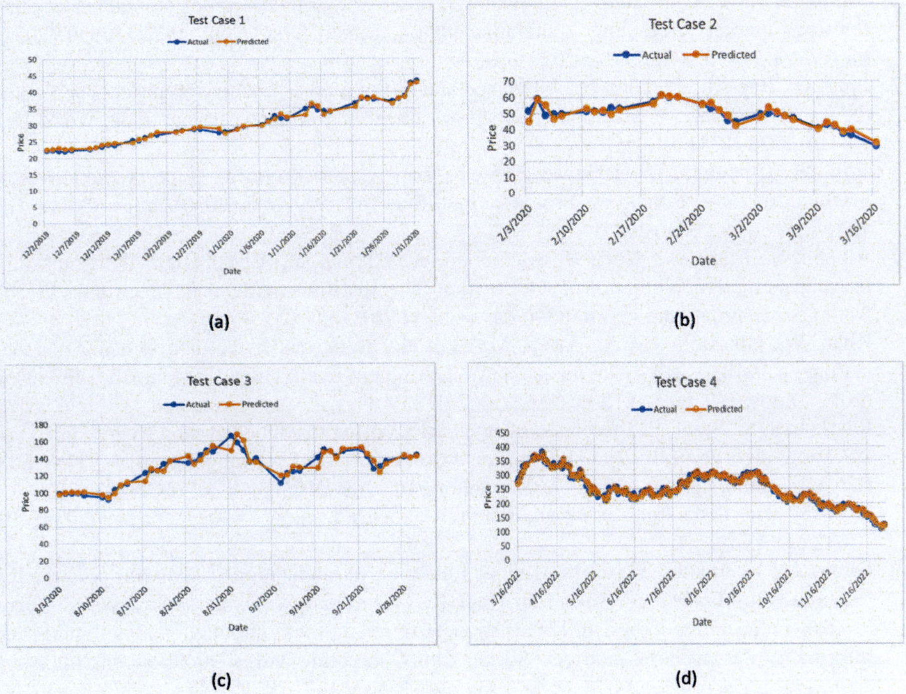

Fig. 2. Results of the proposed approach on the test cases: Case 1 (a), Case 2 (b), Case 3 (C), and Case 4 (d).

4 Conclusion and Future Work

In conclusion, this study utilized an ensemble regression approach that combined multiple data sources to predict stock closing prices. The effectiveness of the approach was evaluated using various measures, and the results demonstrated its ability to predict stock prices under different scenarios. The approach was able to handle fluctuations, anticipate sudden changes, and predict both simple and radical changes in stock prices. The high R^2 values indicate that the approach can explain a significant portion of data variability. However, further optimization or refinement may be required for specific datasets. Overall, the ensemble stacking technique is a powerful tool that can provide accurate predictions and reduce individual model biases in predicting stock prices.

Future work can focus on further improving the accuracy of the proposed approach by incorporating more data sources and exploring different ensemble techniques. Additionally, the proposed approach can be extended to predict other financial indicators, such as stock volume and market capitalization.

References

1. Kumbure, M.M., Lohrmann, C., Luukka, P., Porras, J.: Machine learning techniques and data for stock market forecasting: a literature review. Expert Syst. Appl. **197**, 116659 (2022). https://doi.org/10.1016/j.eswa.2022.116659
2. Jung, H., Raj, M., Riyanto, Y.E.: Finance and trade: a cross-country empirical analysis on the impact of financial development and asset tangibility on international trade. World Dev. **34**(10), 1728–1741 (2006)
3. Vijh, M., Chandola, D., Tikkiwal, V.A., Kumar, A.: Stock closing price prediction using machine learning techniques. Proc. Comput. Sci. **167**, 599–606 (2020). https://doi.org/10.1016/j.procs.2020.03.326
4. Albahli, S., Irtaza, A., Nazir, T., Mehmood, A., Alkhalifah, A., Albattah, W.: A machine learning method for prediction of stock market using real-time twitter data. Electronics **11**(20), 3414 (2022). https://doi.org/10.3390/electronics11203414
5. Khan, W., Ghazanfar, M.A., Azam, M.A., et al.: Stock market prediction using machine learning classifiers and social media, news. J. Ambient Intell. Hum. Comput. **13**, 3433–3456 (2022). https://doi.org/10.1007/s12652-020-01839-w
6. Ravikumar, S., Saraf, P.: Prediction of stock prices using machine learning (regression, classification) algorithms. In: 2020 International Conference for Emerging Technology (INCET), Belgaum, India, pp. 1–5 (2020). https://doi.org/10.1109/INCET49848.2020.9154061
7. Alakwah, H.: Tesla stock price prediction. Mendeley Data, V1 (2023). https://doi.org/10.17632/c7r6ky4xgc.1
8. Kiran, A., Vasumathi, D.: Data Mining: min–max normalization based data perturbation technique for privacy preservation. In: Raju, K., Govardhan, A., Rani, B., Sridevi, R., Murty, M. (eds) Proceedings of the Third International Conference on Computational Intelligence and Informatics. Advances in Intelligent Systems and Computing, Vol. 1090. Springer, Singapore. 2020. https://doi.org/10.1007/978-981-15-1480-7_66
9. Krawczyk, B., Minku, L.L., Gama, J., Stefanowski, J., Woźniak, M.: Ensemble learning for data stream analysis: a survey. Inf. Fus. **37**, 132–156 (2017). https://doi.org/10.1016/j.inffus.2017.02.004
10. Souhaila, C., Mohamed, M.: Ensemble methods comparison to predict the power produced by photovoltaic panels. Proc. Comput. Sci. **191**, 385–390 (2021). https://doi.org/10.1016/j.procs.2021.07.049
11. Brnabic, A., Hess, L.M.: Systematic literature review of machine learning methods used in the analysis of real-world data for patient-provider decision making. BMC Med. Informat. Decis. Making **21**(1),(2021). https://doi.org/10.1186/s12911-021-01403-2
12. Doreswamy, H.K.S., Yogesh, K.M., Gad, I.: Forecasting air pollution particulate matter (PM2.5) using machine learning regression models. Proc. Comput. Sci. **171**, 2057–2066 (2020). https://doi.org/10.1016/j.procs.2020.04.221
13. Abdelghafar, S., Goda, E., Darwish, A., Hassanien, A.E.: Satellite lithium-ion battery remaining useful life estimation by coyote optimization algorithm. In: 2019 Ninth International Conference on Intelligent Computing and Information Systems (ICICIS), Cairo, Egypt, pp. 124–129 (2019). https://doi.org/10.1109/ICICIS46948.2019.9014752

14. Abdelghafar, S., Darwish, A., Ali, A.: Short-term forecasting of GDP growth for the petroleum exporting countries based on ARIMA model. In: Hassanien, A.E., et al. (eds.) The 3rd International Conference on Artificial Intelligence and Computer Vision (AICV2023), 5–7 March 2023, AICV 2023, Lecture Notes on Data Engineering and Communications Technologies, vol. 164, pp. 399–406. Springer, Cham (2023). https://doi.org/10.1007/978-3-031-27762-7_37
15. Colin Cameron, A., Windmeijer, F.A.G.: An R-squared measure of goodness of fit for some common nonlinear regression models. J. Econometrics **77**(2), 329–342 (1997). https://doi.org/10.1016/S0304-4076(96)01818-0

The Impact of Data Normalization on KNN Rendering

Hassan I. Abdalla(✉) and Aneela Altaf

College of Technological Innovation, Zayed University, Dubai, United Arab Emirates
{Hassan.Abdalla,Aneela.Altaf}@zu.ac.ae

Abstract. Data normalization is a vital preprocessing technique in which the data is either scaled or converted so features will make an equal contribution. The success of classifiers, like K-Nearest Algorithm, is highly dependent on data quality to generalize classification models. In its turn, KNN is the simplest and most widely-used model for different machine learning-based tasks, including text classification, pattern recognition, plagiarism and intrusion detection, ranking models, sentiment analysis, etc. While the core of KNN is basically based on similarity measures, its performance is also highly contingent on the nature and representation of data. It is commonly known in literature that to secure competitive performance with KNN, data must be normalized. This raises a key question about which normalization method would lead to the best performance. To answer this question, the normalization of data with KNN, which has not yet been given good attention, is investigated in this work. We provide a comparative study on the significant impact of data normalization on KNN performance using six normalization methods, namely, Decimal, L2-Norm, Max/Min, Std Norm, TFIDF and BoW. On eight publicly-available datasets, experimental results show that no method dominates the others. However, the L2-Norm, Decimal, and TFIDF methods were shown to obtain the best performance (measured by accuracy, precision, and recall) in most evaluation metrics. Moreover, run time analysis shows that KNN is working efficiently with BoW, followed by TFIDF.

Keywords: KNN · Normalization · Text Classification · Machine Learning · Performance Evaluation

1 Introduction

Text classification is a vital issue, chiefly with big data, in machine learning. The K-nearest neighbor (KNN) is among the most commonly used models for data and text classification [1]. KNN has been investigated over the last three decades and is commonly implemented in so many fields [2]. Mostly, in literature, KNN also poses as the baseline model against which the performance of other models can be measured, such as text classification [3] and object recognition [4]. KNN is a non-parametric model which depends only on data and similarity measures [5]. The k-NN model finds the k-nearest neighbor(s) by walking through all of the training data, and then classifies test data by

A. Hassanien et al. (Eds.): AISI 2023, LNDECT 184, pp. 176–184, 2023.
https://doi.org/10.1007/978-3-031-43247-7_16

computing the distance between test data and all of the training data using the relative distance similarity measure [6]. Despite its being characterized by a lazy algorithm due to its nature of scanning the whole training data just to classify one data point, KNN is comprehensively used for big data classification due to its simplicity and acceptable accuracy [7, 8].

On the other hand, while the performance of similarity measures is heavily investigated in the literature [5], the impact of data normalization on KNN performance has not yet been given much attention. So, the focus of this work is to draw considerate attention to the normalization effects on KNN performance, which would lead to obtaining the highest possible performance.

The rest of this paper is planned as follows. In Sect. 2, we present the most recent work on KNN. Section 3 and 4 draw the KNN and normalization methods used in our experiments. In Sect. 5, we cover the experimental results and discussion. Finally, conclusions are provided in Sect. 6.

2 Literature Review

In several fields, including text classification [1], pattern recognition [2], data mining, and machine learning [4], the k-nearest neighbor (kNN) technique is a widely used machine learning model. The conventional kNN approach is well recognized for being a lazy algorithm because while building the model is not expensive, the searching phase is expensive and becomes unacceptably expensive when dealing with large amounts of data. To find the closest neighbors, it is typically necessary to compute the distances between a test point and all of the training points.

Many different searching methods have been put out in the literature to enhance kNN performance, chiefly in high dimensions, such as kd-trees [4], quadtrees [5], metric trees [6, 7], and R-trees [8]. The fundamental problem with such algorithms is that as the complexity of the target dataset increases, so does their computing cost. The relative searching method may wind up examining the majority of the training points, which makes it slow and lazy and even worse than ordinary k-NN [9], especially for large numbers of k values [9, 10]. Numerous methods for looking for approximations of solutions have been presented to address the efficiency problem in high-dimensional space. For instance, approximation-based techniques [12, 13] and the hashing technique [11]. However, these methods are still plagued by the dimensionality curse [9], particularly when using various K value counts [14]. Moreover, being a rarely explored topic, the impact of data normalization on kNN performance has not received convincing attention.

The role of data normalization for enhancing data quality and, consequently, the classifier's performance has been investigated in many earlier works. The key role of data normalization to build highly accurate models has been investigated for many machine learning models like k-Nearest Neighbors (k-NN) [1], Support Vector Machines (SVM) and Artificial Neural Networks (ANN) [15]. Other studies have verified and validated the effects of data normalisation on classification performance in a variety of domains, including vehicle detection [12], medical data classification [16], and stock market prediction [17].

3 K-Nearest Neighbor Classifier (K-NN)

In this sub-section, we briefly explore the KNN and the necessity for normalization. The main idea of k-NN is based on finding the distances (Euclidean, Manhattan, etc.) between the test data and training data to classify test points via identification of their neighborhood, which is selected as a subset from the training data. The test data is then assigned to the nearest class based on majority rule [5]. The key steps of the k-NN algorithm are as follows; first, the number of nearest neighbors (K) is decided. Second, the distance between the test data and the whole training data is computed. Third, the distances are sorted, and the neighborhood is defined depending on the Kth minimum distance. Forth, the classes of the nearest neighbors are grouped. Finally, based on the majority rule, the prediction of the new data is done so it will be assigned to the closest class.

On the other hand, when the scale of features differs from one to another, KNN performance is degraded significantly. The key reason is that KNN is basically based on the distance computation between the feature values. So, when some values are larger than others, those features of larger values will dominate the similarity and distance, leading to misleading performance. Hence, the normalization comes to overcome this problem and force KNN to give the best performance. In fact, when feature values are normalized, the similarity and distances between data samples are highly likely to be significantly different than when no normalization is applied, chiefly if the scaled feature values are largely different. This could explain the misleading "high/low" performance of KNN when no normalization is done. Thus, KNN performance will differ in both cases. Figures 1 and 2 show the effects of normalization on data spread on both axes of 2D space. Assume having two class classification issues, where a Class 1 data is selected (black color) along with its 10-nearest neighbors (filled green)[1]. The benefits of normalization can be noted by looking at all the nearest neighbors in both figures. For example, in Fig. 1, all neighbors are aligned in the vertical direction of both axes with a smaller range scale; making KNN performance misleading.

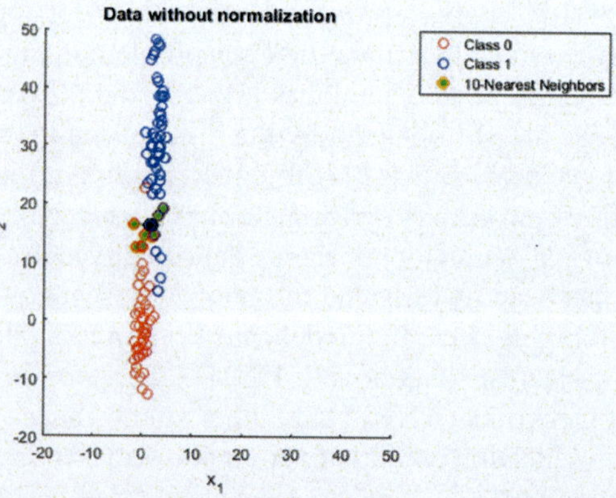

Fig. 1. Classification without normalization

[1] https://stats.stackexchange.com/questions/287425/why-do-you-need-to-scale-data-in-knn.

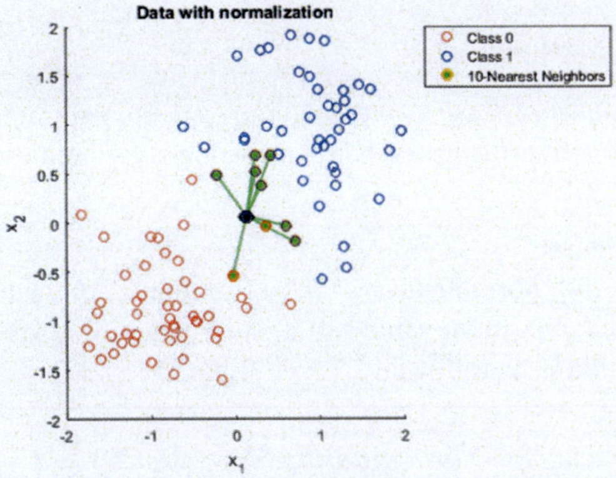

Fig. 2. Classification with normalization

4 Experimental Setup

This work was created using an Intel Core i5 processor, 8 GB of RAM, and Windows 10 Pro (32 bit).

4.1 Datasets

This work has used eight publicly available datasets whose descriptions are given in Tables 1 and 2.

Table 1. Non-IR Datasets Description

Dataset	#attributes	#instances
Glass	10	214
Iris	4	150

Table 2. IR Datasets Description

Dataset	#documents	#classes	#words
Hitech	2301	6	12,942
Classic4	7095	4	7749
Dmoz-Science	6000	12	4821
Opinion	6457	51	2693
Review	4069	5	22,927
WebACE	3900	8880	21

4.2 Normalization Methods

Using the cosine similarity measure, we have implemented six methods for normalization just to record the performance of KNN and identify which method led to the best performance.

A. Decimal Normalization

In decimal scaling normalization, the decimal point of each attribute's values is moved based on the maximum value among all other values for each target attribute. The decimal formula is as follows;

$$Decimal - normalize(attribute - value) = \left(\frac{V^i}{10^j}\right) \tag{1}$$

where v is the attribute value.

B. L_2 Normalization

With this normalization technique, the dataset values are modified in a pattern such that in each row the sum of the squares is always up to 1. The L_2 norm computes the distance of the vector coordinate from the origin of the vector space, and the result is a positive distance value. If we have a vector V, it has three values a_1, a_2 and a_3 ($\|V\| = |a_1| + |a_2| + |a_3|$), then L_2 norm will be computed as the square root of the sum as follows;

$$\|V\| = |a_1| + |a_2| + |a_3| \tag{2}$$

$$L_2(V) = \|V\|^2 = |a_1|^2 + |a_2|^2 + |a_3|^2 \tag{3}$$

C. Min/Max Normalization

In Min-max normalization, the minimum value of each feature is converted into a zero value, and the maximum value is one. Then, each other value would be converted into a decimal value between 0 and 1. The max-min normalization formula is as follows:

$$Min - Max(value) = \{value - min\}/\{max - min\} \tag{4}$$

For example, if 10 is the minimum value of the target feature and 50 is the maximum value, then 30 would be about 0.5.

D. Std Normalization

In the normalized standard deviation, the standard deviation is just divided by the mean i.e.:

$$S_n = \frac{S}{\dddot{X}} \tag{5}$$

E. TFIDF

Simply, TF-IDF is the result of multiplying term frequency (TF) with the inverse document frequency (IDF), as follows;

F. BoW

It is the term frequency weighting, which simply draws the occurrence of each word in the relative document.

5 Results and Discussion

We have used two of the most widely used metrics for performance evaluation. We have used accuracy and MAE, which are described by Eqs. (6–7).

$$ACC = \frac{\text{TruePositive} + \text{TrueNegative}}{\text{TruePositive} + \text{TrueNegative} + \text{FalsePositive} + \text{FalseNegative}} \quad (6)$$

The Mean Absolute Error (MAE) is calculated by the next Eq. 7, in which n is the total number of the documents retrieved, while v_j' and v_j are the predictive class and the true class of document j, respectively.

$$\text{MAE} = \frac{1}{n}\sum_{j=1}^{n}\left|v_{j'} - v_j\right| \quad (7)$$

In general, MAE finds the average absolute distance between the predicted and target classes. The smaller the MAE, the more accurate the normalization method, and so the better KNN performance. Tables 3, 4 and 5 show the recorded performance for KNN under each method, with bolded values implying the higher performance.

As it can be seen from Table 3, the order of best normalization techniques is: L2-Norm, Decimal, and TFIDF. As it can be seen from Table 4, the order of best normalization techniques is: TFIDF, L2-Norm, and BoW.

Table 3. Performance Comparison of Normalization Methods - Accuracy

Dataset/Norm	BoW	TFIDF	Decimal-Norm	L2-Norm	Max/Min	Std-Norm
Hitech	0.69731	0.67510	**0.69729**	**0.69729**	0.60272	0.55331
Iris	0.97401	0.33333	0.97401	**0.98148**	0.84074	0.78144
Classic4	0.93503	**0.93823**	0.93502	0.93502	0.82527	0.86379
Glass	0.95897	0.51282	**0.95897**	**0.95897**	0.86410	0.87179
Dmoz-Science	0.61527	**0.65657**	0.61519	0.61565	0.44676	0.52880
Opinion	0.58815	0.49759	**0.58875**	0.58867	0.36988	0.35148
Review	**0.91319**	0.90158	0.91318	**0.91319**	0.84767	0.81996
WebACE	0.79031	**0.83818**	0.79031	0.79031	0.62151	0.80128

Run Time Analysis

For run-time analysis, we run normalization methods on the Hitech and Iris datasets. The purpose of this analysis is to recognize the normalization method with which KNN is more efficient (Table 5).

It is clear that with BoW, KNN is more efficient and the reason behind that features are used without any transformation with BoW. So, transformation feature values contribute significantly to the speed of KNN. The trade-off is abundantly clear with L2-Norm and

Table 4. Performance Comparison of Normalization Methods - MAE

Dataset/Norm	BoW	TFIDF	Decimal-Norm	L2-Norm	Max/Min	Std-Norm
Hitech	**0.52920**	0.93174	0.71580	0.77617	1.04990	1.06898
Iris	0.02593	0.66666	0.02592	**0.01851**	0.15926	0.21851
Classic4	0.12376	**0.10506**	0.12768	0.10827	0.33780	0.23391
Glass	0.09743	1.64358	0.09231	**0.06410**	0.28205	0.20256
Dmoz-Science	1.73562	**1.54537**	1.71815	1.68481	2.28101	1.91083
Opinion	6.60087	8.78293	**5.96852**	6.36593	11.7555	10.4826
Review	**0.12053**	0.17172	0.16435	0.16011	0.31818	0.46383
WebACE	1.28888	**1.15199**	1.56453	1.42464	2.65199	1.38177

Decimal as they were shown to be superior regarding accuracy, and L2-Norm and TFIDF regarding MAE. Given the results drawn in the tables, it could be deduced that both TFIDF and L2-Norm are the best weighting schemes to be used with KNN. However, this study is not sufficient to be taken as a robust reference. So, with the aim of producing a reliable reference on this topic, future work seeks to investigate more normalization methods with more similarity measures on more middle and large-sized datasets.

Table 5. Run Time in Minutes: Seconds: Milliseconds

Dataset/Norm	BoW	TFIDF	Decimal-Norm	L2-Norm	Max/Min	Std-Norm
Hitech	**10:14.99**	10:19.370	10:15.898	10:29.572	10:35.033	10:21.061
Iris	**00:00.655**	00:01.381	00:00.817	00:00.727	00:01.111	00:00.809

6 Conclusions and Future Work

This work reveals the impact of normalization on KNN performance for text classification. Using accuracy and MAE metrics for performance evaluation, six methods are examined on eight benchmark datasets using cosine similarity measures. Considering the six normalization methods, this study reveals that L2-Norm and TFIDF are the best normalization methods to be used with KNN. When we say the best, it means they led KNN to secure higher performance. However, this study is not sufficient to be taken as a robust reference.

So, with the aim of producing a reliable reference on this topic, future work seeks to propose new normalization methods and investigate more normalization methods with more similarity measures on more middle and large-sized datasets. We will also investigate normalization impact on clustering algorithms' performance [18–21], new similarity measures [22], and advanced neural networks [23].

Acknowledgement. The author would like to thank and appreciate the support they received from the Research Office of Zayed University for providing the necessary facilities to accomplish this work. This research has been supported by the Research Incentive Fund (RIF) Grant Activity Code: R22083—Zayed University, UAE.

Authors' Contributions. Both authors are key contributors in conception and design, implementing the approach and analyzing results of all experiments, and the preparation, writing and revising the manuscript.

Funding. This research has been supported by Research Incentive Fund (RIF) Grant Activity Code: R22083 – Zayed University, UAE.

Data Availability. The dataset used in this work is publicly available.

Competing Interests. The authors declare that they have no competing interests.

References

1. Abdalla, H.I., Amer, A.A.: Towards highly-efficient k-nearest neighbor algorithm for big data classification. In: 2022 5th International Conference on Networking, Information Systems and Security: Envisage Intelligent Systems in 5G//6G-Based Interconnected Digital Worlds (NISS), pp. 1–5. IEEE, March 2022
2. Abdalla, H.I., Amer, A.A.: Boolean logic algebra driven similarity measure for text based applications. PeerJ Comput. Sci. **7**, e641 (2021)
3. Sethi, A., et al.: Supervised enhancer prediction with epigenetic pattern recognition and targeted validation. Nat. Methods **17**(8), 807–814 (2020)
4. Pan, Z., Wang, Y., Pan, Y.: A new locally adaptive k-nearest neighbor algorithm based on discrimination class. Knowl.-Based Syst. **204**, 106185 (2020)
5. Wang, X.: A fast exact k-nearest neighbors algorithm for high dimensional search using k-means clustering and triangle inequality. In: The 2011 International Joint Conference on Neural Networks, pp. 1293–1299. IEEE, July 2011
6. Ciaccia, P., Patella, M., Zezula, P.: M-tree: an efficient access method for similarity search in metric spaces. In: VLDB, vol. 97, pp. 426–435, August 1997
7. Shokrzade, A., Ramezani, M., Tab, F.A., Mohammad, M.A.: A novel extreme learning machine based kNN classification method for dealing with big data. Expert Syst. Appl. **183**, 115293 (2021)
8. Guttman, A.: R-trees: a dynamic index structure for spatial searching. In: Proceedings of the 1984 ACM SIGMOD International Conference on Management of Data, pp. 47–57, June 1984
9. Zhang, S.: Cost-sensitive KNN classification. Neurocomputing **391**, 234–242 (2020)
10. Zhang, S., Li, X., Zong, M., Zhu, X., Cheng, D.: Learning k for knn classification. ACM Trans. Intell. Syst. Technol. (TIST) **8**(3), 1–19 (2017)
11. Gionis, A., Indyk, P., Motwani, R.: Similarity search in high dimensions via hashing. In: VLDB, vol. 99, no. 6, pp. 518–529, September 1999
12. Arya, S., Mount, D.M., Netanyahu, N.S., Silverman, R., Wu, A.Y.: An optimal algorithm for approximate nearest neighbor searching fixed dimensions. J. ACM (JACM) **45**(6), 891–923 (1998)

13. Li, W., et al.: Approximate nearest neighbor search on high dimensional data—experiments, analyses, and improvement. IEEE Trans. Knowl. Data Eng. **32**(8), 1475–1488 (2019)
14. Zhang, S., Li, X., Zong, M., Zhu, X., Wang, R.: Efficient kNN classification with different numbers of nearest neighbors. IEEE Trans. Neural Netw. Learn. Syst. **29**(5), 1774–1785 (2017)
15. Abdalla, H.I., Amer, A.A.: On the integration of similarity measures with machine learning models to enhance text classification performance. Inf. Sci. **614**, 263–288 (2022)
16. Jayalakshmi, T., Santhakumaran, A.: Statistical normalization and back propagation for classification. Int. J. Comput. Theory Eng. **3**(1), 1793–8201 (2011)
17. Pan, J., Zhuang, Y., Fong, S.: The impact of data normalization on stock market prediction: using SVM and technical indicators. In: Berry, M.W., Hj. Mohamed, A., Yap, Bee Wah (eds.) SCDS 2016. CCIS, vol. 652, pp. 72–88. Springer, Singapore (2016). https://doi.org/10.1007/978-981-10-2777-2_7
18. Amer, A.A., Mohamed, M.H., Al Asri, K.: ASGOP: an aggregated similarity-based greedy-oriented approach for relational DDBSs design. Heliyon **6**(1), e03172 (2020)
19. Amer, A.A., Abdalla, H.I.: An integrated design scheme for performance optimization in distributed environments. In: International Conference on Education and e-Learning Innovations, pp. 1–8. IEEE, July 2012
20. Abdalla, H.I.: A brief comparison of k-means and agglomerative hierarchical clustering algorithms on small datasets. In: Qian, Z., Jabbar, M., Li, X. (eds.) WCNA 2021. LNEE, pp. 623–632. Springer, Singapore (2022). https://doi.org/10.1007/978-981-19-2456-9_64
21. Nguyen, L., Amer, A.A.: Advanced cosine measures for collaborative filtering. Adapt. Personal. (ADP) **1**, 21–41 (2019)
22. Abdalla, H.I., Amer, A.A., Amer, Y.A., Nguyen, L., Al-Maqaleh, B.: Boosting the item-based collaborative filtering model with novel similarity measures. Int. J. Comput. Intell. Syst. **16**(1), 123 (2023)
23. Abdalla, H.I., Amer, A.A., Ravana, S.D.: BoW-based neural networks vs. cutting-edge models for single-label text classification. Neural Comput. Appl. 1–14 (2023)

Full Personalized Learning Path Recommendation: A Literature Review

Nesrine Mansouri[1(✉)], Makram Soui[2], and Mourad Abed[1]

[1] LAMIH, Polytechnic University of Hauts-de-France, Valenciennes, France
{nesrine.mansouri,mourad.abed}@uphf.fr

[2] National School of Computer Sciences, University of Manouba, Manouba, Tunisia

Abstract. The widespread usage of online education has led to various learning activities, generating large amounts of data. A set of learning activities that assist users in achieving specific learning objectives make up a learning path. In innovative education applications, learning path recommendation is important because it can help students achieve their learning objectives more quickly, reduce the impact of information overload on students, and provide suitable sequences of learning resources for large numbers of online students. Furthermore, personalizing these learning paths has become a significant endeavor due to the learners' diverse learning styles, backgrounds, and objectives. In this context, many researchers have proposed various approaches and methods for personalizing learning paths using data mining, machine learning, optimization and knowledge graph. The present paper thoroughly investigates and discusses these research works and the theoretical foundations of learning paths. Additionally, we emphasize the importance of introducing online and offline evaluation techniques within learning path approaches to help researchers better evaluate their works and deliver personalized and impactful educational experiences for students. Furthermore, we thoroughly examine the challenges related to learning path recommendation systems, emphasizing the need for significant consideration and resolution. By addressing these challenges, we can not only enhance personalization quality and develop high-quality, reliable, and effective systems but also uncover new avenues for research and innovation in this field.

Keywords: E-learning · Recommender systems · Learning path recommendation · Online evaluation · Offline evaluation

1 Introduction

Over the past few decades, Internet technology has been increasingly adopted in e-learning, associated with rising educational resources. Thus, effective systems are needed to help learners find valuable and adequate resources, among which recommender systems play an essential role. In particular, learning path recommender systems that recommend sequences of educational resources are precious in improving learners learning experiences. However, the traditional form of E-learning systems often triggers learning confusion and cognitive overload by providing the learner with many disorganized

A. Hassanien et al. (Eds.): AISI 2023, LNDECT 184, pp. 185–195, 2023.
https://doi.org/10.1007/978-3-031-43247-7_17

learning resources [1]. Thus, these issues can become nontrivial when the learners have a privileged learning experience, especially when they need to become more familiar with a course or training, or they need more time to learn a lesson. Consequently, during the early 1960s, E-learning systems started using a directional sequence of learning resources [1, 2]. They turned into depending on curriculum sequencing mechanisms which provide users with a learning path based on learning materials [3]. Thus, the system generates specific learning paths which offer a "one size fits all" approach that indeed provides the same educational resources in the same way to all users with different profiles. Consequently, the problem "One size fits all" can cause various issues. Since approaches-based E-learning disregards the users' knowledge background and cognitive abilities, users waste time learning materials they are not qualified to learn. Another problem caused by the "one size fits all" [4] is the lack of ability to convince the users and connect them with the system since the users' preferences (e.g. learning style) are ignored. Besides, disregarding learners' progress during the learning process can negatively impact the system's efficiency according to the E-learning systems [5]. The above issues discourage users from using the system before completing the learning process. This incident is called dropout [6] and its high rate indicates that users are dissatisfied or disagree with the learning process and method [5]. In this way, personalized learning is proposed as an alternative to the "one size fits all" to overcome the mentioned problems [7]. It refers to an approach in education that focuses on providing individualized instruction to meet each student's unique needs and help them achieve their full potential. It has been observed that personalized learning improves student engagement and motivation and increases the student's ownership of the learning process since they deploy individual differences in learning preferences, goals, abilities, knowledge background, etc.

Learning path personalization is a specific approach to personalized learning that focuses on creating customized learning paths for each student based on their unique needs, abilities, and interests, as described in Fig. 1.

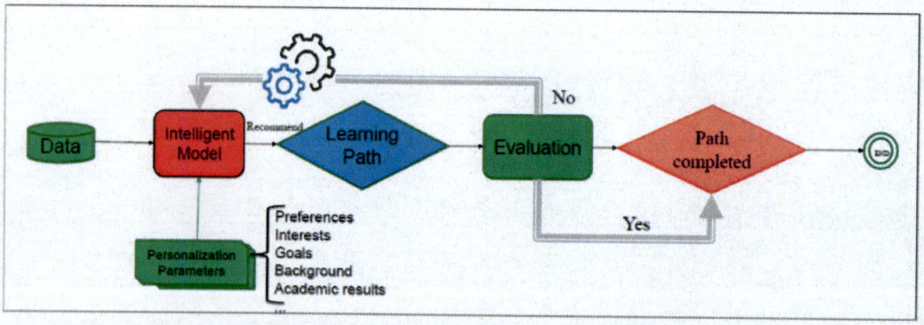

Fig. 1. Learning path recommendation system

This approach is often used in online learning environments, such as MOOCs (Massive Open Online Courses) and other digital educational platforms. Still, it can also be applied to traditional classroom settings. In a learning path personalization system, the student's progress is constantly monitored, and data is used to adjust the content and order of the materials they are presented with. Additionally, the student's interests can be considered to provide them with content that aligns with their specific goals and motivations. For example, a student interested in pursuing a career in a particular field may be presented with resources and opportunities to learn about related topics and gain experience. Learning path personalization can provide many benefits for students, including increased engagement and motivation, better retention of material, and improved outcomes. It can also help to bridge the skills gap and reduce the achievement gap by providing individualized instruction to students of all abilities and backgrounds. However, implementing a learning path personalization system can be a complex and challenging process. It requires significant resources, including technology infrastructure, data collection and analysis, and ongoing professional development for educators. It also requires a clear understanding of learning theories, pedagogy and assessment, and the development of effective learning strategies. Learning path personalization is a specific approach to personalized learning; it creates a customized learning path based on the student's unique needs, abilities, and interests to make the learning experience more engaging, efficient, and effective. Basically, the learning path is established by a teacher, instructor, or expert in the related field and will be similar for all students. Nevertheless, the learning path can be flexible in being adaptable to each learner due to student variances in knowledge level, preferences, interests, and learning goals and needs. These variances refer to a student profile or model, which constitutes the primary information about the learner and then is consumed as parameters to the system on which the recommendation is based. To sum up, learning path recommendation is crucial in innovative education applications. It can provide suitable learning resource sequences for large-scale learners by identifying learners' characteristics and requirements by applying personalization parameters, lessen the effect of information burden on learners, and help learners realize their goals more swiftly.

The main objective of this study is to identify relevant concepts involved in learning path recommendation, discuss the existing challenges and directions of learning path recommendation, cover recommendation techniques trends, and transparently present the evaluation techniques. This review examined electronic databases like Google Scholar and ScienceDirect using keywords like learning path recommendation and personalization. The selection criteria include articles published in English, papers on learning path recommendations, and recent articles.

This paper is organized as follows. Section 1 introduces the main context. Section 2 presents the theoretical background of the recommender system. Section 3 discusses the related works. Section 4 introduces the techniques used to evaluate the learning path recommender. Section 5 discusses the challenges of building a learning path recommender system. Section 6 concludes the work and looks ahead.

2 Theoretical Background

2.1 Recommender System

A recommender system is a technology used to make personalized recommendations to users. It uses data about users' preferences and behavior to suggest items or content they are interested in. Recommender systems can be used in various applications, such as music, movies, news, and e-commerce. The data can be explicit (e.g., ratings or feedback) or implicit (e.g., browsing or purchase history). There are several types of recommender systems, including content-based, collaborative filtering, and hybrid recommender systems. A form of technology called a recommender system provides users with tailored recommendations. Personalized recommender systems in education can be used to deliver customized learning experiences for students by recommending resources, activities, or assessments that align with their individual learning needs and preferences. These systems can be used to suggest content based on a student's performance, interests, or learning style. There are various approaches to building personalized recommender systems in education, including Collaborative filtering, which relies on similar users' past behaviors and preferences to make recommendations [8]. Content-based filtering uses the similarity between the content and the student's past interactions with the content to make recommendations. Hybrid methods combine collaborative and content-based filtering [9].

2.2 Learning Path

According to [10], Learning path (LP) refers to either course sequences in a particular study program or learning object sequences within a course. Indeed, learning object sequences in e-learning are obviously different from those in classical and offline classes because, in e-learning, adaptive sequences are mainly built, in line with each student's learning goals, learning history, learning style, learning needs, and curriculum information. [11] represents the learning path as a route, a student takes in studying the available LOs. More specifically, the learning content sequences that combine learning objects, concepts, and courses are called learning paths [12, 13] and are recommended to individuals matching their prerequisites. In a nutshell, several synonyms have been reported in the literature, including content or learning content sequencing [14], sequence generation [15], curriculum sequence [16], path, and instructional sequence [17].

3 Related Works

This part presents the literature review of the methods used to recommend fully personalized learning paths for learners. Most existing work deployed ontology, knowledge graphs, and collaborative filtering. [18] develop a personalized creativity learning system (PCLS) using data mining. The PCLS includes a series of creativity tasks and a questionnaire regarding several vital variables. Ninety-two college students aged between 18 and 26 are considered as data for this study. This work is based on learner characteristics such as Student ID, Age, College Major, Creativity Learning path, Learning style, and Self-perception of creativity. The main objective of this work is to optimize the invention's performance. [19] used social tags and sequential pattern mining as primary leaner characterization. Collected labels about learning resource sequences earners; these tags are rated and used to recommend learning resource sequences using collaborative tagging techniques. Their aims are based on evaluating the usability of the proposed hybrid method by its implementation in an existing tutoring system for teaching Java programming basics. [20] develop a full-path learning recommendation model, where learners are clustered based on their feature similarity. Then learners' learning paths and learning effects are predicted by Long-Short Term Memory. According to the prediction learning effects, a personalized learning path is recommended for learners based on Edx open datasets WorldUC. Also, They used the track record in past learning experiences and performance as data. This study aims to predict learners' learning paths and performance. [21] implemented a mixed concept map to establish the relationship between learners and learning resources, which described the recommendation as a constraint satisfaction problem and applied the immune algorithm to obtain recommendation results. This work is based on the learner's Path preferences and learning behavior characteristics as selection criteria. Their main aim was to meet their needs and improve the learning efficiency of e-learners. This work considered data consisting of 2150 log items from 156 learners and 57 knowledge units to validate the proposed approach. [22] proposed a learning design recommendation system based on graph theory. The approach is implemented and tested using the concept of cliques to recommend learning paths based on the student's mastery level. This approach serves many purposes, including delivering the shortest path (number of competencies) for learners and describing the ability dependency relationship between learning resources based on graph theory. But, these methods are not considering the diversity of the relationships between the learning resources. The authors used data from 1 to 6 prerequisites (PRE) and provided a gain of one or two competencies (POST). The PRE and POST competencies ranges have been chosen according to qualitative studies of existing learning object datasets (KDD cup 2010 dataset, The Australian Flexible Learning Framework, Globe). [23] developed a group learning path recommendation framework based on a user profile to help each learner in the group efficiently learn the needed new knowledge within the time limitation and to enable the whole group to acquire all the necessary expertise to complete the group learning task, on the whole, using the learning path discovery. Based on this work, learners' knowledge background, learning preference, and group learning tasks are criteria. This work aims to identify effective learning paths for a group of learners. The authors used three different classes of Data, A: self-decided learning path B: group profiling learning path, and C: aggregated prototype learning path. [24] used the data and knowledge graph in the learner's learning history log to first represent the knowledge point learning of learners in different periods as personalized learning to generate a network and then conducted cluster analysis on this customized learning to create a network. In this way, the group learning generation network reflecting the typical characteristics of the learning group is generated. The selected criteria are based on this work using learning logs and curriculum knowledge maps. This work aims at group recommendation in online learning by combining user learning logs and curriculum knowledge maps to generate a personal learning generation network. This work is validated using the log data of the Java course from May 9, 2014, to November 30, 2017, including 11 batches of 247 learners. The knowledge map of the Java course contains 70 knowledge points and 79 edges. A total of 156 learners is used to develop the model. [25] adopted a Competency-Based Approach to offer each learner an individualized learning path for acquiring the competence targeted based on collaborative filtering. The method is based on the learning style, understanding, difficulty of course topics, and course learning materials. The data used in this study is a web service that collects data from the database using SQL language. In general, these works demonstrate various approaches to learning path recommendation, including deep learning-based methods, hybrid recommendation algorithms, and clustering techniques which prove their mettle in providing efficient systems.

4 Evaluation of the Learning Path Recommender System

A learning path recommender system's performance can be assessed in different ways. However, it often involves a mixture of quantitative and qualitative measures. In this way, a holistic view of the system's effectiveness in recommending relevant and diverse learning paths to users will be provided. Basically, this evaluation can be in two forms: online and offline assessment [26].

4.1 Online Evaluation

Online evaluation of learning path recommendations requires assessing the performance of the recommendation system in real time while learners interact with it. Typically, this approach delivers a more accurate assessment of the system's effectiveness in recommending relevant and diverse learning paths to learners. In this sense, A/B testing is one of the widely used ways. It involves randomly assigning learners to two groups. One group obtains recommendations from the learning path recommender system, while the other group obtains recommendations from a totally different source. Then, a performance comparison of the two groups is conducted to distinguish between evaluating the learning path system, thereby assessing the effectiveness [27]. For this type of evaluation, many studies in the literature concentrate on the level of learners' knowledge to define the learning experience [20, 27, 28] deployed the A/B testing to evaluate students' progress using recommended learning paths. The idea consists of dividing the learners into two groups; one group receives a personalized learning path, while the others receive a non-personalized one. After obtaining the necessary recommendations, the learners of the two groups take the final exam. The average score in each group is calculated, and both scores are compared to get a final evaluation of the system. Such an approach is straightforward to implement.

Nevertheless, some things could be improved, such as incorporating human subjects and picking the sample size, and the experiment duration is deemed arduous. Besides, finding similar groups of learners that support the same knowledge is not easy, and it is likely to create between the two tested groups. [23] assess the impact of a learning path by comparing two groups and the rise in performance, which is the difference between the performance before and after using the personalized recommendations. A portion of the literature also suggested evaluating the distance between the recommended and ground-truth learning paths. Specifically, the Edit Distance (ED) is the most frequently utilized measure [21]. In the past, ED has been used to measure how different two Learning paths are [29]. The minimum number of operations required to convert one sequence into the other is all that ED counts. In [21]'s work, determining ground-truth Learning paths by requesting self-organized LP from students via questionnaires is the system's only known interaction with students. The difficulty in deciding whether the self-organized path is correct is one of the main limitations of this work. Even though online evaluation is known to be reliable, these works demonstrate that conducting such an evaluation is pricey in the educational context, primarily due to the amount of time involved in carrying it out and the availability of stakeholders (including learners). Generally, online evaluation experiments are limited in a few general ways; the recommendations made to students may negatively affect their learning experience if the recommendation model's quality is low. In addition to that, when students are divided into groups and receive various recommendations, the online evaluation process cannot ensure equivalence between them. Lastly, the single-variable method cannot assure the split of the test combination because the learning background of each group of students is nearly identical. Online experiments are challenging to replicate. These restrictions apply not only to LP but also to any online educational experiment. Offline evaluation is also a popular evaluation method due to these restrictions.

4.2 Offline Evaluation

This kind of evaluation is often common in learning path recommendation systems which require using historical data to simulate learner profiles and behavior and assess the performance based on specific metrics. Measuring the system's accuracy typically involves the ground truth that can be inferred from the data or through experts in the domain to ensure data reliability [30]. Offline evaluation is a straightforward type of evaluation since it requires only data. Inferring ground truth from the data is quite simple as it involves only the data and no need for experts. However, It relies heavily on the obtained data from the recommender system during the assessment step, which is challenging [31]. It consists primarily in identifying the ground truth of the learning path and then assessing the relation of the recommended learning path with the ground truth [29] Besides, evaluation metrics are frequently used, such as precision, recall, and F1-score [32], along with cross-validation to guarantee that the used metrics are robust and not overfitting the training data.

The offline evaluation procedure aids in enhancing the performance of the algorithms by highlighting weak points and providing insights into the effectiveness of the learning path recommendation system. However, it is essential to remember that offline evaluation is constrained by the quantity and quality of historical data and may not always accurately reflect user behavior in real-time scenarios. Therefore, to verify the effectiveness of the recommendation algorithms in actual situations, online A/B testing with real users is essential.

5 Challenges and Discussions

Learning path recommendation is deemed a daunting task because many factors can indirectly affect the effectiveness of recommendations and should be taken thoroughly to obtain optimal recommendations.

5.1 Personalization

This is the most sensitive factor due to its hardness since each student's profile is unique, with many characteristics, including learning styles, interests, goals, backgrounds, preferences, and interests [33]. In this way, addressing these parameters significantly impacts the recommender system; although challenging, it can ensure that the recommended paths are likely reliable.

5.2 Content Coverage

The learning path should support a broad spectrum of topics and subject areas to ensure the system's effectiveness. In this way, the recommended learning paths can deliver complete coverage of the learners' topics and skills, which can satisfy their interests and needs. Besides, these recommendations should match the pedagogy regarding learning objectives, teaching methods, level of interaction, level of difficulty, and keywords [34].

5.3 Data Quality

To ensure the success of the recommendation system, it is required to check the data quality and ensure its reliability. Generating efficient recommendations is based primarily on the quality of used data that should be accurate and relevant to learners' needs [34].

To cope with these challenges, more efforts should be incorporated into this field and harness the power of combining data analytics, machine learning algorithms, and effective user experience design with building intelligent models to improve the learning path recommendation systems.

6 Conclusion

Learning path recommendation has recently gained significant attention from researchers and experts due to the abundance of online resources and the challenges associated with selecting the suitable learning path to achieve specific goals. The primary purpose of recommending paths is to assist students in navigating vast educational content and identifying the optimal learning path for their specific needs. In this context, several studies developed using hybrid recommendation algorithms, clustering techniques, and deep learning have been reported in the literature. This paper introduces different learning personalization methods and the evaluation techniques employed in this field. Furthermore, we dive into the challenges associated with these approaches and discuss them. The existing methods demonstrated their potential to improve student's learning experiences by providing personalized recommendations to help them efficiently realize their targets. Nevertheless, issues such as data sparsity, cold-start problems, and scalability involve further investigation and research in this field to deliver more efficient personalized recommendations. In future directions, we plan to develop new solutions to mitigate these issues by incorporating techniques like transfer learning to enhance the performance of learning path recommendation systems, especially when dealing with limited data or when faced with cold-start problems.

References

1. Yang, F., Li, F.W., Lau, R.W.: An open model for learning path construction. In: Advances in Web-Based Learning–ICWL 2010: 9th International Conference, Shanghai, China, December 8-10, 2010. Proceedings 9, pp. 318–328. Springer, Heidelberg (2010). Doi: https://doi.org/10.1007/978-3-642-17407-0_33
2. Yang, F., Dong, Z.: Learning Path Construction in e-learning. Lecture Notes in Educational Technology. Springer, Heidelberg (2017)
3. Chi, Y.: Developing curriculum sequencing for managing multiple texts in e-learning system. In Proceedings of International Conference on Engineering Education, pp. 1–8 (2010)
4. Matar, N.: Adaptive unified e-learning approach using learning objects repository structure. Int. J. Digital Soc. **2**(4), 579–587 (2011). https://doi.org/10.20533/ijds.2040.2570.2011.0070
5. Karampiperis, P., Sampson, D.: Adaptive learning resources sequencing in educational hypermedia systems. J. Educ. Technol. Soc. **8**(4), 128–147 (2005)
6. Şahin, Ş, Arseven, Z., Kiliç, A.: Causes of student absenteeism and school dropouts. Int. J. Instr. **9**(1), 195–210 (2016). https://doi.org/10.12973/iji.2016.9115a
7. Bray, B., McClaskey, K.: A step-by-step guide to personalize learning. Learn. Leading Technol. **40**(7), 12–19 (2013)
8. Javed, U., Shaukat, K., Hameed, I.A., Iqbal, F., Alam, T.M., Luo, S.: A review of content-based and context-based recommendation systems. Int. J. Emerg. Technol. Learn. **16**(3), 274–306 (2021). https://doi.org/10.3991/ijet.v16i03.18851
9. Geetha, G., Safa, M., Fancy, C., Saranya, D.: A hybrid approach using collaborative filtering and content based filtering for recommender system. In Journal of Physics: Conference Series (Vol. 1000, No. 1, p. 012101). IOP Publishing, April 2018
10. Rahayu, N.W., Ferdiana, R., Kusumawardani, S.S.: A systematic review of learning path recommender systems. Educ. Inf. Technol., 1–24 (2022). https://doi.org/10.1007/s10639-022-11460-3
11. Premlatha, K.R., Geetha, T.V.: Learning content design and learner adaptation for adaptive e-learning environment: a survey. Artif. Intell. Rev. **44**(4), 443–465 (2015). https://doi.org/10.1007/s10462-015-9432-z
12. Adorni, G., Koceva, F.: Educational concept maps for personalized learning path generation. Lecture Notes in Computer Science (including subseries Lecture Notes in Artificial Intelligence and Lecture Notes in Bioinformatics), vol. 10037 LNAI, pp. 135–148 (2016). https://doi.org/10.1007/978-3-319-49130-1_11
13. Muhammad, A., Zhou, Q., Beydoun, G., Xu, D., Shen, J.: Learning path adaptation in online learning systems. In: 2016 IEEE 20th International Conference on Computer Supported Cooperative Work in Design (CSCWD), pp. 421–426. IEEE, May 2016
14. Seki, K., Matsui, T., Okamoto, T.: An adaptive sequencing method of the learning objects for the e-learning environment. Electron. Commun. Japan (Part III: Fundamental Electronic Science) **88**(3), 54–71 (2005)
15. Rasheed, F., Wahid, A.: Sequence generation for learning: a transformation from past to future. Int. J. Inf. Learn. Technol. **36**(5), 434–452 (2019). https://doi.org/10.1108/IJILT-01-2019-0014
16. de O. C. Machado, M., Bravo, N.F.S., Martins, A.F., Bernardino, H.S., Barrere, E., de Souza, J.F.: Metaheuristic-based adaptive curriculum sequencing approaches: a systematic review and mapping of the literature. Artif. Intell. Rev. **54**(1), 711–754 (2021). https://doi.org/10.1007/s10462-020-09864-z
17. Pepin, B., jan Kock, Z.: Students' use of resources in a challenge-based learning context involving mathematics. Int. J. Res. Undergraduate Math. Educ. **7**(2), 306–327 (2021). https://doi.org/10.1007/s40753-021-00136-x

18. Lin, C.F., Yeh, Y.C., Hung, Y.H., Chang, R.I.: Data mining for providing a personalized learning path in creativity: an application of decision trees. Comput. Educ. **68**, 199–210 (2013). https://doi.org/10.1016/j.compedu.2013.05.009
19. Klašnja-Milićević, A., Vesin, B., Ivanović, M.: Social tagging strategy for enhancing e-learning experience. Comput. Educ. **118**, 166–181 (2018). https://doi.org/10.1016/J.COMPEDU.2017.12.002
20. Zhou, Y., Huang, C., Hu, Q., Zhu, J., Tang, Y.: Personalized learning full-path recommendation model based on LSTM neural networks. Inf. Sci. (N Y) **444**, 135–152 (2018). https://doi.org/10.1016/j.ins.2018.02.053
21. Zhu, H., et al.: A multi-constraint learning path recommendation algorithm based on knowledge map. Knowl. Based Syst. **143**, 102–114 (2018). https://doi.org/10.1016/j.knosys.2017.12.011
22. Durand, G., Belacel, N., Laplante, F.: Graph theory based model for learning path recommendation. Inf. Sci. (N Y) **251**, 10–21 (2013). https://doi.org/10.1016/j.ins.2013.04.017
23. Xie, H., Zou, D., Wang, F.L., Wong, T.L., Rao, Y., Wang, S.H.: Discover learning path for group users: a profile-based approach. Neurocomputing **254**, 59–70 (2017). https://doi.org/10.1016/j.neucom.2016.08.133
24. Zhu, H., Ni, Y., Tian, F., Feng, P., Chen, Y., Zheng, Q.: A group-oriented recommendation algorithm based on similarities of personal learning generative networks. IEEE Access **6**, 42729–42739 (2018). https://doi.org/10.1109/ACCESS.2018.2856753
25. Bendahmane, M., el Falaki, B., Benattou, M.: Toward a personalized learning path through a services-oriented approach. Int. J. Emerg. Technol. Learn. **14**(15), 52–66 (2019). https://doi.org/10.3991/ijet.v14i15.10951
26. Nabizadeh, A.H., Leal, J.P., Rafsanjani, H.N., Shah, R.R.: Learning path personalization and recommendation methods: a survey of the state-of-the-art. Expert Syst. Appl. **159** (2020). https://doi.org/10.1016/j.eswa.2020.113596
27. Nabizadeh, A.H., Gonçalves, D., Gama, S., Jorge, J., Rafsanjani, H.N.: Adaptive learning path recommender approach using auxiliary learning objects. Comput. Educ. **147**, 103777 (2020). https://doi.org/10.1016/J.COMPEDU.2019.103777
28. Choffin, B., Popineau, F., Bourda, Y., Vie, J.J.: DAS3H: modeling student learning and forgetting for optimally scheduling distributed practice of skills. arXiv preprint arXiv:1905.06873 (2019)
29. Zhang, Z., Brun, A., Boyer, A.: New Measures for Offline Evaluation of Learning Path Recommenders (2020). https://doi.org/10.1007/978-3-030-57717–9_19ï
30. Rossetti, M., Stella, F., Zanker, M.: Contrasting offline and online results when evaluating recommendation algorithms. In: RecSys 2016 - Proceedings of the 10th ACM Conference on Recommender Systems, Association for Computing Machinery, Inc, Sep. 2016, pp. 31–34. https://doi.org/10.1145/2959100.2959176
31. Beel, J., Langer, S.: A comparison of offline evaluations, online evaluations, and user studies in the context of research-paper recommender systems. 19th International Conference on Theory and Practice of Digital Libraries, TPDL 2015, Poznań, Poland, September 14-18, 2015, Proceedings 19 (pp. 153-168). Springer, Cham (2015)
32. Su, C.H.: Designing and developing a novel hybrid adaptive learning path recommendation system (ALPRS) for gamification mathematics geometry course. Eurasia J. Math. Sci. Technol. Educ. **13**(6), 2275–2298 (2017). https://doi.org/10.12973/EURASIA.2017.01225A

33. Jeng, Y.L., Huang, Y.M.: Dynamic learning paths framework based on collective intelligence from learners. Comput. Human Behav. **100**, 242–251 (2019). https://doi.org/10.1016/J.CHB.2018.09.012
34. Fennan, A., el Bouhdidi, J., Ghailani, M.: An intelligent architecture for generating evolutionary personalized learning paths based on learner profiles. Article in Journal of Theoretical and Applied Information Technology, vol. 57, no. 2 (2013). https://www.researchgate.net/publication/289069600

Research on the Influence of Recommendation and Reward System on Customer Online Comment Behavior

Wei Liu, Zhipeng Fan(✉), Ming Chen, and Xiang Xie

Harbin University of Commerce, Harbin, Heilongjiang, China
fanzhipeng@hrbcu.edu.cn

Abstract. This article studies consumer online review behaviors based on recommendation reward programs. It mainly studies the influence of different types of rewards in the recommendation reward plan on consumers' online review behavior. Through the questionnaire survey, the results show that consumers are more inclined to choose monetary rewards as their return in the recommendation reward program. At the same time, different rewards in the recommendation reward plan, consumer product satisfaction, and the interaction of the two also have a certain impact on consumers' online review behavior. This article divides the amount of rewards into high rewards and low rewards. When the consumer's satisfaction reaches a certain level, low rewards in the referral reward program will bring more benefits to the merchants than high rewards. High rewards will only have a greater effect on consumers who are less satisfied with the product. When consumer product satisfaction is not high, and the reward limit set by the recommendation reward plan is low, consumers' perception of fairness will cause them not to conduct online evaluation behaviors, even if the reward limit is high, the effect is not significant.

Keywords: recommendation reward · consumer satisfaction · word-of-mouth marketing · reward quota · perceived fairness

1 Introduction

In existing research, most scholars' research results have similar conclusions. Consumers' review behavior not only reflects consumers' opinions on products and goods and services themselves, but to a certain extent, it can also give other buyers' opinions about them. Buying behavior has a very important impact. After the purchaser analyzes the reviews before buying, he can reduce the risks that may occur in the purchase, and then choose the most suitable product to purchase, and these will eventually affect the company's product performance [2]. Enterprises will manage consumers' word-of-mouth, and try to encourage consumers to conduct positive word-of-mouth behaviors through incentive measures, and generate positive reviews, thereby attracting more buyers to purchase products [3]. The main way is to attract consumers to purchase behavior through monetary and material rewards. For example, after the consumer performs the

A. Hassanien et al. (Eds.): AISI 2023, LNDECT 184, pp. 196–211, 2023.
https://doi.org/10.1007/978-3-031-43247-7_18

recommendation behavior, the buyer will return the consumer voucher or cash so that the consumer will have a preferential activity in the next purchase, or they can directly get the cash red envelope [4]. The other is not to provide these material rewards, but to rank the effectiveness of consumer reviews, through other purchasers to like these reviews, so as to generate high-efficiency reviews, and carry out these relatively good reviews. Award. The first is a monetary return, and the other is a non-monetary return [5]. In the research, consumers' review scores and evaluation content of products will be related to the product reputation of the merchants, thereby affecting performance [6]. At the same time, the effectiveness of reviews will also have a certain impact on consumers' buying behavior.

For these reasons, scholars in recent years have focused on the content of comments in word-of-mouth marketing and consumers' willingness to comment [7]. Through the summary of the existing research, there are three main aspects. The first is to consider the overall quantity of consumer reviews and the overall quality of the overall willingness to comment, and analyze the impact of these characteristics on consumers and corporate products [8]. The second is to analyze the buyer's willingness to make comments, and to study the motivation of buyers to produce purchase behavior and comment behavior [9]. The third is to conduct an in-depth analysis and mining of the content of comments generated by buyers [10]. In addition, some studies have also analyzed the influence of reward behavior on consumer review behavior. The results show that low-amount reward behavior is unlikely to affect consumers' willingness to comment. Literature [11] shows that low quotas can affect consumers' review behavior, and will have a certain promotion effect on consumers' review behavior. Some scholars have found through research that reward behavior does not reflect the influence of consumers' willingness to comment. For enterprises, the factors that need to be considered in this reward behavior are costs and benefits. Therefore, we will also pay attention to the effect of this reward behavior on consumer review behavior [12]. In existing research, whether these reward behaviors can promote consumers' online review behavior, and how to promote consumers' online review behavior, there is no unified conclusion. Therefore, this article also focuses on the research on the influence relationship between the recommendation reward program and consumer review behavior.

2 Basic Theoretical Research

2.1 Self-determination Theory

Self-determination theory is a theory of motivation that studies consumers' behaviors when making decisions and self-rewards. Consumers will reduce their self-decision-making behavior when facing internal threats from external rewards, and have a certain impact on the motivation for this behavior [13]. The theory of autonomous learning is based on the consumer's intrinsic motivation, focusing on the reasons for the intrinsic motivation and what kind of behavior and what goals the consumer can achieve as the research purpose theory. Self-determination theory is of great significance to the study of consumer behavior. Literature [14] believes that the theory of self-determination is to combine it with students' learning to explore the motivation of students in the learning process. There are three main aspects, namely, independent needs, competent needs and attribution needs. Autonomous learning is the willingness of students to learn actively and take actions. Competency requirements are students who think they are competent and able to complete their studies or work. The need for belonging refers to the ability of students to integrate into the learning collective and gain a sense of belonging during the learning process. The most important of the three needs is autonomous demand.

2.2 Social Theory

The main content of social cognitive theory is that the whole cannot determine every individual, but some individuals will gradually affect the whole. This also leads to when we want to investigate a certain group, we need to study the nature and characteristics of the individual from the inside out and from the top down [15]. Later, some scholars put forward new viewpoints on the theory of social cognition, thinking that in the theory of social cognition, the very important theory is the theory of cognitive evaluation. That is, before consumers act, if external stimuli will have a bad impact on them, it will make consumers weaken the motivation to make such behaviors, thereby reducing the possibility of behavior. On the contrary, if the external stimulus does not affect the consumer's inner motivation, this motivation will not be reduced.

The theory of social trading is that in the process of social trading, consumers will compare their own payment and the returns obtained through the exchange during the entire trading process, and choose as much as possible to be beneficial to their own behavior. In the transaction process, consumers will be paid. In the referral reward mechanism, consumers will receive certain rewards through certain contributions. And this kind of reward may be an intrinsic reward, for example, it may be recognized and respected by people around it, and a certain honor and fame may be obtained. It may also be an external reward, such as cash rewards and commodity rewards, and what consumers need to pay in this process is their own time and energy [16]. In the process of social transaction theory, it is possible to see consumers' views on the mechanism of recommendation and reward schemes. The social transaction theory can be used to explore whether consumers accept the recommendation reward plan, and whether they are willing to spend a certain amount of time and energy to obtain internal and external rewards from the business. In the traditional word-of-mouth recommendation mechanism. Consumers will recommend products that they consider to be better in use and experience to people who are more familiar and close around them through oral transmission, and help people around them make purchasing decisions.

The research in this article is aimed at the merchant's recommendation reward plan. When consumers are very satisfied with the products they have purchased, the recommendation reward program provided by the merchant is a more active external regulation for consumers, and consumers will be more willing to participate in the recommendation reward program. If for consumers, the products provided by the merchants do not meet their psychological needs, and when they are not satisfied with the purchased goods, the external stimulus provided by the merchant's recommendation and reward program is a negative stimulus, which will lead to consumers' feelings of the merchant. The recommendation reward program is inherently resistant and unwilling to participate in such activities. At the same time, it will also affect consumers' psychological activities and behaviors after participating in the referral reward program.

2.3 Word-of-Mouth Research

Word-of-mouth communication is a behavior in which consumers will comment after purchasing a product through their own purchase experience and their feelings about the quality of the product. It is also a new type of marketing method. Consumer reputation is often a very important means of communication, and it is also very effective. This kind of word-of-mouth influence produced by consumers can attract more buyers to purchase goods, thereby enhancing the company's own product sales and brand awareness. In existing research on word-of-mouth rewards, most of the literature studies the issue of word-of-mouth rewards from two aspects: the sender of word-of-mouth and the recipient of word-of-mouth. Literature [17] verified the influence of the amount of reward, the type of reward, and the distribution ratio in the reward scheme in word-of-mouth rewards on the willingness of word-of-mouth senders to send word-of-mouth. Literature [18] et al. have verified the process of consumers' psychological activities when they send word-of-mouth. That is, how does the sender view the behavior of word-of-mouth rewards? It mainly studies the relationship between psychological effects on word-of-mouth recommendation plans and buyers sending word-of-mouth recommendations. With the rapid development of the Internet, e-commerce platforms are paying more and more attention to the research of potential consumers. By setting certain comment and product feedback platforms, potential consumers can view the use and evaluation of consumers who have purchased products. So as to have a deeper understanding of the product. In general, Internet word of mouth refers to the purchase of products that buyers are interested in or products that they have already purchased, so that other consumers can understand the product experience and product purchase experience, which forms the Internet word of mouth [19]. Currently, platforms that can conduct online word of mouth mainly include Weibo likes and product reviews. At present, word-of-mouth communication is very important. Literature [20] found through research that comparing the word-of-mouth communication between consumers and the advertising communication of businesses, it is found that consumers are more inclined to believe in the authenticity of information transmission between consumers. Another very important feature of word-of-mouth communication is that it has a certain direction. Positive word-of-mouth communication can not only attract other consumers to purchase products, thereby improving corporate performance. It can also reduce a lot of costs in the company's product promotion and further improve performance. Negative word-of-mouth communication will cause consumers to give up buying products because of bad word-of-mouth. In the process of word-of-mouth communication, consumers do not necessarily engage in active word-of-mouth communication, even if they are satisfied with the product. Studies have shown that among the many existing consumers, some of them are "silent customers", and they will not evaluate them regardless of whether they are satisfied with the product. When companies can provide incentives, they can change this to a certain extent. Phenomenon. This is also the reason for the recommendation reward mechanism to be studied in this article.

2.4 Referral Rewards

Literature [21] studied the influence of fairness theory in the distribution of word-of-mouth rewards. The main method is to draw relevant conclusions by analyzing the

US EBATES website and testing consumers. Word-of-mouth rewards mainly include pure word-of-mouth and reward word-of-mouth. Pure word-of-mouth often refers to the behavior of consumers spontaneously and actively evaluating products when they are satisfied or dissatisfied with a product, that is, the behavior of spreading word-of-mouth. Rewarding word-of-mouth means that companies use certain rewards to make consumers spread the word of mouth. Most researchers analyze the pure word-of-mouth of consumers, such as the influence of factors such as consumers' trust in the brand and participation in innovation in the process of consumer's pure word-of-mouth communication. However, in real online shopping, most consumers may not actively promote positively even if they are satisfied with the product. From this, it can be concluded that companies can adopt appropriate methods to stimulate buyers to spread the word of mouth, which is a recommendation reward plan.. Existing research defines the concept of recommendation reward as: in order to obtain positive reviews from consumers, a series of incentive measures taken by enterprises to make consumers carry out positive word-of-mouth communication [22].

2.5 Online Comments

Online reviews often have multiple effects. In addition to providing potential customers with certain product information, they can also provide merchants with product modification suggestions and directions to a certain extent [23]. Taking Amazon as an example, research has found that the emotion of online reviews has an effect on consumers' shopping. Through related theoretical research, it is found that among the customers who browse these online reviews, if the customer's knowledge level is low, they will be more easily affected by online reviews. If the educational level and income level of potential customers are higher, they will be less susceptible to online reviews.

3 Models and Assumptions

3.1 Build a Model

Establish a referral reward plan model to study the impact of reward types and rewards on consumer online review behavior. At the same time, it studies the interaction between rewards and consumer satisfaction, as well as the influence of consumers' perceived fairness in it. The first stage mainly studied the influence of different types of rewards established by merchants on consumer online reviews. In the second stage, the effect of the amount of incentive measures in the word-of-mouth rewards on consumers' willingness to recommend word-of-mouth is studied. Consumer satisfaction is an important research variable in word-of-mouth communication research. Therefore, in the study of word-of-mouth rewards and consumer review behavior, it is necessary to consider the variable of consumer satisfaction. In the referral reward program, different rewards (high limit; low limit) set up by the merchants and different consumer product satisfaction (very satisfied; generally satisfied; dissatisfied) and the interaction of the two have an impact on consumers' online review behavior Influence. In the third stage, based on the theory of perceived fairness, consumers will consider whether their own payments and reports match during the transaction process, which plays a very important role in consumer product, evaluation and directly relates to whether consumers are willing to accept recommendation rewards. How to set up plans and recommendation reward plans can better influence consumers' appraisal willingness and appraisal behavior.

3.2 Propose Hypothesis

Pure taste and rewards in word-of-mouth rewards will have different effects on consumers' review motives. This article studies the influence of different types of rewards (monetary rewards; non-monetary rewards) established by merchants on consumer online review behavior. In the referral reward program, different rewards (high limit; low limit) set up by the merchants and different consumer product satisfaction (very satisfied; generally satisfied; dissatisfied) and the interaction of the two have an impact on consumers' online review behavior Influence. In the recommendation and reward plan, the different rewards set by the merchants and the interaction of different product satisfactions affect the online review behavior of consumers, and perceive the moderating effect of fairness.

According to the recommendation reward type, it is divided into two types: monetary reward and non-monetary reward. Propose a hypothesis:

H1: Different types of rewards have different effects on consumers' online review behavior.

H1a: Monetary rewards have a positive impact on consumers' online review behavior.

H1b: Non-monetary rewards do not have a positive impact on consumers' online review behavior.

Consumer satisfaction is the definition of willingness to recommend as the willingness of consumers to participate in this kind of recommendation reward behavior, and the willingness of the store to recommend products for this purpose. Existing research shows that in pure word of mouth, when buyers are more satisfied with the product, they will actively promote the products of the merchants. At this time, the merchants provide certain incentives to customers by formulating a recommendation and reward plan, which will further strengthen consumers' willingness to comment. In this case, consumers' intrinsic motivation and incentives to participate in rewards have an impact on the product recommendation. The basis for this view is that when consumers' internal and external behaviors are consistent, they will promote their review behavior. When consumers are satisfied with the product or actively willing to comment, adding external stimulus behavior will increase the possibility of consumer behavior. This article speculates that when consumers are more satisfied with purchased products, the incentive plan set up by the merchant will promote the product evaluation of consumers. Therefore, this article defines consumer product satisfaction as the evaluation of consumers' perception and experience of this product after the purchase of the product, and comparison of their pre-purchase and post-purchase feelings and expectations. If it meets expectations, is satisfied. This paper proposes the following hypotheses:

H2: Consumer satisfaction has a positive effect on consumer online review behavior.

H2a: The reward quota of the recommendation reward program has a positive effect on consumers' online review behavior.
H2b: The interaction between consumer satisfaction and the amount of rewards in the recommendation and reward program has a positive effect on consumers' online review behavior.

The level of rewards in the recommendation reward program will have different effects on consumers' online review behavior. When the amount of reward in the referral reward program is high, regardless of the consumer's product satisfaction, the consumer's willingness to comment online will be improved to a certain extent. When the amount of rewards in the referral reward program is low, consumers with higher product satisfaction (very satisfied and general satisfaction belong to the consumer group with higher satisfaction) and consumers with lower product satisfaction will have Different or even opposite willingness to comment. The adjustment effect of the reward amount in the recommendation reward plan mainly puts forward the following three hypotheses:

H3: The positive impact of consumer product satisfaction on consumers' willingness to comment online is regulated by the level of rewards in the recommendation reward plan.
H3a: When the amount of reward in the reward plan is high, regardless of the consumer's product satisfaction, it will improve the consumer's willingness to comment online to a certain extent.
H3b: When the amount of reward in the reward plan is low, consumers with high product satisfaction (very satisfied and general satisfaction belong to the consumer group with high satisfaction) and consumers with low product satisfaction will have Different or even opposite willingness to comment.

In the research of recommendation and reward plan, the theory of perceived fairness is introduced. Consumers' perception of fairness in accepting merchant's reward behavior and the influence of the relationship strength variables between consumers' product recommendations. It is the first time to introduce the perceptual fairness variable in the research on recommendation and reward programs and consumer online review behaviors. By reflecting the psychological variables of consumers when evaluating behaviors in the recommendation and reward behavior, the research on the theory of perceptual fairness has become more important. The comprehensiveness and meticulousness also make the research conclusions on the recommendation reward program and consumer online review behavior more scientific. This article proposes hypotheses:

H4: Customers' perceived fairness plays a moderating role in the influence of consumer satisfaction and rewards on review behavior. When the consumer's satisfaction with the product is not high, consumers with low perception of fairness are more likely to participate in the recommendation reward plan than consumers with high perception of fairness, and the effect is more significant when the amount of reward is low.

4 Research Design

This paper designs research hypotheses based on the research model. The thesis mainly conducts in-depth research on the research through experimental research methods, designs the questionnaires through the design of relevant variables, and analyzes the collected questionnaire data. SPSS software was used for reliability and validity analysis and regression analysis. In this chapter, the research method is designed in detail.

The experimental research method is the method that many scholars choose in empirical research. Through the effective manipulation of related variables in experimental research, the relationship between variables in the research of thesis is studied. The main variables to be measured in this article are four variables: reward type, reward amount, consumer product satisfaction, and perceived fairness.

This article sets the scenario of the questionnaire to buy clothes online. Through interviews, "8 yuan" is set as a high amount H, and "1 yuan" is set as a low amount L, and the reward amount is divided into a high amount and a low amount. Consumer product satisfaction is measured in three dimensions: very satisfied VS, generally satisfied GS, and dissatisfied NS. And set up a group of no rewards in the setting of the reward limit to compare the consumer's recognition of the recommendation reward plan. A material reward questionnaire with consumer product satisfaction as very satisfied and a high amount of reward was set up to compare with consumer product satisfaction as very satisfied and a monetary reward with a high amount of reward. The data from these two questionnaires are used to test the hypothesis of this model. Therefore, the paper formed ten sets of questionnaires. When conducting the experimental survey, the ten sets of questionnaires were randomly distributed to the experimental subjects. Each subject would Randomly obtain a questionnaire scenario. It is mainly divided into the following three scenarios.

Scenario 1: The consumer's satisfaction with the product is VS. The scenario design: buy a long-awaited dress with a discount price of 149 yuan, which meets their expected price. After receiving the product, it was very high-quality in all aspects and exceeded expectations. This purchase of the product is in line with your expectations.

Scenario 2: The consumer's satisfaction with the product is GS's situational design: buy a long-awaited dress with a discount price of 149 yuan, which meets their expected price. After receiving the product, all aspects are very high-quality and meet the expectations, but after careful inspection, the cuffs are a bit dirty. After cleaning, the appearance and quality of the product are not affected. The final result is still satisfactory. This purchase of the product is more in line with your expectations.

Scenario 3: The consumer's satisfaction with the product is not NS design: buy a long-awaited clothing with a discount price of 149 yuan, which meets their expected price. After receiving the product, the color and quality were not good and did not meet expectations. However, after careful inspection, the cuffs are a bit dirty. After cleaning, the appearance and quality of the product are not affected. The final result is still satisfactory. This product purchase is more in line with your expectations.

This article needs to measure five variables in the recommendation reward plan: reward type and reward amount, consumer product satisfaction, consumer perception fairness, and consumer review willingness. In the following, for each variable, the specific setting of variable measurement indicators is carried out. There are mainly the following aspects.

In this article, the types of rewards in the referral reward plan are divided into monetary rewards and non-monetary rewards. The specific measurement indicators are shown in Table 1.

Table 1. Reward Type Scale Measurement Items

Measured variable	Operational definition	Measurement item
Reward type	Research on the impact of monetary rewards and non-monetary rewards on consumers' online review behavior in the recommendation reward program	This referral reward is attractive to me
		This referral reward is valuable to me
		This recommendation reward is very good

In this paper, the rewards in the recommendation reward plan are divided into high rewards and low rewards. The specific measurement indicators are shown in Table 2.

Table 2. Reward amount measurement table

Measured variable	Operational definition	Measurement item
Reward amount	The amount of reward is the amount of reward provided by the company in order to obtain consumer evaluation behavior in the re-recommendation reward plan	This recommendation reward is small
		This referral reward is generous
		This is a high reward for me

In this article, consumer product satisfaction is divided into three dimensions: very satisfied, generally satisfied, and dissatisfied. The specific measurement indicators are shown in Table 3.

In this paper, perceived fairness is divided into two dimensions: self-pay and return matching and self-pay and return mismatch. The specific measurement indicators are shown in Table 4.

Table 3. Items and Basis for the Measurement of Consumer Satisfaction Scale

Measured variable	Operational definition	Measurement item
Customer satisfaction	Consumers' perception and experience of the product after the purchase of the product, and the evaluation obtained after comparing their feelings and expectations before and after the purchase	If I make a purchase selection again, I will not choose this shop
		I chose this store to purchase, and my choice was correct
		The product purchased this time has met my expectations

Table 4. Measurement items of the Perceptual Equity Scale

Measured variable	Operational definition	Measurement item
Perceived fairness	In the referral reward program, consumers' fair perception of their own contributions and reports	Pay and reward match, participate in comments
		Pay and return do not match, do not participate in comments

Table 5. Measurement items of online comment behavior scale

Measured variable	Operational definition	Measurement item
Online comment behavior	Consumers' commodity review behavior in order to obtain rewards from merchants	After shopping, I often comment on products
		After shopping, the product will not be reviewed

This article divides online comment behavior into willingness to comment and unwillingness to comment. The specific measurement indicators are shown in Table 5.

5 Data Analysis

5.1 Analysis of the Results of the Questionnaire

This article selects 860 college students as experimental subjects. The main reason is that college students are more frequent online shopping. At the same time, the recommendation and reward mechanism is a relatively new marketing method. Compared with the previous ones, the acceptance of young people is less Higher, so the experimental object of this article is selected as a college student in a university in Beijing. Questionnaire data collection was carried out in the form of questionnaires distributed on the questionnaire star. 860 questionnaires were distributed through the questionnaire star. A total of 860 questionnaires were returned. Invalid questionnaires were eliminated. A total of 752 valid questionnaires were collected, with an effective rate of 87.44%.

In the process of designing the questionnaire, basic information statistics were conducted on the questionnaire participants, mainly based on the gender, age, and education level of the questionnaire participants, in order to conduct more comprehensive research on the subject and the participants. The basic situation of the questionnaire was analyzed and discussed, and descriptive statistics were performed on the questionnaire data. The statistical results are shown in Table 6.

According to the results of the table data, the proportion of women participating in the questionnaire survey is higher than that of men, but the difference is not significant. The proportion of women is about ten percentage points higher than that of men. The statistical results of the age of the participants are about one-third of the total number of participants under the age of 20, and about 68.08% of the main population aged 21–30. The questionnaire is distributed to college students. One data also reflects the effective use of the questionnaire. In terms of the educational level of the participants, the proportion of bachelor degree holders is 72.60%, the largest proportion. The survey results of the time spent surfing the Internet on a mobile phone each day show that the proportion of people who practice surfing the Internet for 1–2 h a day accounted for the largest proportion, accounting for 64.62%. The survey results on the frequency of online shopping show that the frequency of online shopping is 1–2 times a month the most, accounting for up to one-third. Most of them belong to school students. The online time is 1–2 h per day, and the shopping frequency is 1–2 times a month.

Table 6. Statistics of the distribution of the basic characteristics of the sample

Project	Category	Number of samples	Composition ratio (%)
Gender	Male	344	45.74
	Female	408	54.25
Age	20 years old and below	244	32.44
	21–30 years	512	68.08
	31 years old and above	16	0.012
Education level	Specialist	100	13.29
	Undergraduate	546	72.60
	Postgraduate/PhD student	106	14.09
Daily mobile internet time	Less than 1 h	176	23.40
	1–2 h	486	64.62
	2 h or more	90	11.96
Frequency of online shopping	1–2 times a week	130	17.28

(continued)

Table 6. *(continued)*

Project	Category	Number of samples	Composition ratio (%)
	1–2 times a month	590	78.72
	1–2 times a year	32	04.25
Frequency of shopping reviews	Never	70	09.30
	Almost not	120	15.95
	Occasionally	208	27.65
	Often	354	47.07

The questionnaire survey method used in this article is the five-point method, and the reliability analysis method for this type of data is the Cronbach's α coefficient method. Therefore, in this article, this method is adopted. In this method, under normal circumstances, 0.7 is used as a boundary value for investigation. When the Cronbach's α obtained from the analysis of the questionnaire data is greater than the boundary value, the set value can be obtained. The reliability of the questionnaire is normal. When the Cronbach's α obtained from the analysis of the questionnaire data is less than the boundary value, it can be concluded that the reliability of the questionnaire needs to be further adjusted. In the reliability test of the data in this article, this principle is mainly used to test the reliability. After operating the questionnaire data collected by the formal questionnaire through SPSS24.0 software, perform Cronbach's α analysis on the measurement items of the five scales in the study. When the Cronbach's α obtained by the analysis of the questionnaire data is greater than the boundary value, then it can be concluded that the reliability of the set questionnaire is high. When the Cronbach's α obtained from the analysis of the questionnaire data is less than the boundary value, the reliability is low. After testing, the variables set in this article have a high degree of convergence. The questionnaire has high credibility.

5.2 Hypothesis Testing and Analysis

This article mainly analyzes the collected data through SPSS24.0 software, and mainly conducts regression analysis of reliability and validity test. The results of these analyses are used to verify the hypothesis of the article and draw the research conclusions.

The reward methods in this article are mainly divided into monetary rewards and non-monetary rewards. The relationship between monetary rewards and non-monetary rewards and consumer online review behavior is analyzed. The regression analysis results are shown in Table 7 below.

The data in Table 7 shows that the type of reward has a significant positive impact on consumers' online review behavior. $F = 4.844$, $p < 0.001$, indicating that the regression equation fits well, the regression equation has a significant linear relationship, and different types of rewards will have different effects on consumers' online review behavior. Monetary rewards can promote consumers' recommendation and reward behaviors, while material rewards have a negative impact on consumers' online review behaviors,

Table 7. Regression analysis results of reward type variables

Variable	Non-standardized coefficient (B, standard error)		Standard coefficient (Beta value)	T value	Sig	*AdjustmentR*^{2}	F value
Monetary reward	0.129	0.039	0.230	3.329	0.000	0.015	4.844
Non-monetary rewards	0.000	0.042	0.001	0.009	0.992		

and cannot well trigger consumers' online review behaviors. So suppose H1 is verified. Assume that H1a is supported. Assume that H1b is not supported.

In this paper, the results of the regression analysis of the interactive effect of product satisfaction and the rewards of the recommendation reward program on the review behavior on the online review behavior of consumers are shown in Table 8.

Table 8. Regression analysis results of consumer satisfaction and reward amount

Variable	Non-standardized coefficient (B, standard error)		Standard coefficient (Beta value)	T value	Sig	Adjustment	F value
Customer satisfaction	0.835	0.078	0.289	3.329	0.000	0.742	35.288
Award amount	0.345	0.056	0.127	0.009	0.000		
Interaction between the two	–0.178	0.057	0.278	–5.785	0.000		

The regression analysis results in Table 8 show that consumer product satisfaction has a positive impact on online review behavior, $\beta = 0.835$, $p < 0.001$, assuming that H2 is supported, consumers with higher product satisfaction are willing to participate in recommendation rewards the higher the willingness to plan to comment. Recommendation rewards have a significant positive impact on consumers' online review behavior, $\beta = 0.345$, $p < 0.001$, assuming H2a is supported. The results show that when the amount of reward in the recommendation reward plan set up by the merchant is higher, the more likely it is for consumers to conduct online reviews. Finally, the interaction between consumer product satisfaction and recommendation rewards is verified, $\beta = -0.178$, $p < 0.001$, assuming H2b is supported. The results show that in the face of different product satisfaction, different rewards will have different effects on consumers' online review behavior.

The low-value rewards in the referral reward program can even bring more benefits to the merchants than the high-value rewards. The high-value rewards will only have a greater effect on consumers who are less satisfied with the product. Therefore, it can be concluded that hypothesis H5 is partially verified. Also, assume that H3a is true, and assume that H3b is partially true. When considering the product satisfaction of consumers, when only considering the VS and GS scenarios, there will be different findings, but from the data analysis of the level of rewards alone, the willingness to comment on consumption is mainly affected by the product satisfaction. Consumers who are more satisfied with the product will be more willing to participate in the recommendation reward program. At the same time, for consumers who are dissatisfied with the product, the setting of low rewards may have a worse adverse effect.

The moderating effect of perceived fairness has been initially confirmed. In addition, the results also support that perceived fairness has a negative impact on consumers' online review behavior under the influence of the recommendation reward program. Hypothesis H4 is verified. When the consumer's product satisfaction is not high, consumers with low perception of fairness are more likely to receive the difference in the amount of reward in the recommendation reward program, which will affect their online review behavior.

Through the above-mentioned data analysis and hypothesis testing, certain research conclusions have been reached for the three-phase research set up in the thesis. Through empirical research, different types of rewards (monetary rewards; non-monetary rewards) established by merchants have different effects on consumers' online review behavior, among which monetary rewards have a positive impact on consumers' online review behavior. When the consumer's satisfaction reaches a certain level, the low rewards in the referral reward program will bring even more benefits to the merchants than the high rewards. The high rewards are only for those who are less satisfied with the product. Of consumers will have a greater effect. However, from the data analysis of the level of rewards alone, the willingness to comment on consumption is mainly affected by the satisfaction of its products. The more satisfied consumers are with the product, the more willing they will be to participate in the referral reward program. At the same time, for consumers who are dissatisfied with the product, the setting of low rewards may have a worse adverse effect. Through empirical research, consumers with higher perceived fairness, when faced with high rewards, their perceived fairness will weaken the influence of this kind of money. Especially in the face of lower rewards, the influence of perceived fairness will be greatly enhanced. Therefore, when the amount of reward set by the recommendation reward program is low, consumers with high perception of fairness are often unwilling to participate in the recommendation reward program.

6 Conclusion

In this paper, the research on consumer online review behaviors based on the recommendation reward program is divided into three stages. First of all, the different types of rewards (monetary rewards; non-monetary rewards) established by merchants affect the behavior of consumer online reviews. Through empirical research, monetary rewards will have a positive impact on consumers' online review behavior. The second stage studied the influence of different rewards in the recommendation reward plan, consumer product satisfaction, and the interaction between the two on consumer online review behavior. The reward amount is divided into high rewards and low rewards. Consumer products Satisfaction is divided into three levels: very satisfied, generally satisfied, and dissatisfied. Through empirical research, it is verified that consumer product satisfaction, recommendation reward plan rewards, and the interaction between product satisfaction and rewards have a positive and significant effect on consumer online review behavior. The third stage mainly studied the moderating effect of perceived fairness in the process of consumer satisfaction and rewards affecting consumer online review behavior. Through empirical research, it is believed that consumers with higher perception of publicity, when faced with high rewards, will weaken the influence of this kind of money due to their own perceived fairness, especially when facing lower rewards. The influence of fairness will be greatly strengthened. Therefore, when the amount of reward set by the recommendation reward program is low, consumers with high perception of fairness are often unwilling to participate in the recommendation reward program. Although certain research results have been obtained, there are still deficiencies in the research of the thesis, and there are limitations in the research samples, which can be further studied in other groups in future research.

Acknowledgment. This research was supported by the Heilongjiang Philosophy and Social Science Fund Project (21GLC186).

References

1. Wirtz, J., Chew, P.: The effects of incentives, deal proneness, satisfaction and tie strength onword-of-mouth behavior. Int. J. Serv. Ind. Manag. **13**(2), 141 (2002)
2. Chatterjee, P.: Online Reviews: Do Consumers Use Them? Advances in Consumer Research, 2001,28
3. Park, D.H., Lee, J.: e WOM overload and its effect on consumer behavioral intention depending on consumer involvement. Electron. Commer. Res. Appl. **7**(4), 386–398 (2009)
4. Ngo-Ye, T.L., Sinha, A.P.: The influence of reviewer engagement characteristics on online review help ulness: a text regression model. Decis. Support Syst. **61**, 47–58 (2014)
5. Xuemei, D., Jingyu, D., Zhihong, X., et al.: Research on the influence of online reviews on consumer purchase intention. Manage. Rev. **2**(3), 173–183 (2016)
6. Chen, Y., Xie, J.: Online consumer review: word-of-mouth as a new element of marketing communication mix. Manage. Sci. **54**(3), 477–491 (2008)
7. Peters, F., Pumplun, L., Buxmann, P.: Opening the Black Box: Consumer's Willingness to Pay for Transparency of Intelligent Systems. Publications of Darmstadt Technical University, Institute for Business Studies (BWL) (2020)
8. Yoshida, T., Blair, A., D'Alessandro, A., et al.: Enhancing uniformity and overall quality of red cell concentrate with anaerobic storage. Blood transfusion = Trasfusione del sangue **15**(2), 172–181 (2017)
9. Palm, E., Seubert, C., Glaser, J.: Understanding employee motivation for work-to-nonwork integration behavior: a reasoned action approach. J. Bus. Psychol. **3**, 1–14 (2019)
10. Wich, S.: Former employees' negative online comments are unprotected under the act. Manage. Report Nonunion Organizations **43**(1), 3–4 (2020)

11. Filieri, R.: What makes an online consumer review trustworthy? Annals of Tourism Research 58 (2016)
12. Pramudya, R.C., Seo, H.S.: Hand-feel touch cues and their influences on consumer perception and behavior with respect to food products: a review. Foods **8**(7) (2019)
13. Sheldon, K.M., Prentice, M.: Self-determination theory as a foundation for personality researchers. J. Pers. **87**(1), 5–14 (2019)
14. Schwatka, N.V., Sinclair, R.R., Fan, W.: How does organizational climate motivate employee safe and healthy behavior in small business? a self determination theory perspective. J. Occup. Environ. Med. **62**(5), 1 (2020)
15. Mahadi, N., Abdullah, S.N., Baskaran, S.: A review of social cognitive theory and self-care for type 2 diabetes mellitus patients. Business Management and Strategy, 11 (2020)
16. Tohmé, F., Auday, M.: Social choice situations and aggregation criteria: inductive reasoning in social choice theory. J. Logic Lang. Inf. **28**(4), 551–575 (2019)
17. Park, D.H., Kim, S.: The effects of consumer knowledge on message processing of electronic word-of-mouth via online consumer reviews. Electron. Commer. Res. Appl. **7**(4), 399–410 (2008)
18. Xia, D.S., Lu, N.Q.: Impact of tourism virtual community atmosphere on the willingness of online word-of-mouth recommendation. Scientific and Technological Management of Land and Resources (2019)
19. Yi, X., Siqing, P.: The influence of brand trust and brand emotion on word-of-mouth communication: the role of attitude and attitude uncertainty. Manage. Rev. **26**(02), 80–91 (2014)
20. Wirtz, J., Chew, P.: The effects of incentives, deal proneness, satisfaction and tie strength on word-of-mouth behaviour. Int. J. Serv. Ind. Manag. **13**(2), 141 (2002)
21. Ajorlou, A., Jadbabaie, A.: Dynamic Pricing in Social Networks: The Word-of-Mouth Effect. Management science: Journal of the Institute of Management Sciences (2018)
22. Chevalier, J.A., Mayzlin, D.: The effect of word of mouth on sales: online book reviews. J. Marketing Res. (JMR) **43**(3), 345–354 (2006)
23. Shanks, D.R.: Misunderstanding the behavior priming controversy: comment on Payne, Brown-Iannuzzi, and Loersch (2016). J. Exp. Psychol. Gen. **146**(8), 1216–1222 (2017)

Machine Learning Algorithms for Gold Price Prediction

Duyen Mai Thi Nguyen[1], Narayan C. Debnath[1], Luyl-Da Quach[2], and Vinh Dinh Nguyen[1](✉)

[1] School of Computing and Information Technology, Eastern International University, Thủ Dầu Một, Vietnam
{duyen.nguyen.set18,narayan.debnath,vinh.nguyen}@eiu.edu.vn
[2] Department of Software Engineering, FPT University , Can Tho City, Vietnam
daql@fpt.edu.vn

Abstract. Due to its potential for reliably anticipating gold prices, machine learning for gold price prediction has become a hot study topic in recent years, which is essential for investors and policymakers. This research did a comprehensive survey and experiments on various machine learning algorithms, such as linear regression, support vector machines, decision trees, gaussian process regression, random forest regression, polynomial features, lasso regression, k-nearest neighbor, ridge regression and neural networks, to gold price prediction. These models have been able to provide accurate predictions of gold prices, often outperforming traditional statistical methods. However, the quality of the input data and the model selection procedure affect how accurate these models are. Comprehensive experimental results showed that the support vector machine algorithm achieved the most accurate predictions, with the lowest error rates and the best R2 score.

Keywords: Gold price prediction · Algorithms · Machine learning · Neural networks · Support vector machines · Model selection · Decision trees

1 Introduction

Gold has been one of the most valuable metals throughout history, with a diverse range of uses in currency, jewelry, and industrial applications. The price of gold is influenced by many factors, including global events, economic indicators, and investor sentiment. As a result, accurately predicting the future price of gold is a challenging task, and traditional forecasting models often fail to capture the complex dynamics of gold markets.

Recent advances in machine learning techniques offer a promising approach to predicting financial markets' performance, including gold prices. There has been a lot of research on gold price prediction methods before, as listed in Table 1. In this paper, we compare the effectiveness of ten different machine-learning algorithms for predicting gold prices. Specifically, we examine Linear Regression, Support Vector Regression (SVR) [1], Neural Network [2, 3], Decision Tree, Gaussian Process Regression (GPR), Random Forest Regression, Polynomial Features, Lasso Regression, K-Nearest Neighbors (KNN), and Ridge Regression.

A. Hassanien et al. (Eds.): AISI 2023, LNDECT 184, pp. 212–220, 2023.
https://doi.org/10.1007/978-3-031-43247-7_19

Our study aims to provide insights into the performance of these algorithms for predicting gold prices, comparing their accuracy, robustness, and speed. To achieve this, we use historical data on gold prices, as well as relevant economic and financial indicators, to train and test our models. To determine the best strategy for forecasting gold prices, we compare each algorithm's performance over a range of market scenarios and time periods. Linear Regression is a commonly used technique in financial forecasting, and we use it as a baseline to compare the performance of other algorithms. Support Vector Regression (SVR) is another widely used algorithm that can capture non-linear relationships between variables. Neural Network is a powerful machine learning algorithm that can capture complex patterns and relationships within data. A Decision Tree is a simple yet effective algorithm that can handle both categorical and continuous data. Gaussian Process Regression (GPR) is a probabilistic algorithm that can model uncertainty in data. An ensemble learning technique called random forest regression combines the forecasts of many decision trees to increase accuracy. Polynomial Features is a technique that transforms data to capture non-linear relationships. Lasso Regression is a regularization technique that can reduce overfitting in models. A non-parametric method called K-Nearest Neighbors (KNN) may detect non-linear correlations. Finally, Ridge Regression is another regularization technique that can improve model stability.

By comparing the performance of these algorithms, our study aims to provide insights into the most effective approach for predicting gold prices. The results of our research could have significant implications for investors, traders, and policymakers seeking to make informed decisions about gold investments or gold-backed policies. Ultimately, our goal is to contribute to the growing body of research on machine learning applications in finance and provide insights into the dynamics of gold prices.

Section 2 summarizes the existing research related to gold price prediction and its limitations. Section 3 evaluated and discussed the performance of existing machine learning algorithms for a gold price prediction. Finally, we conclude the paper in Sect. 4.

Table 1. Survey On Existing Gold Price Forecast Method

Authors	Method	Description	Experiments
J. Wang et al. [4]	Decision tree, genetic algorithm, support vector machine (SVM)	A new forecast model named SGRU-AM, based on the special gated recurrent unit (SGRU) and attention mechanism (AM) is proposed in this paper to tackle these challenges	Dataset: daily settlement price, the open position, the trading volume, and the gold spot price in China Accuracy: from 94% to 98.9%

(*continued*)

Table 1. (*continued*)

Authors	Method	Description	Experiments
N. Tripurana et al. [5]	Random Forest regression, Decision tree, Support vector regression, linear regression, and Artificial neural network	The goal of "Gold price prediction" is to forecast gold's price using a variety of Machine learning techniques, considering the relationship between several economic factors that influence gold rates	Dataset: taken from the Kaggle website, dated from January 2008 to May 2018 Accuracy: from 65% to 9%
M. Sravani et al. [6]	Multiple linear regression, Random Forest, and Gradient boosting	To analyze the correlation between the economic variables and the gold prices, there are several studies available. The project "Gold price prediction" aims to predict the price of gold using machine learning	Dataset: taken from the Kaggle website, dated from January 2008 to December 2018 Accuracy: 99.77% and 98.59%
R. Ghule et.al. [7]	Random Forest, and linear regression	Predict future gold rates supported by twenty two market variables victimization machine learning technique	Dataset: Data for this study is collected from (kaggle.com) Dec 2011 to September 2016 Accuracy: 97%
K. A. Manjula et. al. [8]	Random Forest, Gradient boosting, Linear Regression	To understand the relationship between the gold price and selected factors influencing it, namely the stock market, crude oil price, rupee-dollar exchange rate, inflation, and interest rate	Dataset: Spot gold price in rupees per ounce is used to represent gold price. Monthly data from Jan 2000 to Dec 2018 were collected Accuracy: more than 95%

(*continued*)

Table 1. (*continued*)

Authors	Method	Description	Experiments
C. Liu et. al. [9]	BP neural network,	Gold futures' research has important theoretical and practical significance. By using genetic algorithms (GA) to optimize the value of the linking weight of the BP neural network	Dataset: 0804 gold futures of the American New York Futures Exchange and the daily closing prices of 240 days from November 16, 2006, to December 20, 2007 Accuracy: reaches the goal of 0.001 error
S. Patalay et. al. [10]	M5P model tree Machine learning algorithm	Develop a system which is based on machine learning algorithms to predict the gold prices based on the historical data related to other closely related commodities and stock market indicators	Dataset: count of 5046 values consisting of 20 years of data in google accuracy: 85%

2 Related Works and Methods

2.1 Linear Regression

Linear regression [11] is a basic and widely used machine learning algorithm for predicting a continuous outcome variable from one or more input variables. A basic linear regression model with only one input variable has the following formula:

$$y = b_0 + b_1 x \tag{1}$$

where: b_0 is the bias term's intercept, b_1 **is the input variable's coefficient,** x is the input variable, and y is the expected outcome variable. In a multiple linear regression model with multiple input variables, the formula becomes:

$$y = b_0 + b_1 x_1 + b_2 x_2 + \cdots + b_n x_n \tag{2}$$

where: x_i is the ith input variable, b_i **is its coefficient, and** b_0 is the bias term or intercept, y is the predicted output variable. The results of the analysis showed that the Linear regression algorithm had an MAE of 22.47, MSE of 1127.4, MedAE of 14.48, explained variance score of 1.0, and R2 score of 1.0.

2.2 Support Vector Regression (SVR)

SVR [12] seeks to reduce the difference in output between expected and actual output while also maintaining a maximum margin between the predicted output and the upper and lower bounds. The formula for the SVR model can be written as:

$$y = b_0 + \sum_{i=1}^{n} a_i * K(x_i, x) \quad (3)$$

where y is the predicted outcome variable, ai is the coefficients for the support vectors, $K(x_i, x)$ is the kernel function that transforms the input data to a higher-dimensional space, b_0 is the intercept or bias term, and n is the number of support vectors.

2.3 Neural Network

Neural networks [13] is a class of machine learning algorithms modeled after the structure and function of the human brain. The formula for a feedforward neural network with one hidden layer can be written as:

$$y = f * (w_2 * f(w_1 * x + b_1) + b_2 \quad (4)$$

where f is the activation function applied to each hidden layer neuron, y is the predicted outcome variable, x is the input variable, and w_1 and w_2 are weight matrices and bias vectors for the input layer and hidden layer, respectively.

2.4 Decision Tree

Decision trees [14] are machine learning algorithms that create a tree-like model of decisions and their possible consequences. The formula for a decision tree model can be represented as a series of if-then statements:

if (input variable <= threshold value)
then (output = value)
else (output = value)

where the input variable is the input variable being evaluated, the threshold value is the value used to split the tree, the output is the predicted outcome variable value is the output value associated with the branch of the tree.

2.5 Gaussian Process Regression

With fresh input data points, the GPR [15] model may be used to forecast the value of the target variable by modeling the covariance between the input data points.

$$y \sim N\left(K(X, X) + \sigma_n^2 * \mathrm{I})\right)^{-1} * Y \quad (5)$$

where y is the predicted outcome variable, *K(X, X)* is the covariance matrix of *X*, σ_n^2 is the noise variance term, *I* is the identity matrix, and *Y* is the target variable.

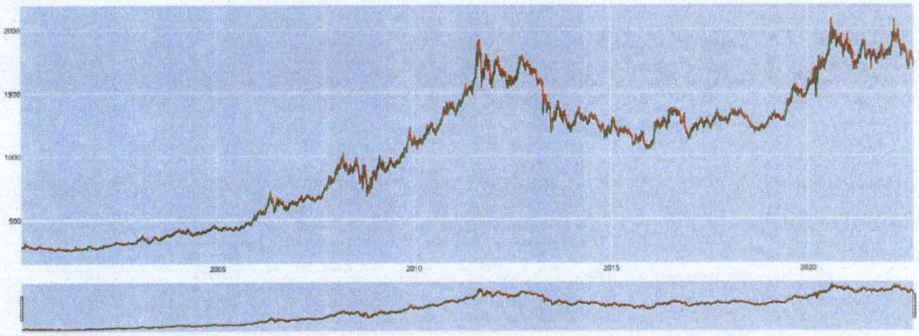

Fig. 1. Gold Futures Candlestick Chart From 2005 to 2020

2.6 Random Forest Regression

Random Forest Regression [16] is a machine learning algorithm that uses an ensemble of decision trees to make predictions. At training time, the algorithm creates a large number of decision trees, and as the final prediction, it produces the mean prediction of each tree.

$$y = \frac{1}{T} * \sum\nolimits_{i=1}^{T} f_i(x) \tag{6}$$

where y is the anticipated result variable, T is the total quantity of decision trees in a forest, and $f_i(x)$ represents the prediction made by the i_{th} decision tree at input variable x.

3 Experimental Results and Discussion

Between 2005 and 2020, the gold market's value rose significantly as shown in Fig. 1, driven by a mix of financial, political, and economic forces. While the price of gold can be erratic in the near term, it is typically regarded as a safe-haven asset and a hedge against inflation, making it an attractive investment for many investors.

Mean Absolute Error (MAE), Mean Squared Error (MSE), Median Absolute Error (MedAE), and R2 Score are the assessment metrics used to evaluate the performance of the models. These indicators are frequently used to gauge a regression model's degree of accuracy in Table 2.

$$MAE = \frac{\sum_{i=1}^{n} |y_i - x_i|}{n}$$

$$MSE = \frac{1}{n}\sum_{i=1}^{n} (|X_i - X^{\sim}|).$$

$$MAD = \text{Median}(|\mathrm{X_i} - X^{\sim}|)$$

Table 2. Experimental Results By Using Various Machine-Learning Methods

Method	MAE	MSE	MAD	R2 Score
Linear Regression [11]	22.47	1127.4	14.84	1.0
Support Vector Regression [12]	22.33	1126.72	14.7	1.0
Neural Network [13]	2048.7	5212109	2370	19.03
Decision Tree [14]	142.88	53735.3	65.1	0.79
Gaussian Process Regression [15]	22.41	1130.42	14.73	1.0
Random Forest Regression [16]	23.51	1223.6	15.22	1.0
Polynomial features [17]	22.47	1127.4	14.84	1.0
Lasso Regression [18]	22.47	1127.4	14.84	1.0
K-Nearest Neighbors [19]	24.48	1336.99	15.37	0.99
Ridge regression [20]	24.48	1336.99	15.37	0.99

$$R^2 = 1 - \frac{SS_{RES}}{SS_{TOT}} = 1 - \frac{\sum_i (y_i - y_i)^2}{\sum_i (y_i - y^-)^2}$$

Both the Support Vector Regression and Linear Regression models perform similarly, with extremely low MAE, MSE, and MAD values and an ideal R2 score of 1.0. These models appear to have quite successfully suited the data. Low MAE, MSE, and MAD values, as well as an ideal R2 score of 1.0, were other indicators of the Gaussian Process Regression model's success.

The MAE, MSE, and MAD values for the Random Forest Regression, Polynomial features, and Lasso Regression models are nearly identical, and all three have a flawless R2 score of 1.0. These models choose the most important attributes and address overfitting using various methods. The Neural Network model performs poorly on this dataset, as seen by its low R2 score of 19.03 and much higher MAE, MSE, and MAD values when compared to the other models. The number of hidden layers, the number of nodes in each layer, and the activation functions all have an impact on how well this model performs since it employs a complex network of interconnected nodes. The Decision Tree model had a lower R2 score of 0.79 and comparatively higher MAE, MSE, and MAD values, which may indicate that the model overfits the data or did not adequately account for its complexity. With greater MAE, MSE, and MAD values as well as lower R2 scores, the K-Nearest Neighbors and Ridge Regression models may not be the best fit for this dataset.

4 Conclusion

This paper delved into the application of ten distinct machine-learning algorithms in predicting gold prices. The algorithms evaluated included Linear Regression, Support Vector Regression (SVR), Neural Network, Decision Tree, Gaussian Process Regression

(GPR), Random Forest Regression, Polynomial Features, Lasso Regression, K-Nearest Neighbors, and Ridge Regression. Each algorithm's performance is evaluated using a variety of standard measures, including Mean Squared Error (MSE), Mean Absolute Error (MAE), Median Absolute Error (MAD), and R2 score.

The results showed that the SVR algorithm achieved the most accurate predictions, with the lowest error rates and highest R2 score. However, the optimization of these models and the incorporation of additional features remains a topic of interest for future research to enhance the precision of gold price prediction using machine learning. Further research is needed to develop more robust models that can capture the complexity and volatility of gold prices, particularly during periods of market instability.

Acknowledgments. The authors would like to thank Eastern International University (EIU) Vietnam for funding this research.

References

1. Nguyen, M.T.-T., Nguyen, V., Dinh, Jeon, J.W.: Real-time pedestrian detection using a support vector machine and stixel information. In: 2017 17th International Conference on Control, Automation and Systems (ICCAS), Jeju, Korea (South), pp. 1350–1355 (2017). https://doi.org/10.23919/ICCAS.2017.8204203
2. Nguyen, V.D., Ngo, N.P., Le, Q.N., Debnath, N.C.: Robust plant leaves diseases classification using efficientnet and residual block. In: So-In, C., Londhe, N.D., Bhatt, N., Kitsing, M. (eds) Information Systems for Intelligent Systems. Smart Innovation, Systems and Technologies, vol 324. Springer, Singapore (2023). https://doi.org/10.1007/978-981-19-7447-2_12
3. Chau, D.H., et al.: Plant leaf diseases detection and identification using deep learning model. In: Hassanien, A.E., Rizk, R.Y., Snášel, V., Abdel-Kader, R.F. (eds.) The 8th International Conference on Advanced Machine Learning and Technologies and Applications (AMLTA2022). AMLTA 2022. Lecture Notes on Data Engineering and Communications Technologies, vol 113. Springer, Cham (2022). https://doi.org/10.1007/978-3-031-03918-8_1
4. Wang, J., Li, Y., Wang, T., Li, J., Wang, H., Liu, P.: A gold futures price forecast model based on SGRU-AM. IEEE Access **9**, 146745–146754 (2021). https://doi.org/10.1109/ACCESS.2021.3122140
5. Tripurana, N., Kar, B., Chakravarty, S., Paikara, B.K., Satpathy, S.: Gold price prediction using machine learning techniques. In: ACI'22: Workshop on Advances in Computation Intelligence, Concepts & Applications at ISIC 2022, May 17–19, Savannah, United States (2022)
6. Sravani, M., Abhilash, Ch., Divya, T., Vasthav, Ch., Priyanka, D.: Gold price prediction. Int. J. Creative Res. Thoughts (IJCRT), 2021 IJCRT|Volume 9, Issue 6 June 2021 | ISSN: 2320-2882
7. Ghule, R., Gadhave, A.: Gold price prediction using machine learning. Int. J. Sci. Res. Eng. Manage. (IJSREM) **06**(06), June 2022
8. Manjula, K.A., Karthikeyan, P.: Gold price prediction using ensemble based machine learning techniques. In: 2019 3rd International Conference on Trends in Electronics and Informatics (ICOEI), Tirunelveli, India, 2019, pp. 1360–1364 (2019). https://doi.org/10.1109/ICOEI.2019.8862557
9. Liu, C.: Price forecast for gold futures based on GA-BP neural network. In: 2009 International Conference on Management and Service Science, Beijing, China, 2009, pp. 1–4 (2009). https://doi.org/10.1109/ICMSS.2009.5302242

10. Patalay, S., Bandlamudi, M.R.: Gold price prediction using machine learning model trees. In: International Conference on Changing Business Paradigm (ICCBP), MDI, Murshidabad (2021)
11. Venkatesh, A., Saravanan, M.S.: An efficient method for predicting linear regression with polynomial regression. In: 2022 3rd International Conference on Smart Electronics and Communication (ICOSEC), Trichy, India, pp. 1603–1606 (2022). https://doi.org/10.1109/ICOSEC54921.2022.9952049
12. Chen, D.: Selecting methods and its application for partners of agricultural supply chain based on Support Vector Regression. In: 2011 International Conference on Computer Science and Service System (CSSS), Nanjing, 2011, pp. 1676–1679 (2011). https://doi.org/10.1109/CSSS.2011.5974669
13. Jeon, H.-J., Nguyen, V.D., Duong, T.T., Jeon, J.W.: A deep learning framework for robust and real-time taillight detection under various road conditions. IEEE Trans. Intell. Transp. Syst. **23**(11), 20061–20072 (2022). https://doi.org/10.1109/TITS.2022.3178697
14. Abdelhalim, A., Traore, I.: A new method for learning decision trees from rules. In: 2009 International Conference on Machine Learning and Applications, Miami, FL, USA, pp. 693–698 (2009). https://doi.org/10.1109/ICMLA.2009.25
15. Sui, Y., Zhang, L.: Visual tracking via locally structured gaussian process regression. IEEE Signal Process. Lett. **22**(9), 1331–1335 (2015). https://doi.org/10.1109/LSP.2015.2402313
16. Dong, P., Peng, H., Cheng, X., Y., Xing, X.Z., Huang, D.: A random forest regression model for predicting residual stresses and cutting forces introduced by turning IN718 alloy. In: 2019 IEEE International Conference on Computation, Communication and Engineering (ICCCE), Fujian, China, 2019, pp. 5–8 (2019). https://doi.org/10.1109/ICCCE48422.2019.9010767
17. Schindler, A., Maier, G.: Object detection in gray scale images based on invariant polynomial features. In: 2010 IEEE International Conference on Image Processing, Hong Kong, China, 2010, pp. 4633–4636 (2010). https://doi.org/10.1109/ICIP.2010.5649524
18. Wei, H., Wang, R., Guo, Z., Yang, Z., Wang, J.: Application of Lasso regression model in the analysis of success probability of electronic experiments. In: 2021 IEEE International Conference on Computer Science, Electronic Information Engineering and Intelligent Control Technology (CEI), Fuzhou, China, pp. 644–649 (2021). https://doi.org/10.1109/CEI52496.2021.9574578
19. Okfalisa, I.G., Mustakim, Reza, N.G.I.: Comparative analysis of k-nearest neighbor and modified k-nearest neighbor algorithm for data classification. In: 2017 2nd International conferences on Information Technology, Information Systems and Electrical Engineering (ICITISEE), Yogyakarta, Indonesia, pp. 294–298 (2017). https://doi.org/10.1109/ICITISEE.2017.8285514
20. Zhang, H., Zhang, S., Yin, Y.: An improved ELM algorithm based on EM-ELM and ridge regression. In: Sun, C., Fang, F., Zhou, ZH., Yang, W., Liu, ZY. (eds) Intelligence Science and Big Data Engineering, IScIDE 2013. LNCS, vol. 8261. Springer, Heidelberg (2013). https://doi.org/10.1007/978-3-642-42057-3_95

Numerical Similarity Measures Versus Jaccard for Collaborative Filtering

Hassan I. Abdalla[1], Yasmeen A. Amer[2], Loc Nguyen[3], Ali A. Amer[1,4](✉), and Basheer Mohamad Al-Maqaleh[2]

[1] College of Technological Innovation, Zayed University, Abu Dhabi, United Arab Emirates
[2] Faculty of Computers and Informatics, Thamar University, Thamar, Yemen
[3] Loc Nguyen's Academic Network, Long Xuyên, Vietnam
[4] Computer Science Department, Taiz University, Taiz, Yemen
aliaaa2004@yahoo.com, aliaaa2004@gmail.com

Abstract. Collaborative filtering (CF) is an important method for recommendation systems, which are employed in many facets of our lives and are particularly prevalent in online-based commercial systems. The K-nearest neighbors (KNN) technique is a well-liked CF algorithm that uses similarity measurements to identify a user's closest neighbors in order to quantify the degree of dependency between the respective user and item pair. As a result, the CF approach is not only dependent on the choice of the similarity measure but also sensitive to it. However, some traditional "numerical" similarity measures, like cosine and Pearson, concentrate on the size of ratings, whereas Jaccard, one of the most frequently employed similarity measures for CF tasks, concerns the existence of ratings. Jaccard, in particular, is not a dominant measure, but it has long been demonstrated to be a key element in enhancing any measure. Therefore, this research focuses on presenting novel similarity measures by combining Jaccard with a multitude of numerical measures in our ongoing search for the most effective similarity measures for CF. Both existence and magnitude would benefit the combined measurements. Experimental results demonstrated that the combined measures are superior, surpassing all single measures across the considered assessment metrics.

Keywords: Collaborating Filtering · Similarity Measure · Jaccard · K-Nearest Neighbor · Recommendation Systems

1 Introduction

With the explosive increase of digital data, recommendation systems (RS) are becoming increasingly necessary to filter the constantly-accumulating flow of information and data online. RS should have the ability to offer the desired recommendations based on the user's preferences [1]. A recommendation system is a system that suggests items for users from a large number of items already available in a database. There are generally three basic methods for making recommendations: collaborative filtering, content-based filtering (CBF), and hybrid-based techniques [2]. CBF exploits the user's information

A. Hassanien et al. (Eds.): AISI 2023, LNDECT 184, pp. 221–229, 2023.
https://doi.org/10.1007/978-3-031-43247-7_20

(gender, age, etc.) to predict his preferences without considering other users' information. On the other hand, CF suggests the item to any user if her/his neighbors are concerned in the same item. Generally speaking, CF-based RS has two models: memory-based and model-based methods [3]. The memory-based approach makes use of the full database to identify a group of users or items that are comparable to the target user or item. While the user-based method seeks to construct a model of user behavior to forecast his selections, On the one hand, the memory-based method usually provides higher precision, whereas on huge datasets, the model-based approach is more effective. The model-based approaches, on the other hand, include training a model and fine-tuning a number of hyper-parameters [4–8].

In general, CF-based RS frequently functions by contrasting the present case with precedent cases. It investigates the actions of a group of users who are comparable to the current user or the traits of a group of objects that are similar to the specific item. The degree of similarity between users increases as they share more common items. User-based and item-based CF are the two primary approaches of CF, respectively. In user-base CF, the flow of incoming items is chosen via a user x user matrix based on the evaluation given by other users who have already expressed an interest in the items. Therefore, if an item has been deemed interesting by a related user, it will immediately be recommended to other people who share that perspective. However, in the item-based model, using an item x item matrix, RS will recommend products that are strikingly similar to those that have already been evaluated and given the highest possible score by the current "target" user. The degree of similarity between the item and its neighbour is then used to forecast the rating. That is, the higher the similarity is, the more the forecasted rating is analogue to the rating's neighbors [10–12].

Therefore, this study's main contribution is that it suggests new similarity measures by combining the Jaccard measure with some numerical measures. The key aim is to experimentally prove that jaccard measures can effectively contribute to enhancing the performance of numerical measures. To this end, on the Movie-lens and Film Trust datasets, numerous experiments have been run using both user-based and item-based models. According to the results, the combined measures are superior, beating all single measures over the assessment metrics. It is worth repeating that the scope of this paper is restricted to illustrating the vital impact of the Jaccard similarity measure on CF.

The remainder of this research is planned as follows. In Sect. 2, the previous works are explored in detail. In Sect. 3, the proposed Jaccard combinations are concisely introduced. Section 3 sets out the experimental setup. Section 4 shows the detailed results of Jaccard combinations. Finally, the conclusion and future work suggestions are stated in Sect. 5.

2 Literature Review

Based on the singularity factor. While [11] proposed Context Based Rating Prediction (CBRP) to determine the context score of each nominated user for the considered "user-item" pair. The mean-jaccard-differences (MJD) were introduced in [9] to use the singularity of user evaluations to improve on traditional similarity measurements. In [13], a composite similarity measure was introduced by combining various well-known similarity metrics (PCC, cosine, and other distance metrics). A similarity measure called (IPWR), short for "improved PCC weighted with user rating preference behavior (RPB)," was drawn in [14] by combining PCC with RPB. The IPWR was seen as superior to the state-of-the-art measures. PIWR was shown to be competitive.

Comparably, [15, 16] created novel methods for choosing neighbors based on the intersection and union of their neighborhoods. These metrics, like JAccard's, relied on shared objects to identify the neighbours of the users of interest. When there are non-shared things between the intended users, the similarity between items would be useless, making these measurements invalid in this situation. On the other hand, employing the Bhattacharyya coefficient, several studies have been dedicated to solving the dilemma of data sparsity [17–19]. Meanwhile, [20] presented a subspace clustering-driven measure to address both the problems of sparsity and dimensionality. On the same page, a fast neighbour user searching (FNUS) approach was proposed in [21] to enhance RS performance via creating item subspaces for interested, uninterested (NINU), and disinterested items. The three subsets of the target User's neighbours were then identified using the co-rated item numbers between the target user and other users. The three neighbour user subsets are combined to find the final neighbor user set. The FNUS was seen to have competitive rendering on huge datasets. In [22], a linear combination is suggested. The preferences and local context of users' behaviour, as well as the proportion of shared ratings between each user pair, were taken into account using PSS, Bhattacharya Coefficient, and Jaccard. New measures were presented in [23, 24] to improve RS accuracy based on global user preference and context information.

A Wasserstein collaborative filtering (WCF) method was put forth in [25]. By minimizing the Wasserstein distance under the user embedding requirement, WCF predicted user preference for cold-start items. Ahn [26] looked at the shortcomings of conventional similarity measurements in CF. The author then created a heuristic similarity measure called PIP, which stands for three semantic heuristics: proximity, impact, and popularity. The author accomplished this by using precise definitions of co-ratings and explanations of user ratings. PIP, however, was demonstrated in [27] to be flawed in several instances of data sparsity. So, to successfully address PIP'S constraints, the heuristic similarity model (NHSM) was developed on the PIP platform.

On the same page, in [28], data sparsity was shown to have a disastrous effect on recommendation system performance. Although the PIP and NHSM measurements [26] offered a better answer to the sparsity issue, the PIP had a relatively wide range of values for each component. As a result, [28] established a modified proximity-impact-popularity (MPIP) similarity measure. The MPIP expression was created to close the gap in the PIP measure, whose range of values for each component was rather wide. It was demonstrated that MPIP is superior to PIP and other competitive measures. Recently, authors in [6] developed the triangle-based cosine measure (TA), the hybrid difference-based similarity measure (HSMD), and the difference-based similarity measure (SMD) to efficiently solve the data sparsity problem. SMD and TA were found to be superior to their competitors. Finally, in [29, 30], a new similarity measure was put forth to fully utilize the limited rating data from cold users.

3 Methodology

The cardinality of the union set divided by the cardinality of the common set is known as the Jaccard measure. The number of common items that both users rate is measured, and is defined as follows:

$$\text{Jaccard}(u_1, u_2) = \frac{|I_1 \cap I_2|}{|I_1 \cup I_2|} \tag{1}$$

Although Jaccard is simple, it is effective within a sparse rating matrix, which has many missing values.

Jaccard Combinations

Concisely, Jaccard is not a dominant measure, but it is an important factor to improve any measure. Given this fact, in this research, we therefore combine the Jaccard measure with other numerical measures in order to take advantage of both the existence and quantity of rating values. The combination is mutual and not resonant, and hence, we use a multiplicative combination of Jaccard and other numerical measures such as cosine, Pearson, PSS, and TA [6]. These numerical measures are typical, having many variants. Furthermore, the advanced version of Jaccard, which was developed by [30], is also combined with cosine, Pearson, PSS, and TA in comparison with Jaccard. This advanced measure is denoted by IJ in Table 1, which shows the combined measures used in this research.

Table 1. Combination of Jaccard and numerical measures

Combined	Jaccard	Numerical measure	Formula
CosineJ	Jaccard	Cosine	Jaccard * cosine
PearsonJ	Jaccard	Pearson	Jaccard * Pearson
PSSJ	Jaccard	PSS	Jaccard * Pearson
TAJ	Jaccard	TA	Jaccard * TA
CosineIJ	IJ	Cosine	IJ * cosine
PearsonIJ	IJ	Pearson	IJ * Pearson
PSSIJ	IJ	PSS	IJ * Pearson
TAIJ	IJ	TA	IJ * TA

The 8th combined measures, such as Cosine J, Pearson J, PSSJ, TAJ, Cosine IJ, Pearson IJ, PSSIJ, and TAIJ are tested and compared with numerical measures such as cosine, Pearson, PSS, and TA.

4 Results

The experimental evaluation makes use of the Movielens dataset. It includes 1682 movies with 100,000 ratings from 943 users, with ratings ranging from 1 to 5. The dataset Movielens is split into 5 folders for the experiments, with a training set and a testing set in each folder. In the same folder, the training set and the testing set are separate sets. The testing parameter r determines the ratio of the testing set to the entire dataset. The testing set, for example, would encompass 10% of the dataset if r = 0.1, which indicates that the testing set has 10,000 ratings (i.e., 10% of 100,000 ratings), while the training set has naturally 90,000 ratings. The nine values of parameter r in the experimental design are 0.1, 0.2, 0.3, 0.4, 0.5, 0.6, 0.7, 0.8, and 0.9. When the KNN method is applied to the training set, the training set grows in size as r decreases, resulting in more accurate measurements as r decreases.

To evaluate CF algorithms, the Precision, recall, and F1 measure are used. The algorithm is better as the quality metric is higher [33]. **The findings of our investigation** are presented in Tables 2, 3 and 4, with grey cells denoting the best performance metrics.

In accordance with the precision metric specified for the recommendation process, Table 2 displays the precision metric for all tested measures over all r = 0.1, 0.2, 0.3, 0.4, 0.5, 0.6, 0.7, 0.8, and 0.9. The last column displays the average precision metrics over all values of r, and the top three good values are indicated by shaded cells.

Table 2. Precision metric within recommendation process

	$r = 0.1$	$r = 0.2$	$r = 0.3$	$r = 0.4$	$r = 0.5$	$r = 0.6$	$r = 0.7$	$r = 0.8$	$r = 0.9$	Avg
Jaccard	0.0056	0.0105	0.0155	0.0207	0.0261	0.0317	0.0377	0.0438	0.0511	0.0270
IJ	0.0062	0.0121	0.0180	0.0240	0.0290	0.0329	0.0321	0.0276	0.0196	0.0224
Cosine	0.0055	0.0104	0.0154	0.0207	0.0262	0.0324	0.0396	0.0508	0.0836	0.0316
Pearson	0.0051	0.0095	0.0141	0.0187	0.0237	0.0291	0.0359	0.0467	0.0803	0.0292
PSS	0.0057	0.0106	0.0157	0.0210	0.0266	0.0329	0.0401	0.0512	0.0834	0.0319
TA	0.0055	0.0104	0.0155	0.0207	0.0263	0.0325	0.0397	0.0509	0.0836	0.0317
CosineJ	0.0056	0.0105	0.0156	0.0209	0.0265	0.0327	0.0399	0.0510	0.0835	0.0318
PearsonJ	0.0052	0.0097	0.0143	0.0190	0.0240	0.0295	0.0362	0.0471	0.0804	0.0295
PSSJ	0.0057	0.0107	0.0158	0.0212	0.0268	0.0331	0.0403	0.0512	0.0831	0.0320
TAJ	0.0056	0.0106	0.0157	0.0209	0.0265	0.0328	0.0400	0.0511	0.0835	0.0319
CosineIJ	0.0055	0.0103	0.0153	0.0205	0.0260	0.0322	0.0394	0.0506	0.0835	0.0315
PearsonIJ	0.0051	0.0095	0.0140	0.0186	0.0236	0.0289	0.0356	0.0465	0.0801	0.0291
PSSIJ	0.0056	0.0106	0.0156	0.0209	0.0265	0.0328	0.0399	0.0511	0.0836	0.0318
TAIJ	0.0055	0.0104	0.0154	0.0206	0.0261	0.0323	0.0395	0.0507	0.0835	0.0316

In the recommendation process, Table 3 displays the recall metric for all tested measures over all r = 0.1, 0.2, 0.3, 0.4, 0.5, 0.6, 0.7, 0.8, and 0.9. The last column displays the average recall metrics over all values of r, and the top three good values are indicated by shaded cells.

Finally, the F1 metric assembles precision and recall. The larger F1 is, the better the measures are.

$$F1 = \frac{2*\text{Precision}*\text{Recall}}{\text{Precision}+\text{Recall}} \tag{2}$$

Table 3. Recall metric within recommendation process

	r = 0.1	r = 0.2	r = 0.3	r = 0.4	r = 0.5	r = 0.6	r = 0.7	r = 0.8	r = 0.9	Avg
Jaccard	0.9266	0.9230	0.9221	0.9191	0.9158	0.9155	0.9073	0.8947	0.8496	0.9082
IJ	0.7928	0.7514	0.6938	0.6142	0.5099	0.3757	0.2249	0.0940	0.0199	0.4530
Cosine	0.9241	0.9208	0.9211	0.9177	0.9150	0.9147	0.9066	0.8937	0.8021	0.9018
Pearson	0.9439	0.9402	0.9388	0.9359	0.9331	0.9309	0.9190	0.8948	0.7834	0.9133
PSS	0.9248	0.9219	0.9215	0.9179	0.9152	0.9143	0.9055	0.8903	0.7936	0.9006
TA	0.9242	0.9211	0.9211	0.9177	0.9149	0.9145	0.9060	0.8928	0.8005	0.9014
CosineJ	0.9266	0.9232	0.9223	0.9193	0.9159	0.9153	0.9066	0.8914	0.7970	0.9020
PearsonJ	0.9429	0.9440	0.9373	0.9351	0.9323	0.9309	0.9186	0.8948	0.7814	0.9130
PSSJ	0.9276	0.9239	0.9232	0.9191	0.9160	0.9142	0.9037	0.8844	0.7860	0.8998
TAJ	0.9265	0.9229	0.9224	0.9191	0.9154	0.9149	0.9060	0.8907	0.7957	0.9015
CosineIJ	0.9198	0.9160	0.9162	0.9127	0.9100	0.9096	0.9009	0.8886	0.8006	0.8972
PearsonIJ	0.9388	0.9334	0.9308	0.9270	0.9214	0.9181	0.9061	0.8844	0.7787	0.9043
PSSIJ	0.9215	0.9189	0.9186	0.9158	0.9132	0.9125	0.9040	0.8905	0.7957	0.8990
TAIJ	0.9207	0.9165	0.9164	0.9131	0.9099	0.9097	0.9009	0.8878	0.7990	0.8971

The averaged values of F1 for all considered metrics are shown in Table 4, which is derived from Tables 2 and 3. The best measurements, whose values are the highest, are indicated by the shaded cells.

Table 4. General MAE and F1 over all measures

	F1
Jaccard	0.052378
IJ	0.042669
Cosine	0.061102
Pearson	0.056653
PSS	0.061638
TA	0.061205
CosineJ	0.061434
PearsonJ	0.057133
PSSJ	0.061781
TAJ	0.061537
CosineIJ	0.060822
PearsonIJ	0.056386
PSSIJ	0.061510
TAIJ	0.060967

5 Conclusions and Future Works

In this work, we combine jaccard with several numerical measures for the purpose of improving CF performance. On the whole, results from experiments demonstrate that the combined measures are preeminent ones that take advantage of data existence and data magnitude. A combined measure has two built-in measures, such as a Jaccard and a numeric measure. With the built-in Jaccard, the missing values are not ignored, and they also contribute to the accuracy of the recommendation process. Real-number values accurately reflect user preferences when using the built-in numeric measure. Suppose, given a numeric measure a, the combined measure of a and Jaccard is called a +. Of course, a + is often better than a, but it is not asserted that a + is better than another numeric measure like B, for example. In other words, whether a + is better than B is dependent on the measure a itself. Similarly, IJ outperforms traditional Jaccard from r = 0.1 to r = 0.6 with precision metric, but PSSIJ outperforms PSSJ from r = 0.1 to r = 0.6 (see Table 3), because the combined measures PSSIJ and PSSJ are primarily determined by PSS. it is possible that we develop a numeric measure as well as possible and then combine it with Jaccard or variants of Jaccard. However, in practice, when a and B are not far different in accuracy, a + is often better than B. In particular, rating data is always incomplete, which which JAccard's good accuracy. Therefore, research on the improvement of Jaccard is necessary, as it is aforementioned that Jaccard is an important factor in enhancing any numeric measure.

Finally, in future work, we aim at developing a comprehensive framework for the Recommendation System (RS) for online retail by presenting new, and enhancing other similarity measures over several datasets. The experimental study is going to be extensively expanded to effectively draw the impact of jaccard combination with trust measures using both user based and item based models [31, 32].

Acknowledgement. The authors would like to thank and appreciate the support they received from the Research Office of Zayed University for providing the necessary facilities to accomplish this work. This research has been supported by the Research Incentive Fund (RIF) Grant Activity Code: R22083—Zayed University, UAE.

Authors' Contributions. All authors have been the key contributors in conception and design, implementing the approach and analyzing results of all experiments, and the preparation, writing and revising the manuscript.

Funding. This research has been supported by Research Incentive Fund (RIF) Grant Activity Code: R22083 – Zayed University, UAE.

Competing Interests. The authors declare that they have no competing interests.

References

1. Shambour, Q.Y., Hussein, A.H., Kharma, Q.M., Abualhaj, M.M.: Effective hybrid content-based collaborative filtering approach for requirements engineering. Computer Systems Science and Engineering **40**(1), 113–125 (2022). https://doi.org/10.32604/CSSE.2022.017221

2. He, W.: Interior design scheme recommendation method based on improved collaborative filtering algorithm. Wirel. Commun. Mob. Comput. **2021**, 1–10 (2021)
3. Amer, A.A., Nguyen, L.: Combinations of jaccard with numerical measures for collaborative filtering enhancement: Current work and future proposal. arXiv preprint arXiv:2111.12202 (2021)
4. Li, T., Li, Y., Yi-Ping Phoebe, C.: Novel multidimensional collaborative filtering algorithm based on improved item rating prediction. Sci. Program. **2021**, 1–14 (2021)
5. Abdalla, H., Amer, A.A.: Boolean logic algebra driven similarity measure for text based applications. PeerJ Comput. Sci. **7**, e641 (2021). https://doi.org/10.7717/peerj-cs.641
6. Amer, A.A., Abdalla, H.I., Nguyen, L.: Enhancing recommendation systems performance using highly-effective similarity measures. Knowledge-Based Systems 217 (2021)
7. Khojamli, H., Razmara, J.: Survey of similarity functions on neighborhood-based collaborative filtering. Expert Syst. Appl. **185**, 115482 (2021)
8. Widiyaningtyas, T., Hidayah, I., Adji, T.B.: User profile correlation-based similarity (UPC-Sim) algorithm in movie recommendation system. Journal of Big Data **8** (2021)
9. Bobadilla, J., Ortega, F., Hernando, A., Bernal, J.: A collaborative filtering approach to mitigate the new user cold start problem. Knowl.-Based Syst. **26**, 225–238 (2012)
10. .Jin, Q., Zhang, Y., Cai, W., Zhang, Y.: A new similarity computing model of collaborative filtering. IEEE Access **8**, 17594–17604 (2020)
11. Chia, E.J., Najafabadi, M.K.: Solving cold start problem for recommendation system using content-based filtering. In: 2022 International Conference on Computer Technologies (ICCTech), pp. 38–42. IEEE (2022, February)
12. Huynh, H.X., et al.: Context-similarity collaborative filtering recommendation. IEEE. Access **8**, 33342–33351 (2020)
13. Choi, K., Suh, Y.: A new similarity function for selecting neighbors for each target item in collaborative filtering. Knowl.-Based Syst. **37**, 146–153 (2013)
14. Ayub, M., et al.: Modeling user rating preference behavior to improve the performance of the collaborative filtering based recommender systems. PLoS ONE 14 (2019)
15. Wang, D., Yih, Y., Ventresca, M.: Improving neighbor-based collaborative filtering by using a hybrid similarity measurement. Expert Systems with Applications **160** (2020)
16. Bag, S., Kumar, S.K., Tiwari, M.K.: An efficient recommendation generation using relevant Jaccard similarity. Inf. Sci. **483**, 53–64 (2019)
17. Koohi, H., Kiani, K.: Two new collaborative filtering approaches to solve the sparsity problem. Clust. Comput. **24**, 753–765 (2021)
18. Cao, H., Deng, J., Guo, H., He, B., Wang, Y.: An improved recommendation algorithm based on Bhattacharyya Coefficient. In: IEEE International Conference on Knowledge Engineering and Applications, pp. 241–244. ICKEA (2016)
19. Patra, B.K., Launonen, R., Ollikainen, V., Nandi, S.: A new similarity measure using Bhattacharyya coefficient for collaborative filtering in sparse data. Knowl.-Based Syst. **82**, 163–177 (2015)
20. Koohi, H., Kiani, K.: A new method to find neighbor users that improves the performance of Collaborative Filtering. Expert Syst. Appl. **83**, 30–39 (2017)
21. Li, Z., Zhang, L.: Fast neighbor user searching for neighborhood-based collaborative filtering with hybrid user similarity measures. Soft. Comput. **25**, 5323–5338 (2021)
22. Saranya, K.G., Sudha Sadasivam, G.: Modified heuristic similarity measure for personalization using collaborative filtering technique. Applied Mathematics and Information Sciences **11**, 307–315 (2017)
23. Al-bashiri, H., Abdulgabber, M.A., Romli, A., Salehudin, N.B.: A developed collaborative filtering similarity method to improve the accuracy of recommendations under data sparsity. Int. J. Adv. Comput. Sci. Appl. **9**, 135–142 (2018)

24. Gazdar, A., Hidri, L.: A new similarity measure for collaborative filtering based recommender systems. Knowledge-Based Systems **188** (2020)
25. Meng, Y., Yan, X., Liu, W., Wu, H., Cheng, J.: Wasserstein collaborative filtering for item cold-start recommendation. In: Proceedings of the 28th ACM Conference on User Modeling, pp. 318–322. Adaptation and Personalization (2020)
26. Ahn, H.J.: A new similarity measure for collaborative filtering to alleviate the new user cold-starting problem. Inf. Sci. **178**, 37–51 (2008)
27. Liu, H., Hu, Z., Mian, A., Tian, H., Zhu, X.: A new user similarity model to improve the accuracy of collaborative filtering. Knowl.-Based Syst. **56**, 156–166 (2014)
28. Manochandar, S., Punniyamoorthy, M.: A new user similarity measure in a new prediction model for collaborative filtering. Appl. Intell. **51**, 586–615 (2021)
29. Osman, N.A., Noah, S.A.M., Darwich, M., Mohd, M.: Integrating contextual sentiment analysis in collaborative recommender systems. PLoS ONE **16** (2021)
30. Pan, R., Ge, C., Zhang, L., Zhao, W., Shao, X.: A new similarity model based on collaborative filtering for new user cold start recommendation. IEICE Trans. Inf. Syst. **E103D**(6), 1388–1394 (2020). https://doi.org/10.1587/transinf.2019EDP7258
31. Cai, X., Tan, W., Zhang, X., Zhou, X.: Using Entropy for Trust Measure in Collaborative Filtering. In: Human Centered Computing: 7th International Conference, HCC 2021, Virtual Event, December 9–11, 2021, Revised Selected Papers, pp. 107–118. Springer Nature Switzerland, Cham (2023, January)
32. Fkih, F.: Enhancing item-based collaborative filtering by users' similarities injection and low-quality data handling. Data Knowl. Eng. **144**, 102126 (2023)
33. Abdalla, H.I., Amer, A.A., Amer, Y.A., Nguyen, L., Al-Maqaleh, B.: Boosting the item-based collaborative filtering model with novel similarity measures. Int. J. Comput. Intell. Syst. **16**(1), 123 (2023). https://doi.org/10.1007/s44196-023-00299-2

Research on the Application of Computer Intelligence Technology in Cost Accounting and Financial Management

Yuan Haiying(✉)

Wenhua College, Wuhan, Hubei Province, China
grapeyuan2022@126.com

Abstract. Cost accounting and financial management play an important role in accounting management, but there is a problem of poor analysis effect. Particle swarm optimization cannot solve the problem of cost accounting and financial management, and the analysis effect is inaccurate. Therefore, this paper proposes a genetic algorithm to carry out a comprehensive analysis of cost accounting and financial management. Firstly, a two-layer chromosome coding method and a decoding method are proposed, and a multiple crossing algorithm and mutation operator and selection strategy are constructed. Then, the optimal individual retention, individual selection times and population disturbance strategies are analyzed, and the precocious convergence of the algorithm is overcome to comprehensively analyze the cost accounting and financial management results. The simulation results show that compared with the particle swarm algorithm, the proposed method has a better effect in improving the accuracy of cost accounting and shortening the cost accounting time.

Keywords: Genetic algorithm · Computer intelligence technology · Cost accounting · Financial management

1 Introduction

Cost accounting is an important content in financial management and has great significance for financial management [1]. However, in the process of cost accounting and financial management, the cost accounting and financial management scheme has the problem of poor accuracy of cost accounting [2], which has a certain impact on financial management. Some scholars have analyzed the application of genetic algorithms to cost accounting and financial management [3], which can analyze cost accounting and financial management schemes and provide support for cost accounting and financial management [4]. Therefore, this paper proposes a genetic algorithm to comprehensively optimize the cost accounting and financial management scheme and verifies the effectiveness of the proposed algorithm [5].

A. Hassanien et al. (Eds.): AISI 2023, LNDECT 184, pp. 230–237, 2023.
https://doi.org/10.1007/978-3-031-43247-7_21

2 Mathematical Description of the genetic Algorithm

The genetic algorithm [6] optimizes the cost accounting and financial management scheme through genetic theory, and the indicators in the cost accounting and financial management scheme are based on the indicators δ_i, which found that the outliers in Cost accounting and financial management were γ_i and adjust the cost accounting and financial management plan, judge the feasibility of the cost accounting and financial management plan, and the judgment result is represented by formula (1).

$$S_i = \sum_{i=1}^{n} \delta_i \cdot \gamma_i \tag{1}$$

Among them, the judgment of cost accounting and financial management outliers is represented by formula (2).

$$\sum x_{ij} = \delta_i \cap \gamma_i \tag{2}$$

Genetic algorithm combined with the advantages of genetic theory, through cost accounting and financial management scheme analysis, can effectively improve the accuracy of cost accounting and financial management.

Suppose I. Cost accounting and financial management objective function is A_i, the judgment variable for costing v_h, satisfaction with costing and financial management programs be x_i, solving the shortest costing mathematical model is represented by Eq. (3).

$$g_h = \sum_{h=1}^{s} v_h \tag{3}$$

B. Choice of Costing and Financial Management options

Hypothesis II The population number of the genetic algorithm is H The total cost of chromosome formation is g_T^x very acidic for the number of central nodes, then, the definition of the fitness function for costing and financial management is shown in Eq. (4).

$$f_x = \left(\max_{x=1}^{H} g_T^x - g_T^x \right) \tag{4}$$

The optimal individual retention strategy and population disturbance strategy are shown by Eq. (5).

$$H = \left(\sum_{x=1}^{H} g_T^x \cdot A_i \right) \tag{5}$$

When the number of algebras is maximum, the crossing probability is represented by Eq. (6).

$$\int_{-\infty}^{+\infty} I_x(i,j) dj = |x(i)|^2 \tag{6}$$

C. Cost Accounting and Financial Management Analysis

Before the genetic algorithm is carried out, the cost accounting and financial management scheme is analyzed, and the basic requirements of cost accounting and financial management are clarified. Combined with the genetic algorithm to process cost accounting and financial management, the empirical distribution function of cost accounting and financial management is represented by formula (7).

$$\begin{cases} y(i) = \sqrt{k}x(ki) \\ I_y(i,j) = I_x(ki, \frac{j}{k}) \end{cases} \tag{7}$$

According to the constraints of genetic algorithm characteristics, the corresponding cost of each chromosome is found, and the basic function of cost accounting and financial management is obtained, which is represented by formula (8).

$$\begin{cases} y(i) = \int_{-\infty}^{+\infty} r(i-v)x(i-v)dv \\ I_y(i,j) = \int_{-\infty}^{+\infty} I_r(i-v,j)x(v,j)dv \end{cases} \tag{8}$$

Carry out comprehensive analysis of cost accounting and financial management, set comprehensive indicators of cost accounting and financial management scheme, and ensure the accuracy of cost accounting and financial management. The cost accounting and financial management scheme is a systematic test solution that must be analyzed innovatively. If cost accounting and financial management can measure the intensity of cost accounting through genetic algorithms, it can be obtained that the iterative equation of cost accounting and financial management is represented by formula (9).

$$Q_{k+1} = a_k \frac{b_k}{b_{k+1}} \psi_k Q_k^T + \frac{1}{b_{k+1}} \delta Q \tag{9}$$

The investigation of the cost accounting and financial management scheme shows that the cost accounting and financial management scheme is distributed and conforms to the objective facts of cost accounting and financial management. Equation (10) can be analyzed from Eq. (9).

$$\begin{cases} b_k = \sum_{i=1}^{n} d_i \\ b_{k+1} = b_k + \delta b_{k+1} \end{cases} \tag{10}$$

Through the genetic algorithm, the amplitude and frequency of cost accounting and financial management can be sensed, and the intersection of financial management can be matched and set, which helps to improve the accurate analysis of cost accounting and financial management.

3 Optimization-Based Genetic Algorithm for Cost Accounting and Financial Management

The genetic algorithm [6] is solved, the chromosome coding and decoding methods of the two-layer distribution segment are constrained, the position of genes in chromosomes is adjusted by using cross-probability and stratified variation algorithms, so as to increase population diversity, while avoiding the optimal solution being changed, the probability is designed during internal mutation, the optimal individual and population disturbance strategy is constructed, and the search ability of genetic computing is improved to achieve cost accounting and financial management. Figure 1 shows the costing and financial management analysis process of the genetic algorithm.

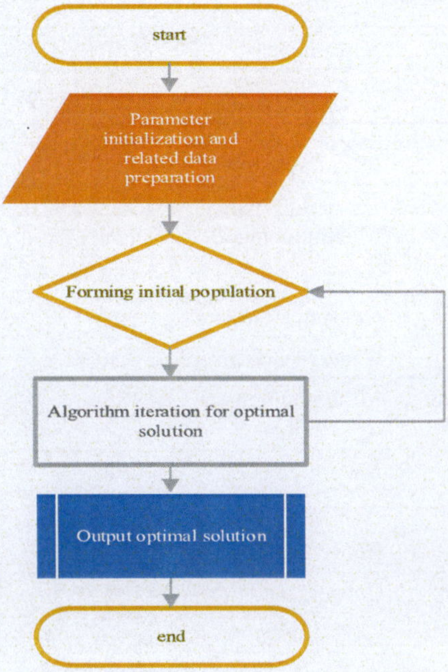

Fig. 1. Genetic algorithm costing and financial management analysis process

Compared with particle swarm algorithm [7], the cost accounting and financial management scheme of genetic algorithm is closer to the basic requirements of cost accounting and financial management of real genetic algorithm. In terms of the rationality and fluctuation range of the cost accounting and financial management analysis process of genetic algorithm, genetic algorithm is superior to particle swarm algorithm. /b20 > From the changes in the cost accounting and financial management scheme of the genetic algorithm in Fig. 2, it can be seen that the stability of the genetic algorithm is better, and the judgment speed is faster. Therefore, the cost accounting and financial management scheme of genetic algorithm of genetic algorithm is faster and more stable.

4 Practical Examples of Cost Accounting and Financial Management

A. Introduction to Cost Accounting and Financial Management

In order to verify the effectiveness of the cost accounting and financial management of the genetic algorithm, it is compared with the particle swarm algorithm. Cost centers and 240 cost center nodes were selected as research objects. The performance comparison of the different algorithms is shown in Table 1.

Table 1. Comparison of the performance of different algorithms

Sample		Performance	Different algorithms	
			Genetic algorithm	Particle swarm arithmetic
1	35 cost center nodes	Minimum number	11	11
		Maximum number	12	12
		Average number	11.15	11.41
		Convergent algebra	7.82	9.47
		Convergence time/s	0.30	0.38
2	1523 cost center nodes	Minimum number	185	206
		Maximum number	208	223
		Average number	197.0	215.51
		Convergent algebra	304.66	351.27
		Convergence time/s	137.37	147.58

B. Cost Accounting and Financial Management

Through the analysis of genetic algorithm, the cost accounting and financial management scheme of genetic algorithm is obtained, and the feasibility of the formed cost accounting and financial management scheme is analyzed. In order to accurately verify the innovative effect of costing and financial management of genetic algorithms, the costing sample is shown in Table 2.

Table 2. Sample costing

Group number	The number of cost centers	The number of cost center nodes	The total number of all cost objects
1	13	128	47811
2	13	244	24301
3	13	775	77401
4	15	1082	108101
5	21	1545	154500

C. Stability of Cost Accounting and Financial Management

To verify the accuracy of the genetic algorithm, the spread convergence curve of the particle swarm algorithm is compared, and the comparison results are shown in Fig. 2.

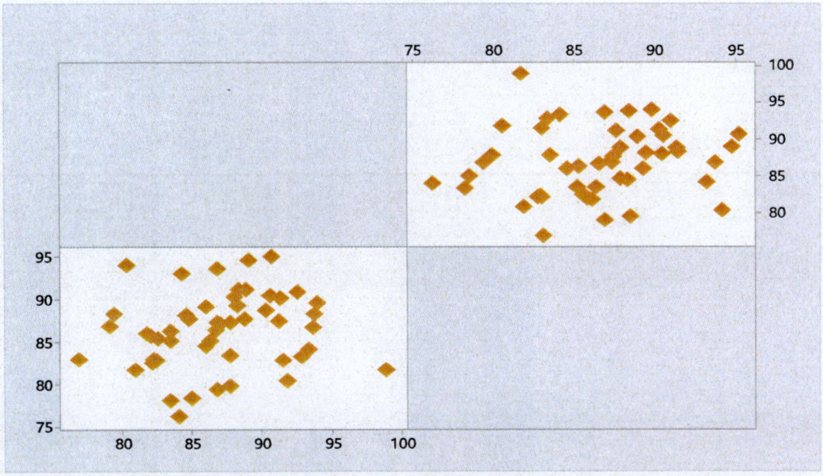

Fig. 2. Convergence curves for different algorithms

It can be seen from Fig. 2 that the effect of cost accounting and financial management of genetic algorithm is better than that of particle swarm algorithm, and the overall convergence effect is better, which shows that the cost accounting and financial management effect of genetic algorithm is better, while the effect of cost accounting and financial management of particle swarm algorithm is poor. Comparison of costing time of different algorithms, the comparison results are shown in Table 3.

Table 3. Comparison of costing time for different algorithms

Group number	Genetic algorithm		Particle swarm arithmetic	
	Number of nodes	Time/s	Number of nodes	Time/s
1	48	13.40	48	16.23
2	80	11.02	80	18.01
3	163	75.29	163	227.73
4	242	179.39	242	419.21
5	301	490.02	301	2320.87

Analysis Table 3 It can be seen that the overall accounting time of the genetic algorithm is faster in terms of cost accounting and financial management, while the overall accounting time of the particle swarm algorithm is slower than that of the genetic algorithm under the condition of the same number of nodes, which indicates that the overall stability of the genetic algorithm is better, and the cost accounting time is faster. The stability of different algorithms is compared, shown in Fig. 3.

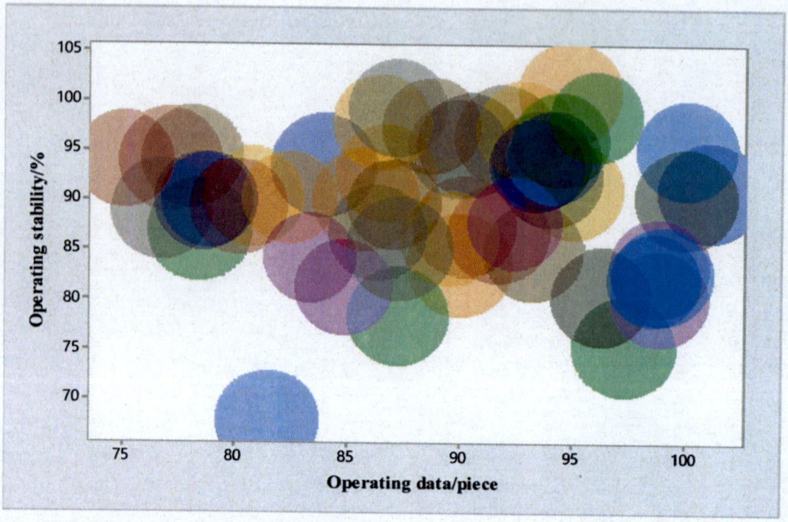

Fig. 3. Comparison of the operation stability of different algorithms

Analysis Fig. 3, the overall search speed of the genetic algorithm is relatively fast, and the optimal solution can be quickly found.

5 Conclusion

Aiming at the poor effect of cost accounting and financial management, this paper proposes a genetic algorithm and combines genetic theory to optimize the cost accounting and financial management scheme. The analysis results show that the genetic algorithm can effectively improve the accuracy of cost accounting and financial management and can also analyze cost accounting and financial management. However, in the process of genetic algorithm operation, too much attention is paid to the optimal solution of search, resulting in certain complexity in the operation process.

References

1. Zhao, L.,, Wang, W., Li, Y., Lou, C., Cheng, W.: Application of equivalence method in cost accounting and performance evaluation of community health service institutions. Chinese Journal of General Practice **26**(13), 1627–1633 (2023)
2. Qian, M., Song, R., Jiao, L.: Research on enterprise environmental costing based on MFCA. Journal of Heihe University **14**(02), 3–45 (2023)
3. Jing, Z.: Financial management early warning optimization model based on chaotic particle swarm algorithm, neural network. Microcomput. Appl. **39**(02), 20–23 (2023)

4. Cai, J.: Simulation study on financial management and decision analysis of colleges and universities under improved decision tree algorithm. J. Jiamusi University (Natural Science Edition) **40**(02), 133–135 and 158 (2022)
5. Xueqin, G.: Financial management early warning method based on chaotic particle swarm algorithmic neural network. Microcomput. Appl. **37**(05), 19–121 (2021)
6. Davis, L.: Handbook of Genetic Algorithms. Van Nostrand Reinhold, New York (1991)
7. Huaping, Z.: Optimization of risk control in financial markets based on particle swarm optimization algorithm. J. Comput. Appl. Math. **368**, 112530 (2020). ISSN 0377-0427

Exploring Factors that Drive FinTech Adoption in Jordan

Mohammad Abuhashesh[1](✉), Ibrahim S. Almashhadani[2], Muhammad Turki Alshurideh[3,4,5], and Mohammad Al-Khasawneh[6]

[1] E-Marketing and Social Media Department, Princess Sumaya University for Technology (PSUT), Amman, Jordan
m.abuhashesh@psut.edu.jo
[2] Business Entrepreneurship, Princess Sumaya University for Technology (PSUT), Amman 11941, Jordan
ibr20198012@std.psut.edu.jo
[3] Department of Marketing, School of Business, The University of Jordan, Amman 11942, Jordan
m.alshurideh@ju.edu.jo, malshurideh@sharjah.ac.ae
[4] Department of Management, College of Business, University of Sharjah, 27272 Sharjah, United Arab Emirates
[5] Applied Science Research Center, Applied Science Private University, Amman 11937, Jordan
[6] E-Marketing and Social Media Department, Princess Sumaya University for Technology (PSUT), Amman 11941, Jordan
m.alkhasaawneh@psut.edu.jo

Abstract. This study aimed to look into FinTech adoption in Jordan. Three predictors were hypothesized to test their impact on behavioral intention: perceived ease of use, financial risks, and privacy risks. The researcher applied a quantitative method. The researcher used convenience sampling to collect data. In conclusion, the sample comprised 296 respondents who were residents of Jordan and users of FinTech applications like mobile banking and mobile wallet. Additionally, statistical tools were used to analyze the data, including structural equation modeling (SEM). This study also contributes theoretically and practically, it attempts to fill a gap in the existing literature on FinTech adoption by advancing a better understanding of determinants that might drive or inhibit behavioral intention to use FinTech services. Also, in an attempt to encourage individuals to use FinTech services and to foster the establishment of the FinTech industry.

Keywords: FinTech · Technology adoption · Behavioral intention · Perceived Ease of Use · Financial Risks · Privacy Risks

1 Introduction

Traditional FinTech services include online and mobile banking, while emerging FinTech services from non-bank financial institutions include mobile wallets, mobile payments, and crowdfunding. FinTech is a new financial system that uses technology to improve

A. Hassanien et al. (Eds.): AISI 2023, LNDECT 184, pp. 238–252, 2023.
https://doi.org/10.1007/978-3-031-43247-7_22

financial activities, financial services, and social welfare. FinTech offers faster transaction rates, no wait times, 24/7 customized services, better information transparency, increased flexibility, and no physical boundaries. Financial technology allows mobile financial transactions (Chawla and Joshi, 2019). FinTech helps people access finance, pay bills, transfer funds, and buy online. Thus, FinTech services are gaining importance across the financial sector. FinTech innovations can improve financial products, lower costs, create new businesses, and create a more inclusive and robust financial system.

FinTech is creative and besides unpredictable, and people are still reluctant to use it, which hinders its growth. Senyo and Osabutey (2020) state that FinTech innovations have been hailed as game-changers for financial inclusion, but they still need to be widely accepted and used. However, the 2020 COVID-19 pandemic has made FinTech services relevant for social isolation, lockdowns, and other pandemic conditions. Thus, this study examines Jordanians' adoption of FinTech like mobile banking (e.g., Arabi Mobile, JKB Mobile, and other banks) and mobile wallets during and after the COVID-19 pandemic (e.g., Zain Cash, Orange Money, and other mobile wallets in Jordan).

Individuals can now pay for products and transfer funds using their portable devices due to the rapid expansion of smartphone usage and technological advancements in portable devices (Jung et al., 2020, Alshurideh et al., 2021a; Al Kurdi et al., 2023). Mobile banking is a service that enables users to do banking transactions without debit/credit access cards, such as tracking money, obtaining alerts, transferring money swiftly, reviewing balances, and more (Al Ketbi and Alshurideh, 2022; Shahid et al., 2022). Mobile wallets are part of the FinTech solutions (Chalik and Faturohman, 2022). The pathway of mobile wallets can broaden the delivery of conventional financial services to neglected parts of society that are unreached by traditional channels (Shaikh et al., 2023).

2 Literature Review

2.1 Perceived Ease of Use

Perceived ease of use is notably essential when users are inexperienced with technology, resulting in increased mental effort and difficulties in understanding complicated information and stimulation. Malaquias and Hwang (2019) indicated that even with a prospective belief in the usefulness of technology, users might express concern about whether or not the technology is too difficult to use. If there is little difficulty in functioning a technology, positive attitudes toward intentions and behavior might evolve to use that technology. According to Suhartanto et al. (2019), a person is more likely to adopt less complicated technology than one that is more complex.

The simplicity of new technology can also increase usage intention and behavior (Alshurideh et al., 2012; Alshurideh et al., 2021b). For instance, people are more likely to use novel technology that is easy to understand and use (Alshurideh et al., 2019; Shaqrah et al., 2023). Users are more likely to adopt a technology if it's easy to use and has a clear interface (Giovanis et al., 2019; Alwan and Alshurideh (2022). User-friendly technology is usually more appealing. According to previous research, digital technology use is often influenced by perceived ease of use (Kar, 2020).

Relevant to FinTech, perceived ease of use is the level to which users would be at ease and endeavor to learn how to use FinTech (Hu et al., 2019). Perceived ease of use denotes that users can easily comprehend or operate FinTech and carry out financial transactions (Lin et al., 2020). A user might feel the ease of using FinTech if it is user-friendly, easy to understand and navigate, and does not require excessive effort (Bashir and Madhavaiah, 2015). Since FinTech services are relatively new, people are more likely to accept them if they are easy to use and understand. Perceived ease of use in financial technology includes time, effort, and comprehension.

Besides, a person may adopt FinTech only when its services are perceived to be more convenient than traditional services and view it as user-friendly, simple to function, and not too difficult to use (Shankar and Datta, 2018). Moreover, when people realize that FinTech services are uncomplicated, their intention to use them will increase (Leong et al., 2020). Nevertheless, users will not accept FinTech if they find it challenging to use, despite the numerous benefits of this innovation (Hassan and Wood, 2020). Hence, the current study proposes the following hypothesis:

H1a. Perceived ease of use has a statistically significant effect on behavioral intention to use FinTech at a significance level of $\alpha \leq 0.05$.

2.2 Financial Risks

Financial risk has been studied for decades as an integral component of perceived risks associated with consumer behavior in offline and online marketplaces. It has been among the principal dimensions of perceived risk in consumer behavior, described as the likelihood that a purchase may result in a loss of funds or other assets. Related to digital technologies, financial risk is a significant facet of perceived risks in adopting digital services, defined as a recurrent possibility of financial loss due to fraudulent activity (Featherman and Pavlou, 2003; Al-Dmour et al., 2001). Moreover, it is an unjustifiable cost generated by digital transactions (Zhao et al., 2019). Financial risk is the possibility of incurring monetary losses due to acquiring a poor digital service or falling victim to digital fraud (Trinh et al., 2020; Awadhi et al., 2021). Generally described, financial risk is a possibility of monetary losses due to financial fraud, collapse in performing transactions, or theft of financial account information while using digital services to obtain things.

Moreover, financial losses in digital services can take different types, such as a loss arising from compromising card details during a particular transaction or activated cybercriminals collecting financial information about the user (Harris et al., 2016). Also, monetary losses in digital services might occur from duplicating a financial transaction due to a technological error, abusing a user's financial information, or purchasing faulty products that do not operate as intended (Mou et al., 2015). A financial loss may appear to be generated mainly through technical glitches or cyberattacks stealing money. Eventually, the phrase alludes to the likelihood of users encountering financial losses due to using digital technologies to conduct particular transactions.

In FinTech, financial risk is the potential of monetary loss due to malfunctions, scams, and transactional problems (Pal et al., 2020). It refers to the likelihood that users of FinTech might experience financial losses during transactions that would not occur if the transactions were completed in traditional forms (Razzaque et al., 2020). Financial risk reflects users' anxiety over the safety of their transactions in using FinTech (Roy et al., 2016). Moreover, financial risk is worries of monetary loss in the use of FinTech owing to the difficulties of demanding reimbursement in the event of transaction failures or a mistake in performing a transaction, in addition to the loss of control over a personal account and, subsequently, losses of funds can be generated. Thus, financial risk is an issue that might discourage individuals from adopting FinTech innovations.

According to Yang et al. (2015), the exchange of funds between accounts in FinTech may raise serious concerns over the theft of financial information, such as account credentials, and the consequent threat of losing money. Also, financial risk is conceivable if FinTech malfunction and consequently cause financial harm, user error occurs in transactions, or users fall victim to cybercrime. Hence, since the an inherent chance of losing money associated with FinTech services, people are naturally apprehensive about using these services (Senyo and Osabutey, 2020). Users may be uncertain about the confidentiality of the information and the verification of the transaction, which can heighten their concerns about the financial risks involved with FinTech. Thus, insecurities over the possibility of incurring financial damage during the execution of transactions may limit the adoption of FinTech services. Therefore, the following hypothesis is suggested:

H1b. Financial risk has a statistically significant effect on behavioral intention to use FinTech at a significance level of $\alpha \leq 0.05$.

2.3 Privacy Risks

Featherman and Pavlou (2003) introduced privacy risk as a fundamental facet of perceived risks that primarily focuses on privacy concerns in adopting digital services. They described it as a possible loss of control over personally identifiable information, for example, when information about a person is exploited without their consent or awareness. Privacy risk is unexpected harm caused by a person's sensitive information becoming compromised through digital transactions (Zhao et al., 2019). Moreover, privacy risk involves a likely loss of privacy and confidentiality of personal information, in addition to the fact that the usage of digital services risks users to data breaches (Trinh et al., 2020). It alludes to concerns that a person might have about the possible violation of their personal data (Johnson et al., 2018). In recent years, the expansion of digital services made possible by digitalization has given rise to privacy concerns among individuals. Remarkably, privacy concerns have become more prominent in the digital revolution.

Concerns over information privacy include collecting a large amount of personal information, its subsequent illegal usage, unacceptable access to obtained information by unauthorized parties, and insufficient protection to prevent incidental or intentional mistakes in keeping personal information (Hsu and Lin, 2016). Mou et al. (2015) mentioned that privacy concerns are associated with transmitting confidential information, such as card details, locations, contact information, and other data that may be revealed. As a result of these concerns, users will feel uncertain and worried about the confidentiality and control of their personal data (Muñoz-Leiva et al., 2017). Additionally, according to Harris et al. (2016), when the personal information of users is leaked, there are psychological problems of disappointment, frustration, and embarrassment (Harris et al., 2016).

Furthermore, As the digital world evolves and digital services expand rapidly, more privacy risks are emerging. According to Yang et al. (2015), the possible violation of privacy arises when personal information is collected, disclosed, transferred, or sold without the knowledge or permission of the user or when cybercriminals steal such personal data. Users have significant worries about protecting their private information by using digital technologies to perform transactions (Zhang et al., 2018), and they are getting seriously wary about the confidentiality of their activities. Thus, it is reasonable to presume that fears of exploiting personal information or information privacy being vulnerable may seriously affect individuals to accept using digital technologies, including FinTech.

Similarly, in FinTech, privacy risk is a chance of exposure or exploitation of personal data (Pal et al., 2020). It pertains to the likelihood that users' personal information, transaction details, and other private data will be leaked when they use FinTech. According to Yang et al. (2015), privacy risk is a big worry for users when it comes to adopting FinTech, as FinTech usage requires much personal information, which might be disclosed or maliciously used if it falls into the wrong hands. Usually, users fear that the information they reveal when making transactions by FinTech might be exploited or exposed (Roy et al., 2016). In addition, financial transactions completed using FinTech may contain such private data. So, when malicious activities or misbehavior occur to data, the interests and rights of users will be threatened, and the adoption of FinTech by users will be negatively impacted. According to Ozturk et al. (2017), users' behavior is adversely affected by privacy risks since they fear that using FinTech would lose their privacy. Accordingly, the following hypothesis is developed:

H1c. Privacy risk has a statistically significant éffect on behavioral intention to use FinTech at a significance level of $\alpha \leq 0.05$.

2.4 FinTech Usage Behavior

Behavioral intentions to accept new technology precede the adoption and usage of that technology. Hence, understanding the determinants of behavioral intentions to use new technology is fundamental for analyzing the adoption of that technology. Numerous studies have shown that behavioral intentions to adopt new technologies are determined by various variables (Singh et al., 2020). Researchers hypothesize that intention may reflect a variety of behavioral elements that affect people to get involved in a behavior (Patil et al., 2020), while usage behavior is regarded as a direct result of behavioral intention and is indicative of a person's readiness to undertake a particular action (Sharma et al., 2020).

In FinTech adoption, Intention to use refers to the likelihood that users will continue using FinTech services in the future. The intention of people to accept using FinTech is predominantly a result of multiple causes impacting it. Various constructs or factors were used to evaluate persons' attitudes on accepting and using FinTech services and equivalent innovations and their intentions and behavior toward these services (Shaikh et al., 2022). Moreover, technological and behavioral determinants that impact users' attitudes toward FinTech are not sufficiently explored in the existing literature (Senyo and Osabutey, 2020). Furthermore, empirical studies in the recent literature have observed a positive relationship between behavioral intentions and usage behavior toward FinTech services (e.g., Senyo and Osabutey, 2020; Thusi and Maduku, 2020).

In contrast, Singh et al. (2020) found no significant relationship between behavioral intention and FinTech usage. Due to the most recent studies affirming the relationship between behavioral intentions and usage behavior, the current research hypothesizes that behavioral intentions to use FinTech can significantly affect FinTech usage.

3 Methodology

The researcher applied a quantitative approach based on a cross-sectional study. Survey research is a quantitative method that employs a questionnaire to survey the sample. This research method attempts to carefully quantify the target sample in line with the criteria used to define the sample. Additionally, this deductive approach was conducted in 2022 using an online questionnaire form to obtain data. A five-point Likert scale that ranges from one (Strongly disagree) to five (Strongly agree) was used to measure all items. The study applied convenience sampling to find FinTech users. Convenience sampling is a non-probability sampling method that selects respondents who are easily accessible to the researcher. Furthermore, this sampling method does not need a random selection of individuals. Using Google forms, a self-administered survey was conducted. The survey was intended to be submitted by participants. Moreover, the distribution of the questionnaire by the researcher started on November 18, 2022, and ended on December 4, 2022. The questionnaire was distributed online through multiple tools and platforms, including email, WhatsApp, Facebook, and LinkedIn. The questionnaire was sent to FinTech users in Jordan to collect the data required for the study. Individuals in Jordan over 18 years old who use FinTech applications such as mobile banking and mobile wallets have been the target population. Only 310 surveys out of 354 circulated with a convenience sample approach were completed. Fourteen responses were excluded since they did not meet the threshold. Ultimately, just 296 submissions from the number of 310 were verified for processing. The descriptive analysis of the data was performed using SPSS. Eventually, using AMOS, structural equation modeling (SEM), path analysis, and confirmatory factor analysis (CFA) were undertaken. AMOS graphics offer researchers all tools required to create and manipulate SEM path diagrams.

3.1 Research Model

(Fig. 1)

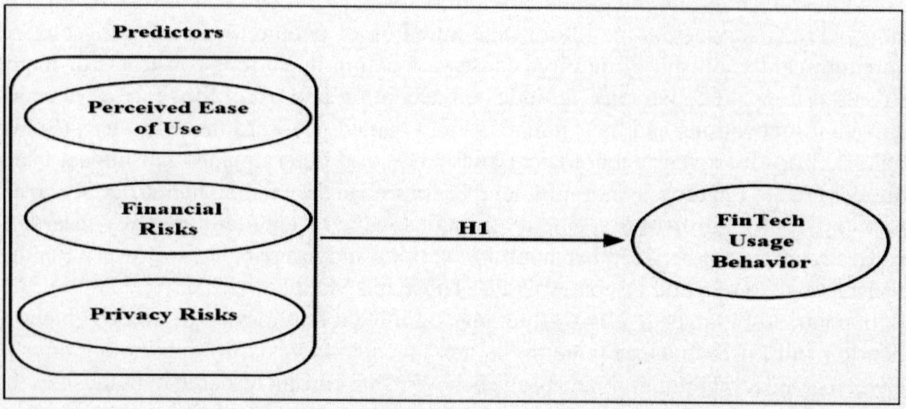

Fig. 1. Conceptual model of FinTech adoption in Jordan.

4 Data Analysis and Findings

4.1 Demographic Analysis

As shown in Table 1, the researcher also asked participants three questions about the usage of FinTech applications. However, the researcher found that mobile banking is the most popular FinTech application, with around 55.4% of respondents using it, and 39.2% of participants revealed that they use both mobile banking and mobile wallet. Additionally, most participants reported using FinTech applications daily or at least 2 to 3 times per week. While 50 percent of respondents have been using FinTech applications for less than two years, the remaining 50 percent have adopted FinTech applications for over two years.

Table 1. Usage of FinTech applications

		Frequency	Percent
Using mobile banking or mobile wallet	Mobile banking	164	55.4
	Mobile wallet	16	5.4
	Both of them	116	39.2
	Total	296	100
		Frequency	**Percent**
Frequency of using FinTech applications	Daily	82	27.7
	From 2 to 3 times a week	60	20.3
	From 4 to 6 times a week	59	19.9

(*continued*)

Table 1. (*continued*)

		Frequency	**Percent**
	Once a week	40	13.5
	Once a month	32	10.8
	A few times a year	23	7.8
	Total	296	100
		Frequency	**Percent**
A number of years in using FinTech applications	Less than 2 years	148	50
	From 3 to 4 years	86	29.1
	From 5 to 7 years	46	15.5
	More than 7 years	16	5.4
	Total	296	100

4.2 Results of Hypotheses Testing

(Table 2)

Table 2. Results of hypotheses testing

Direct Effect Hypothesis		P-Value	Sign	Decision
H1a	Perceived ease of use has a significant effect on behavioral intention to use FinTech	.037	Positively significant	Accepted
H1b	Financial risk has a significant effect on behavioral intention to use FinTech	.008	Negatively significant	Accepted
H1c	Privacy risk has a significant effect on behavioral intention to use FinTech	.338	Insignificant	Rejected

As exhibited in the tables above, Hypotheses (H1a) were accepted where P-Value was less than 0.05. Also, the hypothesis (H1b) was accepted, and this result confirmed that financial risks negatively impacted behavioral intention to use FinTech. While the hypothesis (H1c) was rejected, where P-Value was .338, this result revealed that privacy risks had no significant impact on users' behavioral intention to use FinTech. Furthermore, in this model, the R-squared value was 0.259, indicating that perceived ease of use, financial risks, and privacy risks explained 25.9% of the variation in behavioral intention to use FinTech in Jordan.

5 Discussion and Conclusion

5.1 Discussion

Nowadays, FinTech services in Jordan are growing in popularity. Thus, it is essential to research the behavioral intention of users to discover what determinants encourage them to use FinTech services and what factors cause them to be unwilling to accept and use these solutions. While encouraging people in Jordan to change their behavior from traditional ways of doing financial transactions to using FinTech services is not a simple task, there is a constant need to identify and experimentally assess major causes that might affect persons' intention and behavior to accept FinTech (Alalwan et al., 2018). Therefore, the purpose of this study was to research determinants of FinTech adoption in Jordan. Again, the focus of the study was to assess the determinants of adopting FinTech services by constructing and implementing a conceptual framework using TAM and UTAUT theories. Moreover, the researcher attempted to study users' intention and behavior to use FinTech services in Jordan. The researcher analyzed data sampling using specific software tools. Essentially, structural equation modeling (SEM) was employed to test hypotheses.

The study results were crucial for understanding FinTech adoption in Jordan. Two predictors had a significant impact on behavioral intention to use FinTech. These predictors were perceived ease of use and financial risks. Ultimately, the hypotheses (H1a and H1b) were confirmed. While the impact of privacy risks on behavioral intention to use FinTech was insignificant. Consequently, the hypothesis (H1c) was not supported.

According to study results, the proposed model was effective, and this model empirically has explained how users' behavioral intention to use FinTech applications may be established. Interestingly, the conceptual model demonstrated considerable explanatory power with all predictors and the moderating variable in estimating behavioral intention to use FinTech applications, where $R^2 = 0.384$. However, the subsequent sections elaborate a more in-depth discussion of the empirical results.

The Impact of Predictors on Behavioral Intention to Use FinTech

Similarly, the study looked into the impact of perceived ease of use on behavioral intention to use FinTech. The empirical findings discovered that perceived ease of use significantly predicted behavioral intention to use FinTech. This result was consistent with prior empirical studies that showed the significance of perceived ease of use in determining behavioral intention to use FinTech services (e.g., Shaikh et al., 2020; Singh and Srivastava, 2020). On the other hand, this finding contradicted other empirical findings that have found no relationship between perceived ease of use and behavioral intention to use FinTech (e.g., Daragmeh et al., 2021; Liébana-Cabanillas et al., 2020). This result suggests that simple services foster FinTech adoption among people in Jordan. Due to the contradictory results of previous research, it was thought essential to confirm the effect of the perceived ease of use of FinTech on adoption intention to use FinTech.

Recent results demonstrated that financial risks contribute to a detrimental impact on behavioral intention to use FinTech. This empirical conclusion was similar to previous studies that have found a negative relationship between financial risks and users' behavioral intention to use FinTech services (e.g., Agárdi and Alt, 2022). Therefore, the findings of this study indicate that individuals in Jordan are concerned about losing their money due to their usage of FinTech services, and this uncertainty inhibits their behavioral intention to use FinTech. Lastly, the results demonstrated that privacy risks had no adverse effect on behavioral intention to use FinTech services. This finding contradicted other conclusions that Thusi and Maduku (2020) discovered. This empirical study suggests that people in Jordan who use FinTech services are not apprehensive about losing their personally identifiable data.

The Relationship Between Users' Behavioral Intention and Their Actual Use of FinTech

The study then assessed the relationship between behavioral intention to use FinTech and FinTech usage behavior. This relationship was tested to determine if users' behavioral intention can determine their actual use of FinTech applications like mobile banking and mobile wallet. Usage behavior is the predictable consequence of behavioral intention and indicates a person's behavior to perform a given action (Sharma et al., 2020). Subsequently, empirical findings demonstrated that behavioral intention to use FinTech had a significantly positive impact on the usage behavior of FinTech.

5.2 Theoretical Implications

This study contributed to the body of knowledge and had many theoretical implications. The study assessed multifaceted determinants that might be drivers or inhibitors of behavioral intention to use FinTech. Subsequently, this assessment made a significant contribution to the existing literature by proposing a robust conceptual model that integrated and expanded two of the most widely accepted theories on technology adoption, namely TAM and UTAUT. Additionally, the study has added further findings to the limited number of research in the Jordanian academia on FinTech adoption. Due to the scarcity of research on FinTech in Jordanian academia, a comprehensive methodology for estimating behavioral intention to accept FinTech in Jordan was designed.

This study also helped close a gap in the existing literature on FinTech adoption by better understanding mechanisms that affect users' behavioral intention to use FinTech services, particularly during and post-COVID-19 pandemic.

5.3 Practical Implications

This research has proven that parameters such as functionality and user-friendliness of FinTech services play a significant part in promoting their adoption. Additionally, the opinion of others in Jordan and users' innovation affect behavioral intention to use FinTech. However, this study uncovered perceived concerns that deter users from adopting FinTech applications. In addition to other findings with practical uses, the study results demonstrated how the impact of COVID-19 lockdowns boosted FinTech adoption in Jordan and the extent to which unusual contemporary crises such as the COVID-19 pandemic can disrupt the FinTech industry. In general, governments, policymakers, and developers might employ this knowledge to foster the expansion of the FinTech industry and push individuals to increase their use of FinTech.

To begin, users must remain assured that they can get exposure to dependable, convenient infrastructures that will empower them to use FinTech applications. Furthermore, developers and decision-makers in the FinTech industry should concentrate on creating broad knowledge that will incentivize people to experiment with FinTech services. These objectives can be fulfilled through training courses and tutorial videos that elaborate on the advantages, simplicity, and safety of using FinTech services. Developers of FinTech services should guarantee that users fully comprehend the concept of using FinTech before signing up. They might use assistive techniques like live discussion or robots to create constant guidance that promotes reasons for adopting FinTech. In addition, they can leverage digital channels to illustrate the advantages of using FinTech services to their target market.

According to the study results, the behavioral intention of users to adopt FinTech services has shifted during and post-COVID-19 lockdowns. Despite the spread of COVID-19 disease, individuals were able to conduct financial transactions and stay safe at home by using FinTech applications. The empirical findings of this study discovered that the COVID-19 crisis had a considerable influence on the emergence of FinTech. Accordingly, these findings might be exploited by developers and providers of FinTech services to advocate potential advantages of using FinTech in unforeseen crises such as the COVID-19 pandemic. Additionally, as a consequence of the importance of FinTech during the COVID-19 pandemic, governments and legislators may consider the vital role of FinTech in times of emergency and stimulate the expansion of the FinTech industry in Jordan by mandating people to use FinTech services such as mobile banking and mobile wallet. Finally, recent findings found that financial risks hinder users' behavioral intention to use FinTech. Therefore, developers and marketers of FinTech services should use this conclusion to design strategies and programs to alleviate people's concerns regarding the usage of FinTech. There is a requirement to ensure that the adoption of FinTech enables users to execute their financial transactions in a safe, efficient, and consistent way since this would enhance the user experience and the authenticity of using FinTech.

6 Conclusion

This study was undertaken in Jordan to measure the intentions and behavior of users to adopt FinTech applications, namely mobile banking and mobile wallet. Fundamentally, the study's conceptual model was derived from thoroughgoing research and theories in the existing literature.

Additionally, using a cross-sectional study and non-probability sampling, the researcher gathered data from respondents. Multiple online tools were used to design and circulate an online questionnaire to the data needed. Three hundred ten users consented to fill out the survey and participate in the experiment. However, the researcher eliminated 14 replies because they needed to meet the threshold. In the result, 296 responses were valid for analysis. Then, multiple statistical approaches were used to analyze the data. The data assessments exhibited adequate discriminant and convergent validity, along with sufficient reliability and validity.

The findings showed that perceived ease of use and financial risks significantly impacted the behavioral intention to use FinTech applications such as mobile banking and mobile wallet. At the same time, privacy risks did not impact behavioral intention to use FinTech. Finally, the researcher looked into the relationship between behavioral intention to use FinTech and FinTech usage behavior. The study's findings demonstrated that users' behavioral intention significantly affected their actual use of FinTech. To conclude, recent research provided a foundation for a conceptual model and methodology that integrates multiple components of researching FinTech and understanding determinants that affect its acceptance and adoption by users in Jordan.

References

Agárdi, I., Alt, M.A.: Do digital natives use mobile payment differently than digital immigrants? A comparative study between generation X and Z. Electronic Commerce Research (2022). https://doi.org/10.1007/s10660-022-09537-9

Alalwan, A.A., Dwivedi, Y.K., Rana, N.P., Algharabat, R.: Examining factors influencing Jordanian customers' intentions and adoption of Internet banking: Extending UTAUT2 with risk. J. Retail. Consum. Serv. **40**, 125–138 (2018). https://doi.org/10.1016/j.jretconser.2017.08.026

Al-Dmour, A., Al-Dmour, H., Al-Barghuthi, R., Al-Dmour, R., Alshurideh, M.T.: Factors influencing the adoption of e-payment during pandemic outbreak (COVID-19): Empirical evidence. The Effect of Coronavirus Disease (COVID-19) on Business Intelligence, pp. 133–154 (2021)

Al Ketbi, S.J., Alshurideh, M.: The Impact of Corporate Entrepreneurship on Corporate Performance: The Moderating Role of Digital Readiness in the Banking Sector. In: Proceedings of the 8th International Conference on Advanced Intelligent Systems and Informatics 2022, pp. 499–509. Springer International Publishing, Cham (2022)

Al Kurdi, B., Antouz, Y.A., Alshurideh, M.T., Hamadneh, S., Alquqa, E.K.: The impact of digital marketing and digital payment on financial performance. In: 2023 International Conference on Business Analytics for Technology and Security (ICBATS), pp. 1–5. IEEE (2023)

Alshurideh, M., Nicholson, M., Xiao, S.: The effect of previous experience on mobile subscribers' repeat purchase behaviour. Eur. J. Soc. Sci. **30**(3), 366–376 (2012)

Alshurideh, M., Al Kurdi, B., Salloum, S.A.: Examining the main mobile learning system drivers' effects: A mix empirical examination of both the Expectation-Confirmation Model (ECM) and the Technology Acceptance Model (TAM). In: International Conference on Advanced Intelligent Systems and Informatics, pp. 406–417. Springer, Cham (2019, October)

Alshurideh, M.T., Al Kurdi, B., Masa'deh, R.E., Salloum, S.A.: The moderation effect of gender on accepting electronic payment technology: a study on United Arab Emirates consumers. Review of International Business and Strategy **31**(3), 375-396 (2021a)

Alshurideh, M.T., et al.: Factors affecting the use of smart mobile examination platforms by universities' postgraduate students during the COVID-19 pandemic: An empirical study. In: Informatics, Vol. 8, No. 2, p. 32. MDPI (2021b)

Alwan, M., Alshurideh, M.: The effect of digital marketing on value creation and customer satisfaction. Int. J. Data and Netw. Sci. **6**(4), 1557–1566 (2022)

Awadhi, J., Obeidat, B., Alshurideh, M.: The impact of customer service digitalization on customer satisfaction: Evidence from telecommunication industry. Int. J. Data and Netw. Sci. **5**(4), 815–830 (2021)

Bashir, I., Madhavaiah, C.: Consumer attitude and behavioural intention towards Internet banking adoption in India. J. Ind. Bus. Res. **7**(1), 67–102 (2015). https://doi.org/10.1108/jibr-02-2014-0013

Chalik, F.R., Faturohman, T.: Customer satisfaction of e-wallet user: An adoption of information system success model.Quantitative Analysis of Social and Financial Market Development, 61–83 (2022). https://doi.org/10.1108/s1571-038620220000030005

Chawla, D., Joshi, H.: Consumer attitude and intention to adopt mobile wallet in India – an empirical study. Int. J. Bank Mark. **37**(7), 1590–1618 (2019). https://doi.org/10.1108/ijbm-09-2018-0256

Daragmeh, A., Lentner, C., Sági, J.: FinTech payments in the era of COVID-19: Factors influencing behavioral intentions of "Generation X" in Hungary to use mobile payment. J. Behav. Exp. Financ. **32**, 100574 (2021). https://doi.org/10.1016/j.jbef.2021.100574

Featherman, M.S., Pavlou, P.A.: Predicting e-services adoption: A perceived risk facets perspective. Int. J. Hum Comput Stud. **59**(4), 451–474 (2003). https://doi.org/10.1016/s1071-5819(03)00111-3

Giovanis, A., Athanasopoulou, P., Assimakopoulos, C., Sarmaniotis, C.: Adoption of mobile banking services. Int. J. Bank Mark. **37**(5), 1165–1189 (2019). https://doi.org/10.1108/ijbm-08-2018-0200

Harris, M.A., Brookshire, R., Chin, A.G.: Identifying factors influencing consumers' intent to install mobile applications. Int. J. Inf. Manage. **36**(3), 441–450 (2016). https://doi.org/10.1016/j.ijinfomgt.2016.02.004

Hassan, H.E., Wood, V.R.: Does country culture influence consumers' perceptions toward mobile banking? A comparison between Egypt and the United States. Telematics Inform. **46**, 101312 (2020). https://doi.org/10.1016/j.tele.2019.101312

Hsu, C.L., Lin, J.C.C.: An empirical examination of consumer adoption of Internet of things services: network externalities and concern for information privacy perspectives. Comput. Hum. Behav. **62**, 516–527 (2016). https://doi.org/10.1016/j.chb.2016.04.023

Hu, Z., Ding, S., Li, S., Chen, L., Yang, S.: Adoption intention of fintech services for bank users: An empirical examination with an extended technology acceptance model. Symmetry **11**(3), 340 (2019). https://doi.org/10.3390/sym11030340

Johnson, V.L., Kiser, A., Washington, R., Torres, R.: Limitations to the rapid adoption of m-payment services: understanding the impact of privacy risk on M-Payment services. Comput. Hum. Behav. **79**, 111–122 (2018). https://doi.org/10.1016/j.chb.2017.10.035

Jung, J.H., Kwon, E., Kim, D.H.: Mobile payment service usage: U.S. consumers' motivations and intentions. Computers in Human Behavior Reports **1**, 100008 (2020). https://doi.org/10.1016/j.chbr.2020.100008

Kar, A.K.: What affects usage satisfaction in mobile payments? modelling user generated content to develop the "digital service usage satisfaction model." Inf. Syst. Front. **23**(5), 1341–1361 (2020). https://doi.org/10.1007/s10796-020-10045-0

Leong, C.M., Tan, K.L., Puah, C.H., Chong, S.M.: Predicting mobile network operators users m-payment intention. European Business Review **33**(1) (2020). https://doi.org/10.1108/ebr-10-2019-0263

Liébana-Cabanillas, F., Molinillo, S., Japutra, A.: Exploring the determinants of intention to use P2P mobile payment in Spain. Inf. Syst. Manag. **38**(2), 165–180 (2020). https://doi.org/10.1080/10580530.2020.1818897

Lin, W.R., Wang, Y.H., Hung, Y.M.: Analyzing the factors influencing adoption intention of Internet banking: Applying DEMATEL-ANP-SEM approach. PLoS ONE **15**(2), e0227852 (2020). https://doi.org/10.1371/journal.pone.0227852

Malaquias, R.F., Hwang, Y.: Mobile banking use: A comparative study with Brazilian and U.S. participants. Int. J. Inf. Manage. **44**, 132–140 (2019). https://doi.org/10.1016/j.ijinfomgt.2018.10.004

Mou, J., Shin, D.H., Cohen, J.F.: Trust and risk in consumer acceptance of e-services. Electron. Commer. Res. **17**(2), 255–288 (2015). https://doi.org/10.1007/s10660-015-9205-4

Muñoz-Leiva, F., Climent-Climent, S., Liébana-Cabanillas, F.: Determinants of intention to use the mobile banking apps: An extension of the classic TAM model. Spanish Journal of Marketing - ESIC **21**(1), 25–38 (2017). https://doi.org/10.1016/j.sjme.2016.12.001

Ozturk, A.B., Bilgihan, A., Salehi-Esfahani, S., Hua, N.: Understanding the mobile payment technology acceptance based on valence theory: A case of restaurant transactions. Int. J. Contemp. Hosp. Manag. **29**(8), 2027–2049 (2017). https://doi.org/10.1108/ijchm-04-2016-0192

Pal, A., Herath, T., De', R., Rao, H.R.: Is the convenience worth the risk? An investigation of mobile payment usage.Information Systems Frontiers **23**(4), 941–961 (2020). https://doi.org/10.1007/s10796-020-10070-z

Patil, P., Tamilmani, K., Rana, N.P., Raghavan, V.: Understanding consumer adoption of mobile payment in India: Extending Meta-UTAUT model with personal innovativeness, anxiety, trust, and grievance redressal. Int. J. Inf. Manage. **54**, 102144 (2020). https://doi.org/10.1016/j.ijinfomgt.2020.102144

Razzaque, A., Cummings, R.T., Karolak, M., Hamdan, A.: The propensity to use FinTech: Input from bankers in the Kingdom of Bahrain. J. Inf. Knowl. Manag. **19**(01), 2040025 (2020). https://doi.org/10.1142/s0219649220400250

Roy, S.K., Balaji, M., Kesharwani, A., Sekhon, H.: Predicting Internet banking adoption in India: A perceived risk perspective. J. Strateg. Mark. **25**(5–6), 418–438 (2016). https://doi.org/10.1080/0965254x.2016.1148771

Senyo, P., Osabutey, E.L.: Unearthing antecedents to financial inclusion through FinTech innovations. Technovation **98**, 102155 (2020). https://doi.org/10.1016/j.technovation.2020.102155

Shaqrah, M., et al.: The Impact of Branded Mobile Applications on Customer Loyalty. In: The Effect of Information Technology on Business and Marketing Intelligence Systems, pp. 2215–2229. Springer International Publishing, Cham (2023)

Shahid, S., Islam, J.U., Malik, S., Hasan, U.: Examining consumer experience in using m-banking apps: A study of its antecedents and outcomes. J. Retail. Consum. Serv. **65**, 102870 (2022). https://doi.org/10.1016/j.jretconser.2021.102870

Shaikh, A.A., Alamoudi, H., Alharthi, M., Glavee-Geo, R.: Advances in mobile financial services: A review of the literature and future research directions. International Journal of Bank Marketing **41**(1), 1–33 (2022). https://doi.org/10.1108/ijbm-06-2021-0230

Shaikh, A.A., Glavee-Geo, R., Karjaluoto, H., Hinson, R.E.: Mobile money as a driver of digital financial inclusion. Technol. Forecast. Soc. Chang. **186**, 122158 (2023). https://doi.org/10.1016/j.techfore.2022.122158

Shaikh, I.M., Qureshi, M.A., Noordin, K., Shaikh, J.M., Khan, A., Shahbaz, M.S.: Acceptance of Islamic financial technology (FinTech) banking services by Malaysian users: An extension of technology acceptance model. Foresight **22**(3), 367–383 (2020). https://doi.org/10.1108/fs-12-2019-0105

Shankar, A., Datta, B.: Factors affecting mobile payment adoption intention: An Indian perspective. Global Business Review **19**(3_suppl), S72–S89 (2018). https://doi.org/10.1177/0972150918757870

Sharma, R., Singh, G., Sharma, S.: Modelling Internet banking adoption in Fiji: A developing country perspective. Int. J. Inf. Manage. **53**, 102116 (2020). https://doi.org/10.1016/j.ijinfomgt.2020.102116

Singh, S., Sahni, M.M., Kovid, R.K.: What drives FinTech adoption? A multi-method evaluation using an adapted technology acceptance model. Manag. Decis. **58**(8), 1675–1697 (2020). https://doi.org/10.1108/md-09-2019-1318

Suhartanto, D., Dean, D., Ismail, T.A.T., Sundari, R.: Mobile banking adoption in Islamic banks. Journal of Islamic Marketing **11**(6), 1405–1418 (2019). https://doi.org/10.1108/jima-05-2019-0096

Thusi, P., Maduku, D.K.: South African millennials' acceptance and use of retail mobile banking apps: An integrated perspective. Comput. Hum. Behav. **111**, 106405 (2020). https://doi.org/10.1016/j.chb.2020.106405

Trinh, H.N., Tran, H.H., Vuong, D.H.Q.: Determinants of consumers' intention to use credit card: A perspective of multifaceted perceived risk. Asian Journal of Economics and Banking **4**(3), 105–120 (2020). https://doi.org/10.1108/ajeb-06-2020-0018

Yang, Y., Liu, Y., Li, H., Yu, B.: Understanding perceived risks in mobile payment acceptance. Ind. Manag. Data Syst. **115**(2), 253–269 (2015). https://doi.org/10.1108/imds-08-2014-0243

Zhang, T., Lu, C., Kizildag, M.: Banking "on-the-go": Examining consumers' adoption of mobile banking services. Int. J. Qual. Serv. Sci. **10**(3), 279–295 (2018). https://doi.org/10.1108/ijqss-07-2017-0067

Zhao, H., Anong, S.T., Zhang, L.: Understanding the impact of financial incentives on NFC mobile payment adoption. Int. J. Bank Mark. **37**(5), 1296–1312 (2019). https://doi.org/10.1108/ijbm-08-2018-0229

Exploration of the Training Mode of Application-Oriented Engineering Management Talents Based on the Goal of Innovation and Entrepreneurship

Changjiang Deng(✉)

The International College of NIDA, Bangkok, Thailand
changjiang.deng@stu.nida.ac.th

Abstract. Education is the foundation of talent training, in order to strengthen the cultivation of innovation and entrepreneurship talents in engineering management, the concept should focus on the cultivation of student's professional skills and innovation and entrepreneurship ability, starting from reforming the theory of engineering management and building an innovation and entrepreneurship guidance service system for college students. The traditional education model cannot solve the problem of comprehensive quality evaluation of talents in the innovation and entrepreneurship guidance service system of college students, and the evaluation is unreasonable. Therefore, this paper proposes a big data algorithm to evaluate and analyze the comprehensive quality of innovative and entrepreneurial talents. First of all, the engineering management theory is used to evaluate the engineering management talents, and the indicators are divided according to the requirements of the comprehensive quality evaluation of talents, so as to reduce the comprehensive quality evaluation of talents Disturbing factors. Then, the engineering management theory evaluates the comprehensive quality of college students' innovative and entrepreneurial talents, forms a comprehensive talent quality evaluation plan, and comprehensively analyzes the comprehensive quality evaluation results of talents. MATLAB simulation shows that under certain evaluation criteria, the accuracy of talent comprehensive quality evaluation and talent comprehensive quality evaluation time of college students' innovation and entrepreneurship are better than those of traditional education models.

Keywords: engineering management theory · big data algorithms · innovation and entrepreneurship · comprehensive quality

1 Introduction

Entrepreneurship is to cultivate college students to better adapt to society after graduation and improve innovation and entrepreneurship ability, which is of great significance for reflecting the personal value of college students [1]. However, in the process of talent comprehensive quality evaluation, there is a problem of poor accuracy in the talent comprehensive quality evaluation program, which brings certain reputation losses to the

A. Hassanien et al. (Eds.): AISI 2023, LNDECT 184, pp. 253–260, 2023.
https://doi.org/10.1007/978-3-031-43247-7_23

evaluation of college students' innovation and entrepreneurship ability [2]. Some scholars believe that the application of big data algorithms to the analysis of college students' innovation and entrepreneurship guidance service system [3] can effectively analyze the comprehensive quality evaluation scheme of talents and provide corresponding support for the comprehensive quality evaluation of talents [4]. On this basis, this paper proposes a big data algorithm to optimize the comprehensive quality evaluation scheme of talents and verify the effectiveness of the model [5].

2 Related Concepts

2.1 Mathematical Description of Big Data Algorithms

The big data algorithm uses big data theory to optimize the comprehensive quality evaluation scheme of talents, and finds the unqualified values in the innovation and entrepreneurship guidance service system of college students according to the indicators in the comprehensive quality evaluation of talents is p_i, and integrates the comprehensive quality evaluation scheme of talents is y_i, and finally judges the feasibility of the college students' innovation and entrepreneurship guidance service system is $ZH(x_{ij})$, calculated as shown in formula (1).

$$ZH(x_{ij}) = p_i \geq \max(y_i) \tag{1}$$

Among them, the judgment of outliers is shown in Eq. (2).

$$\max(y_i) = (x_{ij} + a) \succ \sum x_{ij} \cdot (p \cdot h) \tag{2}$$

The big data algorithm combines the advantages of big data theory and uses the college students' innovation and entrepreneurship guidance service system to quantify, which can improve the entrepreneurial ability of talent comprehensive quality evaluation.

Hypothesis I. The comprehensive quality evaluation requirements of talents is x_i, the comprehensive quality evaluation scheme of talents is g_i, the satisfaction of the comprehensive quality evaluation program of talents is f, and the judicial function of the comprehensive quality evaluation scheme of talents is $F(g_i)$, as shown in Eq. (3).

$$F(g_i) = \sum x_i \bigcap \rightarrow \lim_{x \to \infty} f \tag{3}$$

2.2 Selection of Entrepreneurship Programs

Hypothesis II. The function of the college student innovation and entrepreneurship guidance service system is $R(x_i)$, and the weight coefficient is d_i, then, the talent comprehensive quality evaluation requires the unqualified college student innovation and entrepreneurship guidance service system as shown in Eq. (4).

$$R(x_i) = \mathrm{d}_i \cdot \prod F(d_i) - t_i \tag{4}$$

Based on hypotheses I and II, a comprehensive function of entrepreneurial ability can be obtained, and the result is shown in Eq. (5).

$$R(x_i) + F(g_i) \leq \max(y_i) \tag{5}$$

In order to improve the effectiveness of the comprehensive quality assessment of talents, it is necessary to standardize all data, and the results are shown in Eq. (6).

$$R(x_i) + F(g_i) \leftrightarrow \sum x_{ij} \cdot (p \cdot h) \tag{6}$$

2.3 Analysis of Talent Comprehensive Quality Evaluation Program

Before carrying out the big data algorithm, it is necessary to conduct a multi-dimensional analysis of the talent comprehensive quality evaluation plan and map the talent comprehensive quality evaluation requirements to the college student innovation and **entrepreneurship guidance service system database and eliminate the unqualified talent** comprehensive quality evaluation plan is $D(x_i)$. According to Eq. (6), the anomaly evaluation scheme can be proposed, and the results are shown in Eq. (7).

$$D(x_i) = \frac{R(x_i) + F(g_i)}{\sum x_{ij} \cdot (p \cdot h)} \tag{7}$$

Among them, it is $\frac{R(x_i)+F(g_i)}{\sum x_{ij}\cdot(p\cdot h)} \leq 1$, stated that the scheme needs to be proposed, otherwise the scheme needs to be integrated into it $Nc(x_i)$, and the result is shown in Eq. (8).

$$Nc(x_i) = \min[\sum R(x_i) + F(g_i)] \tag{8}$$

The college students' innovation and entrepreneurship guidance service system is comprehensively analyzed, and the threshold and index weight of the talent comprehensive quality evaluation plan is set to ensure the accuracy of the big data algorithm. The innovation and entrepreneurship guidance service system for college students is a comprehensive quality evaluation scheme for systematic testing of talents, and innovative analysis is required. If the college students' innovation and entrepreneurship guidance service system is in a non-normal distribution, its talent comprehensive quality evaluation scheme will be affected is $e^{i\theta}(x_i)$, reducing the accuracy of the overall talent comprehensive quality evaluation is $\arcsin(x_i)$, and the calculation result is shown in Eq. (9).

$$\arcsin(x_i) = \frac{\min[\sum R(x_i) + F(g_i)]}{\sum R(x_i) + F(g_i)} \times 100\% \tag{9}$$

The comprehensive quality evaluation scheme of the survey shows that the entrepreneurial ability program presents a multi-dimensional distribution, which is in line with the objective facts. The innovation and entrepreneurship guidance service system for college students is not directional, indicating that the entrepreneurship ability program has strong randomness, so it is regarded as high analytical research. If the random function of the college student innovation and entrepreneurship guidance service system is $\cos(x_i)$, then the calculation of formula (9) can be expressed as formula (10).

$$\arcsin(x_i) = \frac{\min[\sum R(x_i) + F(g_i)]}{\sum R(x_i) + F(g_i)} \times 100\% + \cos(x_i) \tag{10}$$

Among them, the college student innovation and entrepreneurship guidance service system meet the normal requirements, mainly because the big data theory adjusts the college student innovation and entrepreneurship guidance service system, removes duplicate and irrelevant schemes, and supplements the default scheme so that the dynamic correlation of the entire talent comprehensive quality evaluation scheme is strong.

3 Optimization Strategy of Innovation and Entrepreneurship Guidance Service System for College Students

The big data algorithm adopts the random optimization strategy for the college students' innovation and entrepreneurship guidance service system and adjusts the project management talent parameters to realize the program optimization of the college students' innovation and entrepreneurship guidance service system. The big data algorithm divides the innovation and entrepreneurship guidance service system of college students into different talent comprehensive quality evaluation levels and randomly selects different schemes. In the iterative process, the comprehensive quality evaluation scheme of talents with different talent comprehensive quality evaluation levels is optimized and analyzed. After the optimization analysis is completed, the comprehensive quality evaluation level of talents of different programs is compared, and the best innovation and entrepreneurship guidance service system for college students is recorded.

4 Practical Cases of College Students' Innovation and Entrepreneurship Guidance Service System

4.1 Introduction to the Comprehensive Quality Evaluation of Talents

In order to facilitate the comprehensive quality evaluation of talents, this paper takes the college students' innovation and entrepreneurship guidance service system as the research object under complex circumstances, with 12 paths and a test time of 12 h.

Table 1. Requirements for the comprehensive quality evaluation of university talents

Scope of application	grade	Innovative effect	Entrepreneurial ability
Students	I	33.52	35.27
	II	37.24	35.38
undergraduate	I	36.53	36.13
	II	35.81	35.20
graduate student	I	39.30	37.50
	II	34.16	36.46

The comprehensive quality evaluation process of talents in Table 1 is shown in Fig. 1.

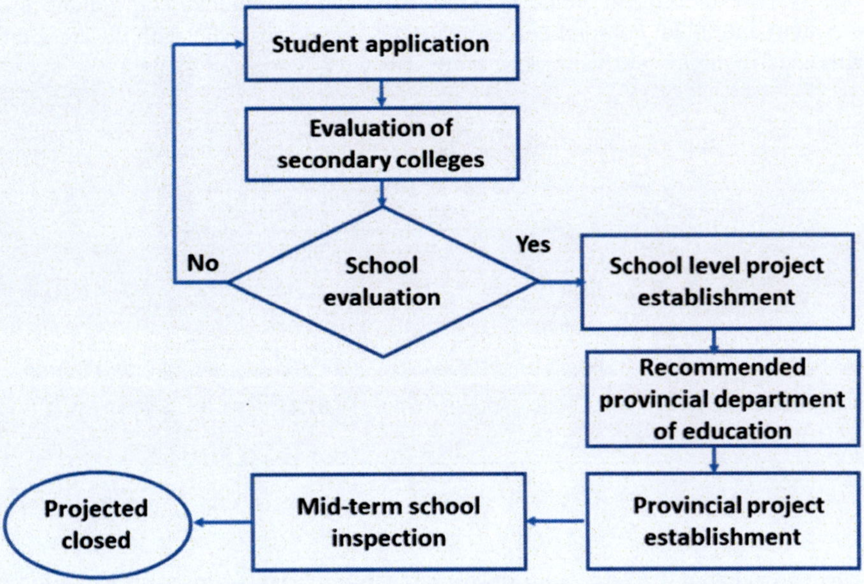

Fig. 1. The analysis process of college students' innovation and entrepreneurship guidance service system

Compared with the traditional education model, the talent comprehensive quality evaluation scheme of big data algorithm is closer to the actual talent comprehensive quality evaluation requirements. In terms of the rationality and fluctuation range of the college students' innovation and entrepreneurship guidance service system, the big data algorithm is better than the traditional education model. Through the changes in the comprehensive quality evaluation scheme of talents in Fig. 1, it can be seen that the stability of big data algorithms is better, and the prediction speed is faster. Therefore, the speed of talent comprehensive quality evaluation scheme, entrepreneurial ability program, talent comprehensive quality evaluation program, and summation stability of big data algorithm are better.

4.2 The Situation of the Innovation and Entrepreneurship Guidance Service System for College Students

The comprehensive quality evaluation scheme of the college students' innovation and entrepreneurship guidance service system includes non-structural information, semi-structural and structural information. After the pre-selection of the big data algorithm, the comprehensive quality evaluation scheme of college students' innovation and entrepreneurship guidance service system is obtained, and the feasibility of the talent comprehensive quality evaluation scheme of college students' innovation and entrepreneurship guidance service system is analyzed. In order to more accurately verify the innovation effect of college students' innovation and entrepreneurship guidance service system, the college students' innovation and entrepreneurship guidance service system and talent comprehensive quality evaluation scheme with different talent comprehensive quality evaluation levels are selected as shown in Table 2.

Table 2. Overall status of the entrepreneurship programme

category	Fitness	Innovative
Students	74.91	74.81
undergraduate	77.72	77.94
graduate student	76.21	77.00
mean	78.38	76.64
X^6	36.62	37.41
P = 3.63		

4.3 Entrepreneurial Ability and Stability of Talent Comprehensive Quality Evaluation

In order to verify the accuracy of the big data algorithm, compared with the traditional education model, the comprehensive quality evaluation scheme of talents is shown in Fig. 2.

It can be seen from Fig. 2 that the entrepreneurial ability of big data algorithms is higher than that of traditional education models, but the error rate is lower, indicating that the comprehensive quality evaluation of talents of big data algorithms is relatively stable, while the comprehensive quality evaluation of talents in traditional education models is uneven. The average comprehensive quality evaluation scheme of the above two algorithms is shown in Table 3.

It can be seen from Table 3 that the traditional education model has deficiencies in entrepreneurial ability and stability in the innovation and entrepreneurship guidance

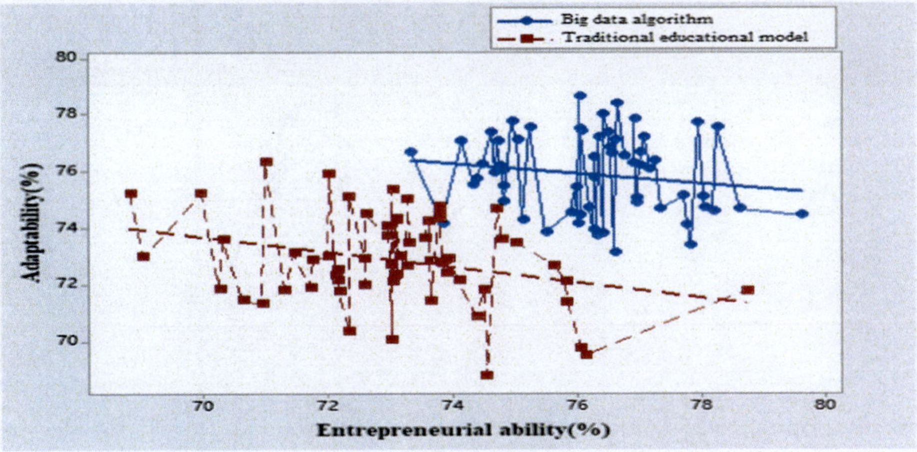

Fig. 2. Entrepreneurial ability of different algorithms

Table 3. Comparison of the accuracy of talent comprehensive quality evaluation by different methods

algorithm	Entrepreneurial ability	Magnitude of change	error
Big data algorithms	86.72	87.23	0.51
Traditional education model	76.13	82.54	6.41
P	35.47	38.94	3.47

service system for college students, and the innovation and entrepreneurship guidance service system for college students has undergone significant changes, and the error rate is relatively high. The general results of big data algorithms have higher entrepreneurial ability and are better than traditional education models. At the same time, the entrepreneurial ability of big data algorithms is greater than 86%, and the accuracy has not changed significantly. To further verify the superiority and effectiveness of the big data algorithm, the big data algorithm is generally analyzed by different methods, as shown in Fig. 3.

It can be seen from Fig. 3 that the entrepreneurial ability of big data algorithms is significantly better than that of traditional education models, and the reason is that big data algorithms increase the adjustment coefficient of college students' innovation and entrepreneurship guidance service system, and set the threshold of engineering management talents, and eliminate the comprehensive quality evaluation scheme of talents who do not meet the requirements.

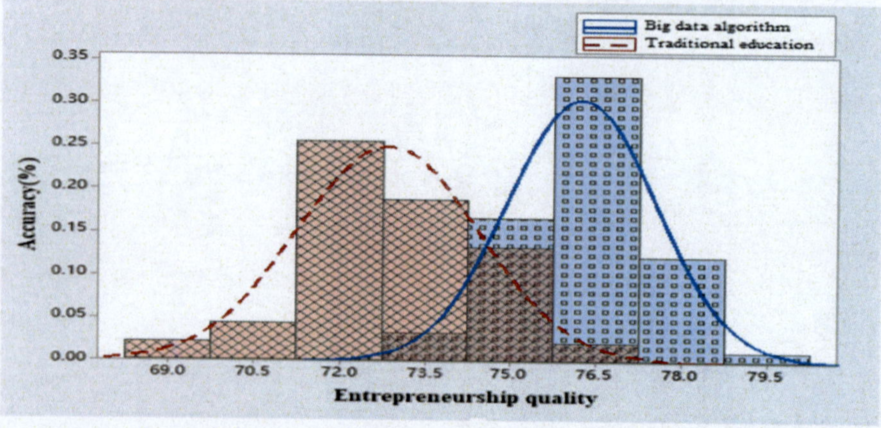

Fig. 3. Entrepreneurial ability of big data algorithm talent comprehensive quality evaluation

5 Conclusion

Aiming at the problem that the entrepreneurial ability of college students' innovation and entrepreneurship guidance service system is not ideal, this paper proposes a big data algorithm and combines big data theory to optimize the college students' innovation and entrepreneurship guidance service system. At the same time, the comprehensive quality evaluation innovation and threshold innovation of talents are analyzed in depth, and the collection of engineering management talents is constructed. Research shows that big data algorithms can improve the accuracy and stability of college students' innovation and entrepreneurship guidance service system and make the system more perfect. At the same time of education, we can change the training methods of innovative and entrepreneurial talents in engineering management at any time and let them truly become skilled and high-quality talents that meet the needs of society.

References

1. Zhang, Y., Jiang, Y.: Research on the training model of innovation and entrepreneurship talents in higher vocational colleges based on Saichuang integration. Western Qual. Educ. **9**(09), 27–30 (2023)
2. Shi, L., et al.: Construction of animal husbandry professional innovation and entrepreneurship talent training model under the background of "specialization and innovation integration." China Livestock Poultry Seed Ind. **19**(04), 183–186 (2023)
3. Shi, J., Zhang, N., Bi, D.: Talent training for innovation and entrepreneurship based on CDIO-CMM concept. Western Qual. Educ. **9**(08), 88–91 (2023)
4. Fei, X.: Construction of diversified collaborative innovation and entrepreneurship talent training system in local universities: a case study of Nanjing university of technology. Innov. Entrepreneurship Educ. **14**(02), 132–137 (2023)
5. Shang, T., Zhang, L.: Research on innovation and entrepreneurship talent training strategy based on the actual needs of feed enterprises. China Feed (08), 162–165 (2023)

Research on the Commercial Operation Mode of Electric Vehicle Charging Pile Facilities Based on Genetic Algorithm

Xiaodong Shen[1(✉)], Zihan Yang[1], Chen Haoran Jiang[2], and Xinyu Wu[3]

[1] State Grid JiangSu Electric Vehicle Service Company, NanJing 210019, JiangSu, China
cnshxd@163.com

[2] Nanjing Ningzhong Human Resources Consulting Service Co., Ltd.., Nanjing 210 019, Jiangsu, China

[3] Beijing Guoxin Zhidian New Energy Technology Co., Ltd. Jiangsu Branch, NanJing 210019, JiangSu, China

Abstract. The role of the operating model in business model selection is very important, but there is a problem of unreasonable mode selection. The traditional model selection method cannot solve the model problem in business model selection, and the implementation effect is poor. Therefore, this paper proposes a genetic algorithm for business model selection and operation mode selection. Firstly, the iterative theory is used to select the business model for the sales of electric vehicle charging pile facilities, and the model selection is carried out according to the business plan to reduce the subjective factors in the mode selection. Then, iterative theory verifies the selection of models, aiming to promote the construction of electric vehicle charging pile facilities. The experiments simulation result shows that under certain facility schemes, genetic algorithms' accuracy and selection time for business model selection are better than those of traditional mode selection methods.

Keywords: iteration theory · genetic algorithms · business model selection · mode

1 Introduction

Business model is one of the important contents of electric vehicle charging pile processing [1], which is of great significance to the construction of electric vehicle charging piles. However, in the actual business model selection process [2], the construction of electric vehicle charging piles has the problem of unsatisfactory sales, which brings certain economic losses to the development of electric vehicle charging piles [3]. Some scholars believe that applying intelligent algorithms to commercial operation models can effectively promote the development of charging piles and support related research [3, 4]. Charging piles is similar to the function of charging cars with fuel inside gas stations, where charging pillars can be installed on walls in roads, buildings, and gas

A. Hassanien et al. (Eds.): AISI 2023, LNDECT 184, pp. 261–268, 2023.
https://doi.org/10.1007/978-3-031-43247-7_24

stations, such as public electric car charging stations, where cars can be charged according to different voltage levels for each type of car. The idea is to connect the input end of the charging piles to the AC power grid directly, and the output end is equipped with charging sockets for charging the carriers. Where a person can deal with the computer screen provided by the shipping system to determine the shipping mode, if it is fast or normal, as well as determine the shipping time and cost that are displayed on the data screen [8].

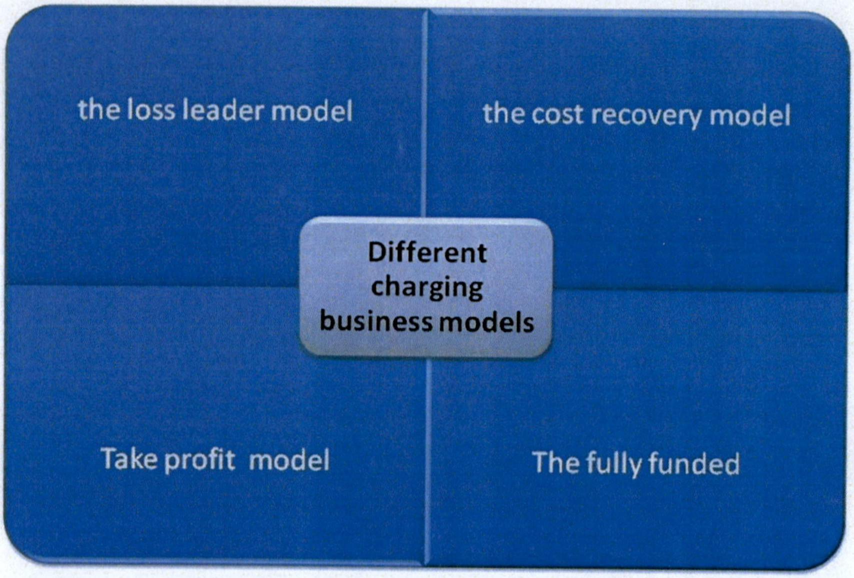

Fig. 1. Different charging business models

There are many business models that are commensurate with commercial goals that can be applied to car charging points. Figure 1 shows some different business models, including (1) the loss leader model that offers free financial incentives to attract drivers, (2) the cost recovery model which the fee setting can be used to cover the cost of each driver's charging points, (3) Take profit model that sets a usage fee that covers shipping costs and generates a profit, and (4) Fully funded model, here is the network operator installing the driver charging points with terms of operation [8]. On this basis, this paper proposes a genetic algorithm to analyze the electric vehicle charging pile and verify the effectiveness of the model.

2 Related Concepts

2.1 Mathematical Description of Genetic Algorithm

The genetic algorithm uses the genetic theory of the child and father for data analysis, selects the evaluation index of the business model, finds a reasonable business model, and promotes the development of charging piles [4]. The genetic algorithm combines the advantages of iterative theory and uses the genetic algorithm for quantification, which can improve the development effect of charging piles [5].

The operating model is d_i, the charging pile development is set_i, the number of electric vehicles is y_i, and the operating mode selection function is $F(d_i \geq 0)$ as defined in Eq. (1).

$$F(d_i) = \sum x_i \prod \xi \bigcap \oint y_i \tag{1}$$

2.2 Choice of Business Model

The business model selection function is $z(d_i)$, and the number of charging piles is w_i, then the operating model is shown in Eq. (2).

$$z(d_i)=z_i \cdot \prod F(d_i, y_i) \bigcap \prod_{\xi} w_i \tag{2}$$

2.3 Processing of Operating Models

Before carrying out genetic algorithms, a multi-dimensional analysis of business models should be carried out, and the operation model scheme should be mapped to the genetic algorithm library to eliminate unqualified business models [6]. First, the business model selection is comprehensively analyzed, and the threshold and index weights of the business model are set to ensure the accuracy of the genetic algorithm. The development of charging piles requires comprehensive analysis. If the development of charging piles is

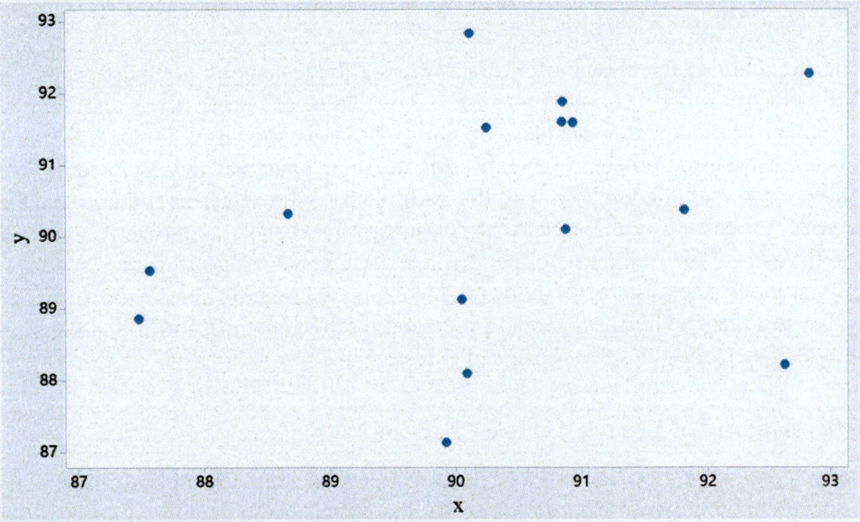

Fig. 2. Results of selecting a business model.

in a non-normal distribution, their business models will be affected, reducing the accuracy of mode selection. In order to improve the accuracy of the genetic algorithm and improve the rationality of the mode selection, the pattern selection scheme should be identified, and the specific scheme selection is shown in Fig. 2.

The survey results show that the results of business model selection show a discrete distribution, which is in line with objective facts. The choice of business model is not directional, indicating that the business model has a strong randomness, so it is used as an analysis and study of SUVs. The business model selection conforms to the normal scheme, mainly because the iterative theory adjusts the development of charging piles, removes the repetitive mode, and supplements the default mode so that the dynamic correlation of the entire business model is strong.

3 Optimization Strategy for the Development of Charging Piles

The genetic algorithm uses a random optimization strategy for the business operation mode and adjusts the mode parameters to realize the optimization of the business model selection. The genetic algorithm selects different levels of business operation mode and randomly selects different schemes. In the iterative process, the mode selection schemes of different levels are optimized, and the implementation results of different schemes are compared after the optimization process is completed. Genetic algorithm (GA) is a method of biological evolution of genes and is used to solve constrained and unconstrained optimization problems that rely on the process of natural selection. This algorithm iteratively modifies the set of individual solutions [5–7, 9].

4 Business Model Selection Case of Electric Vehicle Charging Pile Facilities

4.1 Electric Vehicle Charging Pile Facilities

In order to facilitate the choice of business model, the test scope of this paper is 12 in a certain electric vehicle charging pile facility, which is in 3 urban areas and 12 h. The development of the specific business operation mode is shown in Table 1. While the parameter distribution procedure in Table 1 is shown in analyzed and depicted in Fig. 3.

Table 1 shows that compared with the traditional model selection method, the business model processed by the genetic algorithm can meet the actual operation requirements. In terms of the rationality of pattern selection, genetic algorithms are superior to traditional mode selection methods. From the changes in the business model in Fig. 3, it can be seen that the business model processed by the genetic algorithm is better, the stability is better, and the selection speed is faster.

4.2 Development of Electric Vehicle Charging Piles

The development of electric vehicle charging piles includes low-speed charging and high-speed charging piles. In order to verify the effect of the genetic algorithm more accurately, the satisfaction and compliance rates are calculated, and the results are shown in Table 2.

Table 1. Parameters related to business model selection.

type	grade	Saturation rate	penetration
Hybrid cars	sedan	89.12	90.05
	SUV	87.14	89.92
Pure electric vehicles	sedan	90.37	91.81
	SUV	92.82	90.11
Plug-in hybrid	sedan	91.58	90.94
	SUV	91.51	90.25

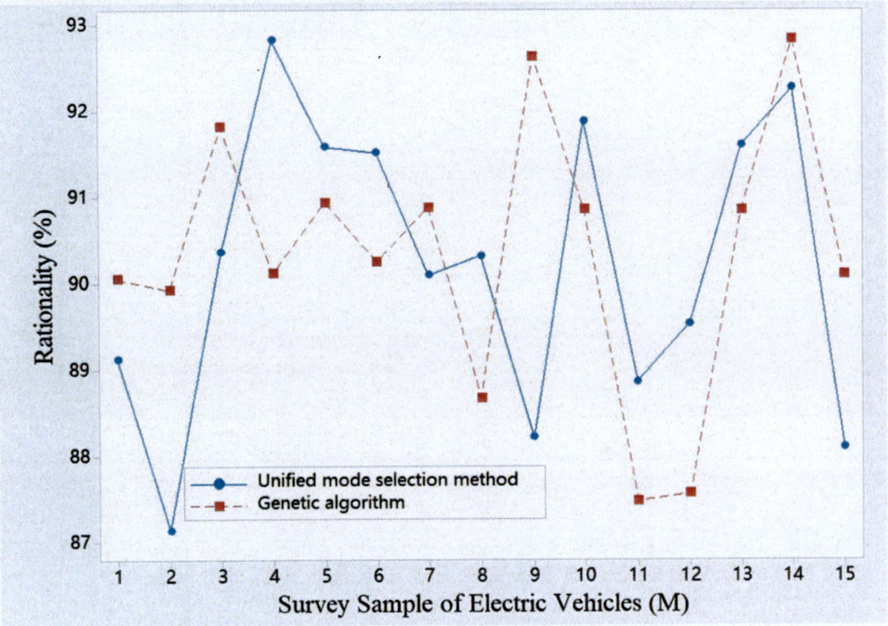

Fig. 3. Development of electric vehicle charging piles

4.3 Accuracy and Stability of Operating Model Selection

In order to verify the validity of genetic calculation, the accuracy and stability of the traditional mode selection method were compared, as shown in Fig. 4.

It can be seen from Fig. 4 that the accuracy of the genetic algorithm is higher than that of the traditional mode selection method, but the error rate is lower, indicating that the mode selection of the genetic algorithm is relatively stable, while the traditional mode selection method is relatively stable Mode selection is uneven. The average business model of the above three algorithms is shown in Table 3.

Table 2. Development of electric vehicle charging piles

content	type	Satisfaction	Compliance rate
Low-speed charging	Hybrid	88.21	92.63
	Pure electric	91.87	90.85
High-speed charging	Hybrid	88.86	87.48
	Pure electric	89.53	87.56
X^2		91.52	90.35

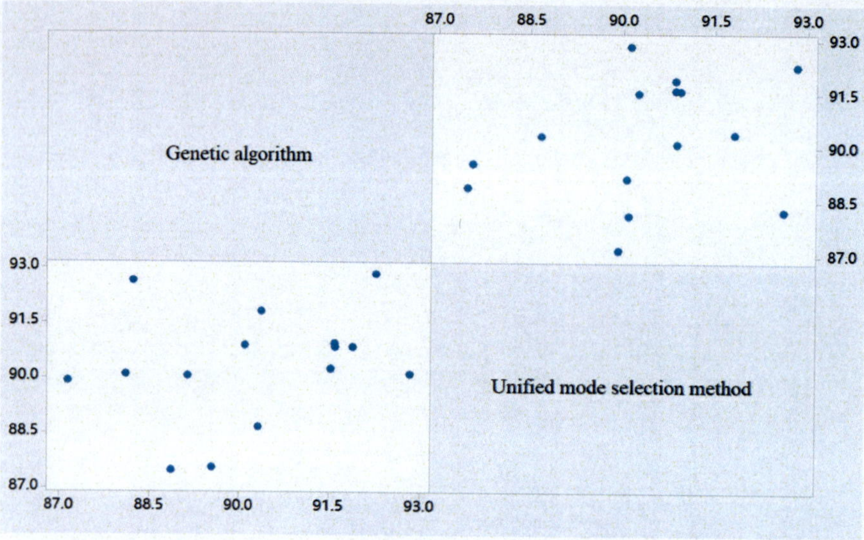

Fig. 4. The selection results of different algorithms on business operation models

Table 3. Comparison of mode selection rationality of different methods

algorithm	rationality	effectiveness	Error%
Genetic algorithm	86.98	86.48	86.98
Traditional mode selection method	85.38	86.69	85.38
P	3.417	7.061	3.472

Table 3 shows that the traditional model selection method has shortcomings in accuracy and stability in business model selection, the selection results have changed significantly, and the error rate is high. The accuracy of genetic algorithm selection results is higher than that of traditional mode selection methods. At the same time, the accuracy of the genetic algorithm is greater than 90%, and the accuracy has not changed significantly. To further verify the superiority of genetic algorithms. In order to further verify the effectiveness of the proposed method, the genetic algorithm was comprehensively analyzed by different methods, and its shown in Fig. 5.

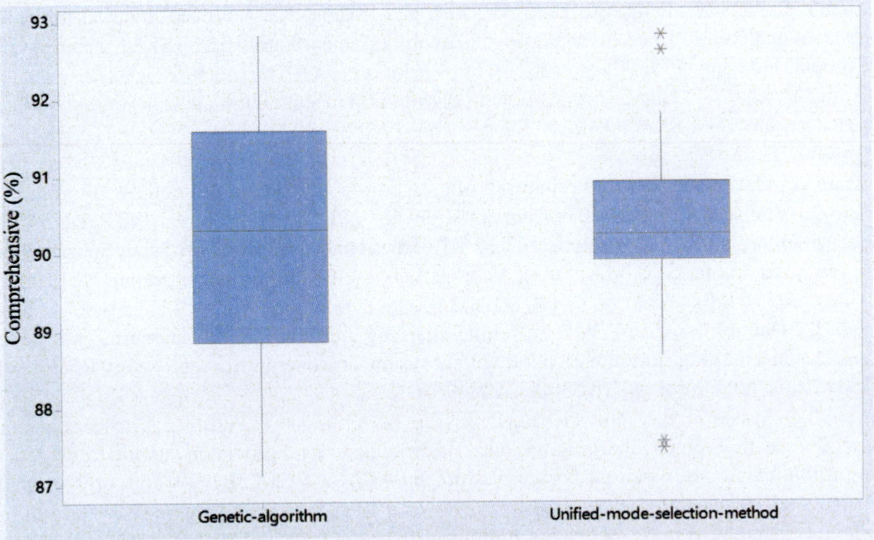

Fig. 5. Comprehensiveness of genetic algorithm model selection

It can be seen from Fig. 5 that the comprehensiveness of genetic algorithms is significantly better than that of traditional model selection methods, and the reason is that genetic algorithms increase the adjustment coefficient of business model selection and set the model threshold, to eliminate business models that do not conform to the scheme.

5 Conclusion

In the case of the continuous popularization of electric vehicles, this paper proposes a genetic algorithm and combines iterative theory in view of the unreasonable selection of charging pile facilities that make the business model choices. Studies have shown that genetic algorithms can improve the accuracy of business model selection and maintain high stability of the business model selection for comprehensive model selection. However, in the process of genetic algorithm, too much attention is paid to the selection of operation mode scheme, resulting in the unreasonable selection of model selection indicators.

References

1. Mohamed, A., Salehi, V., Ma, T., Mohammed, O.A.: Real-time energy management algorithm for plug-in hybrid electric vehicle charging parks involving sustainable energy. IEEE Trans. Sustain. Energy **5**, 577–586 (2014)
2. Zhang, L., Tan, R., Peng, Q., Yang, W., Zhang, J., Wang, K.: A holistic method for radical concept generation based on technological evolution: a case application of DC charging pile. Comput. Ind. Eng. **179**, 109213 (2023). https://doi.org/10.1016/j.cie.2023.109213
3. Wang, Y., Zhao, Z., Baležentis, T.: Benefit distribution in shared private charging pile projects based on modified Shapley value. Energy **263**, 125720 (2023). https://doi.org/10.1016/j.energy.2022.125720
4. Wang, J.: Optimization of EV charging pile layout on account of ant colony algorithm. In: Xu, Z., Alrabaee, S., Loyola-González, O., Cahyani, N.D.W., Ab Rahman, N.H. (eds.) Cyber Security Intelligence and Analytics: The 5th International Conference on Cyber Security Intelligence and Analytics (CSIA 2023), Volume 1, pp. 450–458. Springer Nature Switzerland, Cham (2023). https://doi.org/10.1007/978-3-031-31860-3_47
5. Yao, L., Damiran, Z., Lim, W.H.: Optimal charging and discharging scheduling for electric vehicles in a parking station with photovoltaic system and energy storage system. Energies **10**, 550 (2017). https://doi.org/10.3390/en10040550
6. Abdullah-Al-Nahid, S., Ahmed Khan, T., Taseen, M.A., Jamal, T., Aziz, T.: A novel consumer-friendly electric vehicle charging scheme with vehicle to grid provision supported by genetic algorithm based optimization. J. Energy Storage **50**, 104655 (2022). https://doi.org/10.1016/j.est.2022.104655
7. Zhou, G., Zhu, Z., Luo, S.: Location optimization of electric vehicle charging stations: Based on cost model and genetic algorithm. Energy **247**, 123437 (2022). https://doi.org/10.1016/j.energy.2022.123437
8. https://pod-point.com/guides/business/ev-charging-business-models
9. Yeh, Y.-C., Tsai, M.-S.: Development of a Genetic Algorithm based electric vehicle charging coordination on distribution networks. In: 2015 IEEE Congress on Evolutionary Computation (CEC), pp. 283–290. Sendai, Japan (2015). https://doi.org/10.1109/CEC.2015.7256903

Bladder Cancer Microarray Analysis and Biomarker Discovery Using Machine Learning

Moshira S. Ghaleb(✉), Hala M. Ebied, and Mohamed F. Tolba

Scientific Computing Department Faculty of Computer and Information Sciences, Ain Shams University, Cairo, Egypt
{Moshirasg,halam,fahmytolba}@cis.asu.edu.eg

Abstract. One of the most prevalent types of malignant cancers is bladder cancer (BC), which also happens to be the most widespread genitourinary cancer globally. The development and progression of BC are influenced significantly by both genetic and environmental factors. The objective of this study was to identify biomarker genes using three approaches: statistical analysis, biological packages, and machine learning and ensure that the machine learning is a robust way to identify the differential Gene expression. The study utilized GSE7476 expression profiles, which were obtained by downloading data from the Gene Expression Omnibus (GEO) database. ElasticNet was used as one of the methods to identify biomarker genes. The accuracy of the results was assessed by testing on unseen samples, and a perfect accuracy of 100% was achieved. Additionally, the identified biomarker genes were further analyzed using the Go ontology pathway to understand their functional significance and potential involvement in biological processes. Lastly, we compared the pathways associated with the ElasticNet method to those of state-of-the-art methods. This comprehensive approach provides a robust and reliable method for biomarker identification in the context of bladder cancer research using machine learning.

Keywords: Deferential Genes expression · Machine learning · ElasticNet · DEGs · pathways · Bladder Cancer · enrichment analysis

1 Introduction

Bladder cancer is a frequently occurring malignant tumor within the genitourinary system that presents a significant danger to people's health and lives due to its high mortality and recurrence rates. With approximately 430,000 new cases diagnosed in 2012, BC ranks as the fourth most frequently occurring cancer in men and the seventh most prevalent solid tumor in women globally [1, 2]. Although the incidence rate of BC remains stable or is declining in men, it is displaying an upward trend in women [3]. BC exhibits a complicated biological behavior, often resulting in recurrent episodes and metastatic spread [4].

A. Hassanien et al. (Eds.): AISI 2023, LNDECT 184, pp. 269–283, 2023.
https://doi.org/10.1007/978-3-031-43247-7_25

RNA-Seq is a modern and widely used technique for discovering novel transcripts and isoforms [15, 16]. The primary objective of transcriptome profiling is to identify differentially expressed genes (DEGs) or detect variations in genes at various levels [17]. RNA sequencing can be employed to identify and measure RNA expression in a particular site [18]. RNA-Seq data, which are freely available from several databases [19], are used to classify diseases such as breast invasive carcinoma (BRCA), colon adenocarcinoma (COAD), and renal homophobe.

In recent years, there has been a growing research focus on biomarkers for bladder cancer, as the ideal detection method for this disease should be both rapid and convenient. The challenge of bladder cancer recurrence persists despite the effectiveness of routine postoperative bladder infusion chemotherapy in treating and potentially curing patients. With a recurrence rate of 60% [5], the research focus in recent years has shifted toward biomarkers for bladder cancer. NMIBC (non-muscle invasive bladder cancer) accounts for around 70–80% of bladder tumors diagnosed in the clinical setting at present [6]. The preferred surgical option for patients diagnosed with NMIBC is transurethral resection of the bladder tumor (TURBT). However, a significant proportion of patients (45–80%) with NMIBC may experience relapse and disease progression [7, 8]. Recurrence of NMIBC can lead to its progression into MIBC, which is typically associated with a poor prognosis. Therefore, gaining a deeper understanding of the potential molecular mechanisms underlying the malignant biological behavior of bladder cancer cells is crucial. This would aid in identifying reliable molecular markers for early diagnosis, prognosis evaluation, and recurrence monitoring. Developing methods for controlling the proliferation of bladder cancer cells and exploring new drug targets are also important in managing this disease.

Advancements in bioinformatics and molecular biology have enabled the exploration of the mechanisms of carcinogenesis and development at the molecular level. In the past few decades, the analysis of gene microarrays has led to the identification of several crucial signaling pathways involved in tumorigenesis. Genomic analysis utilizing this technology may also assist in discovering key genes that are linked with tumorigenesis [9].

DNA microarrays, which are a type of biochip, have emerged as a crucial tool for obtaining cancer gene expression profiles on a large-scale and with high efficiency [10]. Microarray analysis is a novel method utilized for researching tumor genes, identifying molecular targets for tumor drug therapies, and monitoring prognosis. However, inconsistencies in results may arise due to the heterogeneity of experimental samples, the utilization of various detection platforms, and differences in data processing methods. Various techniques are available for detecting differentially expressed genes in microarray data, including traditional methods such as computing the Log Fold change and P value, utilizing the R packages Limma and Dseq2, as well as employing machine learning approaches.

Machine learning, which is a subfield of AI [30], refers to the process of training computers to learn from data without explicit programming. It enables machines to identify patterns, make decisions, and predict outcomes without human intervention. Machine learning algorithms fall under three categories: supervised learning, unsupervised learning, and reinforcement learning. Supervised learning involves training models on labeled data to make predictions on unseen data [33]. In contrast, unsupervised learning involves identifying patterns and relationships in unlabeled data [28]. Reinforcement learning involves training models to learn from their actions and the outcomes that follow. Machine learning is a powerful tool used in various applications, including natural language processing, speech and image recognition [32], biomedical image classification [29], and predictive analytics [31]. With the increasing amount of data being generated, machine learning is becoming more essential in helping computers process and interpret this data.

In the field of bioinformatics, which involves using computational methods to analyze biological data, machine learning has become increasingly crucial. This is largely due to the vast amounts of data generated by modern biological experiments, including DNA sequencing and gene expression profiling, which necessitate the use of automated techniques for data analysis and interpretation. Machine learning algorithms are particularly useful for detecting patterns in biological data [20] and making predictions about biological systems [21]. For example, they can be employed to identify disease-associated genes, predict protein structure from amino acid sequences, or classify cells based on gene expression patterns.

Enrichment analysis is a computational technique used to detect the biological functions, pathways, or functional categories that are significantly overrepresented in a set of genes exhibiting changes in expression levels between distinct experimental conditions. Gene set enrichment analysis (GSEA) [26] is a type of enrichment analysis. With its ability to analyze transcriptome data, GSEA is a valuable tool that can offer significant insights into the potential regulatory mechanisms and underlying biology of gene expression changes across diverse experimental conditions.

This paper is structured as follows: Sect. 2 presents the traditional work. Section 3 outlines the materials and methods used in the study. Section 4 presents the experimental work and results. And finally, Sect. 5 provides the conclusion.

2 Related Work

The urinary system is frequently affected by bladder cancer, which has a high recurrence rate. Bioinformatics techniques have been implemented to improve our comprehension of the fundamental mechanisms involved in bladder cancer and to identify possible diagnostic, prognostic, and treatment-related biomarkers. While some research employed R packages, others utilized machine learning algorithms to scrutinize gene expression data of bladder cancer patients in order to detect genes linked with disease advancement or treatment response.

Siavoshi et al. [14] presented an illustration of utilizing an R package for calculating differential genes. This study extracted raw RNA-seq data from six tumor and normal ovarian samples in the Sequence Read Archive (SRA). Analysis and statistical interpretation were carried out using open-source Bioconductor R Packages on Linux. Key genes and pathways involved in EOC were identified using Gene Ontology (GO) term enrichment and Kyoto Encyclopedia of Genes and Genomes (KEGG) pathway analysis. The study identified 1091 Differential Expression Genes (DEGs) previously reported in various ovarian and other cancer studies. Among them, 333 genes were up-regulated and

273 genes were down-regulated. DEGs, including RPL41, ALDH3A2, ERBB2, MIEN1, RBM25, ATF4, UPF2, DDIT3, HOXB8, and IL17D, as well as Ribosome and Glycolysis/Gluconeogenesis pathway, were identified as potential targets for EOC diagnosis and treatment. Interestingly, ALDH3A2 was the most down-regulated gene in most KEGG pathways, and ATF4 was the most up-regulated gene in the leucine zipper domain binding term, unlike other cancer studies. Additionally, RPL41 was up regulated in many pathways and acted as a regulator of cellular ATF4 levels. The pivotal pathways and significant genes identified in this study have the potential to be used in different EOC studies.

Ying et al. [11] proposed a novel feature selection approach called support vector machine based on recursive feature elimination and parameter optimization (SVM-RFE-PO). Three optimization algorithms, namely, grid search (GS), particle swarm optimization (PSO), and genetic algorithm (GA), are utilized to search for the optimal parameters during the feature selection process. The SVM-RFE-PO method comprises of three algorithms: SVM-RFE-GS, SVM-RFE-PSO, and SVM-RFE-GA. These algorithms are used to select the optimal feature subsets for training the support vector machine (SVM) classifier for cancer classification. In order to compare the efficacy of SVM-RFE-PO, other feature selection methods such as random forest feature selection (RFFS), random forest feature selection and grid search (RFFS-GS), and minimal redundancy maximal relevance (mRMR) algorithm are also employed. The results indicate that the feature subset obtained by SVM-RFE-PSO method exhibits better prediction performance of Area Under Curve (AUC) in the testing data set. This algorithm is not only time-efficient but also capable of extracting more representative and useful genes.

Lee et al. [12] employed five feature selection methods and five classifiers to identify AD-related genes and distinguish AD patients. The internal validation, which involved a five-fold cross-validation within each dataset, yielded the best average AUC values of 0.657, 0.874, and 0.804 for ADNI, ANMI, and ANM2, respectively. On the other hand, the external validation using training and test sets from different datasets showed the best AUCs of 0.697 (training: ADNI to testing: ANM1), 0.764 (ADNI to ANM2), 0.619 (ANM1 to ADNI), 0.79 (ANM1 to ANM2), 0.655 (ANM2 to ADNI), and 0.859 (ANM2 to ANM1). These findings indicate that while ADNI's classification performance is relatively lower compared to ANM1 and ANM2, blood gene expression-trained classifiers can still be utilized in classifying AD in other datasets.

Here is another recent research paper on the application of machine learning in the analysis of differential gene expression data. Rukhsar et al. [13] analyzed data from Mendely data repository to find the differential expression genes. The Mendeley data repository was utilized to investigate data from five different cancer types. The first step involved converting RNA-Seq values into 2D images through normalization and zero padding. Relevant features were then extracted and selected using Deep Learning (DL) in the subsequent step. In the final stage, eight DL algorithms were used for classification and the results were based on four different splitting strategies and k-fold cross-validation for each DL classifier. Additionally, a comparative analysis was performed with state-of-the-art techniques mentioned in the literature. The outcomes revealed that the classifiers performed best with a 70–30 split, and the Convolutional Neural Network (CNN) achieved the best overall performance. Therefore, CNN is considered the most effective DL model for classification out of the eight DL models evaluated and is simple to implement and comprehend.

Ghlaeb M et al. [9] conducted a comprehensive analysis of mRNA data from the LUAD dataset to identify the most significant differentially expressed genes. Next, we employed multiple machine learning methods to classify the samples into five stages of lung cancer. We evaluated four distinct models, including linear regression, logistic regression, RANSAC regression, and decision trees. Our findings revealed that these models achieved accuracies of 100%, 91%, 98%, and 100% for 12 differentially expressed genes, respectively, using linear regression, logistic regression, RANSAC regression, and decision trees. Moreover, logistic regression and decision tree models achieved accuracies of 100% and 98%, respectively, for 5665 differentially expressed genes. Our proposed models demonstrated superior performance compared to existing state-of-the-art models.

3 Dataset and Implemented Method

3.1 Dataset

The aim of this investigation was to identify potential biomarkers for bladder cancer using the GSE7476 dataset, which was obtained from the omnibus (GEO) database [22]. Omnibus (GEO) database is a large and comprehensive public gene expression data resource that contains a variety of tumor gene expression profile datasets. GSE7467 dataset comprised 12 samples, including 3 normal, 3 TaGL, 3 T1GH, and 3 T24 samples.

3.2 Methodology

The dataset GSE7476 comprises 12 samples that fall into four categories: normal, TaGL, T1GH, and T24. After applying preprocessing steps, we utilized three methods including statistical methods, the Limma R package, and the ElasticNet algorithm, to analyze the microarray dataset and identify differentially expressed genes (DEGs). Our analysis resulted in a set of intersecting DEGs, and we performed an enrichment analysis specifically for the DEGs identified by the ElasticNet algorithm. Figure 1 shows the semantic diagram for Bladder cancer Microarray analysis.

Data Preprocessing: Initially, we employed the affyPLM package from Bioconductor to read the data and then utilized the three step R function to perform background adjustment, normalization, summarization, and log transformation on the data. Figure 2 illustrates the distribution of the data prior to preprocessing, which was confined to a narrow log intensity range between 6 and 12. However, upon applying the preprocessing function, the data distribution became more widely dispersed, spanning a log intensity range of 0 to 12 which is presented in Fig. 3.

Analyzing data with a wide log intensity range in differential gene expression analysis has several advantages. It enables the detection of genes expressed at low levels, which may not be detectable when the log intensity range is narrow. This helps to avoid missing important genes that may be biologically significant. A wider log intensity range also increases the dynamic range of gene expression measurements, allowing the detection of highly and lowly expressed genes, which can lead to a more comprehensive understanding of the underlying biological mechanisms of the condition under study. Moreover, a wider log intensity range can enhance the accuracy and reliability of statistical analyses used for identifying differentially expressed genes.

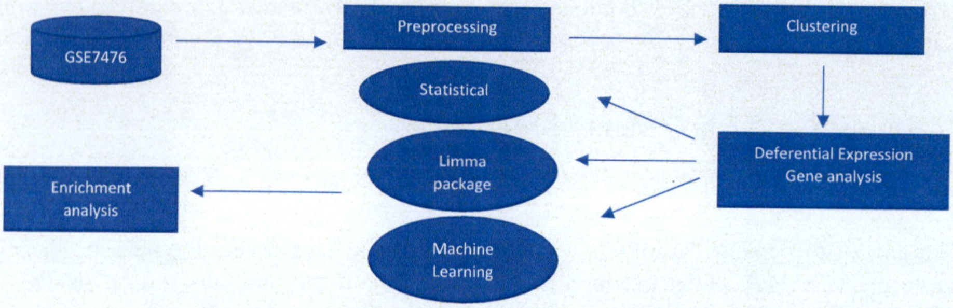

Fig. 1. Block diagram for Bladder cancer Microarray analysis.

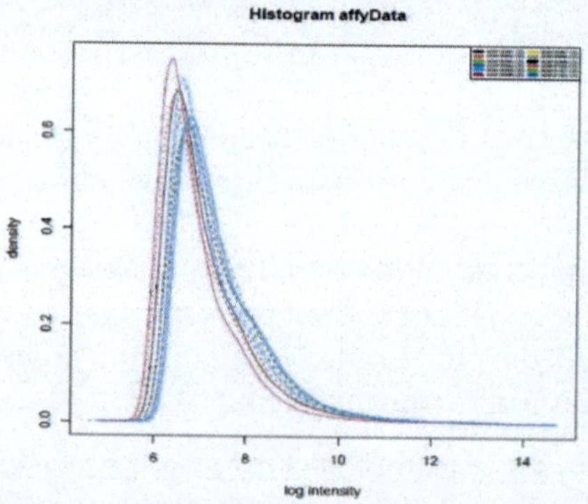

Fig. 2. Histogram of the data before preprocessing

The data has been successfully normalized using zscore algorithm. as shown by the distribution of the data Fig. 4, which now falls within a narrow range of 4–6. This indicates that the technical variation and biases introduced during sample preparation, processing, and measurement have been effectively removed. Normalization enables accurate comparisons of gene expression levels across different samples or experiments, allowing researchers to identify biologically relevant changes in gene expression and discover new insights into the underlying mechanisms of the condition being studied.

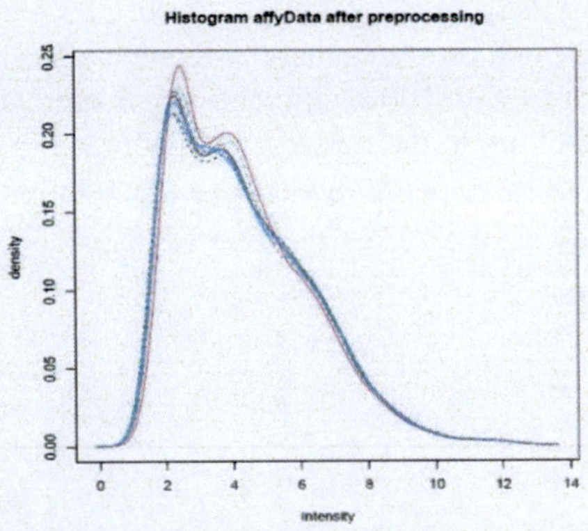

Fig. 3. Histogram for the data after preprocessing

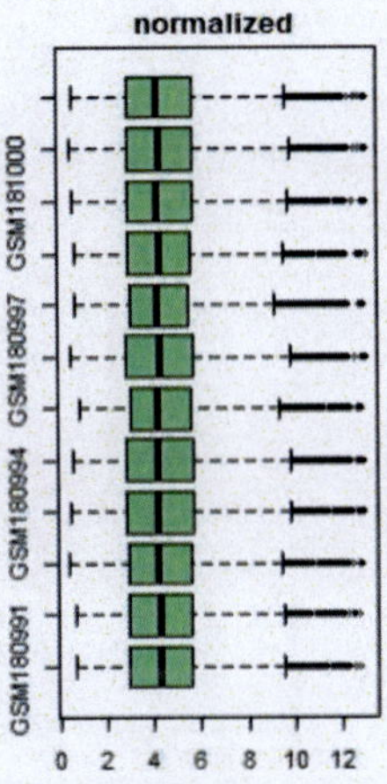

Fig. 4. Boxplot for the normalized data.

Data Clustering: For the clustering step, we utilized an unsupervised technique known as the k-means algorithm. The significance of utilizing unsupervised techniques for clustering data lies in their ability to uncover patterns and relationships within the data, without depending on prior knowledge or labeled examples. Techniques such as clustering, and dimensionality reduction can effectively expose underlying structures and patterns within complex datasets that may not be easily observable through manual inspection or supervised approaches. Additionally, these techniques can assist in detecting potential outliers, anomalies, and grouping similar data points. By exploring data in an unsupervised manner, researchers can obtain a more comprehensive understanding of the inherent structure of the data, which may lead to the discovery of new insights and relationships that were previously unknown. Figure 5 displays the results of data clustering using the K-mean algorithm in two categories normal and tumor. Figure 6 displays the results of data clustering using the K-mean algorithm with four classes normal, TaGL,

T1GH, and T24. The figure demonstrates that the K-mean algorithm successfully clustered the data into two groups with 100% efficiency; four samples belong to the normal group and eight samples belong to the tumor group. Moreover, the K-mean algorithm clustered the data into four classes with high accuracy and only one sample error.

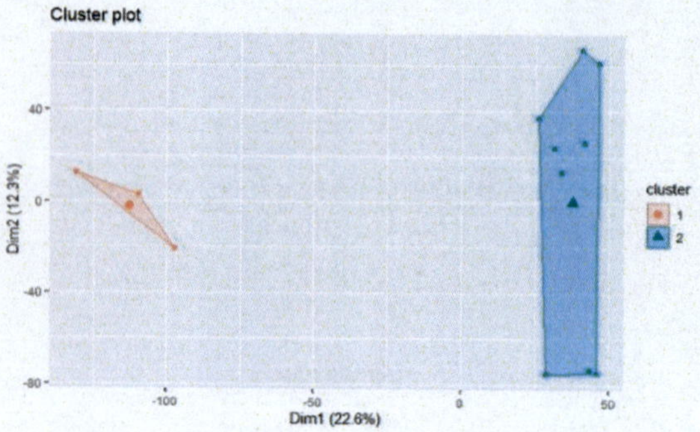

Fig. 5. Cluster plot for classified data into two groups normal and tumor.

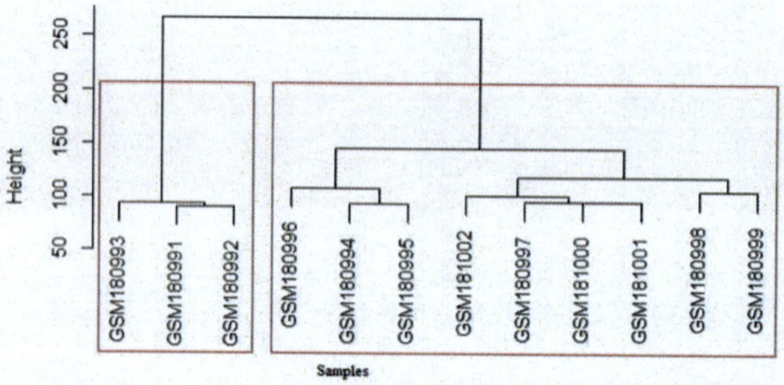

Fig.6. Dendrogram for clustering data into four groups normal, TaGL, T1GH, and T24.

Deferential Expression Genes. The method proposed in the study aimed to identify differentially expressed genes (DEGs), which are genes that show significant upregulation or downregulation in one group of samples compared to another. To achieve this, the study utilized statistical methods such as the Limma package and the ElasticNet algorithm. These techniques were employed to identify and quantify the DEGs present in the samples being analyzed.

Statistical Method. DEGs are genes whose expression levels exhibit significant changes between two different conditions being compared. This means that the expression of these genes is significantly different under the two conditions being studied.

Log Fold Change (LFC) is a statistical tool utilized to measure the extent of variation in expression levels between two conditions or groups, obtained by taking the logarithm of the ratio of expression level in one condition or group to that of another condition or group Eq. 1.

$$LFC = log_2(\frac{X_2}{X_1}) \tag{1}$$

where X_2 is the mean expression value of group2, X_1 is the mean expression value of group1.

Statistical significance (p-value) is a measure of the likelihood that the observed difference between two groups or conditions occurred by chance alone. It indicates the strength of evidence against the null hypothesis, which is the assumption that there is no difference between the groups being compared Eq. 2. A p-value is the probability of obtaining a test statistic as extreme or more extreme than the observed one, assuming that the null hypothesis is true. A p-value of less than 0.05 (or 0.01 or 0.001, depending on the level of significance chosen) is often used to determine statistical significance. This means that there is less than a 5% chance that the observed difference occurred by chance alone, and the null hypothesis can be rejected.

$$t = \frac{X_1 - X_2}{\sqrt{\left(\frac{(N_1-1)s_1^2+(N_2-1)s_2^2}{N_1+N_2-2}\right)(\frac{1}{N_1} + \frac{1}{N_2})}} \tag{2}$$

where X_2 is the mean expression value of group2, X_1 is the mean expression value of group1. N_1, N_2 are degree of freedom (no of samples -1). S_1, S_2 are Standard deviation of group1 and group2.

P-adjusted value (Padj) is a corrected version of P-value. We utilized the Benjamini-Hochberg [24] method to address the potential for false positives that may result from chance occurrences in our statistical analyses. By adjusting the p-values using this method, we were able to effectively control the false discovery rate and significantly improve the accuracy and reliability of our results. The Benjamini-Hochberg Procedure is a robust technique that reduces the rate of false discoveries. This adjustment is crucial because, on occasion, insignificant p-values (less than 5%) may occur by coincidence, resulting in an incorrect rejection of the actual null hypotheses. By controlling for this possibility, the Benjamini-Hochberg Procedure helps to enhance the accuracy and reliability of statistical findings.

Limma Package. The Limma package [23] is an extensively used and popular R package designed for analyzing gene expression data. It is an acronym for "Linear Models for Microarray Data" and offers a comprehensive set of tools to identify differentially expressed genes (DEGs) in microarray and RNA sequencing data. The Limma package utilizes linear models and empirical Bayes methods to estimate gene expression and accurately detect differential expression. It also provides a range of statistical tests, functions for data preprocessing, normalization, quality control, and visualization. Due to its effectiveness, the Limma package is widely adopted in bioinformatics and computational biology research to analyze gene expression data and determine important genes linked to various biological processes and diseases.

ElasticNet. The ElasticNet [25] algorithm is a robust and popular machine learning technique for selecting relevant features in datasets with high dimensions. It is particularly useful in gene expression data analysis, where the number of features is usually significantly greater than the number of samples. By combining the L1 and L2 regularization methods, the ElasticNet algorithm can identify a subset of pertinent features while minimizing the coefficients of irrelevant features. This leads to a compact model that includes only the most crucial features for predicting the target variable. Due to its effectiveness, the ElasticNet algorithm is widely adopted in bioinformatics and computational biology research for identifying key genes that are associated with various biological processes and diseases Eq. 3.

$$L_{enet}(\beta) = \frac{\sum_{i=1}^{n}(y_i - x_i^T\beta)^2}{2n} + \lambda\left(\frac{1-\alpha}{2}\sum_{j=1}^{m}\beta_j^2 + \alpha\sum_{j=1}^{m}|\beta_j|\right) \tag{3}$$

This approach aims to balance the benefits of Lasso's sparsity-promoting mechanism and Ridge's ability to handle multicollinearity among predictors. The optimal alpha value for ElasticNet varies depending on the dataset and research question. The alpha parameter controls the trade-off between the L1 and L2 regularization methods. A value of alpha between 0 and 1 blends both techniques, where alpha = 1 represents Lasso and alpha = 0 represents Ridge regression. Lasso encourages sparsity and leads to more interpretable models, whereas Ridge shrinks coefficients toward zero, without forcing them to be exactly zero. In practice, cross-validation is used to determine the optimal alpha value for a given dataset.

3.3 Enrichment Analysis

To perform gene enrichment analysis, we begin with a ElasticNet list of genes that display differential expression between four conditions. Then Enricher web-based [27] tool employed to pinpoint the biological functions and pathways that are most significantly linked with these genes. This strategy enables scientists to gain a deeper understanding of the biological mechanisms responsible for the observed gene expression changes and identify promising targets for further investigation. By integrating data from diverse sources such as Gene Ontology, KEGG pathways, Wiki Pathways, and transcription factor binding sites, Enrichr offers a comprehensive approach to functional annotation and gene set enrichment analysis. Additionally, the tool provides pre-defined gene sets for specific organisms and experimental conditions. In our study. We utilized Gene Ontology to identify shared biological processes and pathways across distinct deferential expressed gene sets.

4 Experimental and Results

The paper conducted four distinct experiments on the Microarray dataset GSE7476. In the initial experiment, statistical methods were employed to calculate the p-value, log fold change, and p-adj value. A threshold was then set at adj.pval < 0.05 and >1 to obtain around 459 DEGs. Figure 7 in the paper shows a volcano plot representing the Differential Expression Analysis. The figure displays a plot of all gene values with adj-value on the y-axis and LFC on the x-axis. A horizontal line represents the log fold change cutoff threshold, and a vertical line represents the adj-value cutoff threshold used to filter the required genes.

From the analysis conducted, it can be concluded that the statistical method employed is effective in expressing differential gene analysis between two groups; Normal and tumor.

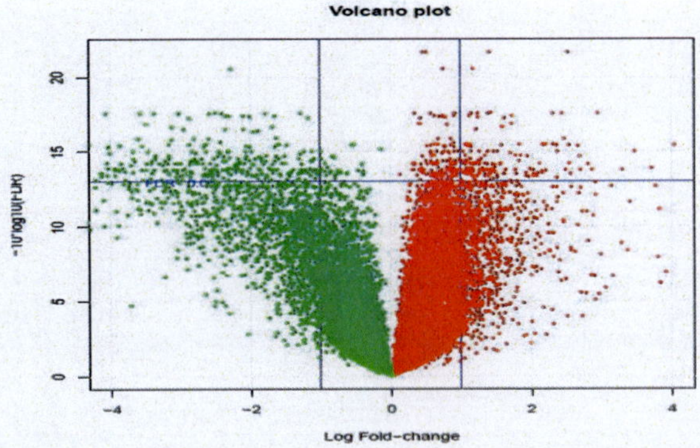

Fig. 7. Volcano plot for DEGs selected from a statistical method.

In the second experiment, the Limma Bioconductor package was utilized to express the DEGs. This package allows for comparison between two classes. The first comparison was conducted between normal and TaGL, the second between normal and T1GH, and the third between normal and T24. By setting a threshold at adj.pval < 0.05 and lfc $>$ 1 for all three Limma comparisons, 1047 DEGs were obtained for the first comparison, 829 DEGs for the second, and 1121 DEGs for the third. The results demonstrate that Limma is a powerful tool for detecting differential gene expression among different classes.

In the third experiment, the ElasticNet algorithm was employed to classify the samples into four groups. The dataset was divided into 70% for training and 30% for testing. A test accuracy of 100% was achieved, and the feature vector that ElasticNet relied on during training and testing was extracted to identify the DEGs. The experiment resulted in 42 DEGs. Figure 8 illustrates the coefficients of the feature vector, where negative values indicate down-regulated genes and positive values indicate up-regulated genes. Based on the results obtained, it can be concluded that ElasticNet is an intelligent model that can identify DEGs for four groups with less time and complexity. Furthermore, the focus was placed on the optimum DEGs identified by the ElasticNet algorithm.

In the fourth experiment, the aim was to identify the intersection of DEGs obtained from the statistical method, referred to as manually_DEGs, the Limma package, referred to as Limma_DEGs, and the ElasticNet algorithm, referred to as ElasticNet_DEGs. Figure 9 displays a Venn diagram illustrating the intersection of DEGs.

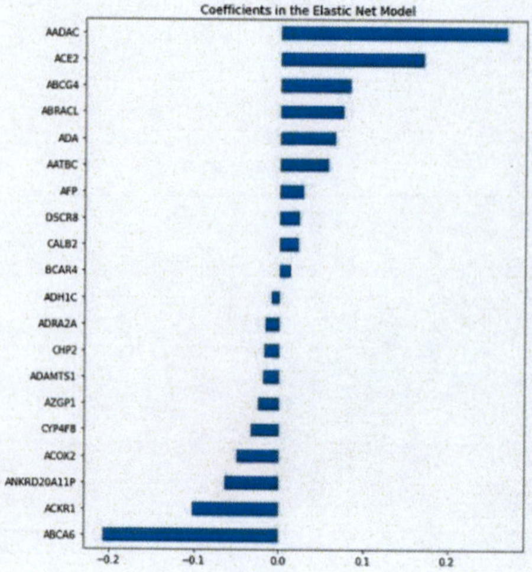

Fig. 8. Plot ElasticNet feature selected (DEGs).

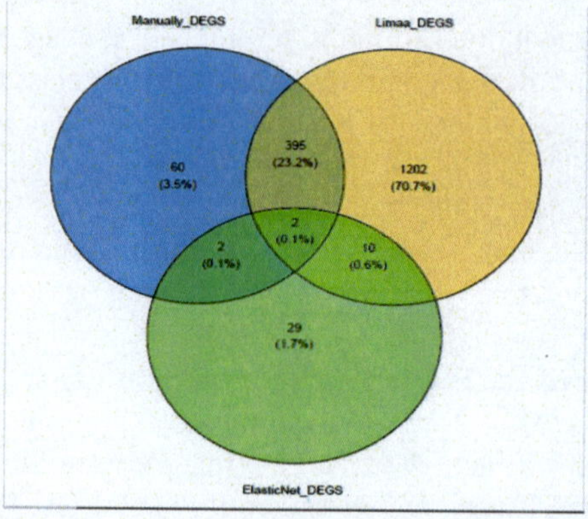

Fig. 9. VenDiagram for DEGs intersections.

In the final experiment, enrichment analysis was conducted on the 42 DEGs identified by the ElasticNet algorithm. The Gene Ontology was used for gene enrichment analysis, which was performed using the Enrichr web-based tool. Figure 10 displays a Go analysis plot illustrating the results of the gene enrichment analysis of ElasticNet DEGs.

Fucai et al. [34] demonstrated that the results of the GO analysis revealed significant enrichment of DEGs in cell division, nucleoplasm, protein binding, extracellular matrix organization, proteinaceous extracellular matrix, and heparin binding, when compared to the state-of-the-art. In contrast, our GO analysis showed that DEGs were significantly enriched in muscle contraction, extracellular matrix organization, extracellular structure organization, external encapsulating structure organization, smooth muscle contraction, and cell migration. Based on these findings, we can infer that our proposed model's GO analysis is more reliable than the state-of-the-art.

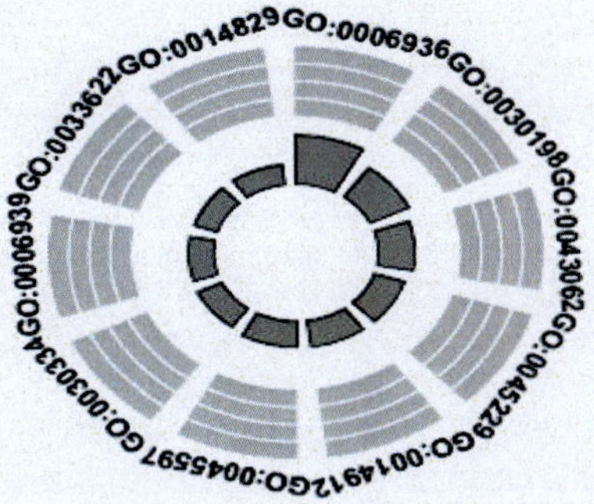

ID	Description
GO:0006936	muscle contraction
GO:0030198	extracellular matrix organization
GO:0043062	extracellular structure organization
GO:0045229	external encapsulating structure organization
GO:0014912	negative regulation of smooth muscle cell migratio
GO:0045597	positive regulation of cell differentiation
GO:0030334	regulation of cell migration
GO:0006939	smooth muscle contraction
GO:0033622	integrin activation
GO:0014829	vascular associated smooth muscle contraction

Fig. 10. Go Term pathway enrichment for ElasticNet DEGs

5 Conclusions

To commence the research, we applied three approaches to the microarray GSE7476 dataset to identify differentially expressed genes: a statistical method, the Limma package, and the ElasticNet algorithm. Preprocessing was conducted for all three methods. The statistical method divided the samples into only two groups: normal and tumor, resulting in 459 DEGs. For limma, three comparisons were made, yielding 1047 DEGs for the first, 829 DEGs for the second, and 1121 DEGs for the third comparison. ElasticNet was trained using 70% of the data and 30% for testing, the clustered it into four categories, which produced 42 DEGs. The DEGs identified by ElasticNet were subjected to GO term analysis and compared with the state-of-the-art. The study's conclusion demonstrated that artificial intelligence algorithms like ElasticNet can accurately identify regulated genes with minimal complexity and reliable outputs, as evidenced by the GO term analysis.

References

1. Antoni, S., Ferlay, J., Soerjomataram, I., Znaor, A., Jemal, A., Bray, F.: Bladder cancer incidence and mortality: a global overview and recent trends. Eur. Urol. **71**, 96–108 (2017)
2. Torre, L.A., Bray, F., Siegel, R.L., Ferlay, J., Lortet-Tieulent, J., Jemal, A.: Global cancer statistics. CA Cancer J. Clin. **65**, 87–108 (2015)
3. Pinto, I.G.: Systemic therapy in bladder cancer. Indian J. Urol. **33**, 118–126 (2017)
4. Burger, M., et al.: Epidemiology and risk factors of urothelial bladder cancer. Eur. Urol. **63**, 234–241 (2013)
5. Liu, S., et al.: The evaluation of the risk factors for non-muscle invasive bladder cancer (NMIBC) recurrence after transurethral resection (TURBt) in Chinese population. PLOS ONE **10**(4), e0123617 (2015)
6. Alfred, W.J., et al.: Updated 2016 EAU guidelines on muscle-invasive and metastatic bladder cancer. Eur. Urol. **71**(3), 462–475 (2017)
7. Clark, P.E., et al.: NCCN guidelines insights: bladder cancer, version 2.2016. J. Nat. Compr. Canc. Netw. **14**(10), 1213–1224 (2016)
8. Kukreja, J.B., Shah, J.B.: Advances in surgical management of muscle invasive bladder cancer. Indian J. Urol. **33**(2), 106 (2017). https://doi.org/10.4103/0970-1591.203416

9. Ghaleb, M.S., Ebied, H.M., Tolba, M.F.: Lung Cancer stages classification based on differential gene expression. In: Hassanien, A.E., et al. (eds.) The 3rd International Conference on Artificial Intelligence and Computer Vision (AICV2023), March 5–7, 2023, pp. 272–281. Springer Nature Switzerland, Cham (2023). https://doi.org/10.1007/978-3-031-27762-7_26
10. Lu, X., Zhang, X.: The effect of GeneChip gene definitions on the microarray study of cancers. BioEssays **28**(7), 739–746 (2006). https://doi.org/10.1002/bies.20433
11. Zhang, Y., Deng, Q., Liang, W., Zou, X.: An efficient feature selection strategy based on multiple support vector machine technology with gene expression data. BioMed Res. Int. **2018**, 1–11 (2018). https://doi.org/10.1155/2018/7538204
12. Lee, T., Lee, H.: Prediction of Alzheimer's disease using blood gene expression data. Sci Rep. **10**(1), 3485 (2006). https://doi.org/10.1038/s41598-020-60595-1. PMID: 32103140; PMCID: PMC7044318
13. Rukhsar, L., Bangyal, W.H., Ali Khan, M.S., Ag Ibrahim, A.A., Nisar, K., Rawat, D.B.: Analyzing RNA-seq gene expression data using deep learning approaches for cancer classification. Appl. Sci. **12**, 1850 (1850). https://doi.org/10.3390/app12041850
14. Siavoshi, A., Taghizadeh, M., Dookhe, E., Piran, M.: Gene expression profiles and pathway enrichment analysis to identification of differentially expressed gene and signaling pathways in epithelial ovarian cancer based on high-throughput RNA-seq data. Genomics **114**(1), 161–170 (2022)
15. Zararsiz, G., Goksuluk, D., Korkmaz, S., Eldem, V., Goksuluk, I.P., Unver, T.: MLSeq Machine Learning Interface to RNA-Seq Data. https://bioconductor.org/packages/release/bioc/vignettes/MLSeq/inst/doc/MLSeq.pdf. Accessed on 1 July 2021
16. Waseem, Q., Alshamrani, S., Nisar, K., Din, W.W., Alghamdi, A.: Future technology: software-defined network (SDN) forensic. Symmetry **13**, 767 (2021)
17. Wesolowski, S., Birtwistle, M.R., Rempala, G.A.: A comparison of methods for rna-seq differential expression analysis and a new empirical bayes approach. Biosensors **3**, 238–258 (2013)
18. Conesa, A., et al.: A survey of best practices for RNA-seq data analysis. Genome Biol **17**, 13 (2016)
19. Urda, D., Montes-Torres, J., Moreno, F., Franco, L., Jerez, J.M.: Deep Learning to Analyze RNA-Seq Gene Expression Data. In: Rojas, I., Joya, G., Catala, A. (eds.) IWANN 2017. LNCS, vol. 10306, pp. 50–59. Springer, Cham (2017). https://doi.org/10.1007/978-3-319-59147-6_5
20. Ciaburro, G., Iannace, G.: Machine-learning-based methods for acoustic emission testing: a review. Appl. Sci. **12**, 10476 (2022). https://doi.org/10.3390/app122010476
21. Xu, C., Jackson, S.A.: Machine learning and complex biological data. Genome Biol. **20**, 76 (2019). https://doi.org/10.1186/s13059-019-1689-0
22. Jiang, P., Liu, X.S.: Big data mining yields novel insights on cancer. Nat. Genet. **47**(2), 103–104 (2015). https://doi.org/10.1038/ng.3205
23. Ritchie, M.E., et al.: limma powers differential expression analyses for RNA-sequencing and microarray studies. Nucleic Acids Res. **43**(7), e47 (2015). https://doi.org/10.1093/nar/gkv007
24. Haynes, W.: Benjamini–hochberg method. In: Dubitzky, W., Wolkenhauer, O., Cho, K.-H., Yokota, H. (eds.) Encyclopedia of Systems Biology, pp. 78–78. Springer New York, New York, NY (2013). https://doi.org/10.1007/978-1-4419-9863-7_1215
25. De Mol, C., De Vito, E., Rosasco, L.: Elastic-net regularization in learning theory. J. Complexity **25**(2), 201–230 (2009). https://doi.org/10.1016/j.jco.2009.01.002
26. Subramanian, A., et al.: Gene set enrichment analysis: a knowledge-based approach for interpreting genome-wide expression profiles. Proc. Natl. Acad. Sci. U.S.A. **102**(43), 15545–15550 (2005). https://doi.org/10.1073/pnas.0506580102

27. Chen, E.Y., Tan, C.M., Kou, Y., et al.: Enrichr: interactive and collaborative HTML5 gene list enrichment analysis tool. BMC Bioinform. **14**, 128 (2013). https://doi.org/10.1186/1471-2105-14-128
28. Ghaleb, M.S., Ebied, H.M., Shedeed, H.A., Tolba, M.F.: Image Retrieval based on self-organizing feature map and map and Multilayer perceptron Neural Networks Classifier. In: Ninth International Conference on Intelligent Computing and Information science (ICICS), pp. 189–193. Cairo, Egypt (2019)
29. Ghaleb, M.S., Ebied, H.M., Shedeed, H.A., Tolba, M.F.: COVID-19 x-rays model detection using convolution neural network. In: Hassanien, A.E., Haqiq, A., Tonellato, P.J., Bellatreche, L., Goundar, S., Azar, A.T., Sabir, E., Bouzidi, D. (eds.) AICV 2021. AISC, vol. 1377, pp. 3–11. Springer, Cham (2021). https://doi.org/10.1007/978-3-030-76346-6_1
30. Ghaleb, M.S., Ebied, H.M., Shedeed, H.A., Tolba, M.F.: Content based image retrieval based on convolutional Neural Network. In: Tenth International Conference on Intelligent Computing and Information science (ICICS), pp. 149–153. Cairo, Egypt (2021)
31. Ghaleb, M.S., Ebied, H.M., Shedeed, H.A., Tolba, M.F.: Weather classification using fusion of deep convolutional neural networks and traditional classification methods. Int. J. Intell. Comput. Inform. Sci. **22**, 84–96 (2022)
32. Ghaleb, M.S., Ebied, H.M., Shedeed, H.A., Tolba, M.F.: Image retrieval based on deep learning. J. Syst. Manag. Sci. **12**, 477–496 (2022)
33. Ghaleb, M.S., Ebied, H.M., Shedeed, H.A., Tolba, M.F.: Content-based image retrieval using fused convolutional neural networks. In: Hassanien, A.E., Snášel, V., Tang, M., Sung, T.-W., Chang, K.-C. (eds.) Proceedings of the 8th International Conference on Advanced Intelligent Systems and Informatics 2022, pp. 260–270. Springer International Publishing, Cham (2023). https://doi.org/10.1007/978-3-031-20601-6_24
34. Tang, F., He, Z., Lei, H., Chen, Y., Lu, Z., Zeng, G., Wang, H.: Identification of differentially expressed genes and biological pathways in bladder cancer. Mol. Med. Rep. **17**(5), 6425–6434 (2018)

Robotic and Control Applications

Autonomous Robot Navigation and Exploration Using Deep Reinforcement Learning with Gazebo and ROS

Ahmad Taher Azar[1,2,3](✉), Muhammad Zeeshan Sardar[2], Saim Ahmed[2], Aboul Ella Hassanien[4], and Nashwa Ahmad Kamal[5]

[1] College of Computer and Information Sciences, Prince Sultan University, Riyadh, Saudi Arabia

[2] Automated Systems and Soft Computing Lab (ASSCL), Prince Sultan University, Riyadh 11586, Saudi Arabia
{msardar,sahmed,aazar}@psu.edu.sa

[3] Faculty of computers and Artificial Intelligence, Benha University, Banha, Egypt
ahmad.azar@fci.bu.edu.eg, ahmad_t_azar@ieee.org

[4] Faculty of computers and Artificial Intelligence, Cairo University, Giza, Egypt
aboitcairo@cu.edu.eg

[5] Faculty of Engineering, Cairo University, Giza, Egypt
nashwa.ahmad.kamal@gmail.com

Abstract. In this paper, robot navigation and exploration methodologies are presented using Deep Reinforcement Learning (DRL). The methodology has two parts. Firstly, the waypoint selection towards the global goal is done from generated points of interest (POI) in the environment. Secondly, a local navigation and obstacle avoidance policy is learned using DRL with a single low-cost 2D LIDAR. Deep Deterministic Policy Gradient (DDPG) is the DRL algorithm used for local navigation policy training and the simulation environment is built in Gazebo with Robot Operating System (ROS). After training the robot, the learned policy and waypoint selection are integrated together to develop a complete autonomous navigation and exploration system. This overall system not only mitigates the necessity of building the map of the environment but also gives similar or better performance in comparison to planning-based methods.

Keywords: Deep Reinforcement learning · Robot navigation · Mobile robots · Robot exploration · Neural networks

1 Introduction

Mobile robot navigation and exploration is a challenging task and due to the recent advancements and adaptability of robotics in almost every field, solving these challenges has garnered a lot of attention [1,2,14,17]. Simultaneous

A. Hassanien et al. (Eds.): AISI 2023, LNDECT 184, pp. 287–299, 2023.
https://doi.org/10.1007/978-3-031-43247-7_26

localization and mapping (SLAM) is a technique widely used for robot navigation and exploration. Typically, SLAM requires a map of the environment to be built manually using onboard robot sensors like cameras, LIDAR, and sonar. In SLAM, the robot builds a map of the environment and localizes itself in the map simultaneously [4,9,37]. Since to build the map, an operator is required to move the robot in unknown areas which is not possible all the time due to environmental constraints, the cost of labor, and limited resources. Therefore, autonomous navigation and exploration is the solution and this is the reason it is having significant attention in the research community.

Deep Reinforcement Learning (DRL) which is a subset of Machine Learning (ML) is a technique that can address the challenges in mobile robot navigation and exploration [3]. DRL is based on training the robot to take action in its environment through trial and error and maximizing a cumulative reward. In DRL there is a neural network (NN) that is trained to receive an environment input, make the agent perform an action, and get a reward for performed action to determine how good or bad that particular action was given the current environment state. DRL for mobile robot navigation and exploration can solve the complex challenges of navigation and exploration. It allows the robot to learn the complex behaviors and adapt to changing environments. With the help of DRL, a robot can learn to navigate through unknown environments, avoid obstacles and explore new areas with minimal prior knowledge [26,31].

The open-source software tools Gazebo and ROS (Robot Operating System) are widely used in robotics study and development. Gazebo is a 3D simulator that allows for the simulation of complex physical systems such as robots and their environments. In contrast, ROS is a software framework that offers libraries and tools for developing and running robot applications. Gazebo and ROS can be used together to simulate mobile robot navigation and exploration in a virtual world. This enables to test and improve algorithms for robot navigation and exploration without the need for a physical robot. In Gazebo, a virtual environment is created that includes the robot and its surroundings to mimic mobile robot navigation and exploration using DRL. A DRL agent then controls the robot by receiving sensor inputs from the surroundings and generating commands to direct the robot's motion. ROS is used to perform the simulation because it offers a framework for managing the robot's state and communicating with the DRL agent [8,34].

In this paper, two problems are being solved, one is autonomous exploration and the second is autonomous navigation in which the robot is required to move to a goal location. The goal is to build a system that can perform both the tasks of exploration and navigation and does not require human control and previous knowledge of the environment. Points of interest (POI) extraction plays an important role in finding the best path to reach the goal. POIs are estimated in the robot's surrounding and one of them is chosen as a waypoint for the robot to follow. Waypoints direct the DRL algorithm's motion policy toward the end goal, so alleviating the local optimum dilemma. The motion is eventually done in accordance with the policy, without the need for creating a map and having prior knowledge of the surroundings. In this paper, DRL based algorithm named Deep Deterministic Policy Gradient (DDPG) is implemented in ROS and Gazebo along with waypoint selection and navigation strategies.

The organization of the paper is such that the similar related work is highlighted in Sect. 2. Section 3 gives details about the proposed methodology for both global and local navigation including the waypoint selection strategy and DDPG training policy. Results and discussion is carried out in Sect. 4. Lastly, the conclusion is communicated in Sect. 5.

2 Related Work

Previously many approaches to SLAM have been studied and developed to perform environmental exploration and mapping. SLAM is performed on mobile robots using onboard sensors including LIDARs, cameras, and some other sensors [7,22,38]. SLAM can be two dimensional (2D) or three dimensional (3D) depending on whether the sensors are 2D or 3D [16,20,25]. It helps the robots build the environment map for their localization within it to perform navigation and exploration. The limitation of existing SLAM techniques heavily rely on the human operators to build the environment map which involves moving the sensor-equipped robot in the entire area of interest [12,18,35]. In fact, many autonomous navigation and exploration approaches involve previous availability of the map [13,23,30].

The rise in deep neural networks (DNN) and their capabilities allowed the development of different techniques to help with the task of robot navigation and exploration and mitigate the dependence on pre-built maps [10,15,21,27]. In [28] although the navigation is done through a planner and the environment mapping is carried via learned predictions of frontier areas to help the robot explore. In [33] a hierarchical structure is presented that combines multiple convolutional layers along with decision-making process for indoor exploration of mobile robots. Similarly another paper [29] is utilizing internal analysis algorithm to fuse decision making and multiple convolutional layers to achieve these goals. RGBD Depth camera is used to train the end-to-end system. [19] propose a two-step process. First, the detection of target objects is done using a DNN. Secondly, Deep Q-Network (DQN) Double Deep Q-Network (DDQN) is used to achieve the navigation to the target object. [32] employ deep Q-learning for exploratory navigation that allows the robot to learn obstacle avoidance in unknown settings. Also, [36] shows how the robot first learns the navigation in simulation and then is converted to real settings using the sim2real approach. DDPG based networks are used to train the robots in continuous action space. In [5] the authors present DRL-based system for autonomous navigation and exploration. The robots are trained using Gazebo and ROS and have the ability to navigate in unknown environments. It also highlights the strategies for waypoint selection to achieve the global goal without falling into local minima. Similarly, in [11], and [24], both the local planner and global planner are coupled together to avoid obstacles. In this case, a local path is created using a neural network in a previously mapped region without any importance to exploration. [39] discusses a method to utilize visual data of the close vicinity to generalize the target-driven navigation. Robot navigation commands are learned with the help of a neural network to reach the provided goal.

Neural networks can perform well in modular-based tasks but they may struggle in a complete end-to-end solution. Therefore an end-to-end solution is presented in here which utilizes DDPG that is one of the state-of-the-art DRL methods. The simulation environment made in a such a way to help in generalizing the solution and easier transfer to real systems. An overall system is described in the Fig. 1.

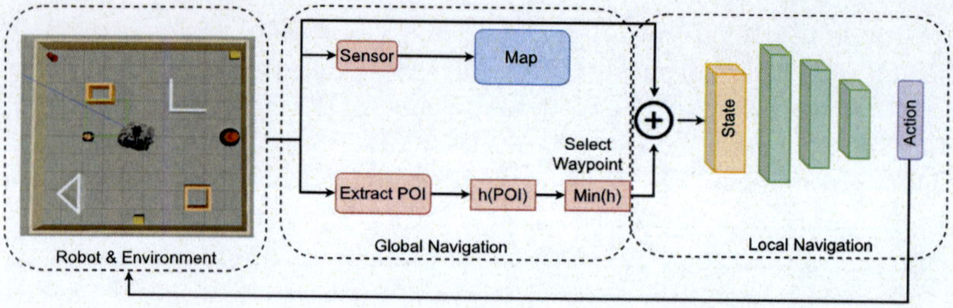

Fig. 1. Overall system for autonomous mobile robot navigation and exploration. On the left side a mobile robot is shown in a Gazebo training environment. The middle part shows the Global navigation flow and right side depicts the local navigation flow

3 Proposed Methodology

The proposed methodology has two parts, one is the optimal selection of way-points for global navigation. The waypoint selection using an evaluation criteria. The second is local navigation which also includes obstacle detection and avoidance. Local navigation of the robot is trained using DRL based method. During each step of navigation, a waypoint is given to DRL based algorithm in the form of polar coordinates relative to the current robot pose. Mapping is also done while navigating.

3.1 Global Navigation

To navigate the robot to the goal point, a waypoint is chosen from the available POIs. Since there is no environment map or pre-available information so the robot can not find an optimal path to the goal so it do so by following chosen POIs. The selection of POI can only be done in the immediate vicinity of the robot because of no available information of the environment and the map. The process which is used to obtain POIs is further divided into two methodologies. The first includes obtaining a POI using a threshold of two consecutive laser scan readings and making the robot move in the presumed gap. The second method is related to the maximum range of the laser and a POI is found if the readings are larger than the maximum range. The criteria of POI selection and deletion include that POI should not be previously visited or near the obstacles or unreachable within the defined number of steps.

Information-based Distance Limited Exploration (IDLE) is used as an evaluation method to select an optimal waypoint from the generated POI list using the criteria explain above [6]. The smallest IDLE score is selected as an optimal waypoint. Equation 1 defines the POI evaluation strategy

$$f(\mu_i) = \tanh\left(\frac{e^{\left(\frac{d(p_t-\mu_i)}{\lambda_2-\lambda_1}\right)^2}}{e^{\left(\frac{\lambda_2}{\lambda_2-\lambda_1}\right)^2}}\right)\lambda_2 + d(\mu_i, g) + e^{I_{i,t}} \quad (1)$$

where score f is the score of each candidate POI μ with index i determined by **the sum of three different components. The very first component from all three** is the Euclidean distance denoted by $d(pt, \mu i)$. This is the distance between the robot position p at a particular time t and the candidate POI μ expressed in the form of $tanh$ function:

$$\tanh\left(\frac{e^{\left(\frac{d(p_t-\mu_i)}{\lambda_2-\lambda_1}\right)^2}}{e^{\left(\frac{\lambda_2}{\lambda_2-\lambda_1}\right)^2}}\right)\lambda_2 \quad (2)$$

where, $\lambda 1$ and $\lambda 2$ are two-step distance limits and e denotes the Euler's number.

The second major component of the Eq. 1 is the Euclidean distance between the goal g and candidate POI μ represented by $d(\mu i, g)$. The third and the final component $e^{I_{i,t}}$ stores the map information score at time t. The $I_{i,t}$ are defined as

$$I_{i,t} = \frac{\sum_{w=-\frac{k}{2}}^{\frac{k}{2}} \sum_{h=-\frac{k}{2}}^{\frac{k}{2}} \mu_{(x+w)(y+h)}}{k^2} \quad (3)$$

where k is kernel size with width w and height h which is responsible for estimating candidate points with x and y coordinates. This way, POI with the lowest value is selected as an optimal waypoint ready to be used in local navigation.

3.2 Local Navigation

In the local planner if compared to the classical planner-based navigation stack, DDPG which is DRL based method is utilized to train the local navigation **policy. DDPG has two different neural networks named actor and critic.** The actor-network takes in the environment state and outputs a continuous action while the critic network tries to estimate the expected reward for taking a particular action a in a specific state s. In the case of mobile robot navigation and exploration, DDPG takes in the LIDAR data points, the robot's current position, and goal coordinates in the form of a state vector. The actor-network has two fully connected (FC) layers comprising. After each layer, a rectified linear unit (ReLU) is applied. The last network layer is connected to the output layer through two action parameters, which indicate the robot's linear velocity v and angular velocity w. To confine the output to the range (−1, 1), a tangent hyperbolic *tanh* activation function is used. An action a is confined by maximum linear and angular velocities denoted by v_{max} and w_{max} respectively before performing it in the environment. Equation 4 shows the action scaling process before applying it to the robot.

$$a = \left[v_{\max} \left(\frac{a_1 + 1}{2} \right), \omega_{\max} a_2 \right] \tag{4}$$

An important thing to note down here is that the linear velocity is adjusted to be positive only because the robot is only allowed to move in the forward direction.

The critic network takes in state s and action a as inputs and gives the Q-value. It consists of a fully connected layer, which is then activated by the ReLU activation function having output layer ls. This layer's output, along with the action, is transformed by two different Transformation Fully Connected Layers (TFC) having same size $m1$ and $m2$. These two layers are combined using Eq. 5.

$$l_c = l_s W_{m1} + aW_{m2} + b_{m2} \tag{5}$$

where l_c is the combined fully connected layer (CFC). W_{m1} and W_{m2} denotes the weights of size $m1$ and $m2$, respectively for two TFC layers. b_{m2} is bias in layer $m2$. Lastly, ReLU activation is applied to l_c. Afterwards, it is connected to the output with 1 parameter representing the Q value. The complete architecture is illustrated in Fig. 2

The reward r is calculated based on three conditions. Firstly, the reward is big positive (r_p) if the distance $d^{'}$ to the goal is less than a threshold $\xi_{d'}$, secondly, the reward is big negative (r_n) if the robot has a collision with an obstacle and lastly it gets the reward based upon the linear and angular velocities. The last condition encourages the robot to explore. The reward is given in Eq. 6

$$r(s,a) = \begin{cases} r_p & if\ d^{'} < \xi_{d'} \\ r_n & if\ collision \\ v - |\omega| & otherwise, \end{cases} \tag{6}$$

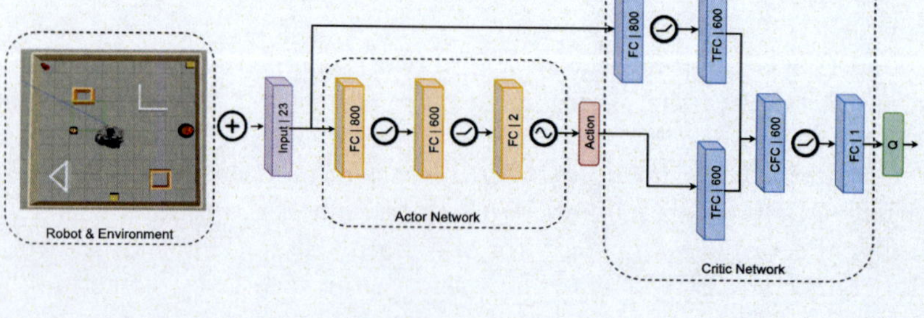

Fig. 2. A complete architecture consisting of robot and its environment (left), actor network (middle) and critic network (right)

To encourage the robot to move towards the goal, a delayed reward is utilized as described in Eq. 7

$$r_{t-i} = r(s_{t-i}, a_{t-i}) + \frac{r_p}{i}; \forall_i = \{1, 2, 3, 4, \ldots, n\}, \quad (7)$$

The rewards are estimated and updated with the previous steps going from 1 upto n. This allows the robot to get the positive goal reward in a descending order over n steps before it can reach the goal. In this way, the robot has learned the robust local navigation policy allowing the robot to read a local goal while avoiding the obstacles using 2D laser scan values.

3.3 Mapping and Exploration

The purpose of the waypoint selection and navigation to it is to reach the final global goal. It is easier for the robot to navigate to the global goal when it is closer to it. During this process of local and global navigation, the robot explores and maps the environment. This mapping is achieved using laser scanner and robot odometry as information sources. The overall algorithm for robot navigation and exploration is presented in Algorithm 1.

4 Results and Discussion

The experimental setup consists of a simulation environment in Gazebo integrated with Robot Operating System (ROS). Rviz is used as a visualization tool to show the LIDAR points, Goals, and robot movement. The Gazebo world of size 10×10 m and the corresponding Rviz view used to train the robot are shown in the Fig. 3 and 4 respectively. The robot for simulation is Turtlebot3 Waffle mounted with LSD-01 360-degree LIDAR. The training is done on a computer equipped with AMD Ryzen 7 5700G with Radeon Graphics Processor, 16GB RAM and Nvidia GeForce RTX3060 GPU.

In order to reduce the training time and keep the state space limited, LIDAR points are down-sampled to 20 points only and cover the frontal 180-degree view of the robot because the robot is only allowed to move in the forward direction. The maximum linear (v_{max}) and angular (w_{max}) velocities are set to 0.5 meters per second and 1 rad per second respectively. The DDPG network parameters used for training are shown in the Table 1.

The actor-network learning is lower than the critic network to not update the action quickly. To make the network learn robust and general features, Gaussian noise is added to LIDAR data and actions. In order to further reinforce the generic learning robot pose and goal are randomly initialized after each episode. One episode is concluded when a goal is reached, or the robot collides or the maximum number of steps per episode are exhausted. One epoch is considered when the steps reach the evaluation frequency value. The reward values are set to +150 for reaching the goal, −100 for colliding and positive for moving straight, and negative for moving into sharp turns. Large positive reward encourages the

Algorithm 1: Mobile robot navigation and exploration algorithm.

```
Initialize the global goal and corresponding thresholds;
while goal_reached do
    Get data from the sensor;
    Get POI from equation 1;
    if D less than n then
        if waypoint = global_goal then
            goal_reached = True;
        else
            if d(pt,g) less than delta then
                waypoint = global_goal;
            else
                for i in POI do
                    Estimate h(i) using equation 1;
                end
                waypoint = POI;
            end
        end
    end
    Get action from DDPG;
    Perform action;
end
```

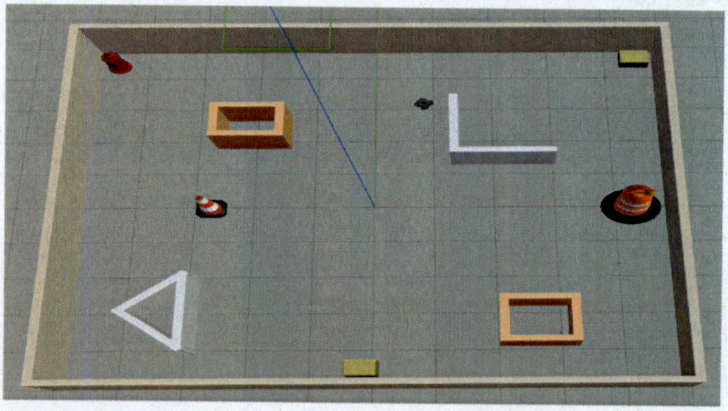

Fig. 3. Gazebo world with turtlebot3 waffle spawned in it.

robot to reach the goal, a large negative to avoid obstacles, and positive for moving straight, and a negative for sharp turns to avoid the robot rotating along its axis of rotation.

The simulation is run for 400 Epochs and we are able to get good results in terms of positive reward value, lower losses, and high Q values. The Q value plot with respect to the number of steps is illustrated in Fig. 6. The loss curve with

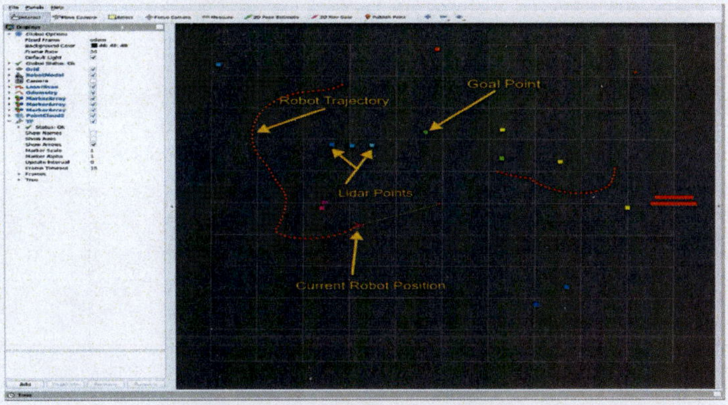

Fig. 4. Rviz visualization of the robot movement in Gazebo world. It shows the robot's position along with trajectory, LIDAR points, and goal points.

Table 1. Parameters used in training the robot navigation and exploration algorithm

Parameter	Value
optimizer	SGD
actor learning rate	0.0001
critic learning rate	0.001
actor and critic momentum	0.5
batch size	64
max steps per episode	500
evaluation frequency	$5\times^{6}$
buffer size	$5\times^{6}$
τ	0.001

respect to the number of steps is shown in Fig. 7. The positive trend of the Q-value plot shows that the model has learned and Q-values are optimized. The loss plot is another depiction of this claim that the robot has learned the features. The evaluation experiments highlight that the reward which was negative at around −80 at the beginning of the training has become positive at the averaging around +50 which means the robot is able to avoid obstacles and reach the goal in a reasonable time frame. Figure 5 shows the reward values with epoch numbers. In the Fig. 5, for epoch number 411 the reward is 69.26 and for epoch number 418 the reward is 106.

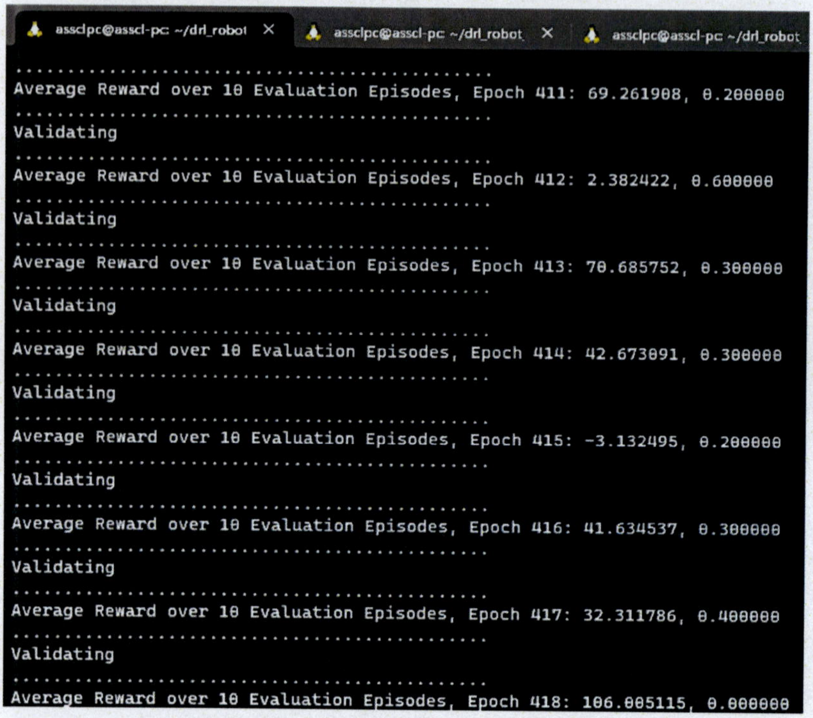

Fig. 5. DDPG evaluation results showing the epoch numbers and corresponding reward. For the epoch 411 the reward is 69.2 and for the epoch 418 the reward is 106.0

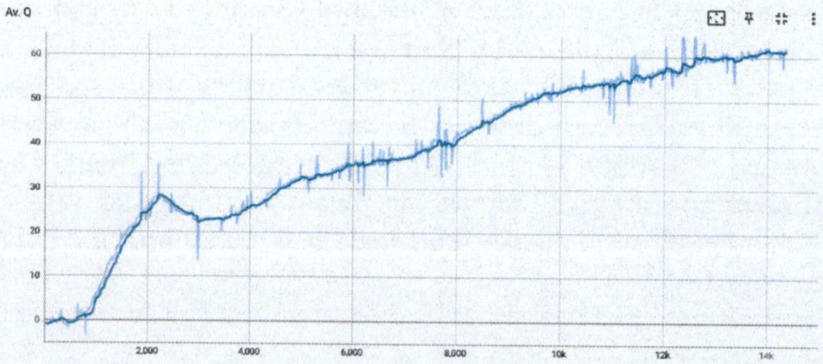

Fig. 6. Value of average Q value. It can be seen average Q value continued to increase with the number of steps until it gets saturated.

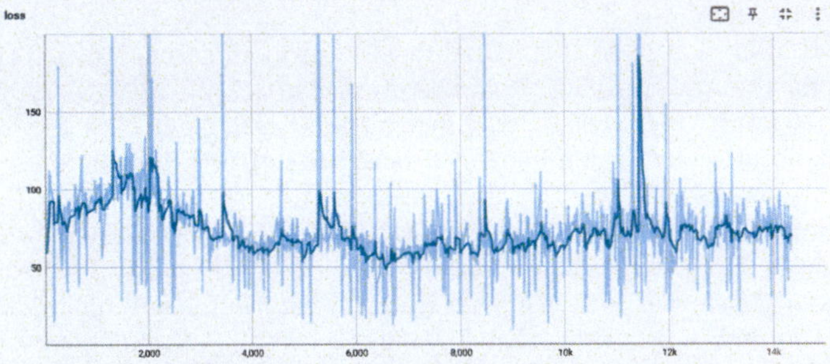

Fig. 7. Loss curve with the number of steps. The loss is decreasing with the training time and epochs to pass depicting the robot learning.

5 Conclusion

This work provides autonomous mobile robot navigation and exploration methodology utilizing DDPG which is a deep reinforcement learning-based algorithm. The exploration and navigation task is divided into a global goal-reaching strategy and local navigation to avoid obstacles. The local navigation policy is trained using DDPG algorithm in a simulated environment built using Gazebo and ROS. This expands the potential uses of mobile robots to include search and rescue, and industrial, and environmental monitoring. The RL-DDPG algorithm has proven substantial potential as a technique for facilitating robots' ability to learn from their experiences and improve their decision-making abilities. The experimental results show that the robot has learnt robust features which allow it to not only navigate to the goal autonomously but also avoid obstacles along the way.

Acknowledgement. The authors would like to thank Prince Sultan University, Riyadh, Saudi Arabia for their support. Special acknowledgement to Automated Systems & Soft Computing Lab (ASSCL), Prince Sultan University, Riyadh, Saudi Arabia.

References

1. Ajeil, F.H., Ibraheem, I.K., Azar, A.T., Humaidi, A.J.: Autonomous navigation and obstacle avoidance of an omnidirectional mobile robot using swarm optimization and sensors deployment. Int. J. Adv. Rob. Syst. **17**(3), 1–15 (2020)
2. Ammar, H.H., Azar, A.T., Tembi, T.D., Tony, K., Sosa, A.: Design and implementation of fuzzy PID controller into multi agent smart library system prototype. In: Hassanien, A.E., Tolba, M.F., Elhoseny, M., Mostafa, M. (eds.) AMLTA 2018. AISC, vol. 723, pp. 127–137. Springer, Cham (2018). https://doi.org/10.1007/978-3-319-74690-6_13
3. Azar, A.T., et al.: Drone deep reinforcement learning: a review. Electronics **10**(9) (2021)
4. Bailey, T., Durrant-Whyte, H.: Simultaneous localization and mapping (SLAM): Part II. IEEE Robot. Autom. Mag. **13**(3), 108–117 (2006)

5. Cimurs, R., Suh, I.H., Lee, J.H.: Goal-driven autonomous exploration through deep reinforcement learning. IEEE Robot. Autom. Lett. **7**(2), 730–737 (2021)
6. Cimurs, R., Suh, I.H., Lee, J.H.: Information-based heuristics for learned goal-driven exploration and mapping. In: Proceedings of 18th International Conference on Ubiquitous Robots, pp. 571–578 (2021)
7. Cui, L., Ma, C.: SOF-SLAM: a semantic visual slam for dynamic environments. IEEE Access **7**, 166528–166539 (2019)
8. DeepSim: Deepsim: a reinforcement learning environment build toolkit for ROS and Gazebo. arXiv preprint arXiv:2205.08034 (2022)
9. Durrant-Whyte, H., Bailey, T.: Simultaneous localization and mapping: part I. IEEE Robot. Autom. Mag. **13**(2), 99–110 (2006)
10. Elkholy, H.A., Azar, A.T., Magd, A., Marzouk, H., Ammar, H.H.: Classifying upper limb activities using deep neural networks. In: Hassanien, A.-E., Azar, A.T., Gaber, T., Oliva, D., Tolba, F.M. (eds.) AICV 2020. AISC, vol. 1153, pp. 268–282. Springer, Cham (2020). https://doi.org/10.1007/978-3-030-44289-7_26
11. Faust, A., Zhang, C., Yang, Y., Krause, J., Darrell, T.: PRM-RL: long-range robotic navigation tasks by combining reinforcement learning and sampling-based planning. In: Proceedings of IEEE International Conference on Robotics and Automation, pp. 5113–5120 (2018)
12. Filipenko, M., Afanasyev, I.: Comparison of various slam systems for mobile robot in an indoor environment. In: 2018 15th International Conference on Control, Automation, Robotics and Vision (ICARCV), pp. 400–407. IEEE (2018)
13. Gao, H., Zhang, X., Wen, J., Yuan, J., Fang, Y.: Autonomous indoor exploration via polygon map construction and graph-based slam using directional endpoint features. IEEE Trans. Automat. Sci. Eng. **16**(4), 1531–1542 (2019)
14. Ibraheem, G.A.R., Azar, A.T., Ibraheem, I.K., Humaidi, A.J.: A novel design of a neural network-based fractional PID controller for mobile robots using hybridized fruit fly and particle swarm optimization. Complexity **2020**, 1–18 (2020)
15. Ibrahim, H.A., Azar, A.T., Ibrahim, Z.F., Ammar, H.H.: A hybrid deep learning based autonomous vehicle navigation and obstacles avoidance. In: Hassanien, A.-E., Azar, A.T., Gaber, T., Oliva, D., Tolba, F.M. (eds.) AICV 2020. AISC, vol. 1153, pp. 296–307. Springer, Cham (2020). https://doi.org/10.1007/978-3-030-44289-7_28
16. Jiang, G., Yin, L., Liu, G., Xi, W., Ou, Y.: FFT-based scan-matching for slam applications with low-cost laser range finders. Appl. Sci. **9**(1), 41 (2018)
17. Khamis, A., et al.: Robotics and intelligent systems against a pandemic. Acta Polytechnica Hungarica **18**(5), 13–35 (2021)
18. Ko, D.W., Kim, Y.N., Lee, J.H., Suh, I.H.: A scene-based dependable indoor navigation system. In: 2016 IEEE/RSJ International Conference on Intelligent Robots and Systems (IROS), pp. 1530–1537. IEEE (2016)
19. Lee, M.F.R., Yusuf, S.H.: Mobile robot navigation using deep reinforcement learning. Processes **10**(12), 2748 (2022)
20. Li, B., Wang, Y., Zhang, Y., Zhao, W., Ruan, J., Li, P.: GP-SLAM: laser-based slam approach based on regionalized gaussian process map reconstruction. Auton. Robot. **44**, 947–967 (2020)
21. Mohamed, N.A., Azar, A.T., Abbas, N.E., Ezzeldin, M.A., Ammar, H.H.: Experimental kinematic modeling of 6-DOF serial manipulator using hybrid deep learning. In: Hassanien, A.-E., Azar, A.T., Gaber, T., Oliva, D., Tolba, F.M. (eds.) AICV 2020. AISC, vol. 1153, pp. 283–295. Springer, Cham (2020). https://doi.org/10.1007/978-3-030-44289-7_27

22. Mur-Artal, R., Tardós, J.D.: ORB-SLAM2: an open-source slam system for monocular, stereo, and RGB-D cameras. IEEE Trans. Rob. **33**(5), 1255–1262 (2017)
23. Palomeras, N., Carreras, M., Andrade-Cetto, J.: Active slam for autonomous underwater exploration. Remote Sens. **11**(23), 2827 (2019)
24. Pokle, A., Karaman, S., Rus, D.: Deep local trajectory replanning and control for robot navigation. In: Proceedings of International Conference on Robotics and Automation, pp. 5815–5822 (2019)
25. Ren, Z., Wang, L., Bi, L.: Robust GICP-based 3D LiDAR SLAM for underground mining environment. Sensors **19**(13), 2915 (2019)
26. Ruan, X., Ren, D., Zhu, X., Huang, J.: Mobile robot navigation based on deep reinforcement learning. In: 2019 Chinese Control and Decision Conference (CCDC), pp. 6174–6178. IEEE (2019)
27. Sayed, A.S., Azar, A.T., Ibrahim, Z.F., Ibrahim, H.A., Mohamed, N.A., Ammar, H.H.: Deep learning based kinematic modeling of 3-RRR parallel manipulator. In: Hassanien, A.-E., Azar, A.T., Gaber, T., Oliva, D., Tolba, F.M. (eds.) AICV 2020. AISC, vol. 1153, pp. 308–321. Springer, Cham (2020). https://doi.org/10.1007/978-3-030-44289-7_29
28. Shrestha, R., Tian, F.P., Feng, W., Tan, P., Vaughan, R.: Learned map prediction for enhanced mobile robot exploration. In: Proceedings of International Conference on Robotics and Automation, pp. 1197–1204 (2019)
29. Sleaman, W.K., Hameed, A.A., Jamil, A.: Monocular vision with deep neural networks for autonomous mobile robots navigation. Optik **272**(170), 162 (2023)
30. von Stumberg, L., Usenko, V., Engel, J., Stückler, J., Cremers, D.: From monocular slam to autonomous drone exploration. In: Proceedings of European Conference on Mobile Robots, pp. 1–8 (2017)
31. Sugiyama, M.: Statistical Reinforcement Learning: Modern Machine Learning Approaches. CRC Press, Boca Raton (2015)
32. Tai, L., Liu, M.: A robot exploration strategy based on q-learning network. In: Proceedings of IEEE International Conference on Real-Time Computing and Robotics, pp. 57–62 (2016)
33. Tai, L., Li, S., Liu, M.: Autonomous exploration of mobile robots through deep neural networks. Int. J. Adv. Robot. Syst. **14**(4), 1729881417703571 (2017)
34. Takaya, K., Asai, T., Kroumov, V., Smarandache, F.: Simulation environment for mobile robots testing using ROS and Gazebo. In: 2016 20th International Conference on System Theory, Control and Computing (ICSTCC), pp. 96–101. IEEE (2016)
35. Taketomi, T., Uchiyama, H., Ikeda, S.: Visual slam algorithms: a survey from 2010 to 2016. IPSJ Trans. Comput. Vis. Appl. **9**(1), 16 (2017)
36. Xie, L., Wang, S., Markham, A., Trigoni, N.: Towards monocular vision based obstacle avoidance through deep reinforcement learning. CoRR abs/1706.09829 (2017). https://arxiv.org/abs/1706.09829
37. Yousif, K., Bab-Hadiashar, A., Hoseinnezhad, R.: An overview to visual odometry and visual slam: Applications to mobile robotics. Intell. Ind. Syst. **1**(4), 289–311 (2015)
38. Yu, C., et al.: DS-SLAM: a semantic visual slam towards dynamic environments. In: 2018 IEEE/RSJ International Conference on Intelligent Robots and Systems (IROS), pp. 1168–1174. IEEE (2018)
39. Zhu, Y., et al.: Target-driven visual navigation in indoor scenes using deep reinforcement learning. In: Proceedings of IEEE International Conference on Robotics and Automation, pp. 3357–3364 (2017)

Check for updates

Building Your Own Geminoid: A DIY Guide to Creating an Affordable Social Robot

Ibrahim A. Hameed(✉) and Andreas Glomsrud(✉)

Department of ICT and Natural Sciences, Norwegian University of Science and Technology (NTNU), Larsgårdsvegen 2, 6009 Ålesund, Møre og Romsdal, Norway
{ibib,andreglo}@ntnu.no

Abstract. Social humanoid robots are at the forefront of cutting-edge research, but they come with a hefty price tag and a sluggish pace of development. These complex robots are often proprietary, built with prohibitively expensive parts and methods, and require a significant amount of time and resources to create. Fortunately, a team of researchers from the Creative Machines Lab at Columbia University (NY, USA) has developed a solution to this problem by creating an affordable and straightforward open-source robotic face. In this paper, we present a revised version of their system that enables the use of standardized metric parts to build a budget-friendly Geminoid. Although we faced constraints on parts acquisition and other factors, we were able to establish a solid foundation that will pave the way for future developments that will make this technology more accessible.

Our work highlights the adaptability of design methods such as 3D printing, which allowed us to create a cost-effective and flexible solution to a longstanding challenge in the field of robotics. With our findings, we hope to inspire others to explore new avenues of research that will lead to the creation of advanced social humanoid robots that are both affordable and accessible.

Keywords: Social robots · Human-robot interaction · Open-source robotics · Social humanoid robots · Affordable robots · 3D printing

1 Introduction

The idea of robots and artificial intelligence has fascinated mankind for centuries. From the mythical tales of the bronze automaton Talos to the sophisticated machines of today, robots have captured our imagination and become a part of our everyday lives. However, while robots have been used in many industries, most of them lack the human-like qualities that make social interaction possible [1].

Social robots that can understand and communicate with humans have the potential to revolutionize many industries. For example, they could act as

I. A. Hameed and A. Glomsrud—These authors contributed equally to this work.

A. Hassanien et al. (Eds.): AISI 2023, LNDECT 184, pp. 300–315, 2023.
https://doi.org/10.1007/978-3-031-43247-7_27

surrogates for human interaction in hazardous environments or provide support for people with disabilities. In healthcare, social robots are utilized as personal trainers to promote activities among older adults in long-term care facilities [2], for rehabilitation purposes [3], remote interventions [4], and as companions capable of simulating emotions [5]. In education, they serve as second language tutors, teaching assistants, and storytellers, assisting teachers in repetitive tasks [6,7]. Social robots can also be used in shopping malls and as receptionists [8,9]. The COVID-19 pandemic has further emphasized the importance of social robots in filling the void left by limited human interaction due to enforced social distancing measures [10–12]. To address the technical and artistic challenges of designing and staging believable care scenarios with robots, theater has been employed as a means to dramatize the interactions and enhance the perception and acceptance of this new technology [13], as illustrated in Fig. 1.

Fig. 1. The play "Cornell" [13].

However, the development of humanoid robots is slow due to the prohibitive cost associated with building them. As a result, most of these robots are proprietary and not available for purchase at a reasonable price. This poses a challenge for researchers who want to build on existing work and improve the methods used in this field. To address this challenge, this paper proposes a simple and **affordable testable platform for social robots that uses open-source and readily available solutions.** It is based on the Eva robot, shown in Fig. 2, developed by the Creative Machines Lab at Columbia University in NYC, US [14]. Eva is a robotic face designed to realistically replicate human facial expressions and movements. It is equipped with 14 facial skin muscles, 6 neck muscles, and 5 eyeball and eyelid muscles. Eva serves as a platform for further research into AI for Human-Computer interfacing, specifically exploring the possibility of emotional intelligence in machines. In this paper, the same design concept will be utilized to manufacture a geminoid capable of replicating human facial expressions.

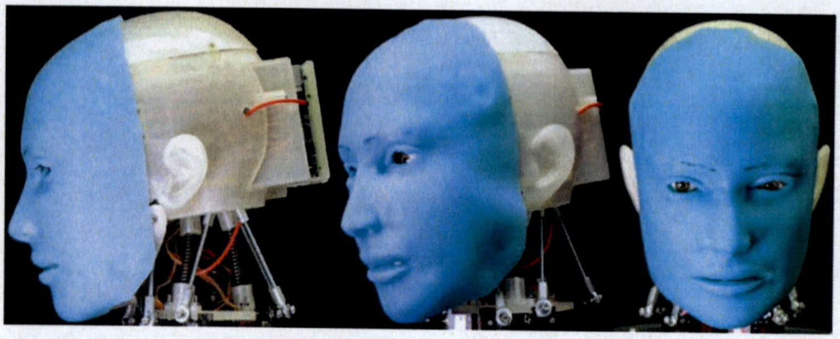

Fig. 2. Eva Robot developed by the Creative Machines Lab at Columbia University in NYC, US [14].

By reducing the cost and complexity of building social robots, we hope to promote further research and development in this field. This, in turn, will facilitate the reiteration and improvement of previous work and methods, enabling us to realize the full potential of social robots in the future.

The paper is organized as follows: the theory and related work are presented in Sect. 2. The implementation details are introduced in Sect. 3. The developed robot is briefly presented in Sect. 4. Concluding remarks and suggestions for future work are presented in Sects. 5 and 6, respectively.

2 Theory and Related Works

2.1 Geminoid

The geminoid project is a remotely controlled lifelike android that attempts to mimic the appearance of the researcher Hiroshi Ishiguro, as it is shown in Fig. 3(a). It is defined as a teleoperated android of an existing person [15]. The system has no internal computing power and relies solely on facial tracking and commands from an extraneous system. It is built to relative realistic size and proportions. The robot is actuated by pneumatic systems and has been proven to be quite lifelike at distance [16] and [17].

2.2 Sophia

Sophia, shown in Fig. 3(b), is an AI driven social robot, it uses text to speech and speech to text algorithms to communicate with humans. Its movements are servo driven, and it has movable limbs depending on its configuration. Sophia is developed by Hanson Robotics and is fully proprietary, even the rubber mixture used for its face is patented by Hanson Robotics. Sophia is the only robot to have been granted a citizenship. That being in Saudi Arabia. Sophia is one of the few robotics projects that has gotten mainstream acknowledgement, being featured on talk shows and interviews all over the world [18].

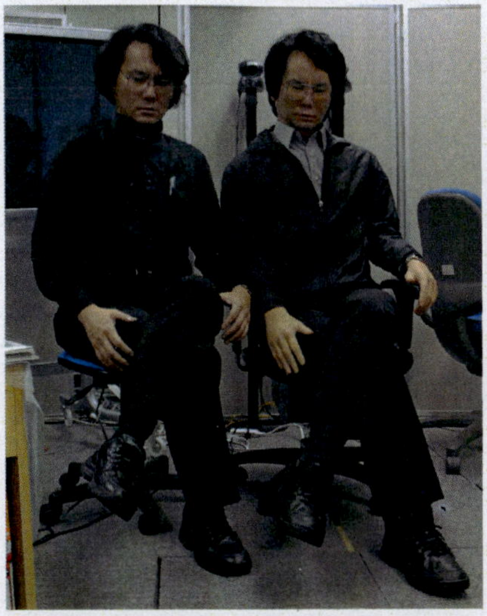

(a) Geminoid and its human source, Hiroshi Ishiguro, designed by ATR Intelligent Robotics and Communication Laboratories)

(b) Sophia Robot

Fig. 3. Humanoid companions: Hiroshi Ishiguro' Geminoid and Sophia Robot [17,18].

2.3 Furhat Robot

Furhat Robotics, as shown in Fig. 4, has developed a fully digital AI face system using a projector. This system can be used as a stand-in, such as a receptionist. The "face" of the robot is displayed on a curved LCD panel on the front, and there is no physical actuation involved. The system is designed for the development of AI algorithms. However, one limiting factor for research using this system is its high cost, which ranges from around $10,000 to $20,000. Additionally, the system relies on proprietary software and hardware developed by Furhat Robotics, which might hinder further development [19].

2.4 Eva Robot

The main drawback of many previous systems is their proprietary and costly nature. In order to address these issues, the Creative Machines Lab at Columbia University in NYC, US [14] developed the Eva robot, as shown in Fig. 2. Their

Fig. 4. Furhat Robot developed by Furhat Robotics [19].

goal was to create a low-cost open-source solution for a social robotic face that can be replicated at an affordable cost.

The system is built using Raspberry Pi, servos, and 3D printed parts, which are easily accessible and inexpensive. This enables the creation of a fully-featured testable robotic face at a fraction of the price compared to previous commercial robots. The open-source nature of the robot also allows for wider utilization in research and education.

The objective of the project was to develop a human-like social robot capable of generating intelligent and generalizable facial expressions. This enables the robot to adapt its behavior in real-time to different interaction situations with humans. The aim was to build a robot face with soft skin that can be easily controlled to express various facial expressions without the need to hardcode each expression individually [20].

The aim of this paper is to utilize the Eva robot [14] as the foundational platform for constructing an affordable robotic face. This robotic face can be controlled using a vision-based AI algorithm to accurately mimic various facial expressions in real time.

The proposed AI-based algorithm is designed to analyze images or video frames of a human face and replicate the facial expression depicted in those visual inputs. This is achieved by mapping a point system onto the human face and distorting a reference image to match the input. Subsequently, the distorted image is used to calculate commands for the servo motors that drive the robotic face [21].

2.5 Uncanny Valley

One key factor when developing or reviewing robots that mimic human appearance is the Uncanny Valley effect. People tend to humanize most everything we interact with. Like being able to see a face in a cloud formation. Because of this pattern seeking nature we can easily recognize when something is not human. Humanoid robots and other created depictions of humans (animations) need to be realistic enough, if they look realistic but act in a way that doesn't align with how a human acts people start to feel uneasy. It is normal to feel disgust or a general feeling that something is wrong. The uncanny valley, shown in Fig. 5, is hard to overcome when you are creating and developing something that aims to mimic humans. An often utilized strategy when developing systems that could closely resemble a humanoid is to deliberately add features that make it distinct from a human. People are usually more accepting and less judgmental towards systems when they don't stride to achieve hyper realism [21].

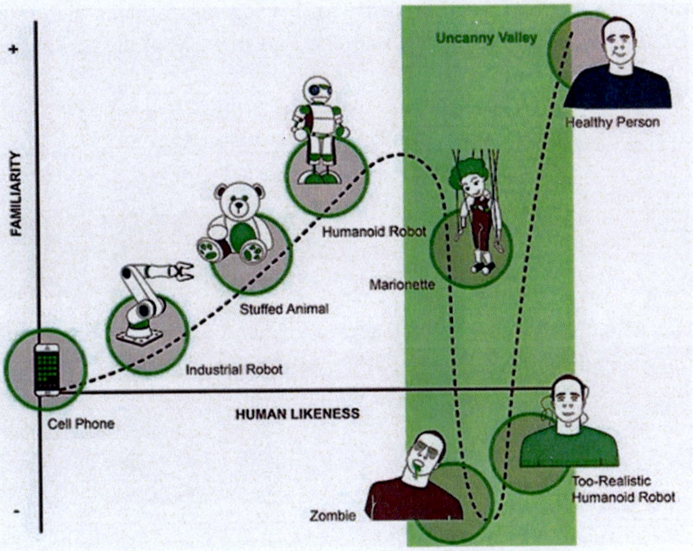

Fig. 5. Uncanny valley (based on M. Mori) [22]

3 Implementation

With the goal of the project being to create a testable prototype utilizing as many predefined systems as possible. The most logical choice of platform is the Eva 1.0 platform. The Eva platform is built in a manner that allows for easy modification of features by adding or removing whole modules. The system is however not designed to be built using the metric system. Some refactoring has to be done to allow for the use of metric parts.

3.1 Refactoring

To allow for assembly using metric parts 3 steps were done.

3.1.1 Modification of 3D Model

The model, shown in Fig. 6, is designed to be assembled using Heat Set Inserts and machine screws. Heat set inserts need correctly sized holes such that there is enough plastic to fasten them. The hole sized on the model are modelled after inserts designed in the Imperial system, making them not fit with any metric inserts you commonly find in Europe. All holes for inserts and their respective through holes for screws were therefore scaled to fit either M3 or M2 Sizing. M3 Sizing was favored when possible due to higher availability of M3 sized inserts.

3.1.2 Removal of Extraneous Systems

Further refactoring of the system was done by removing the eye module. The module is actuated fully mechanically and with parts that were not readily available at the time of planning. Given the complexity of the system as a whole it was removed to simplify.

Eyes can always be added back in as it is a separate assembly. Removing the eyes also frees up space for the possibility of other features.

Fig. 6. 3D model and Eye assembly from Eva 1.0 paper

3.1.3 Simplification of Parts List

Because of refactoring and removal of eye system, the parts list, shown in Table 1, for the project shrunk quite significantly. This is because of two factors. The eye assembly required quite a lot of unique parts that were not utilized in any other assembly within the robot. When refactoring to metric a lot of imperial sizes that are quite similar all got converted to one standard sizing. Within the Eva 1.0 paper the part list included around 20 different metal screws and fastener pieces. This was simplified to 10 screws and fasteners with refactoring.

Some costs of parts do not match up with the amount necessary for building. This is because many parts have a minimum order quantity. Price is calculated according to purchasing minimum order, meaning there will be spare parts. Given this, some money can be saved by finding suppliers with lower order minimums. The price of a Raspberry Pi 3B+ or any other kind of raspberry pi board is also quite inflated, this is because of the silicone shortage of 2020/2021 causing shortage of supply.

Table 1. Items list and cost

Item	Amount (Min Order)	Cost [NOK[a]]
Raspberry Pi 3B+	1	1600
Adafruit 16-Channel PWM/Servo HAT	2	420
GPIO Stacking Header	1	71.05
3x4 Right Angle Male Headers	8(10)	147.29
Brass M2.5 Standoffs	8	45.92
MG90S Servo Motor	19(24)	1374.00
5 V 10 A Switching Power Supply	1	433.90
Barrel Plug	2(3)	37.89
Cooling Fan	1	89.88
Speakers	1	229.9
0.7 mm Steel Wiring (50 m)	1	36.9
1.2 mm Teflon Tube 50 m	1	456.6
Silicone for casting	1	509
PLA 3d Print Fillament 2.5 Kg	1	699
Compression Spring	3(10)	79.8
M2x3.4x4 Brass Heat Set Insert	50	37.5
M3x4.6x5.7 Brass Heat Set Insert	85(100)	223.28
M2x8 mm Machine Screws	42(100)	45.3
M2x14 mm Flat Head	6(100)	112.16
M3x8 mm Machine Screws	41(100)	50.61
M3x12 mm Flat Head	16(100)	45.1
M3x16 mm Flat head	20(100)	37.08
M3x25 mm Flat Head	12(100)	49.74
M3 Locking nut	12(100)	224.5
M3x1000 mm rod	11(5)	195.96
M3 Rod Ends	12	85
SUM:		7579

[a] NOK stands for Norwegian Krone, 1 NOK = $0,095 as in Jun 2023

3.2 Assembly

After refactoring, the skull was 3D printed using the Prusa i3 MK2 3D Printer in Generic PLA plastic. PLA was chosen because of its strength and high melting temperature along with its low cost. PLA is one of the most used 3D printing materials because of these factors. The printing process took about 3 days because parts had to be printed twice due to layer shifting and other printing errors. These kinds of errors are normal on big prints with a print time nearing 24 h. The skull comprises of 4 main parts, as it is shown in Fig. 7, along with some additional smaller pieces for connections to the base of the robot.

To connect these pieces together heat set inserts were melted into all necessary holes with a soldering iron. The holes are designed to have a 2 mm deeper crevice than necessary to allow for the soldering iron to stick through the inset.

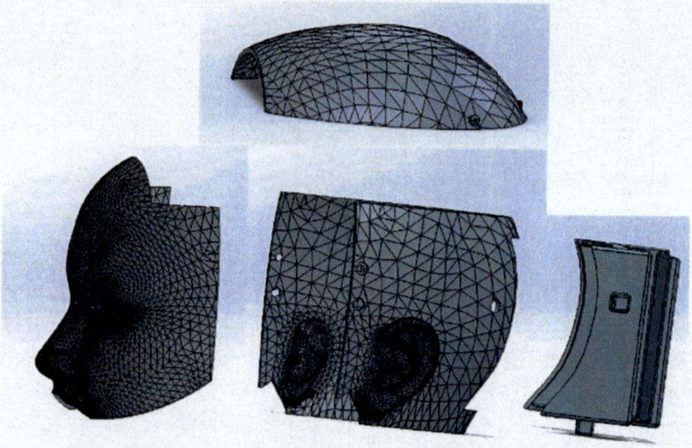

Fig. 7. Four pieces of the skull assembly

The system is designed to be controlled using a Raspberry Pi 3B system. It is however also compatible with any Raspberry pi B board that was released after, shown in Fig. 8(a). For the raspberry pi 3B utilized in the project to connect to all the necessary servo motors, add-on boards needed to be assembled. The utilized add-on boards for these servo connections come disassembled and requires the user to solder some key components to the board. This includes a power connector, connection pins and if wished there are places for capacitors if you do not have clean input power. The two HATS had pin risers soldered to them together with connectors for the servo wires. The first servo has 45° pins soldered along with pass-through GPIO pins. This allows for a secondary board to be stacked on top.

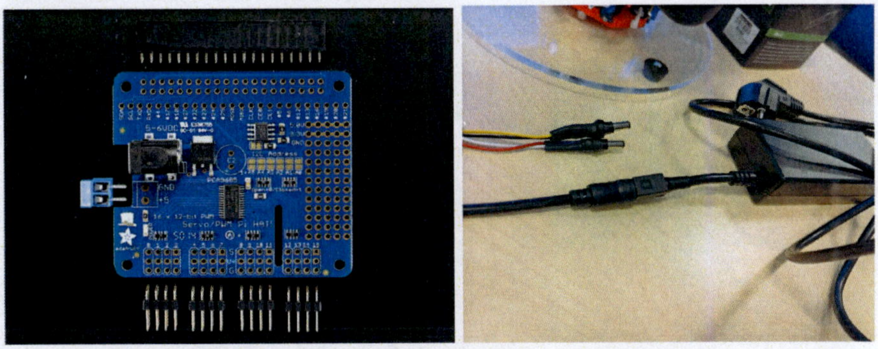

(a) Adafruit bare board. (b) PSU with split wire.

Fig. 8. Adafruit bare board and power supply

The servo motor control boards can each draw up to 50 W (10 V) each. We can supply this by splitting a 100W power supply unit (PSU) (10 V), shown in Fig. 8(b), in such a way that the boards are connected in parallel. According to Ohms law when connecting two components in parallel the voltage across both will be equal and the amperage will be divided along them according to their needs summing up to 10 A. The servo-Hat's each have a 2.1 mm AC Barrel jack as input. The power supply we are using (name and reference) has a 4 pin power connector. The connector has 2 pairs of wires each delivering 10V 5A maximum. These can be split into 2 separate plugs delivering power to both boards from one source.

The base, shown in Fig. 9, was constructed first since it is required to assemble the rest of the skull. It is comprised of two 3D printed pieces that hold the servos at the bottom and one piece that mounts to the skull. In between there are 3 springs with a compression force of 6.69 N/mm. The spring model used is not the same dimension as the one in paper but it has the same compression force and should therefore be about equal. The servos are connected to the base with M3 rods and rod ends.

Following this the back of the skull was assembled, including inserting the raspberry pi into its place. The jaw was also assembled and connected and all wires were connected to the raspberry pi. Because of the way the system is built it is paramount to connect all wires from the lower systems before you install the servo bank. After installing all servos into the system and routing all wires, servo control was implemented with python and ROM (range of motion) was tested.

Fig. 9. Assembled base of the developed robot head.

Every servo has its own separate range of motion depending on its position and orientation within the Robot. Table 2 will describe control board ID, port and angular limits for every servo on the robot. This can be used while programming the behaviour of the robot.

The system is reliant on a pliable mask to convey emotions. This mask is cast from silicone rubber. The mould was 3D printed on a Prusa i3 MK2 3D Printer in Generic PLA and prepped by sanding with 200 grit sand paper on the inside casting surfaces. Followed by two layers of priming spray and a single layer of black lacquer. This was done because the 3D printing process does not create a smooth surface. Casting in a mould with a rough surface would make it nearly impossible to separate from the mold. A non smoothed face would also change the characteristics of both the visual aspect along with the behaviour when deforming.

Before casting a layer of mould release was applied to both the positive and negative side of the mould and left to cure for 12 h. Afterwards about 100g of px30 two part silicone was mixed along with a gray color pigment. This silicone was then poured into the mould and the mould was closed. A piece of flat wood was laid on top of the positive mold piece and a heavy metal block was laid on top to flatten it out. The mould has tabs that make sure the moulds interact correctly.

Table 2. Control board ID ports description

Board	Port	Min angle	Max angle	Neutral angle
0x40	0	135	45	90
0x40	1	45	135	90
0x40	2	135	45	90
0x40	3	45	135	90
0x40	4	135	45	90
0x40	5	45	135	90
0x40	8	120	90	120
0x40	9	60	90	120
0x40	10	0	180	0
0x40	11	180	0	180
0x40	12	0	180	0
0x41	0	0	180	0
0x41	1	0	180	0
0x41	2	0	180	0
0x41	3	0	180	0
0x41	4	0	180	0
0x41	11	0	180	0
0x41	12	0	180	0
0x41	13	0	180	0
0x41	14	0	180	0
0x41	15	0	180	0

After casting of face was finished, lengths of 1.2mm plastic tubing was fitted between the face plate and the servo bank allowing for simple routing of pull wires from silicone mask to driving servos. Tubes were hot-glued onto the face plate and fastened by a set screw on the servo bank. Wires were inserted and tested without the silicone mask connected.

4 Result

The assembled robot, as depicted in Fig. 10, showcases a high level of advancement and sophistication. It is capable of performing a wide range of tasks, including expressing various facial expressions and mimicking phonetics.

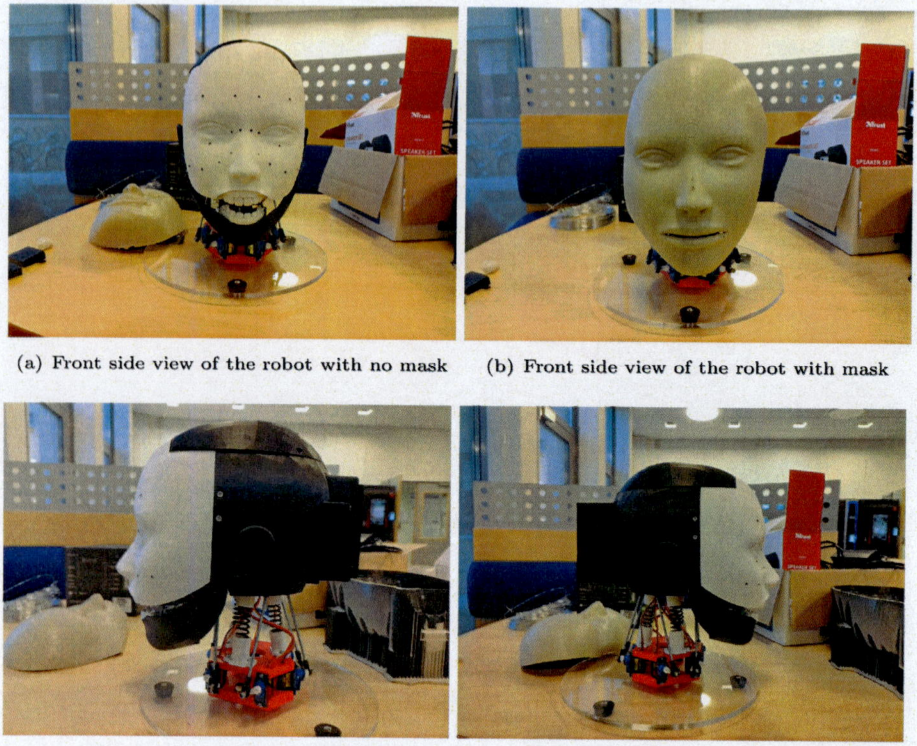

(a) Front side view of the robot with no mask

(b) Front side view of the robot with mask

(c) Left side view of the robot

(d) Right side view of the robot

Fig. 10. Assembled robot.

5 Conclusion

The plan for this project was to complete it as a course report within one semester using by one student using very limited budget. Due to delays in parts acquisition and the need to refactor the model, a fully-featured system could not be developed within the allotted time frame. However, a solid foundation has been established, and with minimal additional work, the system can become fully operational. The total cost of development was also relatively low compared to other platforms, making this design adaptable and ideal as a simple testable platform. In conclusion, it is possible for one person to build a functional Geminoid using limited resources for a course report.

Due to an error in the 3D printing slicer software, a gap was found in the mold. This gap made it challenging to cast the silicone face since parts of the face were not cast as the silicone leaked out of the hole, as shown in Fig. 11. This issue can be easily fixed either within the slicing software or by using another modeling tool. Furthermore, by the end of the designated time frame, there was no suitable method for connecting the servo strings to the silicone mask. While using tape could be a potential solution, it is important to test its compatibility with the specific silicone and wiring used.

When testing the robot, the projected steel wire was too stiff and could not be fastened to the mask. Fishing line was also tested, but it did not work either. A jewelry beading wire made of steel would be optimal since it is flexible yet rigid. It can be riveted together, making it easy to fasten to the mask using tape or other means. The servos used in the system are appropriately sized and powered for their intended use. However, they are equipped with plastic gears, which were selected due to their availability. Replacing these plastic gears with metal ones would be advisable to enhance the durability of the servos. The connections between the servos and the top of the base platform are made using two plastic end rods attached with an m3 rod between them. However, these plastic end rods may wear out quickly over time. They were chosen due to the high cost of metal end rods in the m3 size. While they work fine currently, replacing them with metal end rods would be recommended to increase durability.

Fig. 11. Hole in the back of the mold due to incorrect slicing settings, insufficient infill density, improper support structures, or a problem with the model itself.

6 Future Work

A low-cost Geminoid could have several potential uses in the future. One of the most promising areas of application is in the field of social robotics, where affordable humanoid robots could be used for educational purposes, therapy, and social interaction. Such robots could also be used in research settings to study human-robot interaction, and to develop new methods for controlling and programming robots. Additionally, low-cost Geminoids could be used in entertainment settings, such as theme parks, museums, and other public spaces, where they could provide interactive and engaging experiences for visitors. When equipped with a chatbot such as ChatGPT [23], low-cost Geminoids could be used to provide instructions, information, or assistance in settings such as airports and help desks in other public places. In addition to expressing various facial expressions in real-time, one of the objectives of this project was to mimic phonetics. Mimicking phonetics, or reproducing speech patterns produced by humans, is indeed a complex task for a robot. It requires the integration of advanced technologies and algorithms to achieve accurate speech reproduction.

References

1. Henschel, A., Laban, G., Cross, E.S.: What makes a robot social? A review of social robots from science fiction to a home or hospital near you. Curr. Robot. Rep. **2**(1), 9–19 (2021). https://doi.org/10.1007/s43154-020-00035-0
2. Blindheim, K., Solberg, M., Hameed, I.A., Alnes, R.E.: Promoting activity in long-term care facilities with the social robot pepper: a pilot study. Inf. Health Soc. Care **48**(2), 181–195 (2023). https://doi.org/10.1080/17538157.2022.2086465. PMID: 35702818
3. Martín, F.A., Castillo, J.C., Malfáz, M., Castro-González, A.: Applications and trends in social robotics. Electronics **11**(2), 212 (2022). https://doi.org/10.3390/electronics11020212
4. Guemghar, I., et al.: Social robot interventions in mental health care and their outcomes, barriers, and facilitators: scoping review. JMIR Mental Health **9**(4), 36094 (2022). https://doi.org/10.2196/13322
5. Scoglio, A.A., Reilly, E.D., Gorman, J.A., Drebing, C.E.: Use of social robots in mental health and well-being research: systematic review. J. Med. Internet Res. **21**(7), 13322 (2019). https://doi.org/10.2196/13322
6. Hameed, I.A., Strazdins, G., Hatlemark, H.A.M., Jakobsen, I.S., Damdam, J.O.: Robots that can mix serious with fun. In: Hassanien, A.E., Tolba, M.F., Elhoseny, M., Mostafa, M. (eds.) AMLTA 2018. AISC, vol. 723, pp. 595–604. Springer, Cham (2018). https://doi.org/10.1007/978-3-319-74690-6_58
7. Said, S., Alkork, S., Beyrouthy, T., Youssef, K.: Social robotics in education a survey on recent studies and applications. Int. J. Emerg. Technol. Learn. (iJET) **18**, 67 (2023). https://doi.org/10.3991/ijet.v18i03.33529
8. Niemelä, M., Lammi, H., Heikkilä, P.: A social service robot in a shopping mall: expectations of the management, retailers and consumers. In: Proceedings of the Companion of the 2017 ACM/IEEE International Conference on Human-Robot Interaction, HRI 2017, pp. 227–228. Association for Computing Machinery, New York (2017). https://doi.org/10.1145/3029798.3038301
9. Chen, Z., Nakamura, Y., Ishiguro, H.: Android as a receptionist in a shopping mall using inverse reinforcement learning. IEEE Robot. Autom. Lett. **7**(3), 7091–7098 (2022). https://doi.org/10.1109/LRA.2022.3180042
10. Shah, S.H.H., Steinnes, O.-M.H., Gustafsson, E.G., Hameed, I.A.: Multi-agent robot system to monitor and enforce physical distancing constraints in large areas to combat Covid-19 and future pandemics. Appl. Sci. **11**(16), 7200 (2021). https://doi.org/10.3390/app11167200
11. Khamis, A., et al.: Robotics and intelligent systems against a pandemic. Acta Polytechnica Hungarica **18**(5), 13–35 (2021). https://doi.org/10.12700/aph.18.5.2021.5.3
12. Khamis, A., et al.: AI and robotics in the fight against COVID-19 pandemic. In: Azar, A.T., Hassanien, A.E. (eds.) Modeling, Control and Drug Development for COVID-19 Outbreak Prevention. SSDC, vol. 366, pp. 57–85. Springer, Cham (2022). https://doi.org/10.1007/978-3-030-72834-2_3

13. Jochum, E., et al.: Using theatre to study interaction with care robots. Electronics **8**(4), 457–470 (2016). https://doi.org/10.1007/s12369-016-0370-y
14. Creative Machines Lab: EVA robot. https://www.creativemachineslab.com/eva.html. Accessed 17 June 2023
15. Nishio, S., Ishiguro, H., Hagita, N.: Geminoid: teleoperated android of an existing person. In: Pina Filho, A.C. (ed.) Humanoid Robots, Rijeka. IntechOpen (2007). https://doi.org/10.5772/4876
16. Ishiguro, H.: Studies on humanlike robots – humanoid, android and geminoid. In: Carpin, S., Noda, I., Pagello, E., Reggiani, M., von Stryk, O. (eds.) SIMPAR 2008. LNCS (LNAI), vol. 5325, p. 2. Springer, Heidelberg (2008). https://doi.org/10.1007/978-3-540-89076-8_2
17. Sakamoto, D., Kanda, T., Ono, T., Ishiguro, H., Hagita, N.: Android as a telecommunication medium with a human-like presence. In: HRI 2007, pp. 193–200. Association for Computing Machinery, New York (2007). https://doi.org/10.1145/1228716.1228743
18. Robot, S.: Hanson Robotics (2023). https://www.hansonrobotics.com/sophia/
19. Furhat Robotics: The world's most advanced social robot. https://furhatrobotics.com/. Accessed 19 Apr 2023
20. Faraj, Z., et al.: Facially expressive humanoid robotic face. HardwareX **9**, 00117 (2020). https://doi.org/10.1016/j.ohx.2020.e00117
21. Chen, B., Hu, Y., Li, L., Cummings, S., Lipson, H.: Smile like you mean it: driving animatronic robotic face with learned models, pp. 2739–2746. IEEE Press (2021). https://doi.org/10.1109/ICRA48506.2021.9560797
22. Mori, M., MacDorman, K.F., Kageki, N.: The uncanny valley [from the field]. IEEE Robotics Autom. Mag. **19**(2), 98–100 (2012)
23. OpenAI: ChatGPT (Mar 14 version) [Large language model] (2023). https://chat.openai.com/chat

The Internet of Robotic Things: A Review of Concept, Challenges and Applications

Magy ElBanhawy(✉), Asmaa Mohamed, Walaa Saber, and Rawya Yehia Rizk

Electrical Engineering Department, Faculty of Engineering, Port Said University, Port Said Governorate, Egypt
magy.elbanhawy@eng.psu.edu.eg

Abstract. The integration of robotics and Internet of Things (IoT) leads to the concept of the Internet of Robotic Things (IoRT). It is an emerging vision that brings together the physical world which includes (sensors and accumulators) with robotics and cloud computing. Cloud computing provides robots with tremendous abilities by offering them faster and powerful computational capabilities through massively computation powers and higher data storage facilities. The recent progress in cloud robotics has led to active research in this area from the development of cloud robotics architectures to its varied applications in different robotics domains. These new researches accelerate the development of autonomous system design paradigms and the proliferation of the IoRT. With Machine to Machine (M2M) communication technology, IoRT services can aim to understand connected robots' reactions in order to optimize services and applications. However, cloud and M2M communication systems are overloaded due to the amount of transferred data. Because of the different IoRT data features as well as the way to manage this data, the complexity of IoRT systems increases. This survey explores how the integration of robotic and IoT technologies will enhance the abilities of both the current IoT and the current robotic systems, thus enabling the creation of new, potential services. This survey discusses some of the new technological challenges created by this merger and concludes that a truly holistic view is needed but currently lacking.

Keywords: Internet of Robotic Things (IoRT) · Cloud Robotics · Robot Operating System (ROS) · M2M Communication Networking

1 Introduction

Robotic systems are seen as critical components for humanity's growth in the new digital era. The robotic systems were turned into IoRT applications when technologies of AI, Virtual/Augmented Reality (VR/AR), Deep learning, intelligent networking, Digital Twins (DT), Distributed Ledger technologies (DLTs), swarm technologies and electric mobility emerged [1]. Robotic things can now be connected to anything and everyone at any time, at any location, via various networks and services. Due to new advancements in intelligent networking. Cloud robotics is an intersection between robotics, cloud

A. Hassanien et al. (Eds.): AISI 2023, LNDECT 184, pp. 316–326, 2023.
https://doi.org/10.1007/978-3-031-43247-7_28

computing, deep learning, big data, and Internet of Things (IoT) and other emerging technologies[2].

More like a robot whose sensing and computation are not integrated into a single system, thus the robot having "an extended or a shared brain". As a result, robots are becoming not only smarter by connecting to the cloud, but also cheaper, smaller and easy to build. A networked robotic system refers to a group of robotic devices that are connected via a communication network either wired or wireless. When connected to the cloud, robots can benefit from the powerful computation, storage, and communication resources of the modern data center in the cloud.

Cloud computing provides on-demand, networked access to a pool of virtualized hardware resources (processing, storage, resources, applications) or higher-level services [3]. Processing data streams performed by IoRT devices in a handful of centralized data centers brings concerns about response time, load balancing, latency, massive ingress bandwidth needs and data privacy. Fog computing (also referred to as Edge computing, cloudlets) brings on-demand and elastic computational resources to the edge of the network, closer to the producers of data [4]. The cloud paradigm can be observed in Fig. 1, was adopted by the robotics community, called cloud robotics for offloading resource-intensive tasks for the sharing of data and knowledge between robots [5].

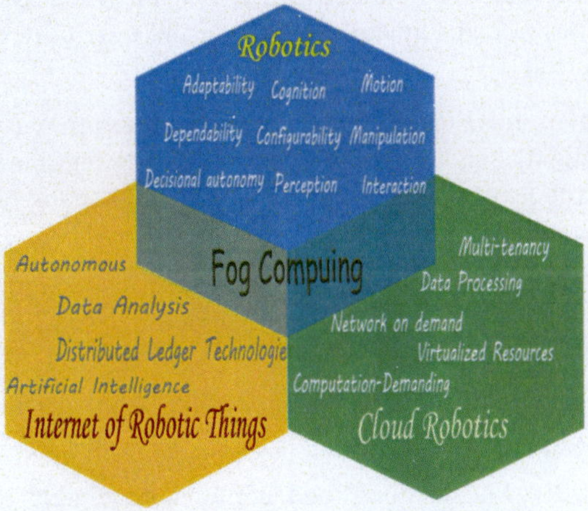

Fig. 1. A scheme of robotics, cloud computing, and IoRT integration.

This survey is intended to shed light on the IoRT applications, technologies and outline the major IoRT challenges in each field. The organization of the remainder of this paper is depicted over a few sections. Section 2 delivers a summary of the recent literature survey. A focus on Cloud Robotics is presented in Sect. 3. Section 4 gives an overview of IoRT techniques, architecture, and abilities. Section 5 highlights some open research challenges in this active field of research. And finally, in Sect. 6 the synopsis and concluding remarks are drawn on the IoRT developments.

2 Literature Review

An intelligent combination of technological advances such as IoRT, Cloud Computing, Smart Grid and Smart Building allows for tracking huge amounts of information. The basic idea of IoRT is to let ‘things’ such as robots to be able to interact and cooperate with each other through wireless communication protocols. The applications for IoRT are widely spread across the world and are far-reaching. For example, IoRT can be used for precision agriculture, manufacturing, health care and much more. In the following section, some of the vast applications are illustrated [6].

2.1 Autonomous Supply Chain

All parts of a supply chain from the production of raw materials to the delivery systems can be built and designed to be automated with the help of IoRT systems [1].

Maintenance: IoRT-empowered robots are capable of autonomously detecting and fixing problems in the system is an important step towards limiting human resources in supply chains as shown in [2] and [6].

Automated E-Commerce: E-commerce is one of the most targets for large-scale IoRT applications [7]. Amazon and Flipkart facilities at the time have no personnel in the lower and middle tiers [1].

Precision Agriculture: IoRT can be used in precision agriculture for the targeted delivery of fertilizers, pesticides, and water [6]. Sensors in fields detect a specific need, which is communicated to robots who deliver in time only the right number of resources to the right place as presented in [8] and [9].

2.2 Healthcare

There are many tasks in a hospital implemented by Humans that could easily be substituted by an intelligent robot [2].

Patient Care: Activities like handling food and other necessities for patients, helping them walk, and assisting them in other ways can be the duty of intelligent and responsive robots as in and [10].

Patient Monitoring: With sensors and other devices connected to a patient’s body, real-time monitoring of their health can be carried out without any problems shown in [11, 12] and [10]. Possibly combining this with an automated drug delivery system could result in a system that treats the patient with accurate and timed dosage alerts [6].

Surgery: Robot-assisted surgery has been popular for some time, and procedures could possibly be performed for the most part by intelligent robots with little assistance from human surgeons as illustrated in [8, 13] and [14].

2.3 Personal Robots

Consumer demands for IoRT personal robots are still only beginning to spread, but it has great potential in the future as the demands could be varied from one human to another [15]. Smart homes can be completely managed by IoRT, with automated systems for household works, security, comfort, communications and convenience, all interconnected to provide humans with a seamless experience as presented [3] and [16].

2.4 Physical Security and Monitoring

IoRT systems can be used to avoid risks to valuable human lives by replacing human personnel in different situations that pose threats [17].

Robot Guards: Guards or security robots can work with camera feeds, inputs from motion sensors, etc., to ensure complete security as in [17] and [18].

Disaster Management: Any area affected by a disaster poses a great amount of risk to rescue personnel. IoRT-supported robots can be used to traverse flood or earthquake-ridden areas, assess the situation, transport supplies, and provide assistance to victims as shown [19].

Military Applications: There have been stories in science fiction about robot wars, where countries would deploy robot forces instead of human forces, to show power without risking millions of human lives presented in [20].

2.5 Automated Vehicles

Driverless cars are used in real time as Tesla Company led this application of IoRT. With the widespread implementation of IoRT, cars can maintain a constant line of communication with other cars, preventing accidents, predicting traffic, and finding comfortable routes [15, 21–23]. The feedback from traffic signals and cameras can also be monitored and processed in real-time by cars to supplement this. Automated Guided Vehicles (AGVs) also have applications in autonomous supply chains as illustrated in [24] and [27].

3 A Cloud Robotics Overview

Cloud Robotics is the way to intelligent robots with higher computing efficiency and lower power consumption. As a result, lower manufacturing costs because there will be less hardware, as well as lower emissions. Cloud robotics innovation become the most traction field due to initiatives by large corporations such as Microsoft, Google and IBM. The need to build and design robots with high performance, accessibility and accuracy has been the primary focus of research activities.

The architecture of Cloud Robotics consists of two types: a machine-to-machine (M2M) level and a machine-to-cloud (M2C) level. The Cloud Robotic architecture mainly consists of M2M or/and M2C architecture. On the M2M level, a group of robots communicate through wireless links to form a computing network (i.e., an ad-hoc cloud) [10]. This unique combination offers us great flexibility in designing computing models to address its specific applications. The following are the three computing models in the robotics environment:

Clone-Based Model: Each robot or virtual machine (VM) has a corresponding system-level clone in the cloud. A task can be performed in the robot or its clone. The set of robotic clones also forms a peer-to-peer network with much higher connectivity than the physical ad-hoc M2M network. Moreover, this model allows for the sporadic outage in the physical M2M network as in Fig. 2 (a).

Peer-Based Model: Fig. 2 (b) shows that each VM or robot in the cloud is considered a computing unit. These robots and VMs form a fully distributed computing mesh. A task can be split into smaller modules (threads) for execution over a subset of the nodes in the computing mesh. By this, the computing load is distributed among the systems [28].

Proxy-Based Model: In the group of networked robots, where one-unit functions as a group leader, communicating with a proxy to bridge the interaction between the robotic network and the cloud as shown in Fig. 2 (c). The set of computing units is organized into a two-tier hierarchy [1].

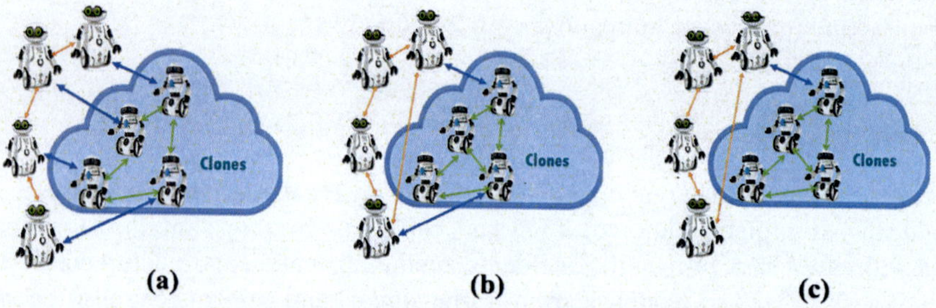

Fig. 2. The different models of the robotics environment.

4 Internet of Robotic Things

IoRT is characterized by "intelligent devices that can monitor events, fuse sensor data from a variety of sources, use local and distributed intelligence to determine the best course of action," this definition provided in [21] focuses more on the robotic cloud concept. Networked Robotics occurs to merge robot system architectures (both hardware and software) and applications that use networks, such as the Internet and cloud computing.

4.1 IoRT Architectures

There is no single architectural design that is agreed upon universally because each organization, company, or each user, for that matter, has different or unique requirements.

Different architectures of IoRT were illustrated in [29]. Moreover, the most hierarchy of architectures includes three-tier architecture, four-tier architecture, five-tier architecture, and seven-layer architecture. The key architecture and the most acknowledge architecture of the as shown in Fig. 3.

Hardware Layer/Physical Layer: The Physical layer represents the lowest level of the IoRT architecture, which includes robots, smartphones, home equipment, sensors, and actuators. IoRT devices may vary from small sensors to a varied range of robotic devices to produce data. Generally speaking, the word robot refers to vehicle, drones, unmanned vessels, and so on. Furthermore, sensors and actuators can be integrated into robot applications, aiming to optimize, monitor, and control various processes such as navigation, calibration, and tuning.

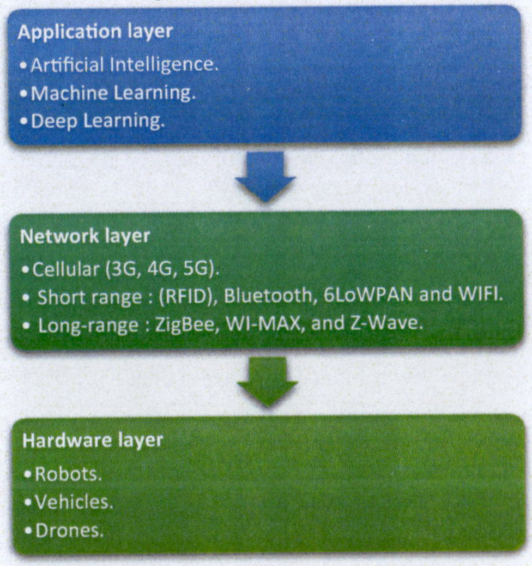

Fig. 3. The architecture of the IoRT systems.

Network Layer: The network layer is the second layer of the IoRT architecture includes controllers, routers, servers, and various communication and control protocols to offer the required connectivity. This layer transfers the sensor data between different layers using different networks type such as: Cellular (3G, 4G, 5G), Short range distance (Radio Frequency Identification (RFID), LAN, Bluetooth, 6LoWPAN and WIFI) and Long-range distance (ZigBee, WI-MAX, and Z-Wave).

Finally, Application Layer: The application layer is the uppermost layer in the IoRT architecture and defines all applications that use IoRT technology. In addition, this layer includes Artificial Intelligence and Machine Learning algorithms, to attain optimized solutions for complex problems in physical environments [30].

5 Challenges

IoRT is a new research field and is in the early stages of development, with many obstacles to overcome. This in-depth and critical investigation of the state of the art in IoRT led to various open research challenges that may be carried out further by researchers in the field of IoRT. The major challenges or gaps that emerged from our study are listed in this section, as well as in Table 1, along with future tasks.

Table 1. Table captions should be placed above the tables.

Challenge	Recommendations/ Practices	Future Opportunities
Overloaded Cloud Resources	• Increasing customer demands in cloud storage systems pose a major challenge. This can be expressed by the imbalance in query handling due to the overload of the system servers in [13] • Traditional cloud systems only achieve a space balance for data storage, but they are limited to simultaneously balancing space resources and system Input-Output, and then minimizing the probability of the overload obtained [3]	• A load balancing scheme for control channel in the Cellular IoT (CIoT) systems in order to support various Internet of Things (IoRT) services. This scheme aims to solve the problems of traffic loads which might be unequal and difficult to release the traffic overload [31]
Fault Tolerance	• Consider using hybrid failure handling method. Developers should incorporate redundant replications and user transparent, fault-tolerant deployment, and execution techniques in orchestration design [25]	• Design a new method for avoiding the unnecessary resource consumption and offer sufficient caching capabilities should be introduced
Data Distribution	• Enable users to make decisions on data sharing [2]. Deploy the social network for service sharing in the Fog	• How to achieve fine-grained data sharing scheme to realize fine-grained data access in Fog Computing
Generalized System Framework	• The present level system framework lacks optimization and theoretical basis. Therefore, it is a challenge to provide a very explicit generalized framework for all robots [26]	•There is an urgent requirement for an algorithm-level system framework that optimizes system performance. A number of robotic tasks in various scenarios can be improved by cloud robotics

(continued)

Table 1. (*continued*)

Challenge	Recommendations/ Practices	Future Opportunities
Communication Constraints	• Ad hoc routing protocols suffer from high latency as a route has to be established before a message can be sent, and are not practical if the network topology is dynamic [14]	•These drawbacks are significant in mobile robotic networks, and may lead to severe performance degradation
Information and learning constraints	• Networked robotics is constrained by the information observed or computed by robots in the network, and by the examples or scenarios that the network encounters, and hence limiting its ability to learn [11, 28]	• On the other hand, the learning process has to be repeated once the environment changes or the team is placed in a new unfamiliar environment
Dynamic adaptability of the system	• Where robots are deployed is dynamic, and this is where the difficulty of robotic tasks lies. It is not easy to make an optimal collaborative computing decision to improve the system in dynamic conditions	• It requires the system to be capable of adapting to a changing condition and it means a group of dynamic parameters rather than statics, which brings more difficulties to the system design
Data Distribution	• Enable users to make decisions on data sharing [5]. Deploy the social network for service sharing in the Fog	• How to achieve fine-grained data sharing scheme to realize fine-grained access. Without such one many real time applications may be blocked
Location privacy	• Deploy anonymous and stateless process which keep user's location and identity private, even from the Fog network as in [21] and [32] • Define some levels of privacy protection in data combination	•Design new effective privacy-preservation methods • Develop mechanisms that allow IoT users to identify their own privacy requirements as well as enforcement mechanism

6 Conclusion

In this survey a background of IoRT and cloud robotics is presented. This survey focused on IoRT definition, abilities, applications, enabling technologies and IoRT architectures. Various methods for navigation of robotic things were reviewed. IoRT issues and challenges have been deeply analyzed. It is clear from this analysis that there is an increasing need for investigation in robot-to-robot cooperation, above all in heterogeneous networks of robots, human–robot interfaces for human-centered interactions. In addition, there are some challenges that must be taken into considerations such as energy management, optimization of efficiency and protection of sensitive data.

References

1. Sayeed, A., Verma, C., Kumar, N., Koul, N.: Approaches and challenges in internet of robotic things. Future Internet **14**(9), 265 (2022). https://doi.org/10.3390/fi14090265
2. Vermesan, O., et al.: Internet of robotic things intelligent connectivity and platforms. Front. Robot. AI **7**(September), 1–33 (2020). https://doi.org/10.3389/frobt.2020.00104
3. Liu, B.: Fog and Cloud Robotics (2022)
4. Hashem, W., Attia, R., Nashaat, H., Rizk, R.: Advanced deep reinforcement learning protocol to improve task offloading for edge and cloud computing. In: Hassanien, A.E., Rizk, R.Y., Snášel, V., Abdel-Kader, R.F. (eds.) The 8th International Conference on Advanced Machine Learning and Technologies and Applications (AMLTA2022), pp. 615–628. Springer International Publishing, Cham (2022). https://doi.org/10.1007/978-3-031-03918-8_51
5. Liu, B., Wang, L., Liu, M.: ElasticROS: an elastically collaborative robot operation system for fog and cloud robotics. **X**(X), 1–20 (2022), http://arxiv.org/abs/2209.01774
6. Romeo, L., Petitti, A., Marani, R., Milella, A.: Internet of robotic things in smart domains: applications and challenges. Sensors (Switzerland) **20**(12), 1–23 (2020). https://doi.org/10.3390/s20123355
7. Zhao, L., Huang, Z.: A moving object detection method using deep learning-based wireless sensor networks. Complexity **2021**, 1–12 (2021). https://doi.org/10.1155/2021/5518196
8. Simoens, P., Dragone, M., Saffiotti, A.: The internet of robotic things: a review of concept, added value and applications. Int. J. Adv. Robot. Syst. **15**(1), 1–11 (2018). https://doi.org/10.1177/1729881418759424
9. Abikoye, O.C., et al.: Application of internet of thing and cyber physical system in industry 4.0 smart manufacturing. In: Singh, K.K., Nayyar, A., Tanwar, S., Abouhawwash, M. (eds.) Emergence of Cyber Physical System and IoT in Smart Automation and Robotics: Computer Engineering in Automation, pp. 203–217. Springer International Publishing, Cham (2021). https://doi.org/10.1007/978-3-030-66222-6_14
10. Afanasyev, I., et al.: Towards the internet of robotic things: Analysis, architecture, components and challenges. In: International, Conference on Developments in eSystems Engineering, vol. October-20, no. May 2020, pp. 3–8 (2019). https://doi.org/10.1109/DeSE.2019.00011
11. Mohammadi, V., Rahmani, A.M., Darwesh, A.M., Sahafi, A.: Trust-based recommendation systems in Internet of Things: a systematic literature review. Hum.-Cent. Comput. Inf. Sci. **9**(1), 21 (2019). https://doi.org/10.1186/s13673-019-0183-8
12. Lins, R.G., Givigi, S.N.: Cooperative robotics and machine learning for smart manufacturing: platform design and trends within the context of industrial internet of things. IEEE Access **9**, 95444–95455 (2021). https://doi.org/10.1109/ACCESS.2021.3094374
13. Chen, W.: Intelligent manufacturing production line data monitoring system for industrial internet of things. Comput. Commun. **151**, 31–41 (2020). https://doi.org/10.1016/j.comcom.2019.12.035
14. Albonico, M., Rohling, A., Santos, J., Varela, P.: Mining evidences of internet of robotic things (IoRT) software from open source projects. In: ACM International Conference Proceeding Services, no. September, pp. 71–79 (2021). https://doi.org/10.1145/3483899.3483900

15. Shao, T., Chowdhury, D., Gill, S.S., Buyya, R.: IoT-Pi: a machine learning-based lightweight framework for cost-effective distributed computing using IoT. Internet Technol. Lett. **5**(3), 1–6 (2022). https://doi.org/10.1002/itl2.355
16. Saurabh, Dhanaraj, R.K.: A review paper on fog computing paradigm to solve problems and challenges during integration of cloud with IoT. J. Phys. Conf. Ser. **2007**(1), 012017 (2021). https://doi.org/10.1088/1742-6596/2007/1/012017
17. Amin, A.: A face recognition system based on deep learning (FRDLS) to support the entry and supervision procedures on electronic exams. Int. J. Intell. Comput. Inf. Sci. **20**(1), 40–50 (2020). https://doi.org/10.21608/ijicis.2020.23149.1015
18. DiLuoffo, V., Michalson, W.R., Sunar, B.: Robot operating system 2: the need for a holistic security approach to robotic architectures. Int. J. Adv. Robot. Syst. **15**(3), 1–15 (2018). https://doi.org/10.1177/1729881418770011
19. Macenski, S., Foote, T., Gerkey, B., Lalancette, C., Woodall, W.: Robot operating system 2: design, architecture, and uses in the wild. Sci. Robot. **7**(66), eabm6074 (2022). https://doi.org/10.1126/scirobotics.abm6074
20. Casamayor Pujol, V., Dustdar, S.: Fog robotics—understanding the research challenges. IEEE Internet Comput. **25**(5), 10–17 (2021). https://doi.org/10.1109/MIC.2021.3060963
21. Ezenwobodo, Samuel, S.: An assessment on the use of mathematical softwares in teaching and learning of mathematics in colleges of education in south-eastern Nigeria: a case study of Anambra and Enugu. Int. J. Res. Publ. Rev. **04**(01), 1806–1812 (2022). https://doi.org/10.55248/gengpi.2023.4149
22. Jeong, J., Yang, J., Baltes, J.: Robot magic show as testbed for humanoid robot interaction. Entertainment Comput. **40**, 100456 (2022). https://doi.org/10.1016/j.entcom.2021.100456
23. Singh, R.B., Zhuang, H.: Measurements, analysis, classification, and detection of gunshot and gunshot-like sounds. Sensors **22**(23), 9170 (2022). https://doi.org/10.3390/s22239170
24. Hassan, A., Attia, R., Rizk, R.: Internet of vehicles and intelligent routing: a survey-based study. In: Hassanien, A.E., Rizk, R.Y., Snášel, V., Abdel-Kader, R.F. (eds.) The 8th International Conference on Advanced Machine Learning and Technologies and Applications (AMLTA2022), pp. 517–531. Springer International Publishing, Cham (2022). https://doi.org/10.1007/978-3-031-03918-8_43
25. Valecce, G., et al.: Robotic-aided IoT: automated deployment of a 6TiSCH network using an UGV. IET Wirel. Sens. Syst. **9**(6), 438–446 (2019). https://doi.org/10.1049/iet-wss.2019.0076
26. Alzoubi, Y.I., Al-Ahmad, A., Jaradat, A., Osmanaj, V.H.: FOG computing architecture, benefits, security, and privacy, for the internet of thing applications: an overview. J. Theor. Appl. Inf. Technol. **99**(2), 436–451 (2021)
27. Attia, R., Hassaan, A., Rizk, R.: Advanced greedy hybrid bio-inspired routing protocol to improve IoV. IEEE Access **9**, 131260–131272 (2021). https://doi.org/10.1109/ACCESS.2021.3114646
28. Patil, A.A., Karande, P.P., Jagtap, N.A., Jamadar, A.R., Patil, A.S.: Manual star delta starter. Int. J. Res. Publ. Rev. **3**(4), 1762–1765 (2022)
29. Shende, S.W.: Artificial intelligence and machine learning for internet of things. J. Phys. Conf. Ser. **1913**, 012151 (2021). https://doi.org/10.1088/1742-6596/1913/1/012151
30. Moussa, W., Nashaat, M., Saber, W., Rizk, R.: Comprehensive study on machine learning-based container scheduling in cloud. In: Hassanien, A.E., Rizk, R.Y., Snášel, V., Abdel-Kader, R.F. (eds.) The 8th International Conference on Advanced Machine Learning and Technologies and Applications (AMLTA2022), pp. 581–592. Springer International Publishing, Cham (2022). https://doi.org/10.1007/978-3-031-03918-8_48

31. El Banhawy, M., Saber, W., Amer, F.: A native enhanced elastic extension tables multi-tenant database. Int. J. Electr. Comput. Eng. **10**(6), 6618–6628 (2020). https://doi.org/10.11591/IJECE.V10I6.PP6618-6628
32. Alkady, Y., Rizk, R.: Location privacy-preserving of vehicular ad-hoc network in smart cities. In: Hassanien, A.E., Rizk, R.Y., Snášel, V., Abdel-Kader, R.F. (eds.) The 8th International Conference on Advanced Machine Learning and Technologies and Applications (AMLTA2022), pp. 532–543. Springer International Publishing, Cham (2022). https://doi.org/10.1007/978-3-031-03918-8_44

Virtual Control System for Presentations by Real-Time Hand Gesture Recognition Based on Machine Learning

NourEldin Osama[1], Yousr Ahmed[1], Hussein Mohamed[1], Seif Eldin Hesham[1], Youssef Ahmed[1], Eman K. Elsayed[1,3], and Dalia Ezzat[1,2,4](✉)

[1] School of Computer Science, Canadian International College (CIC), Cairo, Egypt
dalia_ali@cic-cairo.com
[2] Faculty of Computers and Artificial Intelligence, Cairo University, Cairo, Egypt
[3] Faculty of Science, Alazhar University (Girls) Branch, Cairo, Egypt
[4] Scientific Research Group in Egypt (SRGE), Cairo, Egypt

Abstract. Presentations are a powerful tool for presenters who want to persuade their audiences in today's digital age. This paper exploits advances in hand gesture recognition, and proposes a virtual control system for presentations. The proposed system utilizes a webcam or built-in camera to capture hand gestures. Based on hand gestures, presentations can be controlled virtually and change presentation slides in both forward and backward directions. It is also possible by using the proposed system to get a pointer on the slide, write, or draw virtually on the screen through specific hand gestures. The obtained results show that the proposed system has a high accuracy of 96% in recognizing hand gestures and thus controlling presentations remotely without using any external device.

Keywords: Computer vision · Machine learning · Deep learning · Hand gesture detection · Human computer interaction

1 Introduction

With the fast development of computer technology and artificial intelligence methods, especially deep learning, the accuracy of hand gesture recognition has increased significantly [1–3]. The capacity to recognize hand gestures has been the subject of several research efforts, either to improve the capability or to take use of it in other disciplines. Among the research efforts, this paper [4] proposed a technique for hand gesture recognition. In order to recognize hand gestures in a dataset with a limited amount of gesture photos, an end-to-end fine-tuning strategy of a pre-trained CNN model with a score-level fusion technique is proposed. Massey University (MU) dataset and HUST American sign language (HUST-ASL) dataset are used and the performance of the proposed technique shows a mean accuracy of 98.14% and 64.55% for both datasets, respectively. In this paper [5], a one-hot encoding technique is utilized to transform the categorical data values to binary form. This is followed by the implementation of a crow search algorithm

A. Hassanien et al. (Eds.): AISI 2023, LNDECT 184, pp. 327–335, 2023.
https://doi.org/10.1007/978-3-031-43247-7_29

(CSA) for determining best hyper-parameters for the training of datasets utilizing CNN. The dataset utilized for this experiment is the hand gesture recognition database, which was gathered from the public repository, Kaggle. This database has 10 different folders for hand gesture images for 10 digits (0–9). Each folder has a 2000 collection of images for various hand gestures for the corresponding digits. In this research [5], the proposed model has achieved an accuracy of 100% in the hand gesture recognition database. This paper [6] presents dependable gesture recognition (DGR) approach that identifies the activity of patient monitoring via remote access utilizing a variety of learning methods. According to the analysis, the suggested DGR approach achieves 94.92% high precision for the various gestures and 89.85% high accuracy for various mess factors. The proposed DGR shortens recognition time to 4.97 s and 4.93 s for the varying gestures and mess factor, respectively.

This paper [7] proposes a lightweight hand gesture recognition model leveraging YOLO (You Only Look Once) v3 and DarkNet-53 CNN. The model excelled in extracting hand features, achieving an accuracy, precision, recall, and f1 score of 97.68%, 94.88%, 98.66%, and 96.70% respectively. The model is capable of real-time detection for both static hand images and dynamic video-recorded gestures. This work [8] proposes a method for hand gesture recognition using a CNN model with attention mechanism. The results showed that the method achieved an accuracy of 99.2%, which is significantly higher than the accuracy of other compared methods. This paper [9] proposed a method for hand gesture recognition using an improved single-shot multibox detector (SSD) algorithm, CNN, and support vector machine (SVM) classifier. The findings demonstrated that the proposed method achieved an accuracy of 93.4%.

This research [10] proposes a method for hand gesture recognition using multimodal data fusion and multiscale parallel CNN. This method was evaluated on a self-made dataset containing 10 common hand gestures. The results showed that the proposed method achieved an accuracy of 95%. In this paper [11], CNN-SPP, an integrated CNN with spatial pyramid pooling (SPP), is presented. The CNN-SPP network was designed for hand gesture recognition. The proposed CNN-SPP achieved an average accuracy of 99.34, outperforming other comparative approaches. As a complement to the ongoing development in hand gesture recognition, Google has developed the MediaPipe framework that is a powerful tool for this task [12]. In addition, the OpenCV library [13] can also be used to track and identify hand gestures and hand tips. The OpenCV library and the MediaPipe framework both use machine learning concepts and are both used for real-time computer vision applications.

Hence the field of hand gesture recognition has gained prominence, finding utility in a range of applications such as virtual reality, games, robotics and human-computer interaction [14–19]. In this paper [14], a virtual mouse system based on hand gestures has been proposed. Through hand gestures, the computer can be controlled virtually and can perform many functions such as left-click, right-click, and scroll without using the real mouse. Research efforts have been made to provide virtual systems as an alternative to the real mouse such as the virtual systems proposed in [20, 21].

Recognizing and interpreting hand gestures enables natural and intuitive interactions with devices, including gesture-based presentation virtual controller. In the digital era,

Intelligent presentation creation is crucial as it facilitates coherent and systematic dissemination of information, captivates and engrosses the audience, optimizes temporal allocation, instills confidence and professionalism, fosters adaptability, and ensures efficacious communication. By engaging in meticulous preparation, presenters can enhance the lucidity and organization of their discourse, effectively captivate the audience's attention, adeptly manage temporal constraints, exude self-assurance, adapt deftly to unforeseen circumstances, and communicate their message with eloquence and persuasion. The presenter's interaction with the slideshow plays a significant role in minimizing interruptions from the audience, thereby maintaining the flow and focus of the presentation. Hence, this paper proposes a presentation virtual control system based on hand gesture recognition using the MediaPipe framework.

The remaining portions of this paper are organized as follows: In Sect. 2, the MediaPipe framework is discussed. The proposed system is described in Sect. 3. The results are reported in Sect. 4. Finally, Sect. 5 presents the concluding remarks for this paper.

2 MediaPipe Framework

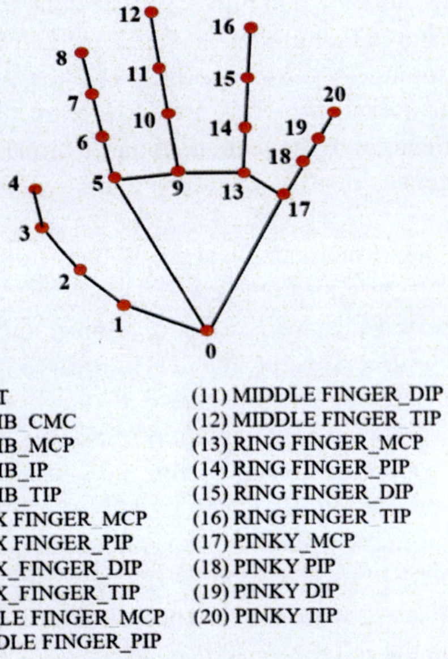

Fig. 1. The coordinates of the landmarks of the hand

MediaPipe is a cross-platform framework, created by Google, for building multimodal applied machine learning pipelines [12]. It provides cutting edge ML models that can be used for many tasks such as face detection, hand tracking, and object detection. For the hand tracking, the MediaPipe Hand Landmarker model package trained on approximately 30,000 real images is used. The hand landmarker model bundle includes a palm detection model and a hand landmarks detection model. The palm detection model finds hands in the input image, and the hand landmarks detection model spots particular hand landmarks on the cropped hand image that the palm detection model specifies. The hand landmarks model can identify 21 hand joints that are recognized and identified by the palm detection model as shown in Fig. 1.

3 Proposed System

The proposed system, presentations virtual control, is a system with a web application interface, which accepts PowerPoint and Portable Document Format (PDF) files as input. Based on hand gestures, the proposed system can control the presentation virtually. The proposed system can move to the next slide in the presentation, move to the previous slide in the presentation, get a pointer on the slide, draw or write text on the slide, and erase any text on the slide. The proposed system performs these functions through two phases: the conversion phase and the hand tracking phase as shown in Fig. 2.

At the conversion phase, the PowerPoint file is converted to Portable Document Format (PDF). Once the file is in PDF format, the system proceeds to the next step of the conversion phase. This step involves converting the PDF file into Portable Network Graphics (PNG) images, as shown in Fig. 2 (a). The hand tracking phase is mainly based on MediaPipe Hand Landmarker model, and it takes the frames captured by the webcam in a laptop or PC as input. As shown in Fig. 2 (b), the second phase of the proposed system begins with the source, which is a video stream from a camera. This real-time video data serves as the input to the hand landmarks model. The hand landmarks model processes the video frames, identifying and recognizing the presence of hands and hand geometry.

The output of the hand landmarks model is then utilized by several components. The hand gesture recognizer interprets the hand landmarks and recognizes specific hand gestures. If the recognized gesture is a hand waving, the slide control component is activated. If the hand gesture is right, the system moves to the next slide. Conversely, if the hand gesture is left, the system moves to the previous slide. Simultaneously, the Pointer component uses the hand landmarks, specifically the INDEX FINGER TIP landmark, to control a pointer on the screen, enabling the user to interact with the system in a more direct and intuitive way. In addition to the pointer, the drawing and eraser components also make use of the hand landmarks. The drawing component is activated when the INDEX FINGER and MIDDLE FINGER are extended, tracking these movements to create drawings on the screen. The eraser component, on the other hand, is activated when the INDEX FINGER, MIDDLE FINGER, and RING FINGER are extended, using these movements to erase drawings on the screen.

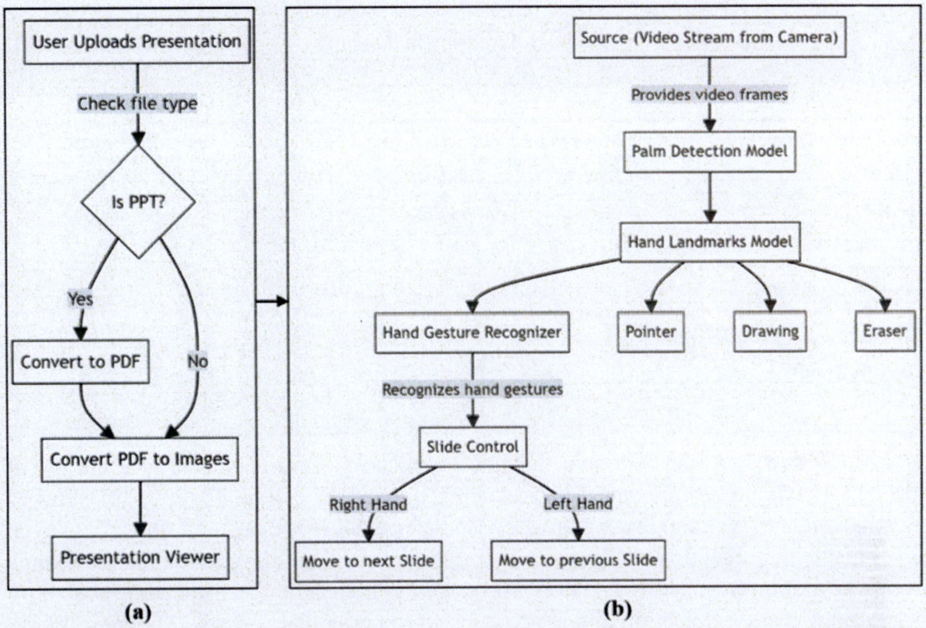

Fig. 2. The illustrative structure of the proposed system, (a) represents the first phase of the proposed system, (b) represents the second phase of the proposed system

4 Results and Discussion

The proposed system is made on the windows operating system. The implemented software of the proposed system utilizes python version 3.11.2, and Django framework version 4.1.7 [22] is used to create an interface for real-time use of the proposed system. The proposed system requires specific hardware: modern multi-core processor such as an Intel Core i5 or AMD Ryzen 5, minimum of 8 GB RAM is recommended for development but 16 GB or more may be beneficial, and Solid-State Drive (SSD) with at least 256 GB.

As described in the Sect. 3, the proposed system can perform five functions: move to the next slide, return to the previous slide, get a pointer on the slide, draw on the slide, and erase the drawing on the slide using different hand gestures. In order to evaluate the effectiveness of the proposed system in performing the five functions, each hand gesture was tested 20 times by three different participants, yielding 60 trials total. The experiment was conducted in a variety of lighting conditions and viewing distances. Each participant tested the proposed system for each hand gesture five times in normal light conditions, five times in dim light conditions, five times at a close distance to the webcam, and five times at a far distance from the webcam. The accuracy of the proposed system is calculated by dividing the number of correctly recognized hand gestures by the total number of hand gestures, as shown in Eq. (1).

$$Accuracy = \frac{Correctly\ detected\ hand\ gestures}{Total\ number\ of\ hand\ gestures} \tag{1}$$

Table 1. Experimental results

Hand gesture	Action	Success	Failure	Accuracy
Right hand gesture	Move to next slide	60	0	100%
Left hand gesture	Back to previous slide	60	0	100%
INDEX FINGER TIP	Get a pointer on the slide	60	0	100%
INDEX FINGER and MIDDLE FINGER	Drawing on the slide	48	12	80%
INDEX FINGER, MIDDLE FINGER, and RING FINGER	Erase the drawing on the slide	60	0	100%
Overall Results	--	**288**	**12**	**96%**

From Table 1, it can be noted that the proposed system achieved an accuracy of 100% in right hand gesture detection and hence move to next slide. Figure 3 (a), shows a person using the right hand gesture to move to the next slide in a presentation. The person raises his hand in the air with his palm facing forward. Then they move their hands in a sweeping motion from left to right. The proposed system achieved an accuracy of 100% in left hand gesture detection and hence back to the previous slide. As shown in Fig. 3 (b), this gesture shows a person using the left hand gesture to go back to the previous slide in a presentation. The person is holding their hand up in the air, with their palm facing down. They are then moving their hand in a sweeping gesture from right to left. The proposed system achieved an accuracy of 100% in detecting INDEX FINGER TIP and hence getting a pointer on the slide.

As shown in Fig. 3 (c), the gesture shows a person using a hand gesture to get a pointer on a slide. The person is holding their hand up in the air, with their index finger extended. They are then pointing at a specific area on the slide. The proposed system achieved an accuracy of 80% in detecting INDEX FINGER and MIDDLE FINGER and hence drawing on the slide. As shown in Fig. 3 (d), this gesture shows a person using two fingers to write on a slide. The person is holding their hands up in the air, with their index fingers extended. They are then moving their fingers in a drawing motion on the slide. The proposed system achieved an accuracy of 100% in detecting INDEX FINGER, MIDDLE FINGER, and RING FINGER and hence erasing the drawing on the slide.

As shown in Fig. 3 (e), this gesture shows a person using a hand gesture to erase writing on a slide. The person is holding their hand up in the air, with their palm facing down. They are then moving their hand in a sweeping motion over the drawing on the slide. The proposed system achieved an average accuracy of 96%. If an undefined hand gesture is used, then no change will occur. As shown in Fig. 3 (f), this gesture shows a person using an undefined hand gesture where the person raises their hand in the air with their hand palm up and then they move their hand in a circular motion.

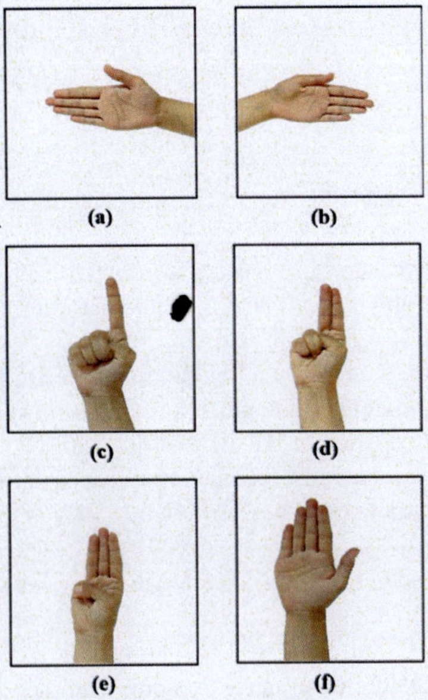

Fig. 3. Hand gestures to control the presentation virtually. (a) represents the hand gesture to move to the next slide, (b) represents the hand gesture to return to the previous slide, (c) represents the hand gesture to get a pointer on the slide. (d) represents the hand gesture to write on the slide, (e) represents the hand gesture to remove the writing on the slide, (f) represents undefined hand gesture

5 Conclusion and Future Work

The main objective of the proposed system presented in this paper is to control presentations virtually using hand gestures rather than using a physical pointer. The proposed system includes two main phases, the conversion phase and the hand tracking phase. Through these phases, the proposed system can be used as a good virtual control tool for presentations. Through the results obtained, it was found that the proposed system has a high ability to accurately recognize hand gestures and then take the correct actions for these gestures. Despite achieving relatively high accuracy, the proposed system may struggle to detect hand gestures accurately if the hand motion is too fast or if the gesture occurs at the edge of the screen. Another limitation of the model is that it may be affected by changes in lighting, which can negatively affect performance. Future work could be to use different image processing techniques to improve the robustness of the proposed system for different lighting conditions.

Code Availability
The code of the proposed system, developed by the authors as part of the graduation project for the 2022–2023 academic year at the School of Computer Science, Canadian International College (CIC), is publicly available at: https://github.com/NourEldin-Osama/Graduation-Project/blob/main/static/js/Hand_Landmarker_Detection.js.

References

1. Oudah, M., Al-Naji, A., Chahl, J.: Hand gesture recognition based on computer vision: a review of techniques. J. Imaging **6**(8), 73 (2020). https://doi.org/10.3390/jimaging6080073
2. Jaramillo-Yánez, A., Benalcázar, M.E., Mena-Maldonado, E.: Real-time hand gesture recognition using surface electromyography and machine learning: a systematic literature review. Sensors **20**(9), 2467 (2020)
3. Parvathy, P., Subramaniam, K., Prasanna Venkatesan, G.K.D., Karthikaikumar, P., Varghese, J., Jayasankar, T.: Development of hand gesture recognition system using machine learning. J. Ambient. Intell. Humaniz. Comput. **12**, 6793–6800 (2021)
4. Sahoo, J.P., Jaya Prakash, A., Pławiak, P., Samantray, Sa.: Real-time hand gesture recognition using fine-tuned convolutional neural network. Sensors **22**(3), 706 (2022). https://doi.org/10.3390/s22030706
5. Gadekallu, T.R., et al.: Hand gesture classification using a novel CNN-crow search algorithm. Complex Intell. Syst. **7**(4), 1855–1868 (2021)
6. Mahmoud, N.M., Fouad, H., Soliman, A.M.: Smart healthcare solutions using the internet of medical things for hand gesture recognition system. Complex Intell. Syst. **7**(3), 1253–1264 (2021)
7. Mujahid, A., et al.: Real-time hand gesture recognition based on deep learning YOLOv3 model. Appl. Sci. **11**(9), 4164 (2021). https://doi.org/10.3390/app11094164
8. Breland, D.S., et al.: Deep learning-based sign language digits recognition from thermal images with edge computing system. IEEE Sensors J. **21**(9), 10445–10453 (2021)
9. Liao, S., et al.: Occlusion gesture recognition based on improved SSD. Concurrency Computation: Pract. Experience **33**(6), e6063 (2021)
10. Gao, Q., Liu, J., Zhaojie, J.: Hand gesture recognition using multimodal data fusion and multiscale parallel convolutional neural network for human–robot interaction. Expert. Syst. **38**(5), e12490 (2021)
11. Tan, Y.S., Lim, K.M., Tee, C., Lee, C.P., Low, C.Y.: Convolutional neural network with spatial pyramid pooling for hand gesture recognition. Neural Comput. Applic. **33**(10), 5339–5351 (2021)
12. Camillo Lugaresi, J.T.: MediaPipe: A Framework for Building Perception Pipelines. https://arxiv.org/abs/1906.08172 (2019)
13. Culjak, I., Abram, D., Pribanic, T., Dzapo, H., Cifrek, M.: A brief introduction to OpenCV. In: 2012 Proceedings of the 35th International Convention MIPRO, pp. 1725–1730. IEEE (2012)
14. Lv, Z., Poiesi, F., Dong, Q., Lloret, J., Song, H.: Deep learning for intelligent human-computer interaction. Appl. Sci. **12**(22), 11457 (2022)

15. Vuletic, T., Duffy, A., Hay, L., McTeague, C., Campbell, G., Grealy, M.: Systematic literature review of hand gestures used in human computer interaction interfaces. Int. J. Hum Comput Stud. **129**, 74–94 (2019)
16. Shriram, S., Nagaraj, B., Jaya, J., Shankar, S., Ajay, P.: Deep learning-based real-time AI virtual mouse system using computer vision to avoid COVID-19 spread. J. Healthcare Eng. **2021**, 1–8 (2021)
17. Hu, B., Wang, J.: Deep learning based hand gesture recognition and UAV flight controls. Int. J. Autom. Comput. **17**(1), 17–29 (2020)
18. Cruz, P.J., et al.: A Deep Q-Network based hand gesture recognition system for control of robotic platforms. Sci. Rep. **13**(1), 7956 (2023)
19. Bora, J., Dehingia, S., Boruah, A., Chetia, A.A., Gogoi, D.: Real-time assamese sign language recognition using mediapipe and deep learning. Procedia Comput. Sci. **218**, 1384–1393 (2023)
20. Tran, D.S., Ho, N.H., Yang, H.J., Kim, S.H., Lee, G.S.: Real-time virtual mouse system using RGB-D images and fingertip detection. Multimed. Tools Appl. **80**, 10473–10490 (2021)
21. Shibly, K.H., Dey, S.K., Islam, M.A., Showrav, S.I.: Design and development of hand gesture based virtual mouse. In 2019 1st International Conference on Advances in Science, Engineering and Robotics Technology (ICASERT), pp. 1–5. IEEE (2019)
22. Forcier, J., Bissex, P., Chun, W.J.: Python Web Development with Django. Addison-Wesley Professional (2008)

Health Informatics

Soft Computing Framework for Digital Healthcare System

Debasis Chanda[1(✉)] and Narayan C. Debnath[2]

[1] MDI Murshidabad, Murshidabad, West Bengal, India
cdebasis04@yahoo.o.in
[2] School of Computing and Information Technology, Eastern International University, Thủ Dầu Một, Binh Duong, Vietnam
Nararyan.debnath@eiu.edu.vn

Abstract. This Paper addresses the problem related to determination of dependency and relationship between diseases. Towards achieving the above objective, we propose Fuzzy Data Mining Approach.

The contribution of this paper lies in the exploration of soft computing framework in addressing the vexing and often unresolved healthcare concerns. In today's age and also in the foreseeable future, the resolution lies in healthcare based on digital systems. The differentiation of the paper lies in proposing a soft computing framework based digital system for healthcare.

The paper not only considers fuzzy clustering mechanism, but also adopts a differentiated and hitherto seldom explored approach of data mining model based on fuzzy relational databases.

We look at various perspectives and dimensions that would facilitate healthcare, such as the geographic penetration of diseases and similarity measures of the same, such that providers of healthcare systems may focus their efforts on incidence of certain diseases, based on dimensions such as geography, demography etc.

Thus, this work provides a novel and hitherto seldom unexplored differentiated approach of digital healthcare system.

Keywords: Fuzzy Data mining · Cluster Analysis · Euclidean Distance · Compatibility Relation · Disease · Relational Databases

1 Introduction

In this work we address problems related to determination of dependency between diseases by looking at Fuzzy Data Mining Techniques.

Fuzzy Data Mining scores over classical data mining techniques since it involves knowledge discovery from data expressed in real world terms.

The paper looks at Fuzzy Cluster Analysis in the context of Dependency between Diseases. Clustering is the process of grouping data into classes or clusters, so that objects within a cluster have high similarity in comparison to one another, but very dissimilar to objects in other clusters. Dissimilarities are assessed based on the attribute

A. Hassanien et al. (Eds.): AISI 2023, LNDECT 184, pp. 339–347, 2023.
https://doi.org/10.1007/978-3-031-43247-7_30

values describing the objects. The paper considers the impact of diseases expressed by the Fuzzy Linguistic Variables.

The paper considers determination of fuzzy compatibility relation in terms of appropriate distance function applied to given data and clustering method based on fuzzy equivalence relations as well as multidimensional classification based on cluster analysis.

From another perspective, Fuzzy Data Mining Model Using Relational Databases is considered. The same looks at multi-dimensional data mining on relational databases,threshold values, fuzzy compatibility relationship and sequential relational joins with relations containing the relevant dimensions (attributes).

It is observed that fuzzy relational databases provide insights (knowledge), which would otherwise not be possible with crisp relational databases. Moreover, search time is faster, and the above methodology also provides us with mechanisms of arriving at strong or weak association rules, based on the threshold values. Thus, it may be possible to ascertain which diseases that have interdependency, based on the defined criteria.

The paper comprises of the following sections: Fuzzy Mathematical Preliminaries are furnished in Sect. 2; the proposed Modeling Framework is discussed in Sects. 3 and 4; Benefits are furnished in Sect. 5; Sect. 6 provides some conclusions.

2 Fuzzy Mathematical Preliminaries [11]

2.1 Mathematical Description of a Data Mining Algorithm

i. A crisp relation $R_c(X,X)$ is reflexive iff $\langle x, x\rangle \in R_c$ for each x $\in X$, that is, if every element of X is related to itself. A fuzzy relation $R_f\ (X,X)$ is reflexive iff $R_f\ (x,x) = 1$.

ii. A crisp relation $R_c\ (X,X)$ is symmetric iff $\langle x, x\rangle \in R_c$, it is also the case that $\langle y, x\rangle \in R_c$, where x,y $\in X$. Thus, whenever an element x is related to an element y through a symmetric relation, y is also related to x. A fuzzy relation is symmetric iff $R_f\ (x,y) = R_f\ (y,x)$

iii. A crisp relation $R_c\ (X,X)$ is transitive iff $\langle x, z\rangle \in R_c$, whenever both $\langle x, y\rangle \in R_c$, $\langle y, z\rangle \in R_c$, for at least one $y \in X$. In other words, the relation of x to y and of y to z implies the relation of x to z in a transitive relation. A fuzzy relation $R_f\ (X,X)$ is transitive if $R_f(x, z) \geq \max_{y \in Y} \min\left[R_f(x, y),\ R_f(y, z)\right]$ is satisfied for each pair $\langle x, z\rangle \in X^2$.

iv. The transitive closure of a crisp relation $R_c\ (X,X)$ is defined as the relation that is transitive, contains $R_c\ (X,X)$, and has the fewest possible members. For fuzzy relations, this last requirement is generalized such that the elements of the transitive closure have the smallest possible membership grades that still allow the first two requirements to be met.

Given a relation $R(X,X)$, its transitive closure $R_T\ (X,X)$ can be determined by a simple algorithms that consists of the following three steps:

1. $R' = R \bigcup (R \circ R)$
2. If $R' \neq R$, make $R = R'$ and go to step 1.
3. Stop: $R' = R_T$

v. A fuzzy binary relation that is reflexive, symmetric, and transitive is known as a fuzzy equivalence relation or similarity relation.
vi. A fuzzy binary relation that is reflexive and symmetric is called a fuzzy compatibility relation.
vii. α-cut of a fuzzy set A is the crisp set α A that contains all the elements of the Universal set X whose membership grades in A are greater than or equal to the specified value of α.

$$\alpha \cdot \mathrm{A} = \{\mathrm{x}|A(x) \geq \alpha\}$$

3 Cluster Based Fuzzy Data Mining Model

Clustering is the process of grouping data into classes or clusters, so that objects within a cluster have high similarity in comparison to one another, but very dissimilar to objects in other clusters. Dissimilarities are assessed based on the attribute values describing the objects.

Cluster Analysis focuses mainly on distance based cluster analysis, objective being to reduce the intra-cluster distance and increase the inter-cluster distance.

A Fuzzy Cluster Analysis in the context of Dependency between Diseases

The impact of diseases may be expressed by the Fuzzy Linguistic Variables viz. Very High, High, Medium, Low & Very Low, which may, in turn, be denoted by a set of integers such as {5,4,3,2,1}. Thus, a point in a multi-dimensional space may represent each record. The groups or clusters, thus formed, reflect in a natural way the structure of given data.

Every fuzzy equivalence relation (a relation that is reflexive, symmetric and max-min transitive) induces a crisp partition in each of its α -cuts. The fuzzy clustering problem can thus be viewed as the problem of identifying an appropriate fuzzy equivalence relation on given data. Although this cannot be done directly, we can readily determine a fuzzy compatibility relation (reflexive and symmetric) in terms of an appropriate distance function applied to given data.

Let us define a fuzzy compatibility relation R, on our data set (n p-tuples of Rp), in terms of appropriate distance function of the Minkowski class by the formula given below, for all pairs $\langle x_i, x_k \rangle \in X$ where q $\in R^+$ (the set of all non-negative real numbers) and δ is a constant that ensures that $R(x_j, x_k) \in [0, 1][0, 1]$. Clearly δ is the inverse value of the largest distance in X.

To illustrate the clustering method based on fuzzy equivalence relations, let us use a small data set consisting of the following five points in R^2, as furnished in Table 1.

Where say X_{k1} represents Cancer, and X_{k2} represents Diabetes.

As the first step, we perform the analysis for q = 2, which corresponds to the Euclidean distance. First, we need to determine the value of δ in the above equation. Since the largest Euclidean distance between any pair of given data points is 4 (between x1 and x5), we have $\delta = 1/4 = 0.25$. Now, we can calculate membership grade of R by the above equation. For example,

$$R(x_1,\ x_3) = 1 - 0.25\{(3-1)2 + (4-1)2\}^{0.5} = 0.1$$

Table 1. Representative Data Set

K	1	2	3	4	5
Xk1	1	2	3	4	5
Xk2	1	2	4	2	1

When determined relation R may conveniently be represented by the matrix

$$R = \begin{bmatrix} 1 & .65 & .1 & .21 & 0 \\ .65 & 1 & .44 & .5 & .21 \\ .1 & .44 & 1 & .44 & .1 \\ .21 & .5 & .44 & 1 & .65 \\ 0 & .21 & .1 & .65 & 1 \end{bmatrix}$$

The relation is not max-min transitive; its transitive closure is $R_T = \begin{bmatrix} 1 & .65 & .44 & .5 & .5 \\ .65 & 1 & .44 & .5 & .5 \\ .44 & .44 & 1 & .44 & .44 \\ .5 & .5 & .44 & 1 & .65 \\ .5 & .5 & .44 & .65 & 1 \end{bmatrix}$

This relation induces four distinct partitions of its α -cuts:

$$\alpha \in [0, .44] : \{\{x_1, x_2, x_3, x_4, x_5\}\}$$
$$\alpha \in (.44, .5] : \{\{x_1, x_2, x_4, x_5\}, \{x_3\}\}$$
$$\alpha \in (.5, .65] : \{\{x_1, x_2\}, \{x_3\}, \{x_4, x_5\}\}$$
$$\alpha \in (.65, 1] : \{\{x_1\}, \{x_2\}, \{x_3\}, \{x_4\}, \{x_5\}\}$$

The cluster tree for Euclidean distance is shown below.

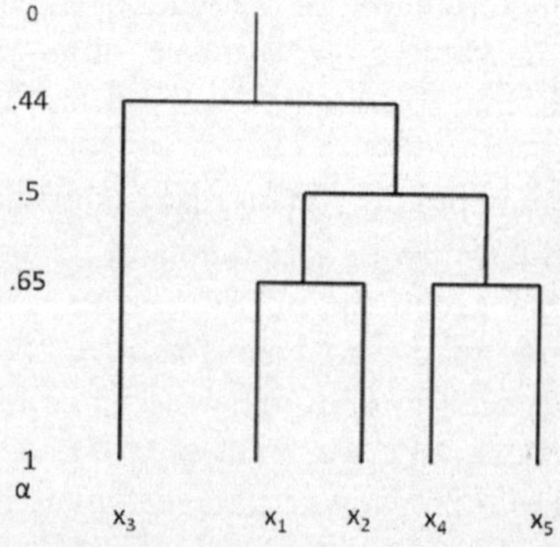

Thus we have multidimensional classification based on cluster analysis. Multidimensional classifications imply similarity of concerned records in a cluster, or in other words, close association. We may observe from the above that there is close association between points (x_1, x_2) and between points (x_4, x_5).

The above exercise considers two attributes. The clustering may be carried out with any number of desired attributes. For example, it may be desired to carry out the analysis with the following four attributes: Cancer, Diabetes, Heart Disease & Kidney Disease. Then if two candidate diseases belong to different clusters as a result of cluster analysis (based on α - cut partitions), we may conclude that these diseases are not interdependent, Whereas, if two candidate diseases belong to the same cluster, they share interdependency.

4 Fuzzy Data Mining Model Using Relational Databases

Let us consider the following fuzzy relations where Fuzzy Linguistic Variables viz. {Very High, High, Medium, Low and Very Low} are denoted by a set of integers such as {5, 4, 3, 2, 1}.

Table 2. Relation disease_name

disease_id	Disease name
1	Disease_1
2	Disease_2
3	
4	

Table 2 is characterized by the attributes disease_id & Disease name. This table provides the ids and the names of the diseases under consideration. For the purpose of codification, we have considered simple numerical codification for the disease_ids. The disease_id uniquely identifies each disease, and is the primary key for this master data table.

Table 3. Relation Disease

disease_id	year	attribute_1	attribute_2	attribute_3
1	2020	5	2	3
2	2020	4	1	5
3	2020	4	4	4
4	2020	2	3	5

The relation Disease captures the attributes of each disease.

Table 4. Relation geography_penetration

disease_id	year	geog_1	geog_2	geog_3
1	2020	5	1	4
2	2020	2	3	2
3	2020	1	5	2
4	2020	4	2	1

In Table 4, the primary key is constituted by a combination of the disease_id & the year. This combination identifies each record uniquely. The attribute values furnished in the Table 3 above are representative in nature.

It may be noted that the relation geography_penetration (Table 4) has been defined keeping the geographic penetration of the concerned disease in mind, which will vary from region to region. Thus region-wise penetration values have been shown as geog_1, geog_ 2, geog_ 3, etc. This relation, therefore, gives us insight into the geographic penetration character of the disease. For example, geog_1 may be Northern Region, geog_2 may be Western Region & geog_3 may be Southern Region. A yardstick of geographic penetration may be the number of patients affected by the disease has in that region. In this table a combination of the disease_id & year uniquely identifies each record. The attribute values furnished in the table above are representative in nature.

We may thus have the following fuzzy compatibility relation defined for domain universal set {1, 2, 3, 4, 5}.

Table 5. Fuzzy Compatibility Relation

5	1	.8	.6	.4	.2
4	.8	1	.8	.6	.4
3	.6	.8	1	.8	.6
2	.4	.6	.8	1	.8
1	.2	.4	.6	.8	1

Table 5 defines the fuzzy compatibility relation. For example, in case of geographies the fuzzy compatibility relation may be defined on the nearness of geographies to the base geography.

Thus, it would be possible to query the above database with the following question: "Which similar diseases have similar penetration in geography 1 for year 2020?".

The answer is obtained by accomplishing a relational join on the relations "disease_name" and "geography_penetration" on the attribute disease_id, selecting the tuples where geog_id = 1 and year = 2020, then projecting the disease_ids as sets, where threshold attribute_1 (e.g. cell_count) $\geq .6$ and threshold penetration for geog_1 $\geq .8$.

On performing the relational join, we have the following relation (relation_1), as illustrated in Table 6.

Table 6. Fuzzy Relational Join

disease_id	attribute_1	geog_1
1	5	5
2	4	2
3	1	2
4	2	4

Now, we observe that attribute_1 (cell_count) for diseases 1 and 2 are similar in the sense that both are > the threshold value of 0.6 (since the compatibility of 5 with 4 from the fuzzy compatibility relation is .8; compatibility of 5 with 2 is .4) However, only for disease 1 the threshold geography penetration for geog_1 ≥ .8 (since the compatibility of 5 with 5 from the fuzzy compatibility relation is 1; compatibility of 5 with 2 is .4). Thus, for the criteria defined by us, only disease_id 1 qualifies.

The result is shown below in Table 7.

Table 7. Fuzzy Relational Join

disease_id	asset
1	5

It may, therefore, be immediately observed that fuzzy relational databases provide us with insights (knowledge), which would otherwise not be possible with crisp relational databases, where we would not be able to define diseases, similar attributes (cell_count) or threshold values. Moreover, search time is faster, as we are dealing with a few integers only when considering the attributes (e.g., attributes like cell_count, blood_sugar_count, etc.) Moreover, the above methodology also provides us with mechanisms of arriving at strong or weak association rules, based on the threshold values. Thus, it may be possible to ascertain which diseases that have interdependency, based on the defined criteria.

As observed, the above methodology leads to knowledge discovery, and is, therefore, an effective data mining process that may be applied on relational databases.

Moreover, multi-dimensional data mining on relational databases using the above approach is also possible by sequential relational joins with relations containing the relevant dimensions (attributes).

5 Benefits of the Proposed Method

5.1 Extraction of Realistic Hidden Patterns

The proposed method is closer to real world expressions (since linguistic variables are used) as compared to existing classical methods, e.g. it would be more realistic to convey that there is strong association between (high cell_count, high blood_sugar_count and low blood_density) instead of exact count.

5.2 Outlier Analysis

There may exist data objects, which do not comply with the general behaviour or model of the data. Such data objects, which are grossly different from or inconsistent with the remaining set of data, are called outliers (or exceptions), and may be of particular interest.

5.3 Fuzzy Relational Databases

Fuzzy relational databases provide us with insights (knowledge), which would otherwise not be possible with crisp relational databases. More Generalized Method.

Fuzzy Mathematics is more generalized than Crisp Mathematics. All methodologies developed with crisp approach can always be addressed with fuzzy approach.

6 Conclusion

The Fuzzy Mathematical Approach discussed in this paper leads to Management Intelligence. Discovery of hidden knowledge, patterns and new association rules in real world terms enable the stakeholders involved in digital health activities, including patients, health care practitioners, researchers, traditional medical device industry firms, gain insights into diseases and their mitigation.

References

1. Chanda, D.: Artificial Intelligence and Data Mining for Mergers and Acquisitions. Chapman and Hall/CRC, London (2021). https://doi.org/10.1201/9780429424571
2. Chanda, D., Dutta majumder, D., Bhattacharya, S.: Soft computing framework for mergers & acquisitions. In: AFOR 2017: Advancing Frontiers in Operational Research: Towards a Sustainable World (2017)
3. Silberschatz, A., Forth, H.K., Sudarshan, S.: Database System Concepts. McGraw Hill, International Edition (2002)
4. Angryk, R.A.: Similarity-driven defuzzification of fuzzy tuples for entropy-based data classification purposes. In: 2006 IEEE International Conference on Fuzzy Systems, pp. 414–422 (2006)
5. EsraAslanertik, B.: Enabling integration to create value through process-based management accounting systems. Int. J. Value Chain Manage. **1**(3), 223 (2007). https://doi.org/10.1504/IJVCM.2007.013302

6. Bouchon-Meunier, B.: Similarity management for fuzzy data mining. In: 2007 International Conference on Intelligent Systems and Knowledge Engineering (ISKE 2007)
7. Dutta Majumder, D., Pal, S.K.: Fuzzy Mathematical Approach to Pattern Recognition. John Wiley & Sons (Halsted), New York (1986)
8. Chiang, D.-A., Chow, L.R., Wang, Y.-F.: Mining time series data by a fuzzy linguistic summary system. Fuzzy Sets Syst. **112**(3), 419–432 (2000)
9. Dutta Majumder, D., Chanda, D.: Datamining & knowledge discovery using a fuzzy mathematical approach for the Indian agricultural system management. In: Fuzzy Logic and Its Application in Technology and Management, Narosa Publishing House, pp. 73–80, June 2006
10. Dutta Majumder, D., Chanda, D.: Study on a framework for agricultural forecasting systems an application of information technology &datamining techniques in the Indian scenario. In: International Conference on Recent Trends & New Directions of Research in Cybernetics & Systems Theory, IASST, Guwahati, India, January 2004
11. Klir, G.J., Yuan, B.: Fuzzy Sets and Fuzzy Logic Theory and Applications. Prentice-Hall of India Private Limited, New Delhi (2002)
12. Ghazavi, S.N., Liao, T.W.: Medical data mining by fuzzy modeling with selected features. Artif. Intell. Med. **43**(3), 195–206 (2008). https://doi.org/10.1016/j.artmed.2008.04.004
13. Yang, G.: The complexity of mining maximal frequent itemsets and maximal frequent patterns. In: Proceedings of the Tenth ACM SIGKDD International Conference on Knowledge Discovery and Data Mining, August 22–25, Seattle (2004)
14. Jin, H., Sun, J., Chen, H., Han, Z.: A fuzzy data mining based intrusion detection model. In: 10th IEEE International Workshop on Future Trends of Distributed Computing Systems (FTDCS 2004), pp. 191–197 (2004)
15. Hu, Y.-C.: A new fuzzy-data mining method for pattern classification by principal component analysis. Cybern. Syst. **36**(5), 527–547 (2005). https://doi.org/10.1080/01969720590944294
16. Han, J., Kamber, M.: Datamining Concepts and Techniques. Morgan Kaufmann Publishers, San Francisco (2001)
17. Han, J., Mining, D.: Concepts and Techniques. Morgan Kaufmann Publishers Inc., San Francisco (2005)
18. Huang, M.-J., Tsou, Y.-L., Lee, S.-C.: Integrating fuzzy data mining and fuzzy artificial neural networks for discovering implicit knowledge. Knowl. Based Syst. **19**(6), 396–403 (2006). https://doi.org/10.1016/j.knosys.2006.04.003
19. Pieter Adriaans and DolfZantinge "Datamining", Addison-Wesley Professional (1996)
20. Chen, Q., Han, J., He, W., Mao, K., Lai, Y.: Utilize fuzzy data mining to find the travel pattern of browsers. In: The Fifth International Conference on Computer and Information Technology. CIT 2005, pp. 228–232 (2005)
21. Subramanyam, R.B.V., Goswami, A.: A fuzzy data mining algorithm for incremental mining of quantitative sequential patterns. Int. J. Uncertain. Fuzz. Knowl. Based Syst. **13**(6), 633–652 (2005)
22. RupaRegeNitsure: Basel II Norms: Emerging Market Perspective with Indian Focus. Economic and Political Weekly, pp. 1162–1166 (2005)
23. Simha, J.B., Iyengar, S.S.: Fuzzy data mining for customer loyalty analysis. In: 9th International Conference on Information Technology, vol. 200, 6, pp. 245–246, 18–21 December 2006

Quantum Optimized AlexNet for Histopathology Breast Image Diagnosis

Hamza Kamel Ahmed, Baraa Tantawi, Malak Magdy, and Gehad Ismail Sayed(✉)

School of Computer Science, Canadian International College (CIC), Cairo, Egypt
Gehad_Sayed@cic-cairo.com

Abstract. Breast cancer is a prevalent and life-threatening disease affecting millions of women worldwide. Timely and accurate detection is crucial for improving patient outcomes. Deep learning techniques have shown remarkable success in image classification tasks, including breast cancer diagnosis. However, the integration of quantum computing into deep learning frameworks remains relatively unexplored. This paper investigates the potential of leveraging quantum computing to enhance image classification, particularly in breast cancer detection. The focus is on utilizing the "breakhis-400x" binary dataset to develop an advanced breast cancer image classifier. The proposed Quantum-Optimized AlexNet (QOA) approach, combines the feature extraction capabilities of the AlexNet model with a quantum layer acting as a linear layer. Experimental results on the BreakHis-400x dataset demonstrate the significant potential of the QOA model, achieving an overall accuracy of 93.67%. These findings highlight the utility of Quantum Computing in improving deep learning models for image classification, particularly in medical imaging analysis, and contribute to the advancement of precision medicine in breast cancer diagnosis.

Keywords: Quantum Computing · Image Classification · Machine Learning · Convolutional Neural Network · Breakhis-400x

1 Introduction

In the advent of computational technology, the landscape of cancer diagnosis has significantly transformed. The integration of artificial intelligence (AI) techniques, particularly deep learning, with medical imaging has been a significant breakthrough in this realm. This paper delves into the confluence of deep learning, quantum computing, and transfer learning in the context of breast cancer diagnosis. Breast cancer, a dominant global health issue, accounts for nearly a quarter of all cancer cases among women worldwide [1]. Histopathological images, which offer diverse and complex morphological features, play a crucial role in the detection and diagnosis of this disease. However, the intricate tissue structures and the subtle differentiation between benign and malignant features in these images pose significant challenges to effective diagnosis [2]. Recent advancements in computational technology have heralded the use of deep learning, a subset of artificial

A. Hassanien et al. (Eds.): AISI 2023, LNDECT 184, pp. 348–357, 2023.
https://doi.org/10.1007/978-3-031-43247-7_31

intelligence, in medical imaging. Deep learning has shown significant promise in image classification tasks, particularly in identifying and distinguishing intricate patterns in histopathological images [3]. Quantum computing, rooted in the principles of quantum mechanics, has surfaced as a promising avenue for the advancement of various computational tasks, including deep learning. The inherent potential of quantum computing lies in its computational speed-ups and its ability to capture complex patterns and features that classical computer struggle to comprehend [8]. The integration of a quantum layer into deep learning models enables the extraction of more meaningful features from less information [10]. This potential for higher representational power stems from the ability of quantum systems to exist in a superposition of states, allowing for the simultaneous processing of a vast amount of information. This attribute is particularly advantageous when dealing with high-dimensional data, such as images, where conventional computational methods often struggle to capture the underlying complexity.

Transfer learning, an approach in deep learning, leverages pre-existing models trained on large-scale image datasets to understand and extract complex features from new datasets [4]. It has been particularly effective in image classification tasks, including the analysis of histopathological breast images. Pre-trained models, such as AlexNet, ResNet18, and InceptionV3, have significantly influenced the field of image recognition, each bringing unique architectural advancements that have proven beneficial in feature extraction [5–7]. Despite the progress made by transfer learning and deep learning models, the quest for improved performance and efficiency in image classification tasks remains an ongoing challenge. The integration of quantum computing into transfer learning offers a compelling solution. Quantum layers, when introduced into pre-existing models, have the potential to speed up the learning process and extract more meaningful features from the vast information available in large datasets. This is particularly pertinent given the high dimensionality of image data and the importance of feature extraction for successful image classification [8]. For instance, this paper demonstrated that the use of transfer learning in the design of thin-film solar cells enhanced quantum efficiencies, underscoring the potential for quantum computing to improve computational speed and efficiency in deep learning tasks [9]. This paper proposed an optimized version of AlexNet deep-learning architecture. It consists of two main phases, namely data preprocessing phase and quantum optimized AlexNet phase. In the data preprocessing phase, several data augmentation techniques are applied to overcome the imbalance problem of the adopted dataset. The processed data fed the proposed quantum optimized AlexNet, where a quantum hybrid layer is introduced. The main contribution of this paper can be summarized as follows:

a) An optimized version of AlexNet based on using quantum hybrid layer is proposed.
b) The integration of quantum layer provides an opportunity to extract more interpretable features from histopathology breast images.
c) This paper highlights the potential of adopting the principles of quantum computing in the field of medical imaging diagnosis.
d) A comparative analysis between different deep-leaning architectures and the state-of-the-art models are introduced and discussed.

The rest of the paper structure is organized as follows; a review of the related work for breast cancer detection is introduced in Sect. 2. The data description is introduced in Sect. 3. Section 4 introduces the overall proposed quantum optimized AlexNet model. Section 5 discusses the obtained results. Finally, in Sect. 5, the main finding of the paper with the future direction is proposed.

2 Literature Review

Several studies have been conducted by researchers in the field of breast cancer detection. These studies have utilized various machine learning approaches such as SVM, random forest, KNN, decision tree, CNN, and logistic regression on different datasets. The authors in [11] examined the performance of decision tree, SVM, naive Bayes, and K-nearest neighbor using the WISCONSIN dataset. The results showed that SVM achieved the highest accuracy of 96.40%. Another study in [12], the authors compared the performance of current BCRAT and BOADICEA models with machine learning models such as random forest, logistic regression, k-nearest neighbor, and MCMC GLMM. The evaluation metric used was accuracy and the machine learning models showed an improvement of 30 to 35 percent over previous models. In [13], the authors employed multiple machine learning approaches to predict and diagnose breast cancer using the WISCONSIN dataset. They used criteria such as sensitivity, accuracy, specificity, precision, and ROC curves. By combining feature extraction, preprocessing, and classifier methods, genetic programming was used to determine the best model.

The authors in [14] investigated the performance of decision tree and random forest on the UMMC dataset in Kuala Lumpur, Malaysia. The evaluation parameter used was accuracy, and the random forest achieved an accuracy of 82.7%. Another study in [15], the authors compared ML models such as K-NN, ANN, SVM, RF, and LR using the UCI dataset. The decision parameters used were accuracy, precision, and f1-score. The results showed that ANN achieved a maximum accuracy of 98.57%, precision of 97.82%, and f1-score of 98.90%. While the authors in [16] employed well-known machine learning classification methods to diagnose and characterize breast cancer using the Wisconsin breast cancer dataset. The evaluation measures used were accuracy and time. The findings indicated that an extreme learning machine achieved a 99% accuracy rate. In [17], the authors investigated the performance of machine learning algorithms CNN, SVM, KNN, logistic regression, naive Bayes, and random forest on the BreaKHis 400X dataset. Accuracy and precision were used as decision parameters, and the results showed that CNN outperformed other machine learning algorithms in terms of accuracy and precision.

On the BreaKHis 400X dataset, the authors in [18] have illustrated that the accuracy of a given transfer learning pipeline could be further increased by optimizing the hyperparameter of the swapped, fully connected classifier. The proposed framework, upon improvement, could facilitate pathologists in discerning the malignancy of breast cancer. and it got an accuracy of 88% using MobileNet-optimized SVM.

3 Dataset Description

In this paper, BreakHis dataset is adopted. It is a comprehensive collection of 9,109 breast cancer histopathological images obtained from 82 patients, captured at various magnification factors [19]. This paper focused exclusively on images taken at 400X magnification. The dataset includes 2,480 benign and 5,429 malignant samples, stored in PNG format with dimensions of 700 $\times$ 460 pixels and three RGB channels. The used 400X subset dataset comprised of 1,232 malignant and 588 benign samples, ensuring balanced representation. Figure 1 shows samples of the adopted dataset.

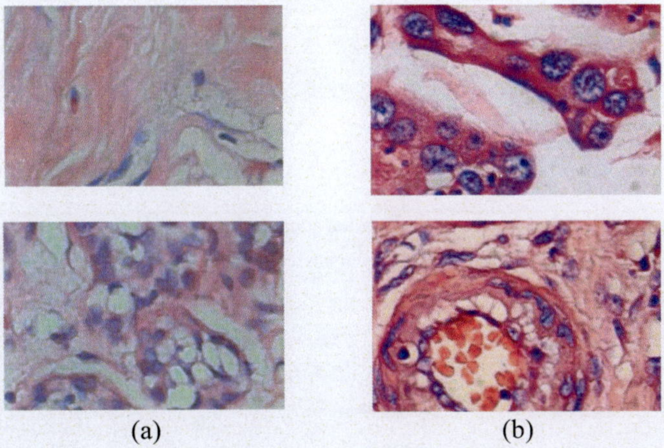

Fig. 1. Samples of BreakHis 400X dataset; (a) benign and (b) malignant images.

4 The Proposed Quantum Optimized AlexNet Approach

The proposed approach integrates quantum computing and deep learning for breast image diagnosis. It consists of two main phases, preprocessing and quantum optimized AlexNet phase. In the preprocessing phase, several data augmentation techniques are applied to tackle the imbalance dataset and to enhance model robustness. Then, a Quantum Optimized AlexNet (QOA) approach is introduced, replacing the last fully connected layer of AlexNet with a quantum hybrid layer. This approach leverages quantum computing's strengths to enhance deep learning performance in histopathology. Figure 2 shows the overall the architecture of the proposed quantum optimized AlexNet approach.

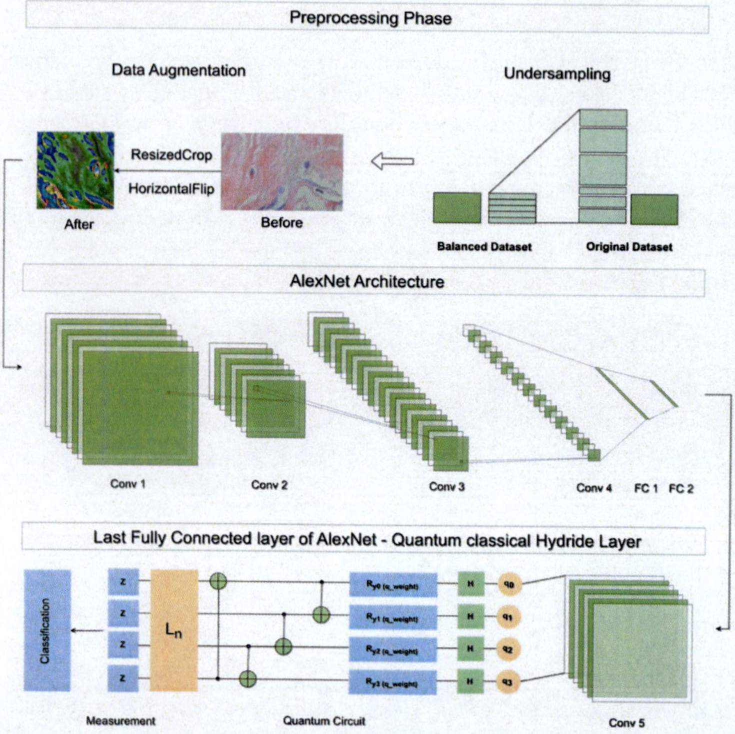

Fig. 2. The architecture of the proposed quantum optimized AlexNet approach.

4.1 The Data Preprocessing Phase

The data preprocessing considers an essential step for ensuring the quality and usability of the data for the Quantum Optimized AlexNet model. Given the often-unbalanced nature of medical datasets, undersampling techniques are applied to create a more balanced representation of classes, which is critical for preventing model bias towards the majority class. Following the balancing of data, data augmentation techniques are implemented, specifically RandomResizedCrop and RandomHorizontalFlip. These techniques artificially increase the diversity of the training data, thereby making the model more robust and capable of better generalizing to unseen data. Through these preprocessing steps, the data fed into the next stage, the Quantum Optimized AlexNet (QOA) approach, is balanced, diverse, and well-suited for effective learning.

4.2 The Quantum-Optimized AlexNet Phase

In the QOA stage of the proposed model, a hybrid approach that leverages both classical deep learning and quantum computing techniques was employed to optimize the AlexNet architecture for breast image diagnosis. The AlexNet architecture, a deep convolutional neural network, is renowned for its capacity to learn complex representations from high-dimensional data. In the proposed model, the last fully connected layer of the standard

AlexNet architecture was replace with the hybrid quantum-classical layer to enhance the network's performance and optimize its computational capabilities.

The quantum circuit is then employed, which involves a sequence of Hadamard layers, parametrized rotations around the y-axis (RY) layers, and entangling layers. The Hadamard layer is applied to each qubit, leading to a superposition of states. The RY layers, which are parametrized, provide rotation angles that can be learned and adjusted during the training process. The entangling layers create entanglement among the qubits through a series of CNOT gates, which is a cornerstone of quantum computing and offers computational advantages over classical methods. This quantum circuit is applied independently to each element of the batch. The quantum states resulting from this process are then collected, producing a four-feature output. These four features serve as the input to a classical linear layer, which performs the final classification of the input breast image. By integrating this hybrid layer into the AlexNet architecture, we aim to enhance its diagnostic capabilities and provide more accurate and computationally efficient breast image diagnoses.

5 Experimental Results and Discussion

In this section, several experiments were conducted to evaluate the robustness and the effectiveness of the proposed quantum optimized AlexNet approach. This study involved four key experiments, focusing on optimizing a quantum layer-integrated image classification approach. The first determined the best pre-trained model by replacing their last layer with a quantum layer. The second investigated the impact of data augmentation on the model performance. The third experimented with various learning rate step sizes, gauging their effects on accuracy. The final experiment addressed class imbalance via undersampling. Performance was evaluated using accuracy and F1 scores, with all experiments conducted in a consistent environment to ensure reliable comparisons. To evaluate the overall performance of the suggested model, different assessment measures are utilized. These measures include accuracy, precision, recall, and f1-score. In the first experiment, the most effective pre-trained model is determined to incorporate into the proposed approach. ResNet18, AlexNet, and InceptionV3 are considered. This is due to the deep-learning architectures have demonstrated reliable performance across a range of tasks in the literature. To infuse quantum computing capabilities into these models, the final layer of those architectures are replaced with a quantum layer. This layer essentially serves as a bridge between classical and quantum computing, capitalizing on the strengths of both paradigms. Upon running each adapted approach for 10 epochs, the performance differences can be observed. As shown in Table 1, AlexNet achieved the highest accuracy of 92.29% in the shortest time of 16m 3s, making it the most suitable model among those tested. Conversely, ResNet18 and InceptionV3 demonstrated lower accuracies and longer computational times.

Table 1. Quantum ResNt18 vs. AlexNet vs. InceptionV3 in terms accuracy and time

Pretrained models	Accuracy (%)	Time (in Min.)
ResNet18	79.08	32
AlexNet	92.29	16
InceptionV3	84.56	52

In the second experiment in Fig. 3, the effect of data augmentation techniques on the performance of the proposed approach is examined. This paper considers random vertical flips and random resize crops are applied. This is due to these techniques are known for their efficacy in improving model generalization by introducing variability to the training data. Before applying these augmentation techniques, the proposed quantum optimized AlexNet achieved an accuracy of 88.94%. However, as indicated in Fig. 3, the application of data augmentation led to a noticeable improvement in model performance. After undergoing the described augmentation techniques and being trained for 20 epochs, the model's accuracy rose to 92.54%.

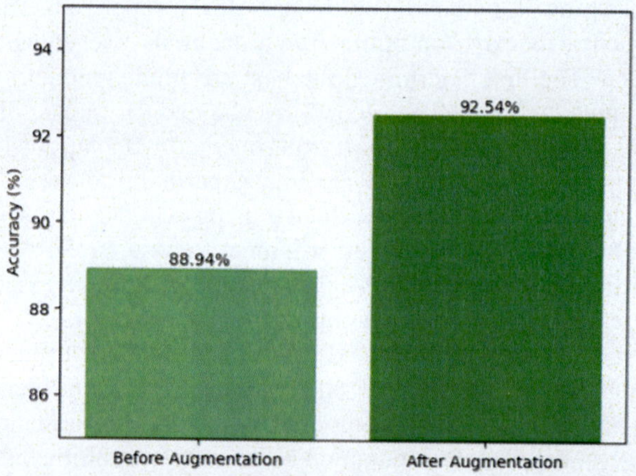

Fig. 3. The accuracy results before and after applying data augmentation techniques.

In the third experiment, the impact of varying the step size of the learning rate for a total of 70 epochs is examined. The results, as presented in Fig. 4, revealed that the step size of 20 yielded the highest accuracy of 93.37%. This experiment emphasizes the importance of proper hyperparameter tuning in enhancing the performance of deep learning models.

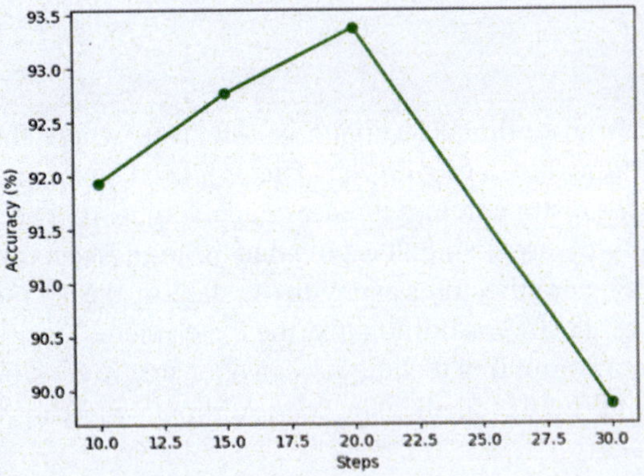

Fig. 4. The performance of the proposed approach using different learning rates

Table 2 provides a comparison between the performance of the proposed approach, QOA, and other state-of-the-art models using the same dataset. The proposed approach demonstrates superior performance, achieving an accuracy of 93.39% and an F1 score of 93.21%, indicating high precision and recall capabilities. The model from [21] achieved an accuracy of 91.85 $\pm$ 1.77% and an F1 score of 0.91 $\pm$ 0.02, coming close to the proposed approach's performance but falling slightly short. The models presented in [18] and [22], despite their unique approaches, reported lower accuracies and F1 scores. This comparison clearly establishes the QOA approach as a competitive choice against state-of-the-art models. Its superior performance underscores the effectiveness of the quantum layer in enhancing the precision and recall of the model, thus highlighting its promise for future exploration in the field of quantum machine learning. However the good performance of the quantum computing, it should be mentioned that due to the limited number of qubits presently available in quantum hardware, exploiting large image sizes in quantum computing is extremely difficult [23]. Quantum computers need exponentially more qubits to process classical data as the amount of input increases.

Table 2. The performance of the proposed QOA approach with the state of the art models

	Accuracy (%)	F1 Score (%)	Recall (%)	Precision (%)
QOA (The proposed Approach)	93	93	93	93
A-MIL [20]	84	–	–	–
MPCS-OP [21]	92	91	–	–
MobileNet-Optimized SVM [18]	88	–	–	–
LBSE [22]	81	86	82	90

6 Conclusion

This paper introduces an optimized version of AlexNet, where the quantum layer is introduced in the building block of AlexNet. The proposed quantum optimized AlexNet approach is applied for histopathology breast image diagnosis. The results showed that proposed approach is very promising. It is obtained the highest accuracy result compared with other deep-learning architectures and with the state-of-the-art models. It is achieved an overall accuracy of 93.67%. Additionally, the experimental results revealed that the integration of quantum computing techniques can open up the possibility for exploration and advancements in the field of medical imaging, ultimately leading to improved patient outcomes.

In the future, the proposed optimized version of AlexNet can be further extended beyond breast cancer diagnosis to other medical imaging tasks. Additionally, the possibility of quantum-optimized AlexNet can be further explored to advance medical diagnostics in lung cancer detection, cardiovascular imaging, or neuroimaging.

References

1. Bray, F., Ferlay, J., Soerjomataram, I., Siegel, R.L., Torre, L.A., Jemal, A.: GLOBOCAN estimates of incidence and mortality worldwide for 36 cancers in 185 countries. Ca Cancer. J. Clin. **68**(6), 394–424 (2018). https://doi.org/10.3322/caac.21492
2. Ayyad, S.M., et al.: Role of AI and histopathological images in detecting prostate cancer: a survey. Sensors **21**(8), 2586 (2021). https://doi.org/10.3390/s21082586
3. Rączkowski, Ł, Możejko, M., Zambonelli, J., Szczurek, E.: ARA: accurate, reliable and active histopathological image classification framework with Bayesian deep learning. Sci. Rep. **9**(1), 14347 (2019). https://doi.org/10.1038/s41598-019-50587-1
4. Antoniou, A., Storkey, A., Edwards, H.: Augmenting image classifiers using data augmentation generative adversarial networks. In: Kůrková, V., Manolopoulos, Y., Hammer, B., Iliadis, L., Maglogiannis, I. (eds.) ICANN 2018. LNCS, vol. 11141, pp. 594–603. Springer, Cham (2018). https://doi.org/10.1007/978-3-030-01424-7_58
5. Krizhevsky, A., Sutskever, I., Hinton, G.E.: ImageNet classification with deep convolutional neural networks. Commun. ACM **60**(6), 84–90 (2017). https://doi.org/10.1145/3065386
6. He, K., Zhang, X., Ren, S., Sun, J.: Deep residual learning for image recognition. In: Proceedings of the IEEE conference on computer vision and pattern recognition, pp. 770–778 (2016). https://doi.org/10.1109/CVPR.2016.90
7. Szegedy, C., Vanhoucke, V., Ioffe, S., Shlens, J., Wojna, Z.: Rethinking the inception architecture for computer vision. In Proceedings of the IEEE conference on computer vision and pattern recognition, pp. 2818–2826. (2016). https://doi.org/10.1109/CVPR.2016.308
8. Tilaye, G.F., Pandey, A.: Investigating the effects of hyperparameters in quantum-enhanced deep reinforcement learning. Quantum Eng. **2023**, 16 (2023). https://doi.org/10.1155/2023/2451990
9. Kaya, M., Hajimirza, S.: Using a novel transfer learning method for designing thin film solar cells with enhanced quantum efficiencies. Sci. Rep. **9**(1), 1 (2019). https://doi.org/10.1038/s41598-019-41316-9
10. Henderson, M., Shakya, S., Pradhan, S., Cook, T.: Quanvolutional neural networks: powering image recognition with quantum circuits. Quantum Mach. Intell. **2**(1), 2 (2020). https://doi.org/10.1007/s42484-020-00012-y

11. Pallathadka, H., Mustafa, M., Sanchez, D.T., Sajja, G.S., Gour, S., Naved, M.: Impact of machine learning on management, healthcare, and agriculture. Materials Today: Proc. **80**, 2803–2806 (2023). https://doi.org/10.1016/j.matpr.2021.07.042
12. Ming, C., Viassolo, V., Probst-Hensch, N., Chappuis, P.O., Dinov, I.D., Katapodi, M.C.: Machine learning techniques for personalized breast cancer risk prediction: comparison with the BCRAT and BOADICEA models. Breast Cancer Res. **21**(1), 1–11 (2019). https://doi.org/10.1186/s13058-019-1158-4
13. Dhahri, H., Maghayreh, E.A., Mahmood, A., Elkilani, W., Nagi, M.F.: Automated breast cancer diagnosis based on machine learning algorithms. J. Healthc. Eng. **2019**, 4253641 (2019). https://doi.org/10.1155/2019/4253641
14. Ganggayah, M.D., Taib, N.A., Har, Y.C., Lio, P., Dhillon, S.K.: Predicting factors for survival of breast cancer patients using machine learning techniques. BMC Med. Inform. Decis. Mak. **19**, 1–17 (2019). https://doi.org/10.1186/s12911-019-0801-4
15. Islam, M.M., Haque, M.R., Iqbal, H., Hasan, M.M., Hasan, M., Kabir, M.N.: Breast cancer prediction: a comparative study using machine learning techniques. SN Comput. Sci. **1**, 1–14 (2020). https://doi.org/10.1007/s42979-020-00305-w
16. Gupta, C., Gill, N.S.: Machine learning techniques and extreme learning machine for early breast cancer detection. Int. J. Innovative Technol. Exploring Eng. (IJITEE) **9**(04), 163–167 (2020) ISSN: 2278–3075
17. Bhise, S., Gadekar, S., Gaur, A.S., Bepari, S., Deepmala Kale, D.S.A.: Breast cancer detection using machine learning techniques. Int. J. Eng. Res. Technol. **10**(7) (2021)
18. Arzmi, M.H., et al.: The classification of breast cancer: the effect of hyperparameter optimisation towards the efficacy of feature-based transfer learning pipeline. In: Arzmi, M.H., et al. (eds.) Deep Learning in Cancer Diagnostics: A Feature-based Transfer Learning Evaluation, pp. 15–19. Springer Nature Singapore, Singapore (2023). https://doi.org/10.1007/978-981-19-8937-7_3
19. Spanhol, F.A., Oliveira, L.S., Petitjean, C., Heutte, L.: A dataset for breast cancer histopathological image classification. IEEE Trans. Biomed. Eng. **63**(7), 1455–1462 (2015). https://doi.org/10.1109/TBME.2015.2496264
20. Chhipa, P.C., Upadhyay, R., Pihlgren, G.G., Saini, R., Uchida, S., Liwicki, M.: Magnification prior: a self-supervised method for learning representations on breast cancer histopathological images. In: Proceedings of the IEEE/CVF Winter Conference on Applications of Computer Vision, pp. 2717–2727. (2023). https://doi.org/10.48550/arXiv.2203.07707
21. Patil, A., Tamboli, D., Meena, S., Anand, D., Sethi, A.: Breast cancer histopathology image classification and localization using multiple instance learning. In: 2019 IEEE International WIE Conference on Electrical and Computer Engineering (WIECON-ECE), pp. 1–4. Bangalore, India (2019) https://doi.org/10.1109/WIECON-ECE48653.2019.9019916
22. Kang, X., Liu, X., Nie, X., Yin, Y.: Learning binary semantic embedding for breast histology image classification and retrieval. In: ICASSP 2021–2021 IEEE International Conference on Acoustics, Speech and Signal Processing (ICASSP), pp. 1180–1184. IEEE (2021). https://doi.org/10.1109/ICASSP39728.2021.9415036
23. Chiang, C.F.: Quantum phase estimation with an arbitrary number of qubits. Int. J. Quantum Inform. **11**(01), 1350008 (2013). https://doi.org/10.1142/S0219749913500081

Quantumedics: Brain Tumor Diagnosis and Analysis Based on Quantum Computing and Convolutional Neural Network

Hamza Kamel Ahmed, Baraa Tantawi, Malak Magdy, and Gehad Ismail Sayed(✉)

School of Computer Science, Canadian International College (CIC), Cairo, Egypt
Gehad_Sayed@cic-cairo.com

Abstract. Before computer-aided diagnosis tools were created, physicians did physical exams, examined patient histories, and gathered patient data. Medical image analysis used to be laborious and subjective when done manually. Variations and diagnostic errors were also possible. Integrating cutting-edge technology like quantum computing and deep learning into medical image analysis offers great promise. The paper introduces Quantumedics, a system that classifies brain MRI images using a quantum convolutional neural network (QCNN) model and segments tumors using DeepLabV3 +. Data preparation, image classification-based QCNN, and tumor extraction and analysis make up its three primary phases. When a tumor is found, a segmentation model built on DeepLabV3 + is used to precisely define the tumor region. Then, significant information is derived from the tumor region, including the tumor's laterality and tumor-to-brain ratio. The proposed Quantumedics system is quite promising, according to the experimental findings. In addition, the results showed that DeepLabV3 + outperformed U-Net and earlier versions in terms of segmentation results. Additionally, the outcomes showed that medical experts could utilize the proposed Quantumedics as a diagnosis and analysis tool. Overall, 99% accuracy was attained.

Keywords: Quantum computing · Deep-learning architecture · Classification · Image analysis · Brain MRI images

1 Introduction

Human physicians' knowledge and experience in the field of medical diagnosis were heavily dependent on them prior to the invention of automated diagnosis technologies. In-depth physical examinations, evaluations of medical histories, and information gathering would all be performed by physicians [1]. Using their knowledge and intuition, they would then determine a diagnosis based on the observed symptoms and patterns. The manual examination of medical images, such as X-rays or MRI scans, by physicians in the past to detect anomalies or possible diseases, often entailed time-consuming and subjective procedures. The physician's skill and expertise in interpreting the photos had a significant impact on the outcomes, which could lead to discrepancies in diagnosis and

A. Hassanien et al. (Eds.): AISI 2023, LNDECT 184, pp. 358–367, 2023.
https://doi.org/10.1007/978-3-031-43247-7_32

errors [2]. Additionally, it may be time-consuming and laborious to consult professionals or get a second opinion. Thus, it may result in lengthening the patient's therapy and possibly affecting their prognosis as a whole.

The interpretation of medical images is crucial to diagnosing and comprehending a wide range of diseases in the realm of healthcare. A potent tool for examining tumors and other brain diseases is magnetic resonance imaging (MRI) [3]. However, successful analysis of these images necessitates specific knowledge and expertise, frequently requiring the assistance of medical experts. Researchers investigated the incorporation of cutting-edge technology into systems for medical image analysis in order to improve the effectiveness of diagnostic procedures and lessen the strain on healthcare practitioners [4]. For the classification of tumors, the authors of [5] presented a 2D CNN architecture with auto-encoder networks. They evaluated the performance of their proposed model against six different conventional machine learning algorithms and various deep learning architectures. The experimental findings demonstrated that the proposed model can significantly and 96% accurately classify the type of tumor. For the diagnosis of Alzheimer's disease in [6], the authors combined five pre-trained transfer-learning architectures. In order to assess the overall effectiveness of their proposed model, 600 participants from the ADNI collection were used. The findings demonstrated that the proposed ensemble transfer-learning approach may effectively identify between participants with Alzheimer's disease and subjects with normal cognitive function. Overall, it had a 96% accuracy rate. Support vector machines (SVM) and the CNN-DNN deep-learning architecture were used by the authors of a different study [7] to classify brain MRI images. They evaluated the performance of LeNet, AlexNet, and ResNet, two other deep-learning architectures. The experimental findings showed that the proposed model utilizing CNN-DNN achieved 100% classification accuracy. Additionally, the results revealed that considering the age and gender of the patient subjects can be very promising for automated diagnosis of the brain.

From all of the previous work, it is clear that deep learning architectures have much promise over traditional machine learning techniques, particularly for image classification problems. Due to its superior performance at tasks like image recognition, natural language processing, and recommendation engines, deep learning has revolutionized a variety of sectors [8]. Despite their success, deep learning algorithms still have several issues that limit their potential. The two main constraints are computational complexity and the capability of efficient scaling. The foundations of quantum computing offer a potential solution to this problem. Quantum computing makes use of the concepts of quantum mechanics to process information fundamentally differently than regular computers. This opens up new possibilities for tackling challenging deep-learning issues that are currently computationally taxing or insurmountable [9]. It is possible to build quantum neural networks that can handle larger datasets and more complex models by combining deep learning and quantum computing techniques. In the training and inference phases, quantum algorithms can enhance performance and reduce computational bottlenecks. As a result, numerous enhanced deep-learning architectures based on quantum computing concepts have been put out in the literature, including [9, 10] and [11].

This paper proposes a classification and analysis system for brain MRI images called as Quantumedics. The proposed Quantumedics combines the principles of quantum computing and deep learning architecture to improve the classification accuracy of brain MRI images for tumor detection. It consists of three main phases: data preprocessing, image classification based on QCNN, and finally, tumor extraction and analysis. By leveraging the quantum nature of the model, it enables more complex computations and potentially superior discrimination between different image patterns. This quantum-powered classification step serves to distinguish medical images as either normal or abnormal, specifically in terms of tumor presence. In cases where a tumor is detected, Quantumedics incorporates a segmentation model to precisely identify and delineate the tumor region within the image. A deep learning-based segmentation model, namely DeepLabV3 + is utilized to accurately segment the tumor from the original brain MRI image. However, providing patients with a comprehensive understanding of their medical images and diagnosis is equally crucial. To address this need, Quantumedics offers an analysis of the extracted tumor region. The type of tumor with the tumor's laterality and the tumor-to-brain-ratio are finally extracted and reported. This extracted information can be further used by physicians as a second opinion in their examination. The main contribution of the paper can be summarized in the following points:

a) This paper study introduces a system for MRI brain tumor analysis and classification based on utilizing QCNN for tumor classification and DeepLab V3 for tumor segmentation.
b) The proposed Quantumedics system offers an analysis for the extracted tumor region including the tumor's laterality and the tumor-to-brain-ratio.
c) A comparative analysis of each component of the proposed Quantumedics system with the state-of-the-art model for brain MRI images.
d) The proposed system can be used as a valuable tool for clinical diagnosis and treatment planning.

The remainder of this paper is structured as follows: Sect. 2 introduces the Quantumedics system, including the utilization of the QCNN approach, and the segmentation and analysis model. Section 3 presents the experimental setup and discusses the results obtained. Finally, Sect. 4 concludes the paper and outlines potential directions for future research in this domain.

2 Proposed System: Quantumedics

The proposed system, called Quantumedics, intends to help patients and medical professionals in the initial evaluation of medical images. The system uses cutting-edge technology to boost the effectiveness and precision of medical image analysis, including quantum computing, deep learning, and natural language processing. A brain MRI image dataset is gathered from two benchmark datasets [12] and [13] to be utilized for tumor classification and segmentation in order to assess the effectiveness of Quantumedics. The different scenarios included in the dataset gathered and used for training and evaluation ensure the generalizability and dependability of the system. The proposed Quantumedics system consists of three main phases, namely data preprocessing, image classification based on QCNN, and tumor extraction and analysis. Figure 1 shows the overall building blocks of the proposed Quantumedics system for brain MRI image analysis and classification. Next, a detailed description of each phase is provided.

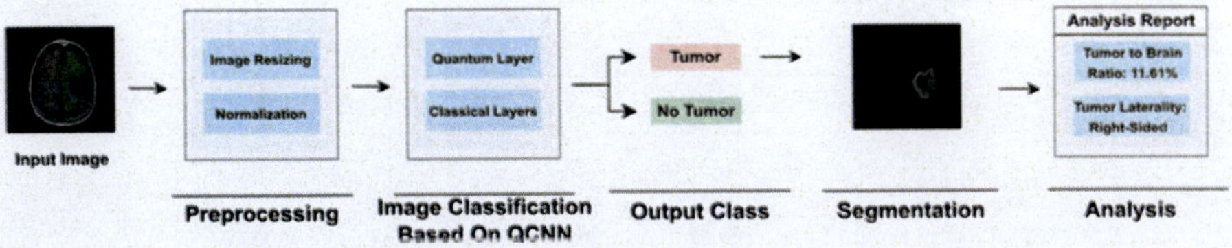

Fig. 1. Building blocks of the proposed Quantumedics system for brain MRI diagnosis and analysis.

2.1 Data Preprocessing Phase

To prepare the original dataset of brain MRI images for further analysis and modeling, there is a preprocessing phase. To standardize the input features, the original images are downsized to 28x28 before data normalization is used. When data are normalized, each data point is scaled down to a value between 0 and 1, reducing the influence of data distribution and scale changes.

2.2 Image Classification-Based on QCNN Phase

The images that have been processed are fed into the Quantum Convolutional Neural Network (QCNN) method in this phase. The quantum part and the classical part make up the two primary components of the QCNN. Then comes a thorough explanation of every part.

2.2.1 Quantum Part

The QCNN's quantum part acts as a preprocessing layer, transforming the input data using the capabilities of quantum computing. Quantum encoding, random quantum circuits, quantum convolution, and quantum measurement make up its four main components.

Using the AngleEmbedding method, which converts classical information into the rotation angles of quantum gates, the input data is embedded into a quantum state throughout the quantum encoding step. This method allows for a compact representation of the input data within the quantum system. The embedded quantum state is subjected to a random quantum circuit in the random quantum circuit step. This circuit comprises identical-angle controlled-NOT gates and single-qubit gates selected at random. The input data is transformed nonlinearly by the random circuit, which could show intricate patterns. Similar to a classical convolutional layer, quantum circuits are used to process the images in the quantum convolution step. Multiple quantum filters are applied to the image before an activation function is applied to the output tensor to add nonlinearity. Finally, each qubit is measured on a computational basis using the Pauli-Z operator in the quantum measurement step, and the expectation values are then employed as features in the classical section of the model. The quantum circuit of the proposed QCNN is shown in Fig. 2.

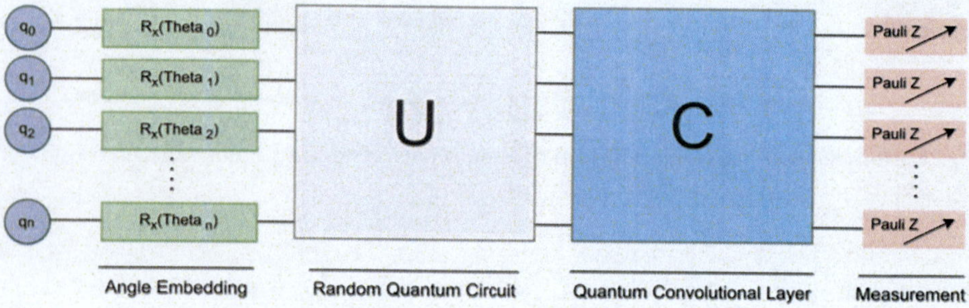

Fig. 2. The quantum circuit of the proposed QCNN

2.2.2 Classical Part

A deep learning architecture that mixes convolutional and fully connected blocks makes up the QCNN's traditional component. Fully connected blocks combine the learned characteristics and make conclusions about the data whereas convolutional blocks identify local patterns and features in the input data. Convolutional layers, max-pooling layers, and dropout layers are included in the classical section to lessen overfitting. A final fully connected layer with a SoftMax activation function serves as the architecture's apex and generates class probabilities for each class. First, the first pooling layer receives as input the expectation values computed from the quantum portion. The classical component makes precise predictions using the features it has derived from the quantum layer after learning intricate patterns. Following the extraction of a number of features using a convolution layer with filters equal to 32, a pooling layer followed by a convolution layer with filters equal to 64, and finally the same convolution layer and pooling layer set but with filters equal to 126 are applied. With 256 filters applied, the final set of pooling and convolution layers is applied. The fully connected layer, where the final class label is reported, is fed by the output from the previous set.

2.3 Tumor Extraction and Analysis Phase

The image is fed into the DeepLabV3 + -based segmentation model if a tumor is found during the classification phase. The DeepLabV3 + deep learning architecture is a powerful and popular tool for segmenting images. It was developed by Google Research and uses arousal spatial pyramid pooling (ASPP) and dilated convolutions to collect fine-grained details and multi-scale contextual data, yielding accurate segmentation results. The segmentation output is further enhanced by the addition of a decoder module. It has accomplished a lot of jobs successfully, demonstrating its adaptability in handling difficult scenarios and creating precise segmentation masks. Due to its outstanding performance and versatility, it is a well-liked choice for computer vision applications like object detection, semantic segmentation, and medical image analysis [14]. The performance of DeepLabV3 + is compared with other older versions of DeepLab and with U-Net deep learning-based segmentation models as it will be seen in the result section.

The tumor's laterality and tumor-to-brain ratio are determined from the tumor region after the tumor has been extracted from the image. The tumor-to-brain ratio shows how much of the brain is occupied by the tumor. This ratio can be used to calculate the tumor burden or determine how much of the brain is affected by the tumor. It can be determined by the tumor's laterality whether it is on the left or right side of the brain. A left-sided tumor, a right-sided tumor, or a bilateral tumor is the value of this measurement. A tumor that affects both the left and right sides of the brain is said to be bilateral.

It is important to understand that the proposed Quantumedics does not intend to take the place of qualified healthcare experts' knowledge. In contrast, it acts as a support tool to help with the first evaluation and give patients easily understandable information about their medical photos. The ultimate objective is to increase the effectiveness of the diagnostic procedure while maintaining the critical importance of medical professionals in evaluating results, making precise diagnoses, and recommending the most effective course of treatment.

3 Experimental Results and Discussion

This section discusses the findings from four experiments conducted to assess the system's effectiveness in detecting and analyzing brain tumors. First, the best embedding type for encoding data was evaluated. Following that, multiple segmentation models were contrasted. In the end, the proposed Quantumedics system's total performance was compared to state-of-the-art alternatives.

In the first experiment shown in Fig. 3, the goal was to determine the best embedding type for converting data from a classical to a quantum state. Considered embeddings included AngleEmbedding, BasisEmbedding, and AmplitudeEmbedding. AngleEmbedding outperformed all other options in the experiment, showing 99.2% accuracy. Comparatively, AmplitudeEmbedding obtained 96% accuracy whereas BasisEmbedding produced 86%. All sorts of embeddings produced the same results over an interval of 700 epochs.

The primary goal of the second experiment in Fig. 4 is to assess how well the retrieved tumors perform utilizing DeepLabV3 + -based image segmentation of brain MRI data. U-Net, Deeplab v1, Deeplab v2, Deeplab v3, and Deeplab v3 + were five different image segmentation models that were evaluated. With a Dice score of 0.89, the results revealed that Deeplab v3 + performed in the most promising manner. The results for the other models were as follows: 0.77 for U-Net, 0.81 for Deeplab v1, 0.82 for Deeplab v2, and 0.82 for Deeplab v3. These findings are illustrated graphically in Fig. 4, which highlights Deeplab v3 + 's advantage in the segmentation of brain MRI images.

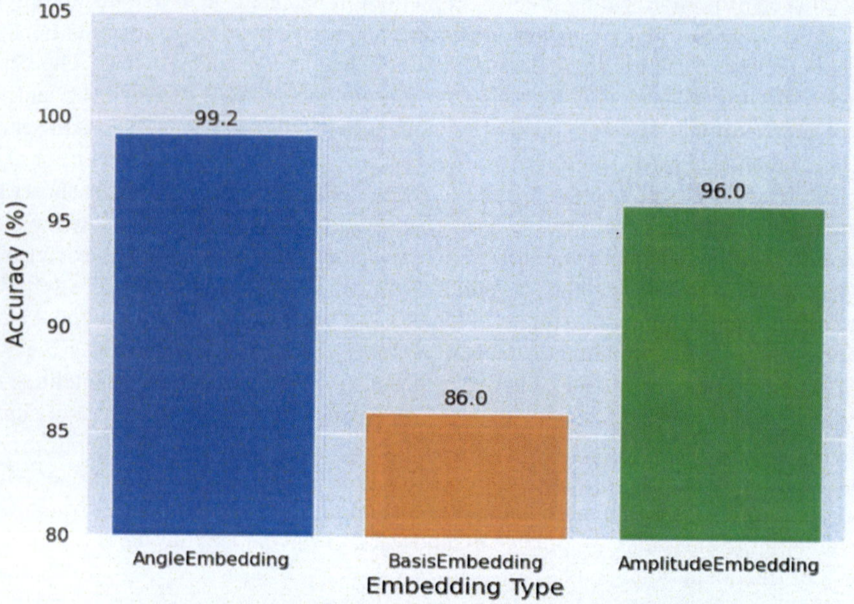

Fig. 3. Embedding Types Comparison over 700 Epochs

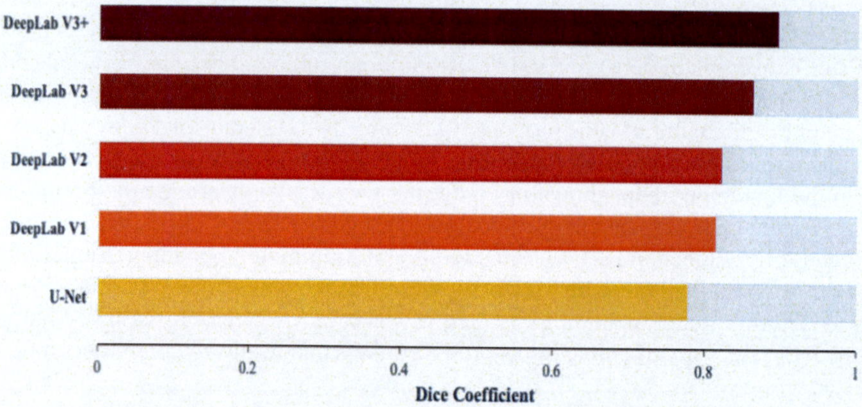

Fig. 4. Dice Scores of Five Image Segmentation Models

Another experiment is carried out to determine how well the quantum computing principle is integrated into CNN. In this experiment, the proposed QCNN's performance is contrasted with that of a classical CNN, whose architecture does not include a quantum layer. The classification accuracy before and after the quantum layer was embedded is shown in Fig. 5. As can be seen, adding the quantum layer, where the proposed QCNN is implemented, significantly improves CNN's performance. Compared to the 90% accuracy of the conventional CNN, the proposed QCNN achieved an accuracy of 99%.

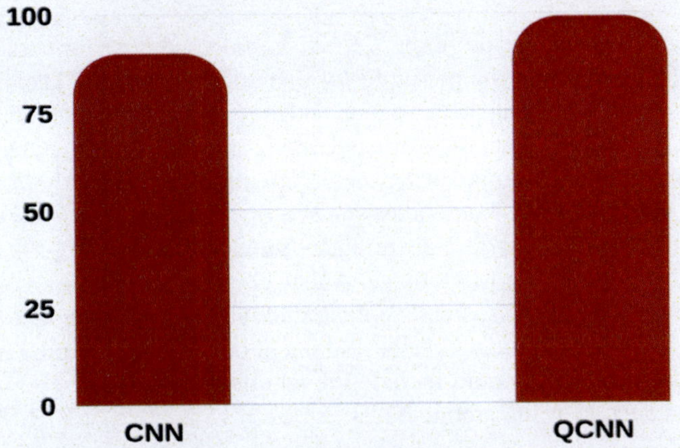

Fig. 5. Classification accuracy of CNN and the proposed QCNN

Using measures like accuracy, sensitivity, precision, and F1-score, the experiment in Table 1 attempted to compare the performance of the proposed system with that of various state-of-the-art models. It is evident that the proposed QCNN approach of Quantumedics is superior to the most recent models. Furthermore, given that the proposed system achieved an overall accuracy of 99%, the results show the potential of the proposed system to provide improved diagnostic processes and contribute to better patient outcomes.

Table 1. The performance of the proposed QCNN approach with the state of the art models.

	Accuracy (%)	F1 Score (%)	Sensitivity (%)	Specificity (%)
The proposed Approach	99.2	99.2	99.2	99.2
RN-OKELM [15]	97.93	–	97.92	97.98
HQC-CNN [16]	97.8	–	–	–
CNN [17]	84	83	76	91
Resnet50 [18]	97.2	–	–	–
CVAE [19]	96.88	96.88	96.88	96.88
PatchResNet [20]	98.10	–	–	–
Deep Belief Network [21]	95.78	–	–	–

4 Conclusion

In this paper, Quantumedics, a comprehensive system for tumor classification and analysis in brain MRI images, is introduced. The proposed system combines the usage of Quantum Convolutional Neural Networks (QCNN) for tumor classification and DeepLabV3

+ for tumor extraction. The proposed QCNN improves the performance of the conventional CNN by applying the principles of quantum mechanics. The analysis of the extracted tumor is also included in this paper, and the tumor's lateralization and tumor-to-brain ratio are determined from the extracted tumor area. According to the experiment's findings, the proposed Quantumedics system correctly classified tumors with an accuracy of 99%. Additionally, the outcomes showed that, in comparison to DeepLabV3 + 's earlier iterations and U-Net deep-learning architectures, DeepLabV3 + can considerably remove tumors from the MRI image. A dice coefficient of 0.89 was obtained. The results also demonstrated the system's potential to help doctors diagnose brain tumors, direct treatment choices, and track the development of the illness. Future enhancements to the system's capabilities could include the addition of other imaging modalities or more advanced deep learning architectures.

References

1. Razali, N.F., Isa, I.S., Sulaiman, S.N., Abdul Karim, N.K., Osman, M.K., Che Soh, Z.H.: Enhancement technique based on the breast density level for mammogram for computer-aided diagnosis. Bioengineering **10**(2), 153 (2023)
2. Darwish, A., Sayed, G., Hassanien, A.: Meta-heuristic optimization algorithms based feature selection for clinical breast cancer diagnosis. J. Egyptian Math. Soc. **26**(2), 321–336 (2018)
3. Hussain, M., Koundal, D., Manhas, J.: Deep learning-based diagnosis of disc degenerative diseases using MRI: a comprehensive review. Comput. Electr. Eng. **105**, 108524 (2023)
4. Sayed, G.I., Solyman, M., Hassanien, A.E.: A novel chaotic optimal foraging algorithm for unconstrained and constrained problems and its application in white blood cell segmentation. Neural Comput. Appl. **31**, 7633–7664 (2019)
5. Saeedi, S., Rezayi, S., Keshavarz, H., R Niakan Kalhori, S.: MRI-based brain tumor detection using convolutional deep learning methods and chosen machine learning techniques. BMC Medical Informatics and Decision Making **23**(1), 1-17 (2023)
6. Yamanakkanavar, N., Choi, J.Y., Lee, B.: MRI segmentation and classification of human brain using deep learning for diagnosis of Alzheimer's disease: a survey. Sensors **20**(11), 3243 (2020)
7. Wahlang, I., et al.: Brain magnetic resonance imaging classification using deep learning architectures with gender and age. Sensors **22**(5), 1766 (2022)
8. Wu, S.B., Li, Z.M., Gao, J., Zhou, H., Wang, C.S., Jin, X.M.: Classification of quantum correlation using deep learning. Opt. Express **31**(3), 3479–3489 (2023)
9. Kamruzzaman, A., Alhwaiti, Y., Leider, A., Tappert, C.C.: Quantum deep learning neural networks. In: Advances in Information and Communication: Proceedings of the 2019 Future of Information and Communication Conference (FICC), Volume 2, pp. 299–311. Springer International Publishing (2020)
10. Kodipalli, A., Fernandes, S.L., Dasar, S.K., Ismail, T.: Computational framework of inverted fuzzy c-means and quantum convolutional neural network towards accurate detection of ovarian tumors. Int. J. E-Health and Medical Commun. (IJEHMC) **14**(1), 1–16 (2023)
11. Ovalle-Magallanes, E., Avina-Cervantes, J.G., Cruz-Aceves, I., Ruiz-Pinales, J.: Hybrid classical–quantum convolutional neural network for stenosis detection in X-ray coronary angiography. Expert Syst. Appl. **189**, 116112 (2022)
12. Navoneel Chakrabarty. Brain MRI Images for Brain Tumor Detection . Kaggle (2019). https://www.kaggle.com/datasets/navoneel/brain-mri-images-for-brain-tumor-detection
13. Mateusz Buda. LGG-MRI Segmentation . Kaggle (2019). https://www.kaggle.com/datasets/mateuszbuda/lgg-mri-segmentation

14. Rogelio, J.P., Dadios, E.P., Vicerra, R.R.P., Bandala, A.A.: Object detection and segmentation using deeplabv3 deep neural network for a portable x-ray source model. J. Advanced Comput. Intell. Intelligent Informatics **26**(5), 842–850 (2022)
15. Devanathan, B., Kamarasan, M.: Automated brain tumor diagnosis using residual network with optimal kernel extreme learning machine. In: 2022 4th International Conference on Smart Systems and Inventive Technology (ICSSIT), pp. 860–865. IEEE (2022)
16. Dong, Y., Fu, Y., Liu, H., Che, X., Sun, L., Luo, Y.: An improved hybrid quantum-classical convolutional neural network for multi-class brain tumor MRI classification. Journal of Applied Physics, **133**(6) (2023)
17. Indraswari, R., Ardan, I.S., Arifin, A.Z., Tjahyanto, A., Rakhmawati, N.A., Kusumawardani, R.: Brain tumor detection on magnetic resonance imaging (MRI) images using convolutional neural network (CNN). In: 2022 9th International Conference on Electrical Engineering, Computer Science and Informatics (EECSI), pp. 367–373. IEEE (2022)
18. Çinar, A., Yildirim, M.: Detection of tumors on brain MRI images using the hybrid convolutional neural network architecture. Med. Hypotheses **139**, 109684 (2020)
19. Salama, W.M., Shokry, A.: A novel framework for brain tumor detection based on convolutional variational generative models. Multimedia Tools and Appl. **81**(12), 16441–16454 (2022)
20. Muezzinoglu, T., et al.: PatchResNet: multiple patch division–based deep feature fusion framework for brain tumor classification using mri images. Journal of Digital Imaging, pp. 1–15 (2023)
21. Doshi, R., Hiran, K.K., Doppala, B.P., Vyas, A.K.: Deep belief network-based image processing for local directional segmentation in brain tumor detection. J. Electron. Imaging **32**(6), 062502 (2023)

Automated Diagnoses Glaucoma Approach in Retinal Fundus Images Using Support Vector Machine

Osama M. Kamara(✉), Ahmed H. Asad, and Hesham A. Hefny

Department of Computer Science, Faculty of Graduate Studies for Statistical Research, Cairo University, Giza, Egypt
osama_kamara@yahoo.com

Abstract. Glaucoma is chronic disease steals sight slowly so causing blindness gradually. Although, it cannot be cured at all but only the early treatment can only slow down its progress. However, minority of doctors can diagnose it accurately, so it is very crucial issue to improve the performance of glaucoma diagnosis. This paper proposed an approach with two phases, phase (1) is the approach without feature selection and in phase (2) we apply feature selection algorithms and compare results. It is based on superpixel algorithm and supervised machine learning then in phase 2 we perform feature selection for selecting the most significant features to improve the performance, reduce data dimensionality and decrease elapsed time, then we performed the support vector machine (SVM) classifier. The proposed approach is evaluated on Rim-One r3 dataset, and the result is that the data dimensionality is decreased by 50% and elapsed time decreased by 42%, on the other hand, we get phase 1 accuracy 99.2% but it is decreased to 98.5% in phase 2 and mean CRD-error is increased by 0.02.

Keywords: Feature Selection · Glaucoma Diagnosis · Machine Learning · Retinal Fundus Image · Segmentation

1 Introduction

Glaucoma disease can't be treated and it is the second cause of blindness after diabetic retinopathy. As glaucoma can't be treated but only, we can slow down its progress so it is very crucial issue to build and improve an automated computer-aided diagnosis (CAD) system that can accurately screens and diagnoses glaucoma to slow down the progression speed of glaucoma in its early stages [1]. Glaucoma disease damages optic nerve head, and the most useful clinical indicator that successfully differentiate between healthy and glaucomatous eye is the measurement of vertical cup-to-disc ratio (CDR) in retinal fundus image [2].

In the few old years there are many approaches tried to detect glaucoma disease in automated way using machine learning and deep learning utilized by artificial intelligence as a computer aid system, we introduce an automated approach for detecting glaucoma based on supervised machine learning and feature selection methods to increase performance and decrease consumed time by the system [2].

A. Hassanien et al. (Eds.): AISI 2023, LNDECT 184, pp. 368–379, 2023.
https://doi.org/10.1007/978-3-031-43247-7_33

This paper is organized as follows: Sect. 1 presents an introduction and background. Section 2 presents a survey of the most recent previous and related works that applied feature selection algorithms. Section 3 presents the proposed approach. Section 4 discusses experiment and result in details. Finally, Sect. 5 discusses the conclusion.

2 Previous Works

In this section, we introduce the most recent approaches which performed feature selection algorithms as an important step to perform classification for glaucoma detection in fundus retinal images as listed in Table 1.

Parashar et al. in [4] proposed glaucoma diagnosis approach based on two-dimensional discrete wavelet transform (2D-DWT) to decompose retinal image into sub-bands then apply the grey level co-occurrence matrix (GLCM) and extract features then to select the most relevant features using Relief algorithm and finally uses K-nearest neighbor classifier (KNN), it is tested on Rim-one dataset and its accuracy is 92.1%.

Zemmal et al. in [5] proposed semi-supervised approach based on features extraction using combination of three methods: Hu Moments, Central Moments and GLCM then select the most relevant features using new proposed genetic algorithm method and finally used Transudative SVM classifier to evaluate the approach on Rim-One dataset then got accuracy 98%.

Septiarini et al. in [6] proposed a statistical approach where apply image preprocessing to retinal images first such as Otsu, Sobel, and Dilation methods then to extract features of the average intensity (mean), standard deviation, smoothness, 3rd moment, uniformity and entropy, then is to apply the correlation feature selection (CFS) method which indicate the level of inter-correlation between the extracted features. His approach was evaluated on 84 fundus images that were collected from hospitals and used Naïve Bayes, SVM classifiers to get accuracy of 93% and accuracy of 95% by using KNN classifier.

Rabbia et al. in [7] proposed novel approach based on deep CNNs to extract LBP, HOG, SURF features and then to apply MRMR feature selection algorithm and finally perform SVM classifier, and it is tested on Rim-One dataset and got accuracy 98.8%. Jose et al. in [8] proposed an approach based on extracting texture features then perform genetic algorithm as search optimization feature selection algorithm, and finally he applies SVM classifier on Rim-One dataset and got accuracy 93.4%.

Romany et al. in [9] proposed an approach based on CNNs and used Weighted Least Square, it used holistic features execution, and it is tested on Rime-One dataset, and it got accuracy 99%.

Sibghatullah et al. in [10] proposed an approach that can clear noise of fundus images based on applying a non-Gaussian bivariate distribution with wavelet coefficients of fundus image then extract features as first step before applying is the Neighborhood Component Analysis (NCA) features selection method and finally apply least square support vector machine for classification, this approach is evaluated on Rim-One dataset and gets accuracy of 91.22%.

Table 1. Shows the most recent approaches based on feature selection algorithms and its results.

Author(s)	Method	Feature selection algorithm	Performance metrics	dataset
Prashar [14]	2D-DWT, GLCM K-nearest neighbor classifier (KNN)	Relief algorithm	Acc. 92.1%	Rim-one
Zemmal [5]	Hu Moments, Central Moments and GLCM SVM classifier	proposed genetic algorithm	Acc. 98%	Rim-one
Septiarini [6]	Otsu, Sobel and Dilation methods KNN classifier	correlation feature selection (CFS)	Acc. 95%	Unknown (collected from hostpitals)
Rabbia [7]	CNNs to extract LBP, HOG, SURF	MRMR feature selection	Acc. 98.8%	Rim-one
Jose [8]	Diversity Indexes	search optimization feature selection	Acc. 93.4%	Rim-one
Romany [9]	CNNs	Weighted Least Square	Acc. 99%	Rim-one
Sibghatullah [10]	wavelet coefficients	Neighborhood Component Analysis (NCA)	Acc. 91.2%	Rim-one
Ali Raza [11]	CNNs of FEDS-Net	Grad-CAM	Acc. 99.7%	Rim-one
Arwa [12]	GLCM	sequential features selection	Acc. 97%	Rim-one
Maila [13]	CNNs	Gain Ratio Algorithm	Acc. 93%	Rim-one + DRISHTI
Deepak [14]	2-D-C-VMD	Relief feature selection	Acc. 96%	Rim-one
Fan Guo [15]	UNet++ neural network algorithm	Chi-square feature selection	Acc. 84%	Origa

(*continued*)

Table 1. (*continued*)

Author(s)	Method	Feature selection algorithm	Performance metrics	dataset
Shruti [10]	Curvelet Empirical Wavelet Transform (EWT)	T-test features selection	Acc. 97%	Rim-one
Nirmala [17]	Adaptive Gamma Correction with Weighting Distribution (AGCWD)	Sequential Floating Forward Selection (SFFS)	SVM Acc 91.5% Naive Bayes Acc 89%	Rim-one

Ali Raza et al. in [11] proposed a novel convolution neural network of FEDS-Net for feature extraction which consists of 9 layers: Convolutional layer (Con), Batch normalization layer (BN), Stride convolutional layer (StCon), Bottleneck layer (BotlNeck), Rapid features downsampling (RFD), Transposed Convolutional layer (TransCon), Grouped Convolutional layer (GC), Efficient Convolutional depth (ECD) and Pixel classification layer (PCL). The used the gradient-weighted class activation mapping (Grad-CAM) technique which determine the most important parts in fundus image for classification as a features selection method which is based on the heat activation maps extracted from many layers and then marks every part of the fundus image with a different color. This approach is evaluated on Rime-One dataset and result is accuracy 99.7%.

Arwa et al. in [12] proposed an approach based on preprocessing image and features extraction of color, shape and texture features using the gray level co-occurrence matrix (GLCM), then apply sequential features selection method to introduce the most 6 significant features to use it in SVM classifier, this approach was evaluated on Rim-One dataset and got accuracy 97%.

Maila et al. in [13] proposed an approach based on using texture and shape descriptors then to use convolutional neural networks CNNs as a pre training then inputs of support vector machine SVM classifier, then the Gain Ratio Algorithm was performed to reduce features vector dimension, this approach was tested on DRISHTI and Rim-One datasets and got accuracy 93%.

Deepak et al. in [14] proposed a novel approach based on 2-D compact variational mode decomposition (2-D-C-VMD) algorithm, and Relief feature selection algorithm was performed, this approach was tested on Rim-One dataset and got accuracy 96%.

Fan Guo et al. in [15] proposed an approach based on a combination of two types of features, clinical measurement features and image features. It uses UNet++ neural network algorithm to segment optic disc and optic cup in addition to selecting the best features by performing variance and correlation analysis to remove low variance features and redundant features which have low contribution then as third step it uses Chi-square feature selection method and it uses also GBDT which is an integrated model of decision trees and finally after selecting best features it uses a Gradient Boosting Decision Tree

(GBDT) classifier, this approach is tested on Origa dataset and got accuracy 84%. Shruti et al. in [16] proposed an approach based on Curvelet Empirical Wavelet Transform (EWT) to compute Correntropy features then perform T-test features selection based on statistical analysis method and apply SVM classifier to detect glaucoma. This approach was tested on Rim-One dataset and got accuracy 97%.

Nirmala et al. in [17] proposed an approach based on the Adaptive Gamma Correction with Weighting Distribution (AGCWD) as preprocessing method for fundus image contrast improvement and localize the ROI, then to extract spatial domain, Local binary pattern (LBP), Histogram of oriented gradient (HOG) and Fractal features then to apply Sequential Floating Forward Selection (SFFS) feature selection method to select the most significant features. This approach was tested on Rim-One and got accuracy 91.5% with SVM classifier, accuracy 89% with Naive Bayes classifier, accuracy 87% with Logistic Regression classifier, so the best performance is SVM with accuracy 91.5%.

3 Proposed Method

We proposed an approach based on supervised classification and superpixels methods, then we experiment the most significant four feature selection algorithms and we compare its results by selecting the most relevant features to reduce data dimensionality. Figure 1 shows the proposed approach diagram and its phases.

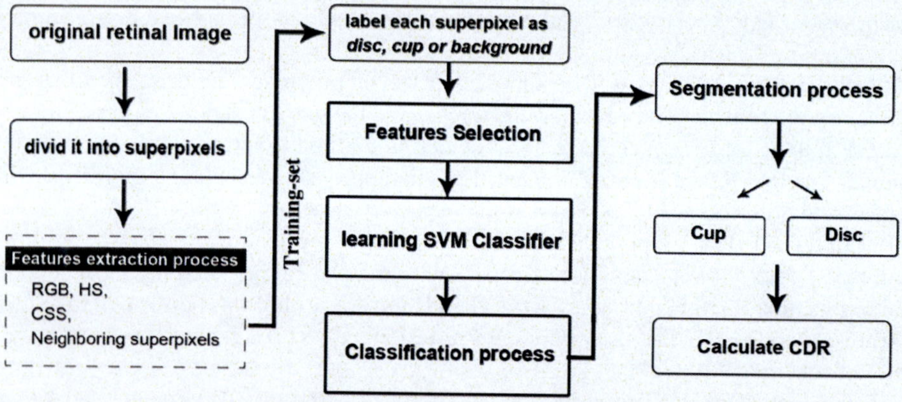

Fig. 1. The proposed approach diagram

3.1 Superpixel Method

Simple Linear Iterative Clustering method (SLIC) algorithm is used to divide fundus image into suitable amount of superpixels [3].

3.2 Feature Extraction

Features is extracted based on color and texture of every superpixel, color features include applied histogram of the five color channels: Red, Green, Blue, Hue, Saturation, also we extract texture features by applying the mean and variance of center surround statistics (CSS) by generating dyadic Gaussian pyramids, finally we have 185 features is extracted from each superpixel [3].

3.3 Feature Selection

Feature selection is a good improvement factor which results in a reduction in the amount of data needed while no loss in significant information or features by clarifying the most relevant features and ranking them based on weight and importance, so it improves classifiers, makes them easily understood, reduces elapsed computation time and reduces data dimensionality [18].

We applied four filtering feature selection algorithms as an important step to increase performance and reduce features vector dimension. Chi-square tests algorithm, Neighborhood Component Analysis algorithm (NCA) and Minimum Redundancy Maximum Relevance algorithm (MRMR) [2].

Every feature selection algorithm is separately experimented and the results was collected for each algorithm. First we want to know the suitable features number we should consider, so, we applied The ReliefF feature selection algorithm to select the most 70 important and significant features, but we got mean CDR-error of 0.45 which is not acceptable result, the we tried to experiment 90 of the most significant features to get satisfied result of mean CDR-error 0.1, hence, we consider the most 90 significant features to experiment every feature selection algorithm.

3.3.1 Minimum Redundancy Maximum Relevance (MRMR)

It finds an optimal set of features that is mutually and maximally dissimilar, can represent the response variable effectively as it minimizes the redundancy of a feature set, and maximizes the relevance of a feature set to the response variable [19] as shown by the following Eqs. (1) and (2):

$$v_s = \frac{1}{|\varsigma|}\Sigma_{xes} I(x, y) \tag{1}$$

$$w_s = \frac{1}{|\varsigma|^2}\Sigma_{x,zes} I(x, z) \tag{2}$$

The MRMR method tries to get the optimum set of features $\boldsymbol{S}$, which maximize $\boldsymbol{v_s}$ and minimize $\boldsymbol{w_s}$ where $\boldsymbol{v_s}$ is relevance of $\boldsymbol{\varsigma}$ to the class label $\boldsymbol{y}$ and $\boldsymbol{w_s}$ is the redundancy of $\boldsymbol{\varsigma}$.

3.3.2 Diagonal Adaptation of Neighborhood Component Analysis (NCA)

It is a non-parametric method with the goal of maximizing prediction accuracy of regression and classification algorithms. It classifies a multi-class using a training set containing $\boldsymbol{n}$ observations [20].

3.3.3 ReliefF

It is an updated version of Relief algorithm whereas ReliefF can work with multiple classes, and it has also higher performance. ReliefF ranks features importance by examining the weight of every predictor that gives a different value to neighbors of the same class [21]. It gives ***0*** value for all predictors weight w_j^i then makes random observation x_r in order to get the K-nearest observations to x_r for each class and updates each nearest neighbor χ_q as shown by the following Eqs. (3) and (4):

If x_r and χ_q are in the same class

$$w_j^i = w_j^{i-1} - \frac{\Delta_j(x_r, \chi_q)}{m} d_{rq} \tag{3}$$

If x_r and χ_q are in the different classes

$$w_j^i = w_j^{i-1} + \frac{P_{yr}}{1 - p_{y_r}} \cdot \frac{\Delta_j(x_r, \chi_q)}{m} d_{rq} \tag{4}$$

w_j^i is the weight of the predictor at the $\boldsymbol{i^{th}}$ iteration step, p_{y_r} is the prior probability of the class to which x_r belongs, P_{yr} is the prior probability of the class to which χ_q belongs, m is the number of iterations specified by 'updates' $\Delta_j(x_r, \chi_q)$ which is the difference in the value of the predictor between observations x_r and χ_q. Let χ_{rj} denotes the value of the jth predictor for observation x_r, and let χ_{qi} denotes the value of the jth predictor for observation χ_q.

3.3.4 Chi-square Tests

It is a feature selection algorithm that depends on X^2 statistic in order to give a value of each feature or attribute [22]. It works in two steps, first one is to start by giving high values to all features e.g., 0.5, and sort them based on that values then perform the X^2 statistic as Eq. (5) on pairs of features based on its intervals, then merge them with the lower value of X^2 and repeat it as need, in the second step, it improves the previous step by continue merging until no features values can be merged

$$\boldsymbol{x}^2 = \sum\nolimits_{i=1}^{2} \sum\nolimits_{j=1}^{k} \frac{(A_{ij} - E_{ij})^2}{E_{ij}} \tag{5}$$

where k is the number of classes,A_{ij}: no. Patterns in the ith interval, jth class, E_{ij} : expected frequency of A_{ij} Chi-square tests is a good method to discretize features of numeric attributes in addition to it can work with multiple classes.

3.4 Classification and CDR Computation

Linear Support Vector Machine (SVM) is effective in high dimensional features vector so we used the linear SVM to classify and label every superpixel in order to indicate if it is belonged to disc, cup or background areas then we fit a circle boundary by applying the circle equation for disc and cup, finally we calculate the vertical diameter of each one and get the CDR = vertical cup diameter/vertical disc diameter [2].

4 Experiments and Results

4.1 Dataset Description

We use the public and common Rim-One r3 dataset to evaluate our approach [23]. Rim-One r3 contains 159 retinal images (Refer to Fig. 2). These images were divided into 85 healthy, 35 suspect glaucomatous and 39 glaucomatous.

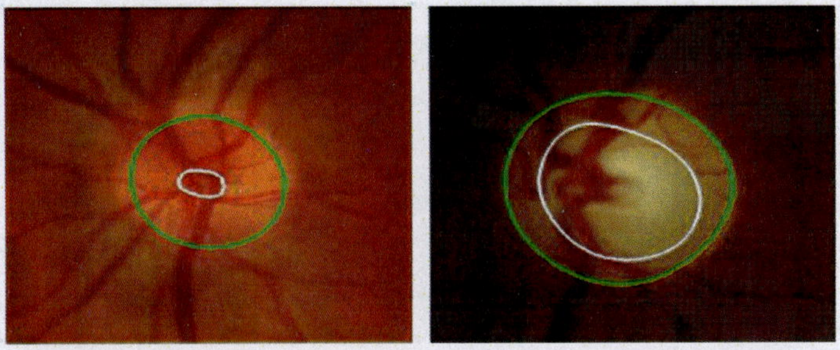

Fig. 2. Two fundus images of Rim-One r3 dataset, the left is healthy and the right is glaucomatous

4.2 Experiments

The proposed approach has two phases, in phase 1 [3] we applied the approach without decreasing features or applying any of feature selection methods and get its results and in phase 2 we apply the four feature selection methods mentioned above and record its results and we compare results of phase1 and phase2, as listed in Table 2 also we compare the accuracy result of applying each feature selection method (Refer to Fig. 4), further we compare our results with the previous work (Refer to Fig. 5.)

First, we divide each fundus image into 1200 superpixels using the simple linear iterative clustering algorithm (SLIC) then extracted color and texture features from each superpixel to get features vector with dimension of 185 features [3],

In phase 2 we apply feature selection methods to reduce features vector dimension and remove redundant and list its result in Table 2, but in phase1 we ignore this step to collect result of each phase to be compared. Finally, we perform linear SVM to classify each superpixel as disc, cup or background and then we collect the labeled superpixels into groups in order to initiate segmentation process of disc and cup. (See Fig. 3) shows segmented OD and OC.

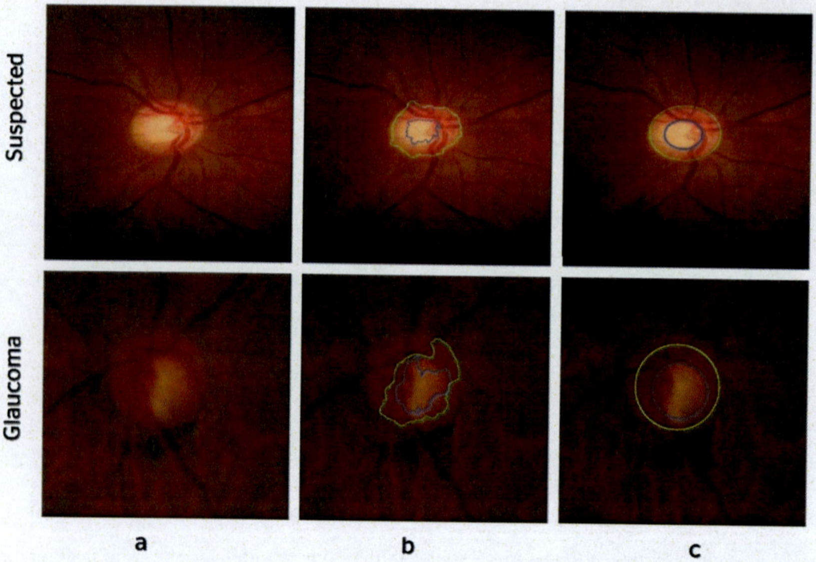

Fig. 3. Fitting circle boundary to disc and cup regions: a. original retinal fundus images, b. classified disc region (green) and cup region (blue), c. fitted circle boundary to disc and cup regions

As we obtained the classified pixels of disc region and cup region, then we fit a circle boundary using the geometrical circle equation and then calculated and divided the vertical diameter of the disc circle (VDD) by the vertical diameter of the cup circle (VCD) to get the Cup Disc Ratio and detect glaucoma in fundus retinal image.

Table 2. shows and compare results of phase 1 and phase 2

		ACC	Time elapsed	Mean CDR error
Phase 1	Kamara [3]	0.9920	404.75	0.088
Phase 2	Proposed chi-square	0.980	265	0.12
	Proposed NCA	0.985	287	0.1
	Proposed Relief	0.985	276	0.12
	Proposed MRMR	0.980	234	0.1

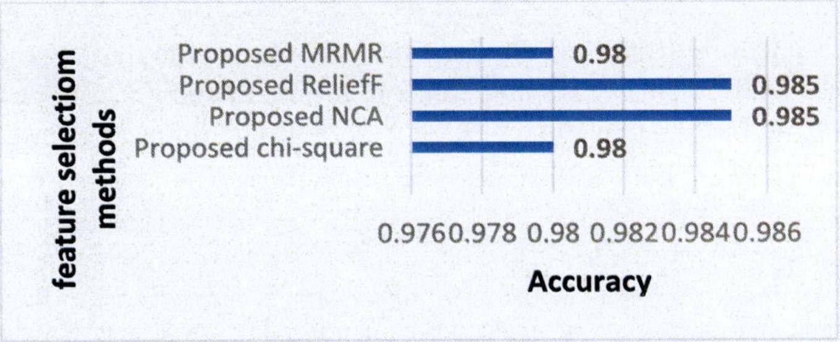

Fig. 4. Show result accuracy of each feature selection method when it is applied.

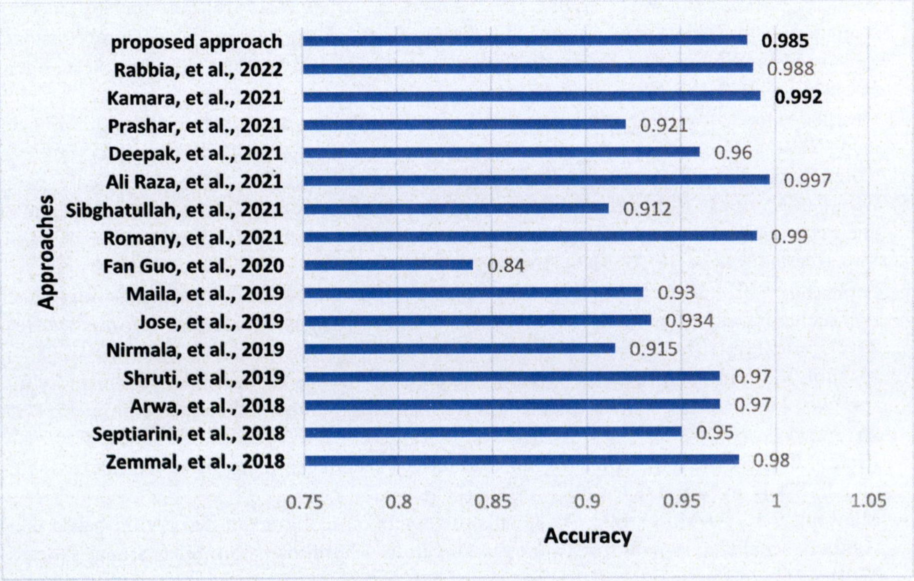

Fig. 5. Show comparison between accuracy in the most recent approaches and our proposed approach.

5 Conclusions and Future Work

We proposed an approach for automated detecting glaucoma disease in retinal fundus images based on machine learning supervised classification and superpixels methods. In addition to, we applied four filtering feature selection algorithms separately in order to increase performance and reduce features vector dimension. Chi-square tests algorithm, Neighborhood Component Analysis algorithm (NCA) and Minimum Redundancy Maximum Relevance algorithm (MRMR). Our proposed approach is evaluated on Rim-One r3 dataset and it has succeeded to detect glaucoma whereas the result is that the number of features vector is decreased to 90 features and elapsed time also decreased by 34%, also, accuracy decreased from 99.2% before feature selection to 98.5% and mean CRD-error increased by 0.02, also accuracy decreased only by 0.007. In the future work we may apply deep learning and Convolutional Neural Networks to the proposed approach and evaluate it with other datasets and compare results to know if can we improve the approach more and detect glaucoma disease in various and different datasets.

References

1. Tham, Y.-C., Li, X., Wong, T.Y., Quigley, H.A., Aung, T., Cheng, C.-Y.: Global prevalence of glaucoma and projections of glaucoma burden through 2040: a systematic review and meta-analysis. Ophthalmology **121**(11), 2081–2090 (2014)
2. Nicolela, M.T., Vianna, J.R.: Optic nerve: clinical examination. Pearls of Glaucoma Management, pp. 17–26 (2016)
3. Kamara, O.M., Asad, A.H., Hefny, H.A.: An enhanced approach for automated glaucoma diagnosis in retinal fundus images. In: Proceedings of the International Conference on Advanced Intelligent Systems and Informatics, pp. 185–197 (2021). Springer International Publishing, 2022
4. Parashar, D., Agrawal, D.: Improved classification of glaucoma in retinal fundus images using 2D-DWT. In: 2021 International Conference on Advances in Electrical, Computing, Communication and Sustainable Technologies (ICAECT), pp. 1–5. IEEE (2021)
5. Zemmal, N., et al.: Robust feature selection algorithm based on transductive SVM wrapper and genetic algorithm: application on computer-aided glaucoma classification. Int. J. Intell. Syst. Technol. Appl. **17**(3), 310–346 (2018)
6. Septiarini, A., Khairina, D.M., Kridalaksana, A.H., Hamdani, H.: Automatic glaucoma detection method applying a statistical approach to fundus images. Healthcare Informatics Res. **24**(1), 53–60 (2018)
7. Mahum, R., Rehman, S.U., Okon, O.D., Alabrah, A., Meraj, T., Rauf, H.T.: A novel hybrid approach based on deep CNN to detect glaucoma using fundus imaging. Electronics **11**(1), 26 (2021)
8. Araújo, J.D.L., et al.: Glaucoma diagnosis in fundus eye images using diversity indexes. Multimedia Tools and Appl. **78**, 12987–13004 (2019)
9. Mansour, R.F., Al-Marghilnai, A.: Glaucoma detection using novel perceptron based convolutional multi-layer neural network classification. Multidimension. Syst. Signal Process. **32**(4), 1217–1235 (2021)
10. Khan, S.I., Choubey, S.B., Choubey, A., Bhatt, A., Naishadhkumar, P.V., Basha, M.M.: Automated glaucoma detection from fundus images using wavelet-based denoising and machine learning. Concurrent Eng. **30**(1), 103–115 (2022)
11. Raza, A., Adnan, S., Ishaq, M., Kim, H.S., Naqvi, R.A., Lee, S.-W.: Assisting glaucoma screening process using feature excitation and information aggregation techniques in retinal fundus images. Mathematics **11**(2), 257 (2023)
12. Elseid, A.A.G., Hamza, A.O., Fragoon, A.: Developing a real time algorithm for diagnosing glaucoma. ICTACT Journal on Image & Video Process. **9**(2) (2018)
13. Claro, M., et al.: An hybrid feature space from texture information and transfer learning for glaucoma classification. J. Vis. Commun. Image Represent. **64**, 102597 (2019)
14. Parashar, D., Agrawal, D.: 2-D compact variational mode decomposition-based automatic classification of glaucoma stages from fundus images. IEEE Trans. Instrum. Meas. **70**, 1–10 (2021)
15. Guo, F., Li, W., Tang, J., Zou, B., Fan, Z.: Automated glaucoma screening method based on image segmentation and feature extraction. Med. Biol. Eng. Compu. **58**, 2567–2586 (2020)
16. Jain, S., Sharma, S.D.: Classification of glaucoma fundus images using curvelet empirical wavelet transform. In: 2019 Fifth International Conference on Image Information Processing (ICIIP), pp. 321–325. IEEE (2019)

17. Krishnamoorthi, N., Chinnababu, V.K.: Hybrid feature vector based detection of Glaucoma. Multimedia Tools and Applications **78**, 34247–34276 (2019)
18. Shardlow, M.: An analysis of feature selection techniques. The University of Manchester **1**(2016), 1–7 (2016)
19. Ding, C., Peng, H.: Minimum redundancy feature selection from microarray gene expression data. J. Bioinform. Comput. Biol. **3**(02), 185–205 (2005)
20. Yang, W., Wang, K., Zuo, W.: Neighborhood component feature selection for high-dimensional data. J. Comput. **7**(1), 161–168 (2012)
21. Robnik-Šikonja, M., Kononenko, I.: Theoretical and empirical analysis of ReliefF and RReliefF. Mach. Learn. **53**, 23–69 (2003)
22. Liu, H., Setiono, R.: Chi2: feature selection and discretization of numeric attributes. In: Proceedings of 7th IEEE International Conference on Tools with Artificial Intelligence, pp. 388–391. IEEE (1995)
23. Fumero, F., Alayón, S., Sanchez, J.L., Sigut, J., Gonzalez-Hernandez, M.: RIM-ONE: an open retinal image database for optic nerve evaluation. In: 2011 24th International Symposium on Computer-Based Medical Systems (CBMS), pp. 1–6. IEEE (2011)

Improving Machine Learning Techniques with Imbalanced Data Treatment for Predicting Diabetes

Radwa Taher[1], Sameh H. Basha[1,2(✉)], and Areeg Abdalla[1]

[1] Faculty of science, Cairo University, Giza 12613, Egypt
rtaher@sci.cu.edu.eg, {samehbasha,areeg}@cu.edu.eg
[2] Faculty of science, Galala University, Suez 43511, Egypt
samehbasha@gu.edu.eg

Abstract. This paper introduces a high-efficiency computational model designed for early predicting diabetes on the PIMA dataset, which contains two interleaved imbalanced classes. Choosing among different machine learning models to classify the data. The extra tree, one of the best features' selection method, is used to improve the results, and the synthetic minority oversampling technique (SMOTE) algorithm is employed to handle the problem of imbalanced data set. The proposed model has achieved superiority in comparing with the state of the art machine learning classification models according to accuracy, precision, recall, f1-score, area under curve (AUC) and specificity measures.

Keywords: Diabetes prediction · Extra Tree · Imbalanced data · Machine Learning

1 Introduction

Diabetes affects about 422 million people around the world according to WHO and causes 1.6 million deaths every year, most of them in low and middle income countries [1]. It prevents human body from producing enough insulin that is type 1; or prevents the body from making the right use of insulin that is type 2. Also there is gestational diabetes which occurs during pregnancy. Diabetes has dangerous complications that affect the whole human body or cause death. Early diagnosis helps avoid some of these complications and predicting the disease gives an alarm for changing life style.

Machine Learning (ML) is a branch of artificial intelligence which intensively embedded in systems for better classifying. It is used in different aspects in the Healthcare field like disease identification, diagnosis, prediction, medical imaging diagnosis and other imaging and many others. It helps the health practitioner to better analyze the big amount of data. ML works very well on different kinds of datasets [2] like clinical data, collected throughout the patient treatment, sensor data like those attached to arms, hands and thighs, and omics data which is high dimensional, huge and very complex for example genomic, transcriptomic and proteomics data.

A. Hassanien et al. (Eds.): AISI 2023, LNDECT 184, pp. 380–391, 2023.
https://doi.org/10.1007/978-3-031-43247-7_34

Many different classification models were used in disease prediction and diagnosis like: random forest algorithm, logistic regression, naive bayes and k-nearest neighbors are used for liver disease prediction [3], J48, decision tree, k-nearest neighbors, random forest and support vector machines for diabetes prediction [4] and artificial neural networks are applied for diabetes diagnosis in [5]. Also deep learning neural network are proposed in [6], deep convolutional neural network (CNN) and CNN-LSTM (convolutional neural network - long short term memory) are used for diabetes classification [7].

The aim of this paper is to build an efficient computational model to predict diabetes mellitus. Four experiments are done, the first investigate the performance of different classification models on the dataset. The second one starts with a preprocessing step. The third experiment adds a feature selection method; extra tree. Finally the SMOTE algorithm is used to handle the problem of imbalanced data set. The results have been gradually improving throughout. And the final model shows superiority when compared to others.

The paper is structured as follows, the literature review is in Sect. 2, the proposed model in Sect. 3, and the results and analysis in Sect. 4 and finally Sect. 5 is the conclusion.

2 Literature Review

Diabetes is one of the most common disease around the world. Goutham Swapna [7] implemented J48 decision tree, k-nearest neighbors random forest and support vector machines to predict diabetes and found that After preprocessing KNN and random forest performance were much better. LDA-ANFIS was used by Esin Dogantekin [2] which is intelligent diagnosis system consists of two stages. First one is linear discriminant analysis (LDA) which is used to separate features between healthy and patient data and second stage is that the features selected by LDA are used as input for adaptive network based fuzzy inference system (ANFIS). The classification accuracy of this model was about 84.6%.

Long short-term memory (LTSM), Markov, and plain recurrent neural network(RNN) on a manually collected diabetes data set from a regional Australian hospital was implemented by Trang Pham [8]. LTSM gives the best performance based on precision is 59.6%. Abir Alharbi [9] used the Saudi type 2-diabetes dataset obtained from King Khalid University hospital and applied GA-ELM algorithm hybrid model consists of genetic algorithm and extreme learning machine algorithm for diabetes diagnosis. Goutham Swapna [7] found that combining statistical models like logistic regression with neural network results in acceptable accuracy and efficiency compared with logistic regression and neural network. The criteria is to minimize error function in neural network training.

PIMA Indians data set attracting researchers' attention for several years. Mamman Mamuda [10] is used Levenberg Marquardt (LM), Bayesian Regulation (BR), scaled conjugate gradient (SCG) and the experimental results show that LM over perform others based on Mean Squared Error. Fareeha Anwar Qurat-Ul-Ain [11] used J48, multi-layer perceptron, hoeffding Tree, JRip, bayes net and random forest. Best precision value equal to 77% and a recall equal to 78% using the hoeffding Tree algorithm. Decision tree, SVM and naive bayes were applied to detect diabetes at early stage and naive bayes gives the highest accuracy [12]. Distributed computing framework like Hadoop which can effectively deal with extremely large data sets. When applying on PIMA Indians data set [13] Random forest algorithm produces highest accuracy than decision tree and naive bayes.

Deep learning also gives promising performance in diabetes diagnosis and prediction. Huaping Zhou [14] proposed deep learning for predicting diabetes (DLPD) model which the experimental results show its effectiveness with high accuracy. It is a deep neural network which uses drop down regularization to handle the over-fitting issue. This model predicts the type of diabetes that a person can face. Results show improvements over the state-of-the-art methods. Ashiquzzaman [15] proposed deep learning neural network composed of fully connected layers followed by dropout layers to address the over-fitting and results were better than other state-of-art methods. Goutham Swapna [7] authors used deep convolutional neural network (CNN) and convolutional neural network - long short term memory (CNN-LSTM) for diabetes classification and found that combination of CNN and LSTM gave the maximum Accuracy.

3 Diabetes Datasets

PIMA Indians diabetes data set is one of the diabetes benchmark data sets which is originally owned by the National Institute of diabetes and digestive and kidney diseases and is publicly published in UCI machine learning repository https://archive.ics.uci.edu/ml/datasets/pima+indians+diabetes [2021]. It consists of 8 diagnostic measurements which are independent variables and one dependent variable recognize if the person has diabetes or not and contains 768 females which are at least 21 years old. The target of study is to classify people into diabetic person with 1 value in outcome column and non diabetic person with

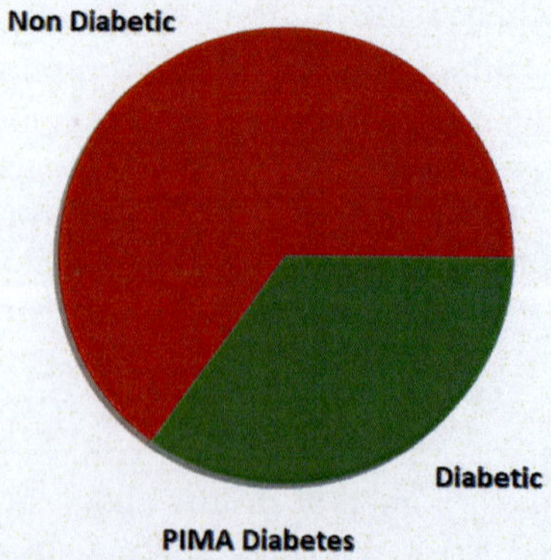

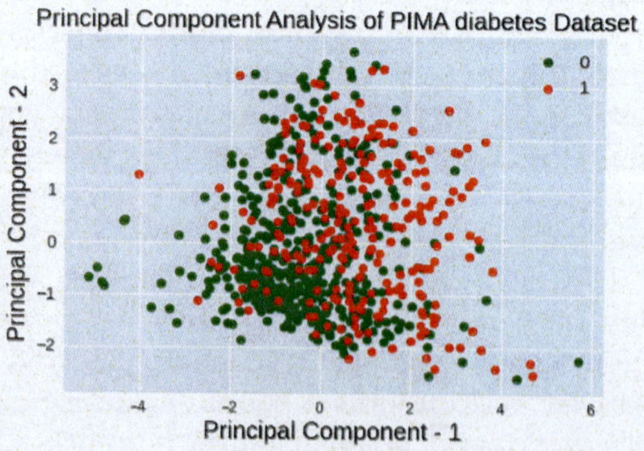

Fig. 1. a)PIMA dataset distribution b)Principals Component Analysis of PIMA

0 value. The two classes are interleaved as shown in Fig. 1 b). It is difficult to identify them separately and they may be considered to have the same center.

PIMA has some problems need to be handled like having some missing values. In glucose column there are 5 missing values, in blood pressure column there are 35 missing values, in skin-thickness there are 227 missing values, in insulin there are 374 missing values, in BMI there are 11. As shown in Fig. 1 PIMA Indians dataset is imbalanced as 268 of 768 are in class 1, the others are 0.

4 Proposed Prediction Model

As shown in Fig. 2, our proposed model consists of five stages. Firstly, the PIMA Indians dataset is preprocessed using different preprocessing techniques to handle noise and missing values, which affect classification results.

Secondly, using extra tree, we did feature engineering extracting new statistical features; mean, standard deviation, count of features and most important feature. To treat the imbalance in the PIMA dataset the SMOTE algorithm is applied at the third step. Fourthly different machine learning models were applied. The performance of classification model was evaluated using accuracy, recall, precision and f1-Score.

4.1 Preprocessing Dataset

To handle missing values, we use the median for non-diabetics and diabetics if the missing values are in classes 0 and 1, respectively. The outliers are handled using the interquartile range. The dataset was then standardized to reduce any other outlier effects.

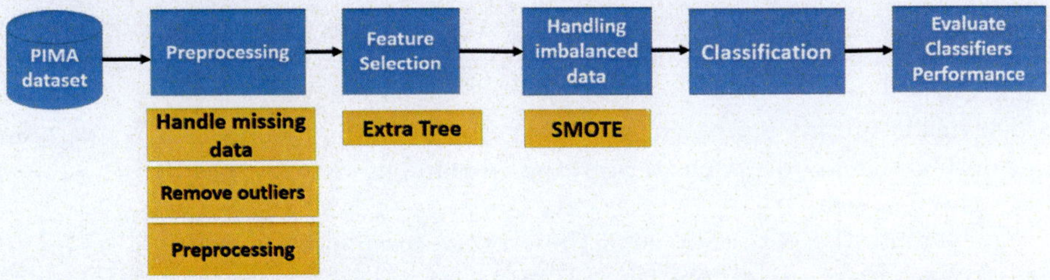

Fig. 2. General phases of proposed model

4.2 Feature Selection

Feature selection reduces training time, over-fitting and improves accuracy. Extremely randomized trees classifier (Extra Trees classifier) is ensampling learning algorithm. Like random forest, It fits number of randomized decision trees to the data. Random splits of all observations are carried out to ensure that the model does not over fit the data [16].

A significant number of distinct decision trees are produced by extra trees from the entire training dataset. Each tree is provided with a random sample of k features from the feature-set [17]. The algorithm selects a split rule for the root node based on k and a cut point that is partially random. This random sample of features leads to the creation of multiple decorrelated decision trees. A majority vote is used to determine the final prediction from all of the trees' predictions.

The output of Extra Tree is Feature Importance. We choose the most important features. Every feature is listed in descending order according to its feature importance The most important seven features of PIMA dataset in descending order are insulin, glucose, skinthickness, BMI, diabetespedigreefunction, age, pregnancies.

4.3 Handling Imbalanced Data

PIMA Indians dataset is imbalanced. Random over sampling is a technique to handle the imbalanced data set. Some random data points are randomly selected with replacement from the minority class and added to the training dataset multiple times to produce more balanced dataset [18]. The drawback of this technique is increasing over-fitting and redundancy. Another algorithm is random under sampling is to select random examples from the majority class and delete them from the training dataset until we reach more balanced dataset.

In 2002 Nitesh Chawla, et al. [19] proposed synthetic minority oversampling technique, or SMOTE to handle imbalanced data. SMOTE works on minority class by finding the k nearest neighbors to the data points in the feature space, draw new synthetic examples in a randomly selected points on line between an example in data set and its neighbors. Synthetic samples are created by taking into account the feature space of the instances. Then, synthetic data will be generated by combining random data with a randomly chosen nearest neighbour [20].

Downside of this approach is that synthetic examples are created without considering the majority class resulting in ambiguous examples if overlapped with other classes.

To evaluate the performance of the proposed model we have use the famous measure; accuracy, recall, precision, f1-score [18, 21].

5 Results and Analysis

For experiments, 10-fold cross-validation was used to average the measures of fitness to get the most accurate performance of the used classifiers. In 10-fold cross-validation, the dataset is divided into 10 equal-sized sub-samples; one single sample is treated as a validation subset for testing, and the remaining sub-datasets (sub-samples) are considered the training dataset. The classification techniques that were used in this study are LR, SVM, KNN, DT, Naïve Bayes (NB), RF and DNN, XGB, and LGMB. Fig. 3 explains the stages of the four experiments.

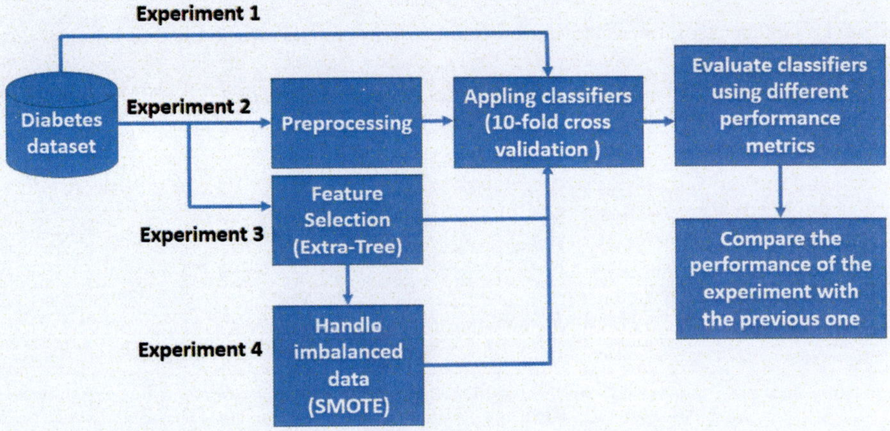

Fig. 3. The work flow of the stages of the four experiments

The first experiment was to run the classifiers on the PIMA dataset without any preprocessing. The results as shown in Table 1. In PIMA dataset LR gave best accuracy 77.8%, followed by ANN with accuracy 76.6%, RF accuracy was 76.4%, XGB accuracy was 76.3%, SVM accuracy was 75.7%, NB accuracy was 75.1% LGMB 73.3%, DT 72.6%, KNN 70.0%.

The second experiment was to preprocess the PIMA dataset. The dataset problems were handled like missing data and outliers. Also, the data was standardized to preserve data quality as a step of preprocessing.

It is remarkably noticeable that the preprocessing step improved the results dramatically with all classifiers. Applying preprocessing on PIMA dataset improved Logistic Regression classifier accuracy by 8.8%, recall by 18%, precision by 8.5% and f1-score by 14.1%, AUC by 10.7% and specificity by 18%. And improved SVM accuracy by 13.1%, recall by 36.5%, precision by 8.9%, s1-score by 25.7%, AUC by 18.6% and specificity by 36.5%. And improved NB by 11.3%, recall by 21.5%, precision by 13.1%, f1-score by 17.6%, AUC by 13.7% and specificity by 21.5%. KNN accuracy was improved by 18.8%, recall by 31.3%, precision by 26.1%, F1-score by 29%, AUC by 21.8%, specificity by 31.3%. RF accuracy was improved by 13.9%, recall by 14.5%, precision by 17.6%, F1-score by 21.3%, AUC by 16.2%, specificity by 23.8%. DT accuracy was improved by 14.5%, recall by 16.4%, precision by 21.8%, f1-score by 18.9%, AUC by 14.9%, specificity by 14.4%. ANN accuracy was improved by 16.7%, recall by 28.7%, precision by 19.7%, F1-score by 26.4%, AUC by 16.4%, specificity by 28.7%. LGMB accuracy was improved by 17%, recall by 25.5%, precision by 23.7%, f1-score by 24.7%, AUC by 19.1%, specificity by 25.5%. XGB accuracy was improved by 14.2%, recall by 22.8%, precision by 18.4%, f1-score by 24.8%, AUC by 16.1%, specificity by 22.8%. So it is obvious that preprocessing stage had a noticeable improvement in the diabetes prediction using PIMA dataset as shown in Fig. 4.

Table 1. Comparison between the four experiments performance on **PIMA diabetes dataset**

Experiment	Accu	Prec	Recall	F-score	AUC	Spec
LR						
Experiment 1	77.8	73.3	58.8	64.7	73.5	58.8
Experiment 2	86.6	81.8	76.8	78.8	84.2	76.8
Experiment 3	86.9	81.8	78.4	79.5	85.0	78.4
Experiment 4	86.1	85.2	87.1	86.1	86.0	87.1
After tuning	**87.6**	**87.0**	**88.0**	**87.4**	**87.6**	**88.0**
SVM						
Experiment 1	75.7	73.4	47.7	57.1	69.2	47.7
Experiment 2	88.8	82.3	84.2	82.8	87.8	84.2
Experiment 3	88.9	83.2	83.7	83.0	87.8	83.7
Experiment 4	91.0	89.7	92.4	91.0	90.8	92.4
After tuning	**94.0**	**94.1**	**93.8**	**94.0**	**93.8**	**93.8**
NB						
Experiment 1	75.1	65.3	59.3	61.7	71.3	59.3
Experiment 2	86.4	78.4	80.8	79.3	85.0	80.8
Experiment 3	86.4	78.1	81.6	79.5	85.2	81.6
Experiment 4	86.5	86.3	86.1	86.1	86.2	86.1
After tuning	**87.5**	**86.8**	**88.0**	**87.4**	**87.5**	**88.0**
KNN						
Experiment 1	70.0	58.1	49.7	53.3	65.0	49.7
Experiment 2	88.8	84.2	81.0	82.3	86.8	81.0
Experiment 3	89.8	86.8	81.8	83.5	88.0	81.8
Experiment 4	89.5	88.2	90.9	89.4	89.3	90.9
After tuning	**92.4**	**88.3**	**97.5**	**92.7**	**92.3**	**97.6**
DT						
Experiment 1	72.9	61.0	61.0	60.9	70.0	61.0
Experiment 2	87.4	82.8	77.4	79.8	84.9	77.4
Experiment 3	88.0	84.4	78.0	80.7	85.4	78.0
Experiment 4	89.5	89.7	89.4	89.5	89.5	89.4
After tuning	**90.3**	**90.3**	**90.2**	**90.3**	**90.3**	**90.2**
RF						
Experiment 1	76.4	69.1	59.8	63.4	72.6	59.8
Experiment 2	90.3	86.7	83.6	84.7	88.8	83.6
Experiment 3	90.1	87.0	82.3	84.2	88.3	82.3
Experiment 4	91.7	91.6	91.7	91.6	91.4	91.7
After tuning	**92.0**	**91.5**	**92.4**	**92.0**	**91.7**	**92.4**
ANN						
Experiment 1	76.6	71.5	59.7	63.0	72.8	59.7
Experiment 2	93.3	91.2	88.4	89.4	92.2	88.4
Experiment 3	92.7	89.6	87.8	88.4	91.5	87.8
Experiment 4	94.6	94.7	94.3	94.5	94.5	94.3
After tuning	**99.3**	**99.2**	**99.5**	**99.3**	**99.3**	**99.5**
LGMB						
Experiment 1	73.3	62.9	58.0	59.9	69.7	58.0
Experiment 2	90.3	86.6	83.5	84.6	88.8	83.5
Experiment 3	89.8	84.8	83.9	84.0	88.5	83.9
Experiment 4	93.0	**92.3**	94.2	93.2	93.0	94.2
After tuning	**93.0**	92.0	**94.5**	**93.2**	**93.0**	**94.5**
XGB						
Experiment 1	76.3	68.7	60.4	63.8	72.6	60.4
Experiment 2	90.5	87.1	83.2	84.6	88.8	83.2
Experiment 3	91.1	91.2	85.6	84.1	85.6	89.5
Experiment 4	92.0	91.1	93.0	92.0	91.9	93.0
After tuning	**92.8**	**92.1**	**94.1**	**93.0**	**92.8**	**94.1**

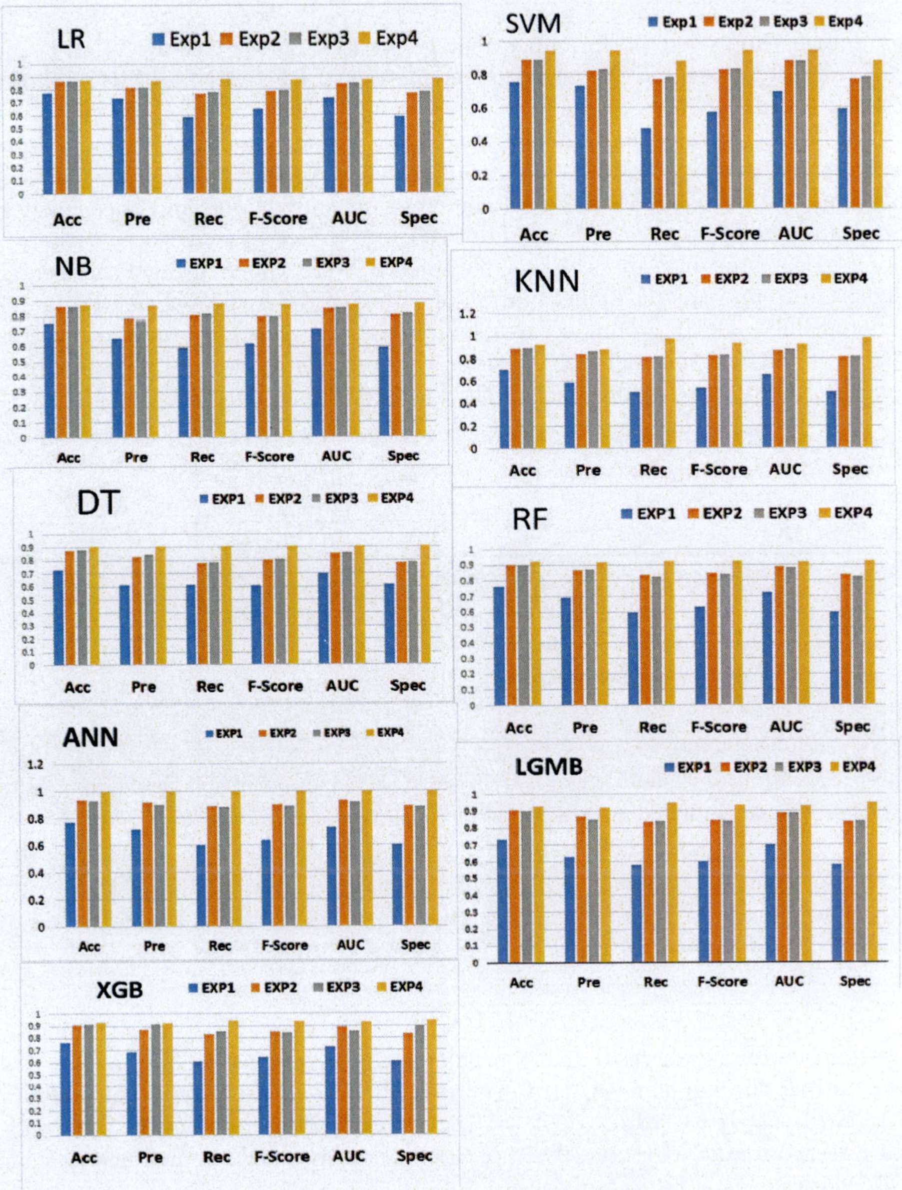

Fig. 4. Comparison between the four experiments per formance on PIMA diabetes dataset

The purpose of our third experiment is to identify the most crucial features, employ them following preprocessing, assess classifier performance, and compare it to the results of our second experiment to determine the impact of this phase. Adding Feature Selection step to the classification model on PIMA Indians dataset improved performance metrics of LR, SVM, NB, KNN, DT and XGB. Slightly decreased the performance metrics of RF, ANN, LGMB as shown in Table 1 and Fig. 4.

The PIMA dataset has imbalanced classes so our fourth experiment is to handle this issue by SMOTE algorithm which dramatically improved all the performance metrics of all the used classifiers specially recall and specificity. It also dramatically improved Precision of NB.

As a last step classifiers' hyper-parameters were tuned using Grid Search Algorithm to find the best combination of values of the hyper-parameters which is compatible with PIMA dataset as shown in Table 2. Accuracy, recall, precision, f1-score, AUC and specificity are performance metrics were used to evaluate the classifiers prediction performance of the four experiments, as shown in Table 1.

Table 2. PIMA dataset hyper-parameters tuned

Classifiers	Hyper parameters
LR	C= 0.01, penalty= 'l1', solver= 'saga'
RF	criterion= 'entropy', maxFeatures='auto', n_estimators= 200
SVM	C= 10, gamma= 1, kernel= 'rbf'
DT	criterion= 'entropy', maxdepth= 11
NB	var_smoothing= 0.0533
KNN	n_neighbors=1
ANN	optimizer='adam', neurons=85, batch_size=16, epochs=50, activation='relu'
LGMB	baggingFraction= 0.8, featureFraction= 0.9, learningrate= 1.0, maxbin= 90, maxdepth= 30, mindatainleaf= 20, minsumhessianileaf= 0, numleaves= 80, subsample= 0.01
XGB	min_samples_split= 10, max_depth= 7, min_child_weight= 5, gamma= 0.1, n_estimators= 180, colsample_bytree= 0.8

After handling diabetes datasets problems, finding the most effective features and handing the imbalanced data sets and getting the best hyper-parameters using Grid search we achieved 99.3% accuracy, 99.2% precision, 99.5% recall, 99.3% f1-Score and AUC and 99.5 specificity at using deep neural network on PIMA dataset.

Table 3 shows the comparison between the most recent studies on the PIMA dataset and our results. It is clear that the proposed model is superior to all previous models according to different performance metrics. For example, the accuracy of the LR has increased by 10 percent to the previous maximum value by Kumar in 2020 [18]. Also the accuracy of the NB has improved by almost 9 percents from the maximum value in 2021 [21]. Among all classifiers, the proposed SVM reached the best accuracy. The four experiments supports our idea of the importance of The preprocessing, feature selection and the method of handling missing data.

As shown in Table 3 the fourth experiment which used ANN gave the best Accuracy 99.3%, best precision 99.2%, recall 99.5%, f1-score 99.3%, AUC 99.3% and specificity 99.5%. Using feature selection and SMOTE with preprocessed date and finding the best hyper-parameters gave the best performance compared to the other experiments and give competitive results with state-of-art models as shown in Table 3.

Table 3. Comparison of the proposed models with previous recent and best results on PIMA dataset

Author	Method	Accuracy	Recall	Precision	F1-Score
Francesco Mercaldoa, Vittoria Nardone 2017[22] (PIMA)	Neural Network (MLP)	–	74	74	74
	Multilayer Perceptron	–	74	74	74
	Hoeffding Tree	–	75	75	75
	JRip	–	76	76	76
	BayesNet	–	74	74	74
	Random Forest	–	76	75	76
Basha S.H., Abdalla A.S. 2018 [6] (PIMA)	NRCS	70.55	77.64	77.04	77.04
Deepti Sisodia, Dilip Singh Sisodia, 2018 [12] (PIMA)	Naive Bayes	76.30	75.9	76.30	77.8
	SVM	65.10	65.1	42.4	51.3
	Decision Tree	73.82	73.8	73.5	73.6
Dilip Kumar Choubey, Prabhat Kumar, [18] 2020 (PIMA)	Logistic regression	78.69	78.7	78.3	77.8
	PCA_Logistic regression	76.022	76	75.5	75
	PSO_Logistic regression	79.56	79.6	79.3	78.7
	KNN	73.47	73.5	72.5	72
	PCA_KNN	75.27	75.3	74.7	74
	PSO_KNN	73.91	73.9	73	72.5
	ID3 DT	75.65	75.7	76.5	72.9
	PCA_ID3 DT	74.34	43	75.8	71
	PSO_ID3 DT	75.65	75.7	76.5	72.9
	C4.5DT	76.52	76.5	76.1	76.2
	PCA_ C4.5 DT	78.25	78.3	81.9	78.7
S Kumari, D Kumar [21] 2021 (PIMA)	LR	74.89	61.25	64.47	61.25
	KNN	71.92	59.75	58.33	59.75
	SVM	74.02	48.75	67.24	56.52
	NB	74.12	65	61.90	63.41
	Soft Voting	79.08	70	73.13	71.56
Chang, V., Bailey, J 2022 (PIMA) [23]	J48 decision tree	74.78	88.43	70.86	78.68
	RF	79.57	81.33	89.40	85.17
	NB	78.67	86.75	81.88	84.24
Poposed Model	LR	**87.6**	**87.0**	**87.0**	**88.0**
	RF	**92**	**91.5**	**92.4**	**92**
	NB	**87.5**	**86.8**	**88.0**	**87.4**
	DT	**90.3**	**90.3**	**90.2**	**90.3**
	KNN	**92.4**	**88.3**	**97.5**	**92.7**
	SVM	**94.0**	**94.1**	**93.8**	**94.0**

6 Conclusion

This article shows the importance of the preprocessing, feature selection stages and the method of handling missing data. We used PIMA Indian dataset. We found that preprocessing is effective because it makes dramatically improvement in all performance metrics which are accuracy, recall, precision and F1-score and with different classification algorithms specially on PIMA dataset because it has a lot of problems need to be handled. Using Feature Selection improved performance metrics of most classifiers, however, it slightly gave a negative effect on the RF, ANN and LGMB. Adding SMOTE technique for imbalanced data treatment noticeably improved all performance metrics of all classifiers used in this study for both of the used datasets specially Recall and precision because the diabetic persons was the minor class.

References

1. www.who.int/health-topics/diabetes#tab=tab_1
2. Dogantekin, E., Dogantekin, A., Avci, D., Avci, L.: An intelligent diagnosis system for diabetes on linear discriminant analysis and adaptive network based fuzzy inference system: LDA-ANFIS. Digital Sig. Process. **20**(4), 1248–1255 (2010)
3. Singh, J., Bagga, S., Kaur, R.: Software-based prediction of liver disease with feature selection and classification techniques. Procedia Comput. Sci. **167**, 1970–1980 (2020)
4. Kandhasamy, J.P., Balamurali, S.: Performance analysis of classifier models to predict diabetes mellitus. Procedia Comput. Sci. **47**, 45–51 (2015)
5. Soltani, Z., Jafarian, A.: A new artificial neural networks approach for diagnosing diabetes disease type II. Int. J. Adv. Comput. Sci. Appl. **7**(6), 89–94 (2016)
6. Basha, S.H., Abdalla, A.S., Hassanien, A.E.: NRCS: neutrosophic rule-based classification system. In: Bi, Y., Kapoor, S., Bhatia, R. (eds.) IntelliSys 2016. LNNS, vol. 15, pp. 627–639. Springer, Cham (2018). https://doi.org/10.1007/978-3-319-56994-9_42
7. Swapna, G., Kp, S., Vinayakumar, R.: Automated detection of diabetes using CNN and CNN-LSTM network and heart rate signals. Procedia Comput. Sci. **132**, 1253–1262 (2018)
8. Pham, T., Tran, T., Phung, D., Venkatesh, S.: Predicting healthcare trajectories from medical records: a deep learning approach. J. Biomed. Inform. **69**, 218–229 (2017)
9. Alharbi, A., Alghahtani, M.: Using genetic algorithm and elm neural networks for feature extraction and classification of type 2-diabetes mellitus. Appl. Artif. Intell. **33**(4), 311–328 (2019)
10. Mamuda, M., Sathasivam, S.: Predicting the survival of diabetes using neural network. In: AIP Conference Proceedings. vol. 1870, p. 040046 (2017). AIP Publishing LLC
11. Anwar, F., Ejaz, M.Y., Mosavi, A., et al.: A comparative analysis on diagnosis of diabetes mellitus using different approaches-a survey. Inform. Med. Unlocked **21**, 100482 (2020)
12. Sisodia, D., Sisodia, D.S.: Prediction of diabetes using classification algorithms. Procedia Comput. Sci. **132**, 1578–1585 (2018)
13. Yuvaraj, N., SriPreethaa, K.: Diabetes prediction in healthcare systems using machine learning algorithms on Hadoop cluster. Cluster Comput. **22**(1), 1–9 (2019)
14. Zhou, H., Myrzashova, R., Zheng, R.: Diabetes prediction model based on an enhanced deep neural network. EURASIP J. Wirel. Commun. Netw. **2020**(1), 1–13 (2020)

15. Ashiquzzaman, A., et al.: Reduction of overfitting in diabetes prediction using deep learning neural network. In: Kim, K.J., Kim, H., Baek, N. (eds.) ICITS 2017. LNEE, vol. 449, pp. 35–43. Springer, Singapore (2018). https://doi.org/10.1007/978-981-10-6451-7_5
16. Alfian, G., et al.: Predicting breast cancer from risk factors using SVM and extra-trees-based feature selection method. Computers **11**(9), 136 (2022)
17. Nadesh, R.K., Arivuselvan, K., et al.: Type 2: diabetes mellitus prediction using deep neural networks classifier. Int. J. Cogn. Comput. Eng. **1**, 55–61 (2020)
18. Choubey, D.K., Kumar, P., Tripathi, S., Kumar, S.: Performance evaluation of classification methods with PCA and PSO for diabetes. Netw. Model. Anal. Health Inform. Bioinform. **9**(1), 1–30 (2020)
19. Chawla, N.V., Bowyer, K.W., Hall, L.O., Kegelmeyer, W.P.: Smote: synthetic minority over-sampling technique. J. Artif. Intell. Res. **16**, 321–357 (2002)
20. Kishor, A., Chakraborty, C.: Early and accurate prediction of diabetics based on FCBF feature selection and SMOTE. Int. J. Syst. Assur. Eng. Manag. (2021), 1–9 (2021). https://doi.org/10.1007/s13198-021-01174-z
21. Kumari, S., Kumar, D., Mittal, M.: An ensemble approach for classification and prediction of diabetes mellitus using soft voting classifier. Int. J. Cogn. Comput. Eng. **2**, 40–46 (2021)
22. **Mercaldo, F., Nardone, V., Santone, A.: Diabetes mellitus affected patients clas**sification and diagnosis through machine learning techniques. Procedia Comput. Sci. **112**, 2519–2528 (2017)
23. Chang, V., Bailey, J., Xu, Q.A., Sun, Z.: Pima Indians diabetes mellitus classification based on machine learning (ML) algorithms. Neural Comput. Appl. **35**, 1–17 (2022)

Federated Learning Assisted Covid-19 Detection Model

Sourav Banerjee[1](✉), Sudip Barik[1], Debashis Das[2], Uttam Ghosh[3], and Narayan C. Debnath[4]

[1] Computer Science and Engineering, Kalyani Government Engineering College, Kalyani, Nadia, India
{mr.sourav.banerjee,debashis.das,ghosh.uttam}@ieee.org, sudipbarik@outlook.com

[2] Computer Science and Engineering, University of Kalyani, Kalyani, Nadia, India

[3] Department of CS and DS, Meharry Medical College, Nashville, TN, USA

[4] School of Computing and Information Technology, Eastern International University, Binh Duong, Vietnam
narayan.debnath@eiu.edu.vn

Abstract. Federated learning (FL) ensures data privacy and security, which has been emerging as a promising approach for developing an accurate and efficient model for COVID-19 detection. This paper highlights the FL-assisted systematic review of the current state-of-the-art COVID-19 detection methods. This work conducts a comprehensive study of the relevant literature and identifies relevant articles. The included articles described various FL-based COVID-19 detection models that utilized different types of data and machine learning algorithms. Furthermore, these models have several advantages over traditional centralized models, including privacy protection, data security, and scalability. The findings of this systematic review suggest that FL-based COVID-19 detection models have great potential for accurate and efficient COVID-19 detection while protecting data privacy and security. FL requires further research to validate the findings on different metrics and enhance the performance of the models.

Keywords: COVID-19 · SARS-CoV2 · Chest X-ray · CT-scan · Deep Learning · Artificial intelligence · Machine Learning · FL

1 Introduction

As of December 2021, the COVID-19 pandemic has caused a worldwide public health crisis, with more than 260 million reported cases and over 5 million deaths confirmed globally. Detecting COVID-19 early and accurately is essential for successfully managing the pandemic. COVID-19 detection has been accomplished through the use of a range of techniques, including RT-PCR, CT scans,

S. Banerjee, S. Barik, D. Das, U. Ghosh and N. C. Debnath—These authors contributed equally to this work

A. Hassanien et al. (Eds.): AISI 2023, LNDECT 184, pp. 392–399, 2023.
https://doi.org/10.1007/978-3-031-43247-7_35

and chest X-rays. These methods are expensive, time-consuming, and necessitate specialized equipment and personnel training [1].

Recently, the application of machine learning (ML) techniques has shown significant promise in the detection of COVID-19. These methods are capable of analyzing vast volumes of data, recognizing patterns and trends, and delivering precise and prompt diagnoses [2]. The utilization of centralized ML models for detecting COVID-19 brings about numerous challenges, including but not limited to data privacy, security, and scalability. FL is an approach to ML that is decentralized in nature, allowing models to be trained on data that is distributed across multiple devices or institutions, without the need to transfer that data to a central location. FL offers several advantages over traditional centralized ML models, such as improved privacy, data security, and scalability [3].

By combining FL and ML techniques for COVID-19 detection, it is possible to overcome the limitations of conventional centralized models, enhance the accuracy and efficiency of COVID-19 detection, and guarantee data privacy and security. This systematic review aims to explore the current work in FL-based COVID-19 detection and provide insights into the potential of this approach for the effective management of the COVID-19 pandemic.

Combining FL and ML techniques improves COVID-19 detection, overcoming limitations and ensuring privacy, and security. This review explores FL-based detection's potential for pandemic management.

1.1 Legal and Ethical Issues

When using FL-based models for COVID-19 detection, important ethical and legal considerations arise [4]. Data privacy and security are critical concerns due to the sharing of model updates instead of raw data. Protecting patient data and preventing the disclosure of sensitive information is crucial. Obtaining informed consent from individuals whose data is used in FL-based models is challenging, particularly in the case of COVID-19 where data collection is often done for public health purposes without individual consent.

Ethical issues in FL bias in training data impact accuracy; data ownership, sharing, and liability pose legal concerns.

1.2 Background and Motivation

The COVID-19 pandemic has greatly impacted public health. Early detection and diagnosis of the disease are crucial due to its rapid spread and lack of effective treatment options. (ML), particularly (FL), has shown promise in disease detection while preserving data privacy. FL-based COVID-19 detection models using various data types have proven accurate and efficient while addressing the challenges of centralized models. This review aims to summarize the current state of FL-based COVID-19 detection, evaluating performance, advantages, and limitations, and suggesting future research directions. Valuable insights can guide the development of more accurate and secure approaches for COVID-19 detection using FL [5].

1.3 The Need for FL over ML

FL enables decentralized training of ML models across multiple devices or servers without requiring a centralized data storage system. This approach is particularly beneficial in situations where data privacy is a significant concern because it enables models to be trained on sensitive data without the need to expose that data to a central repository. Compared to conventional ML techniques, FL offers numerous benefits in the domain of COVID-19 detection. One of the key advantages is that it enables the development of models using data from multiple sources, including data from different hospitals, clinics, and even individual patients. This is important because COVID-19 is a global pandemic, and accurate detection models need to be developed using data from diverse populations [6]. Another advantage of FL in COVID-19 detection is that it allows for the development of models that are more robust to variations in the data. Because the data is distributed across multiple sources, models can be trained on data that is more representative of the overall population. This can lead to models that are more accurate and less prone to overfitting [7]. offers robustness by training models on diverse, representative data, enhancing accuracy and reducing overfitting [7]. FL improves data privacy and security by avoiding centralized storage of sensitive patient data, reducing the risk of breaches, and maintaining patient trust [8]. FL is essential for COVID-19 detection due to its advantages over traditional ML: robust models with diverse data sources, patient privacy protection, and improved data security. FL is crucial for evolving the pandemic's precise detection models [9].

2 Design Searching Strategy

FL has gained popularity for addressing data security and privacy concerns in ML. It is widely used for developing accurate COVID-19 detection models. This article reviews FL-based COVID-19 detection and its potential impact on healthcare [10].

This article reviewed some articles on FL-based COVID-19 detection models that used various data types (medical images, health records, clinical data) and ML algorithms (CNN, SVM, decision tree). The models showed high accuracy and AUC, suggesting their potential for accurate COVID-19 detection.

This review examined FL-based COVID-19 detection studies meeting specific criteria. It excluded traditional ML methods, studies without performance metrics, non-COVID-19 studies, and pre-2020 publications. Figure 1 shows strict selection criteria ensured a thorough evaluation of FL for COVID-19 detection.

This article used strict criteria to select relevant and high-quality studies for review, enabling a comprehensive evaluation of COVID-19 detection with FL.

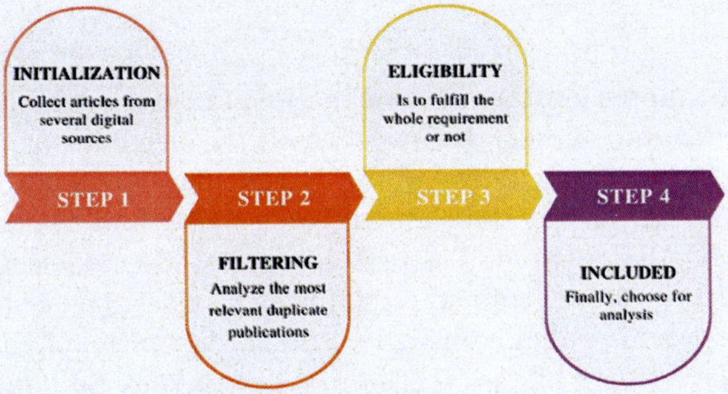

Fig. 1. A flowchart outlining the process for including or excluding articles.

2.1 Data Collection and Analysis of Some Available COVID-19 Datasets

Collecting and analyzing data are fundamental aspects of any research study, including the one on COVID-19 detection using FL. This paper involves collecting relevant COVID-19 datasets and analyzing the data to develop and assess models for COVID-19 detection based on FL.

Several datasets relevant to COVID-19 detection are publicly available, including the COVID-19 Open Research Dataset (CORD-19), COVID-19 Image Data Collection, and the COVID-19 CT segmentation dataset. These datasets contain clinical data, radiology images, and CT scans and can be accessed through repositories like Kaggle, GitHub, and the NIH database. Table 1 provides an overview of the datasets used in previous studies, including test data such as Chest X-rays and CT Images and their sources.

Data analysis involves pre-processing and cleaning data by removing missing or erroneous points. Pre-processed data is split into three sets for training, validation, and testing. FL models train on decentralized data, sharing model updates instead of raw data, ensuring privacy and security.

Table 1. Overview of the used datasets is presented in the previous studies.

Test Dataset	Modalities	Sources
covid-chest x-ray-dataset	X-ray	https://github.com/ieee8023/covid-chestxray-dataset
COVIDx	X-ray	https://github.com/RunwenHu/COVID-19
COVID19CXr	X-ray	https://twitter.com/ChestImaging
Covid19 AR	X-ray	https://nbia.cancerimagingarchive.net/nbia-search/
Muhammedtalo	X-ray	https://github.com/muhammedtalo/COVID-19/tree/master/X-Ray%20Image%20DataSet
Covid19	X-ray	https://www.kaggle.com/bachrr/covid-chest-xray
Pulmonary Abnormalities.	X-ray	http://www.kaggle.com/kmader/pulmonary-chest-xray-abnormalities
BSTI	X-ray and CT .	https://www.bsti.org.uk/covid-19-resources
Kaggle	X-ray and CT	http://medicalsegmentation.com/covid19/
UCSD-AI4H	CT	https://github.com/UCSD-AI4H/COVID-CT
COVID-19	CT	https://github.com/KevinHuRunWen/COVID-19
MedSeg	CT	http://medicalsegmentation.com/covid19
Biorxiv Dataset	CT	https://www.biorxiv.org/
Covid19 Segmentation Data	CT	http://medicalsegmentation.com/covid19/

The advantages and limitations of FL-based models for COVID-19 detection are evaluated by comparing them to conventional centralized models.

So, Data collection and analysis are essential elements of the paper that discusses COVID-19 detection using the FL approach. The analysis involves gathering relevant COVID-19 datasets and developing and evaluating FL-based models for COVID-19 detection. The performance of the models is assessed using standard metrics after being trained on data that is decentralized. The results of the analysis can provide valuable insights into the potential of FL-based models for COVID-19 detection while protecting patient privacy and data security.

3 Results and Discussion

FL-based COVID-19 detection is a promising approach that has gained significant attention in recent years. This systematic review aimed to analyze relevant literature published from 2020 to April 2023 to provide an overview of the current state-of-the-art in FL-based COVID-19 detection. The review applied specific inclusion and exclusion criteria to select relevant studies for analysis. The results of the review highlighted the potential of FL-based models for accurate and efficient COVID-19 detection while protecting data privacy and security. Different ML algorithms (CNN, SVM, decision trees) showcased the adaptability of FL-based models for COVID-19 detection, achieving high accuracy and AUC values despite data limitations and privacy concerns.

The review also highlighted the importance of protecting data privacy and security in healthcare. Traditional centralized models require the aggregation of data from multiple sources, which can pose a risk to the privacy and security of sensitive patient information. FL-based models enable data to remain decentralized, with model updates being shared instead of raw data. This approach ensures that sensitive patient data is kept confidential and secure while still allowing for the development of accurate and efficient COVID-19 detection models.

FL-based COVID-19 detection models overcome the limitations of centralized models by training on distributed data. Collaboration and standardization are needed for adoption. FL models offer accurate and efficient detection while safeguarding data privacy.

Table 2. Some related work based on FL

Ref.	Technology	Modalities	Key Contributions	Limitations
2020 [11]	federated learning	chest X-ray	In the work, we propose the use of FL for COVID-19 data training and deploy experiments to verify the effectiveness	The study did not account for the potential variability in imaging techniques, which may affect the accuracy of the DL models in identifying COVID-19
2021 [12]	5G-Enabled FL (B5G)	chest X-ray	Introducing FL into the 5G-driven. FL solves the problem of decentralising data and protects patients' privacy.	Further research may be required to validate the effectiveness and scalability of the proposed scheme in large-scale IIoT deployments
2021 [13]	blockchain-based FL	CT image	Authors propose using blockchain and FL techniques to develop a DL model for COVID detection using CT imaging, outperforming other models.	More data is needed to validate the model's accuracy and feasibility in real-world settings
2021 [14]	CNN-based federated DL	chest X-rays and CT scans	This study is important for COVID-19 as AI-assisted diagnosis and FL in medical imaging have been crucial for managing the pandemic	The study only focused on COVID-19 lung abnormalities, not other diseases. Findings may not apply to other lung conditions
2021 [15]	Clustered FL	X-ray and Ultrasound	Collaborative learning framework processing visual data at the edge device, using multi-modal ML algorithm for COVID-19 diagnosis	The main issue is varying client sample counts in collaborative learning
2021 [16]	Genetic CFL algorithm	Israel's Covid-19 Dataset	Clustered FL models optimize hyperparameters with genetic algorithms	The efficiency of FL is heavily influenced by client training
2022 [17]	Genetic Clustered FL (Genetic CFL)	X-ray	Introducing Genetic CFL, a superior hybrid algorithm for COVID-19 detection, surpassing traditional AI methods. FL utilizes IoT device data for real-time ML model training.	The algorithm needs significant resources, and may not work well on small datasets
2022 [18]	Federated Learning	CT	The approach was tested on a public COVID-19 CT dataset, outperforming traditional ML methods with high accuracy.	Limited dataset size may impact the generalizability and accuracy of the model due to insufficient clinical data
2022 [19]	Context fuse model, Attention mix module, Residual convolutional module, CTnet-10	CT	This article provides insights on using DL techniques for COVID-19 diagnosis using CT scan images, useful for medical image analysis researchers.	The article lacks comparisons to existing techniques, uses only one metric, and lacks dataset details

4 Limitations of the Current Study

The FL approach for COVID-19 detection is a promising solution for training ML models using data from multiple sources while preserving data privacy. However, there are still several limitations in the current studies on this topic can be described as follows:

- Secondary objectives: Current FL-based COVID-19 detection methods prioritize secondary objectives, limiting accuracy and stability enhancement [20].
- Insufficiency of data: Privacy concerns limit the availability of medical images, impacting the accuracy of FL-based COVID-19 detection [21].
- Potential bias: The model's estimates may be affected by biased observational studies, limiting the quality and effectiveness of the findings.
- Performance time: Federated ML models have slower performance than traditional ML models, limiting their practical use in urgent situations like the COVID-19 pandemic [22].

So, the early studies about "FL-based Covid-19 detection" has limitations in terms of secondary objectives, insufficiency of data, potential bias, and performance time. In Table 2 visualize some previous work and their limitations

5 Conclusion and Future Research Directions

FL has been utilized in the development of a DL-aided architecture for COVID-19 detection based on chest X-ray images. Accessing patients' private data raises concerns about violating patient privacy, and traditional machine-learning models typically necessitate such access. The Internet of Medical Things (IoMT) has the potential to enhance the accessibility of healthcare services, and it may also be leveraged in future research on FL-based COVID-19 detection. Future research in this field may involve exploring the effectiveness of Federated Learning (FL) in other aspects of COVID-19 diagnosis, such as identifying clinical symptoms or using data from different sources. FL-assisted models have shown high accuracy and AUC values, making it a promising method for COVID-19 detection based on chest X-ray images while ensuring data anonymity and security. Future work should focus on evaluating FL models using various metrics and optimizing the prediction of chest lesions caused by COVID-19 infection and pneumonia.

References

1. Wu, F., Zhao, S., Yu, B., Chen, Y., Wang, W., Song, Z., et al.: A novel coronavirus associated with human respiratory disease in China. Nature **579**(7798), 265–269 (2020)
2. Ahmed, S., et al.: Deploying machine and deep learning models for efficient data-augmented detection of COVID-19 infections. Viruses **12**(7), 769 (2020)
3. Kumar, R., et al.: Blockchain-federated-learning and deep learning models for Covid-19 detection using CT imaging. IEEE Sens. J. **21**(14), 16301–16314 (2021)
4. Naz, S., Phan, K.T., Chen, Y.-P.P.: A comprehensive review of federated learning for Covid-19 detection. Int. J. Intell. Syst. **37**(3), 2371–2392 (2022)
5. Faiman, J., et al.: Serious adverse events of special interest following MRNA Covid-19 vaccination in randomized trials in adults. Vaccine **40**(40), 5798–5805 (2022)
6. Li, T., Sahu, A.K., Talwalkar, A., Smith, V.: Federated learning: challenges, methods, and future directions. IEEE Signal Process. Mag. **37**(3), 50–60 (2020)
7. Xu, J., Glicksberg, B.S., Su, C., Walker, P., Bian, J., Wang, F.: Federated learning for healthcare informatics. J. Healthcare Inform. Res. **5**, 1–19 (2021)
8. Li, L., Fan, Y., Tse, M., Lin, K.-Y.: A review of applications in federated learning. Comput. Indust. Eng. **149**, 106854 (2020)
9. Li, X., Gu, Y., Dvornek, N., Staib, L.H., Ventola, P., Duncan, J.S.: Multi-site fMRI analysis using privacy-preserving federated learning and domain adaptation: abide results. Med. Image Anal. **65**, 101765 (2020)
10. Ghimire, B., Rawat, D.B.: Recent advances on federated learning for cybersecurity and cybersecurity for federated learning for internet of things. IEEE Internet Things J. **9**(11), 1–1 (2022)
11. Liu, B., Yan, B., Zhou, Y., Yang, Y., Zhang, Y.: Experiments of federated learning for covid-19 chest x-ray images. arXiv preprint arXiv:2007.05592 (2020)
12. Wang, R., Xu, J., Ma, Y., Talha, M., Al-Rakhami, M.S., Ghoneim, A.: Auxiliary diagnosis of Covid-19 based on 5G-enabled federated learning. IEEE Netw. **35**(3), 14–20 (2021)

13. Kumar, R., et al.: Blockchain-federated-learning and deep learning models for Covid-19 detection using CT imaging. IEEE Sens. J. **21**(14), 16301–16314 (2021)
14. Dou, Q., et al.: Federated deep learning for detecting Covid-19 lung abnormalities in CT: a privacy-preserving multinational validation study. NPJ Digital Med. **4**(1), 60 (2021)
15. Qayyum, A., Ahmad, K., Ahsan, M.A., Al-Fuqaha, A., Qadir, J.: Collaborative federated learning for healthcare: Multi-modal Covid-19 diagnosis at the edge. IEEE Open J. Comput. Society **3**, 172–184 (2022)
16. Kandati, D.R., Gadekallu, T.R.: Genetic clustered federated learning for covid-19 detection. Electronics **11**(17), 2714 (2022)
17. Kandati, D.R., Gadekallu, T.R.: Genetic clustered federated learning for Covid-19 detection. Electronics **11**(17), 2714 (2022)
18. Florescu, L.M., et al.: Federated learning approach with pre-trained deep learning models for Covid-19 detection from unsegmented CT images. Life **12**(7), 958 (2022)
19. Afif, M., Ayachi, R., Said, Y., Atri, M.: Deep learning-based technique for lesions segmentation in CT scan images for Covid-19 prediction. Multimed. Tools Appl. 1–15 (2023)
20. Li, Z., et al.: Integrated CNN and federated learning for covid-19 detection on chest x-ray images. IEEE/ACM Trans. Comput. Biol. Bioinform. (2022)
21. Naz, S., Phan, K.T., Chen, Y.-P.P.: A comprehensive review of federated learning for Covid-19 detection. Int. J. Intell. Syst. **37**(3), 2371–2392 (2022)
22. Salman, F.M., Abu-Naser, S.S., Alajrami, E., Abu-Nasser, B.S., Alashqar, B.A.: Covid-19 detection using artificial intelligence (2020)

Early Diagnosis of Parkinson's Disease Using Hand Drawings Images

Manar Sultan(✉), Ghada Hamed, Manal Tantawi, and Mohamed F. Tolba

Ain Shams University, Cairo, Egypt
ManarSutan@cis.asu.edu.eg

Abstract. This work presented a brief of the most recent technologies that tried to early diagnose Parkinson's Disease (PD). Early diagnosis of PD can help to detect its symptoms and evolution which increases the chances of treatment because there's no treatment for PD. So, we compiled several papers published in some well-established databases, all of which have almost the same diagnosis phases: preprocessing and classification and a few papers applied augmentation before classification. Each one has been mentioned to identify its objective, methodology, and results. Our brief showed that most papers used different approaches that use machine learning mechanisms to perform the automatic PD diagnosis. The main contribution of this paper is to propose an analyzed survey for the most recent proposed computer-assisted diagnosis systems mainly from 2018 to 2021, and how they can affect while handling the PD identification issue, and there are two experiments done using HandPD and NewHandPD datasets: the first experiment done using hybrid classification approach based on both SVM and PCA, while the second done using our modified CNN architecture, the modification done by changing of the number of filters: using 30 convolutional layers instead of 10, 50 pooling layers instead of 20 and 250 neurons at dense layer instead of 100. HandPD gave us the best accuracy using our modified CNN by 100.00% which is much better compared with "hybrid classification approach" mentioned above & "CNN model classification based".

Keywords: Review · Parkinson's diseases · Hand-Drawings · Handwriting · deep learning · SVM · PCA · CNN

1 Introduction

The pigmented neurons in the substantia nigra, which are located in the interior of the middle brain, become degenerated in Parkinson's disease (PD). The loss of these neurons results in a decrease in dopamine, the neurotransmitter responsible for controlling and regulating body movements, decreases when these neurons are destroyed. Millions of people worldwide suffer from this condition, which causes body tremors, slow movements, hypertonia, and balance issues. It is anticipated that this pathology will significantly rise in the coming years. Sadly, Parkinson's disease currently lacks a cure; however, early diagnosis can assist in better managing its symptoms and progression. Even though medicine has made a lot of progress, it's still hard to diagnose PD early.

A. Hassanien et al. (Eds.): AISI 2023, LNDECT 184, pp. 400–410, 2023.
https://doi.org/10.1007/978-3-031-43247-7_36

In fact, the most common warning signs don't show up until five to ten years after the disease started [1]. However, the likelihood of a successful treatment rises when this condition is detected early, allowing for faster support. Hand drawing is an involves the nervous system's motor movements. Using graphics tablets that enable the drawings acquisition in real time, numerous studies have examined hand-drawing disorders in Parkinson's patients. Previous research primarily used Latin languages [3, 4] used some straightforward examples like characters, words, and full sentences [2–6]. They also looked at drawings like loops, superimposed circles, and the Archimedean spiral [2–4, 7]. As a result, researchers concluded that the handwriting alteration varies in significant way even between PD patients taken on medication when symptoms discovered by treatments of dopaminergic [8]. Kinematics changes, such as increasing the movement duration or decreasing velocities while writing [5, 11, 12], or shape changes and an abnormal reduction in size of letter are typically the signs of this deterioration in handwriting abilities, which is known as PD dysgraphia [9].This could be used medically to obtain an automatic diagnostic aid system (DAS) for early Parkinson's disease detection and follow-up [13–15].

In this study we applied a traditional method SVM and PCA, and a modified model CNN based on different two papers which applied their study on drawings of each dataset separately and calculate accuracy foreach shape, that what we have done in addition to merge shapes of the same dataset and calculate accuracy.

Our experiment and results could be found on Sect. 5. Rest of the paper is organized as follow: In Sect. 2: Literature review about recent works, Sect. 3: Datasets used in various recent works, Sect. 4: Evaluation metrics.

2 Literature Review

This section reviewed the most recent proposed methods for PD diagnosis especially since 2018, it contains two sections: the first section lists the most traditional methods while the second one lists deep learning models. For each we discussed both main conducted phases represented in the data preprocessing and classification then the achieved results by each.

2.1 Traditional Algorithms

In this paper they used PD Spiral Drawings depending on Digitized Graphics Tablet then fed them to more than one classifier Logistic Regression, ensemble model Random Forest, K – nearest neighbor classification and SVC C-Support Vector Classification, best accuracy was 91.6% [15].

In this paper they used features extracted from PaHaW dataset, used Mann-Whitney U test to decrease the input data dimensionality and neglect the non-relevant features by removing them then used SVM by the idea of 4-way classification based on age and sex, the best accuracy was with sex female [16].

In this paper they used Signals of NewHandPD dataset then they pre-processed the sensor signals of all tests and subsample each sensor signals channel in uniform way by setting the length by 3000 and finally fed to random forest classifier with stratified 5-fold cross validation which gave accuracy = 89.4% [17].

In this paper they used PDMultiMC dataset then they have two-stages feature selection approaches, first one used statistical exams, the second one depends on a classifier fed them to SVM which gave 90.625% on 1 stage Feature selection and 96.875% on 2 stages Feature selection [18].

In this paper they used extracted features from 2 datasets (PaHaW and NewHandPD), then they tried 4 classifiers 1st: C4.5 an algorithm [33 in paper] based on the ID3 classifier used widely in data mining to obtain a decision using decision trees, 2nd: Random Forest classifier [34 in paper] used to apply a decision trees forest on various subsets of the training set., 3rd: SVMs and Cartesian genetic programming [43 in paper] is used to utilize computer software programs by utilizing the biological properties that occur in the populations natural evolution processes. It proved its effectiveness in various real-world applications. Accuracy with their respective standard deviation for each algorithm respectively using PaHaW dataset is 59.97 ± 15.83%, 79.98 ± 14.53%, 82.63 ± 11.81% and 71.18 ± 15.58%. Using NewHandPD dataset 76.07 ± 14.02%, 90.00 ± 11.32%, 89.31 ± 11.08% and 80.39 ± 15.29% [19].

In this paper they used Hand Spiral and Hand Meander datasets selected features that are extracted from the datasets through PCA and LDA and the obtained results for each dataset have been compared, classifiers used were K-nearest neighbors, Support vector machines, random forest and decision trees, results of hand meander and hand spiral respectively for 4 algorithms 85.26, 85.26, 81.05 and 87.36 and 84.73, 81.57, 77.89 and 80.52 [20] "hybrid SVM & PCA".

In this paper they used images from HandPD and NewHandPD datasets, they perform 4 steps in processing phase: Feature Extraction, F score based features ranking, Training Data LODSO CV and Validation Data LODSO CV then classification phase: applied Logistic Regression (LR), Gaussian Naive Bayes (GNB), SVM and Linear Discriminant Analysis (LDA) then experiment phase: splatted into 2 trials for 4 algorithms respectively 71.21, 89.39, 71.21, 69.69 and 63.04, 19.56, 70.65, 80.43 [21].

2.2 Deep Learning Models

In this paper they used Images from combination of PaHaw, HandPD, NewHandPD and Parkinson's Drawing datasets in addition to using augmentation to maximize number of images by applying flipping, threshold, illumination, rotation and contrast then images splatted to Train and validation set then more than one model had been tried like: GoogLeNet, AlexNet, VGG16, VGG19, ResNet50 and ResNet101, the best result AlexNet fine-tuned gives with 99.22% [14].

In this paper they used images from NewHandPD dataset and processed them by removing outliers through neglecting values less than 5% and greater than 90% on each channel then a normalization is performed using the z-score normalization. Finally, Bidirectional Gated Recurrent Units (BiGRUs) and the one-dimensional convolutions are used for classification which gave results in ranges from 80.00% to 88.57% [13].

In this paper they used Images from Unmentioned dataset name and apply augmentation methods like random horizontal flip, random change of image and random crop and zoom on the image, they used pre-trained Resnet32 and it needs the size of images to be 224×224×3 so nearest neighbor interpolation is used to resize all images, this paper has an extra class compared with others most papers classify patient and control but this classify PD patients from Essential Tremor (ET) patients and from controls, this work gives acc = 98.2 ± 1.35% With Hyper-parameter optimization and acc = 95.3 ± 1.66% Without Hyper-parameter optimization [22].

In this paper they used HandPD and ParkinsonHW images resized to 224 × 224 × 3 before fed to CNN and results as mentioned in table with HandPD dataset using CNN, Proposed model LSTM and CNN + LSTM respectively 99%, 97.8% and 99.3%, and with ParkinsonHW 84.8%, 83.5% and 85.5% [23] "CNN model classification based".

In this paper they used HandPD dataset, and they created two versions using two different image sizes: first using 64 × 64 pixels, and the other one using 128 × 128 pixels. The objective is to evaluate the image resolution influence through the experiments, they used 2 models: ImageNet gave 93.42 ± 3.17 and 95.74 ± 1.60 with image size 1 and 2 respectively while Cifar-10 90.28 ± 6.09 and 76.96 ± 21.6 [24].

In this paper they used images reconstructed from (PaHaW) dataset then they resized them to 150 × 150 pixels and they converted each grayscale image to the RGB format also they used three representations to enhance feature learning and these representations helps increase the complexity of the data Raw images: Median filter residual images and Edge images. Both provide the input to one of three CNNs. Then, the combined feature vectors obtained by the three CNN models. Then input into traditional machine learning models like: "SVM (lin), SVM (RBF), ADA and ET" and this have reported successful results the best accuracy based on various classifiers was 86.67% [25].

In this paper they used Images from PaHaw. Literature showed that fusion techniques can enhance the CNN-based system performance in significant way. So, this paper used two fusion techniques to enhance feature learning and classification: late fusion technique and early fusion technique then the model used pre-trained CNN (AlexNet) then SVM which gives 83% [26].

In this paper they used HandPD dataset, and they faced two problems: 1st: biasedness results from using imbalanced data so they utilized random undersampling to get balanced training data and get rid of the biasedness issues faced in the constructed models. The 2nd problem: classification accuracy low rate with limited clinical significance, so they proposed a learning system which is cascaded and merge Adaboost model with Chi2 model to improve the PD detection accuracy, it gave 72.46% with spiral shapes and 78.04% with meanders [27].

3 Datasets

This section shows the existing public datasets: name, participants number, genders, figures, how they collected the dataset, what did they used to collect data and so on.

Info.	Name							
	HandPD		NewHandPD		PDMultiMC		PaHaW	Parkinson's Drawing (kaggle)
Healthy participants	18		35		16		38	NA
	M	F	M	F	M	F		
	6	12	18	17	5	11		
Patient particpants	74		31		16		37	NA
	M	F	M	F	M	F		
	59	21	21	10	12	4		
#Total samples/dataset	736 images		264 images 792 signals		NA		NA	204 images

3.1 The HandPD Dataset

It collects hand drwings samples from two individuals' groups: 1st is Healthy Group and 2nd is Patient Group. They were obtained from "Botucatu Medical School, São Paulo State University – Brazil". The main objective is filling out a form of 4 meanders and 4 spirals, then cropped out from the form and stored in "jpg" format. The dataset contains 92 individuals.

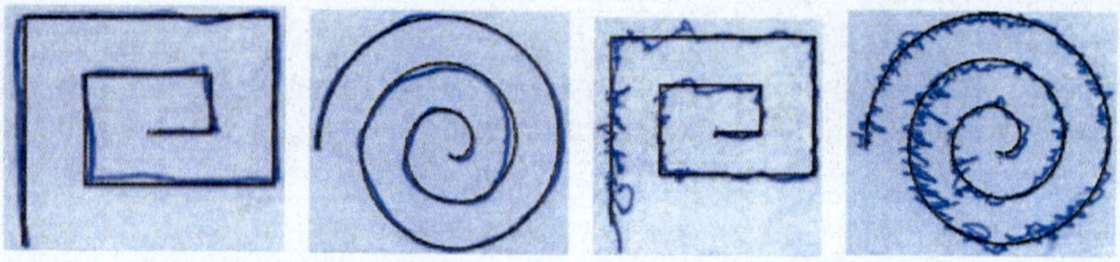

Fig. 1. Meander and spiral images drawn by HC VS PC

3.2 The NewHandPD

An enhanced version of HandPD that is collected from **66 individuals** and is divided into 2 sets: Patient and Healthy. Each person draws 12 samples (4 meanders, 4 spirals, left and right-handed diadochokinesis and 2 circled movements either in the air or on the paper. NewHandPD dataset is more balanced than HandPD.

PDMultiMC

The Parkinson handwriting dataset contains 32 subjects selected at "Saint George Hospital University of Balamand Medical Center (Beirut, Lebanon)".

PD patients are examined in one of two states. The first state is after one hour from taking the regular dopaminergic medication dose which is known as on-state. Another state refers to the patients examined without dopaminergic medication and this state is known as off-state. The handwriting process is performed through two tasks, first writing freely and then copying words and patterns which were written on the left of the paper in 3 different languages:

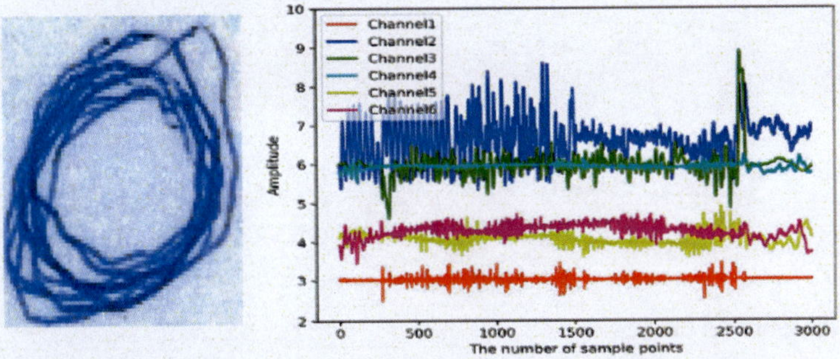

Fig. 2. Spiral image drawn by PC and and respective sampled sensor sig-

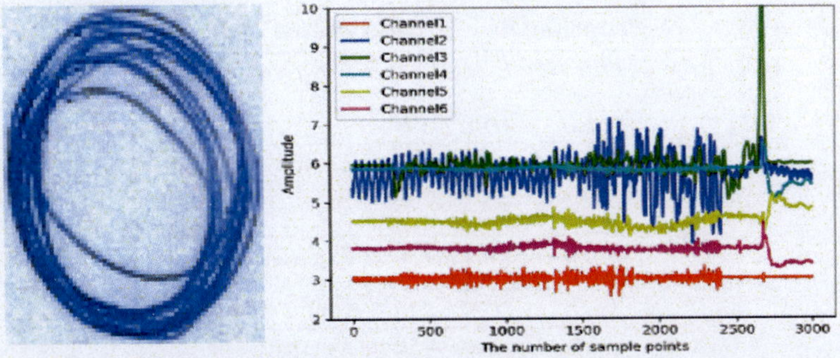

Fig. 3. Spiral image drawn by HC and respective sampled sensor signals.

1. letter l (Repetitive cursive).
2. triangular wave.
3. rectangular wave.
4. Monday word within the Monday–Tuesday sequence (Repetitive).
5. word 'Tuesday' (Repetitive).
6. **first name (Repetitive).**
7. last name (Repetitive).

→Constrains of 7 tasks:

- From 1:3 "copying the patterns from left to right until 10 cycles".
- Writing words in the subject's familiar language and writing sequence 5 times.
- Write 5 full names, each one on a separate line.

The "Parkinson's disease handwriting database" (PaHaW).

This dataset is composed of 37 patients with PD and of 38 years old. Participants completed 8 handwriting & drawings tasks in according to different templates like archimedes spiral drawing (as mentioned in Fig. 1, Fig. 2 & Fig. 3), or writing letters in 2nd, 5th and 8th tasks in a cursive way.

Parkinson's Drawing (Kaggle)

A dataset that contains drawings (Spiral as mentioned in Fig. 1, Fig. 2 & Fig. 3 and waves as mentioned in Fig. 4).

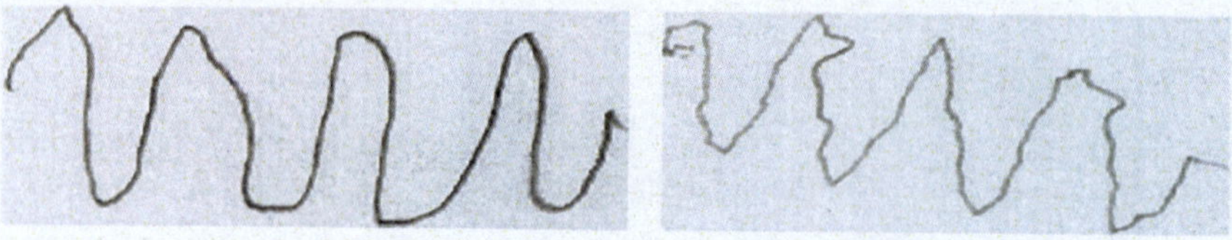

Fig. 4. Wave image drawn by HC VS PC

4 Evaluation Metrics

The different metrics used through the recent proposed methods to evaluate PD diagnosis are:

$$\text{acc} = \frac{True\ Negative + True\ Positive}{False\ Negative + True\ Negative + False\ Positive + True\ Positive} * 100\% \quad (1)$$

$$\text{pre} = \frac{True\ Positive}{False\ Positive + True\ Positive} * 100\% \quad (2)$$

$$\text{rec} = \frac{True\ Positive}{False\ Negative + True\ Positive} * 100\% \quad (3)$$

$$\text{SPE} = \frac{True\ Negative}{False\ Positive + True\ Negative} * 100\% \quad (4)$$

$$\text{SEN} = \frac{True\ Positive}{False\ Negative + True\ Positive} * 100\% \quad (5)$$

$$\text{F1score} = \frac{2 * True\ Positive}{False\ Negative + False\ Positive + 2 * True\ Positive} * 100\% \quad (6)$$

$$\text{AUC} = 0 : 1 \pm \int \text{ROC(t)dt} \quad (7)$$

$$\text{MCC} = \frac{\text{TN} * \text{TP} - \text{FN} * \text{FP}}{\sqrt{(\text{TN} + \text{FN})(\text{TP} + \text{FN})(\text{TN} + \text{FP})(\text{TP} + \text{FP})}}, \quad (8)$$
$$[\text{T : True, F : False, P : Positive, N : Negative}]$$

True and False Positives are the count of the PD subjects classified correctly and the count of healthy subjects discovered as PD, respectively. Also, True and False Negatives are the count healthy controls classified correctly and the PD patients classified incorrectly, respectively [16].

F1-Score is the F-measure that is in range of 0 and 1 such that 1 means that the predictions are perfect and 0 represents that the predictions are worst. The accuracy of the model of prediction measured by MCC which has a value between -1 and + 1, where + 1 means that the prediction is perfect and -1 is the count of disagreements between prediction and observation [27].

5 Methods and Experimental Results

This section discussed our conducted experiments and the obtained results through the next two sections. The first section discussed the used models and the next one showed the achieved results. The experiments are executed on Google co-laboratory using python language and TensorFlow framework.

5.1 Methods

We used one of the most traditional and common algorithms, which is SVM, and the deep learning CNN model following the paper's architecture in [23] shown in Fig. 5.

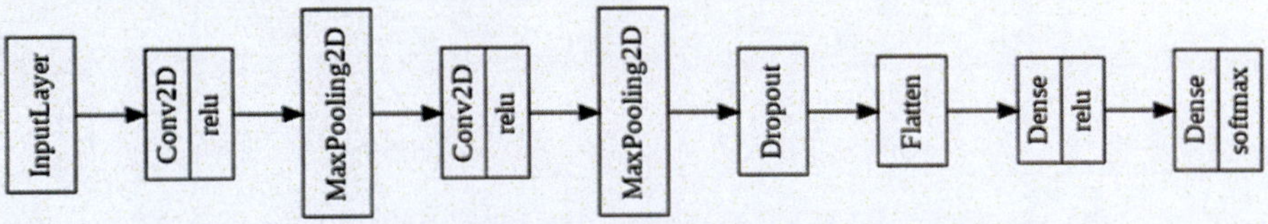

Fig. 5. CNN architecture based on paper [23]

5.2 The Achieved Results

Based on paper [20], they used datasets that contained (meander and spiral) shapes, we tried SVM & PCA algorithm on HandPD and NewHandPD. As mentioned above in Sect. 2.1, the results of hand meander and hand spiral respectively with SVM & PCA are 85.26 and 84.73. While our trial with HandPD gave 91.89% and 94.59% and with NewHandPD gave 76.25% and 82.5%. Those trials are based on the ".csv" file which contains the extracted features (Table 1).

Table 1. Results of paper [20] vs our experiment

#Paper	Dataset		Method	Accuracy	Our experiment	
					Separate shapes	Both shapes
[20]	HandPD	Meander	SVM + PCA	85.26%	91.89%	89.59%
		Spiral		84.73%	94.59%	
	NewHandPD	Meander		NA	76.25%	83.01%
		Spiral			82.5%	
	Both two datasets	Meander			85.26%	84.73%
		Spiral			86.84%	

The 2nd experiment is based on the CNN model architecture in paper [23] with some changes in the number of filters in convolutional, pooling, and dense layers. They used

HandPD and ParkinsonHW datasets while we tried CNN on the NewHandPD to obtain the results shown in Table 2.

Table 2. Experimental results of paper [23] vs our experiment

#Paper	Dataset		Model	Accuracy	Our experiment	
					Separate shapes	Both shapes
[23]	HandPD	Meander	CNN	99%	96.36%	100.00%
		Spiral			96.36%	
	NewHandPD	Meander		NA	17.72%	54.43%
		Spiral			56.96%	
	Both two datasets	Meander			65.61%	63.06%
		Spiral			65.61%	

6 Conclusion

This work discussed and analyzed the most recent proposed methods since 2018 for the PD diagnosis. For each introduced implemented method, we discussed the datasets used, the applied methods, and accuracies they reached. We split them into two groups: papers based on the traditional machine learning methods and others based on the deep learning models. Moreover, the different datasets used for the PD diagnosis are mentioned and compared versus each other with the different metrics used to evaluate the proposed method. Our experiments followed "hybrid SVM & PCA" and "CNN model classification based" using the HandPD and NewHandPD datasets. By following SVM & PCA on HandPD dataset: our experiment achieved accuracies of 91.89% with meander, 96.36% with spiral and 89.59% with both. Also, we used SVM & PCA on NewHandPD dataset: which obtained 76.25% and 82.5% with meander and spiral respectively and 83.01% for both. In the second experiment we followed "CNN model classification based" with some modification in the number of filters like setting the convolutional layer filters to 30 instead of 10, pooling layers filters to 50 instead of 20 and dense layer neurons to 250 instead of 100 which gave 96.36% for meander, spiral and 100.00% for both and this much better than their accuracies. Finally, we used CNN based model on the NewHandPD which gave 17.72% and 56.96% with meander and spiral respectively and 54.43% for both.

References

1. Man, J.H.K., Groenink, L., Caiazzo, M.: Cell reprogramming approaches in geneand cell-based therapies for Parkinson's disease. Corel **286**, 114–124 (2018)
2. Broeder, S., Nackaerts, E., Nieuwboer, A., Smits-Engelsman, B.C.M., Swinnen, S.P., Heremans, E.: The effects of dual tasking on handwriting in patients with parkinson's disease s. Neuroscience **263**, 193–202 (2014)
3. Drotar, P., Mekyska, J., Rektorova, I., Masarova, L., Smekal, Z., Faundez-Zanuy, M.: A New Modality for Quantitative Evaluation of Parkinson's Disease: In-Air Movement, 978–1–4799–3163–7/13/$31.00 ©2013 IEEE
4. Smekal, Z., Mekyska, J., Rektorova, I., Faundez Zanuy, M.: Analysis of Neurological Dis orders Based on Digital Processing of Speech and Handwritten Text, 978–1–4673–6143–9/13/$31.00 ©2013 IEEE
5. Rosenblum, S., Samuel, M., Zlotnik, S., Erikh, I., Schlesinger, I.: Handwriting as an Objective Tool for Parkinson's Disease Diagnosis. Springer-Verlag, Berlin Heidelberg (2013)
6. Drotar, P., Mekyska, J., Rektorova, I., Masarova, L., Smekal, Z., Faundez-Zanuy, M.: Analysis of In-Air Movement in Handwriting: A novel Marker For Parkinson's Disease, Elsevier Ireland Ltd, 2014 0169–2607/©
7. Akyol, K.: A study on the diagnosis of Parkinson's disease using digitized wacom graphics tablet dataset. Int. J. Inf. Technol. Comput. Sci. **12**(12), 45–51 (2017)
8. Tucha, O., et al.: Kinematic analysis of dopaminergic effects on skilled handwriting movements in Parkinson's disease. J. Neural Transm. **113**(5), 609–623 (2006)
9. Letanneux, A., Danna, J., Velay, J.-L., Viallet, F., Pinto, S.: From micrographia to Parkinson's disease dysgraphia. Mov. Disord. **29**(12), 1467–1475 (2014)
10. Lang, A.E., Lozano, A.M.: Parkinson's disease. N. Engl. J. Med. **339**(15), 1044–1053 (1998)
11. Ponsen, M.M., Daffertshofer, A., Wolters, E.C., Beek, P.J., Berendse, H.W.: Impairment of complex upper limb motor function in de novo Parkinson's disease. Parkinsonism Relat. Disord. **14**(3), 199–204 (2008)
12. Van Gemmert, A., Teulings, H.L., Contreras-Vidal, J.L., Stelmach, G.E.: Parkinsons disease and the control of size and speed in handwriting. Neuropsychologia **37**(6), 685–694 (1999)
13. Diaz, M., Moetesum, M., Siddiqi, I., Vessio, G.: Sequence-based dynamic handwriting analysis for Parkinson's disease detection with one-dimensional convolutions and BiGRUs. Expert Syst. Appl. **168**, 114405 (2021)
14. Kamran, I., Naz, S., Razzak, I., Imran, M.: Handwriting dynamics assessment using deep neural network for early identification of Parkinson's disease. Futur. Gener. Comput. Syst. **117**, 234–244 (2021)
15. Kamble, M., Shrivastava, P., Jain, M.: Digitized spiral drawing classification for Parkinson's disease diagnosis. Measurement: Sensors **16**, 100047 (2021)
16. Gupta, U., Bansal, H., Joshi, D.: An improved sex-specific and age-dependent classification model for Parkinson's diagnosis using handwriting measurement. Comput. Methods Programs Biomed. **189**, 105305 (2020)
17. Shoujiang, X., Pan, Z.: A novel ensemble of random forest for assisting diagnosis of Parkinson's disease on small handwritten dynamics dataset. Int. J. Med. Informatics **144**, 104283 (2020)
18. Taleb, C., Likforman-Sulem, L., Khachab, M., Mokbel, C.: Feature Selection for an Improved Parkinson's Disease Identification Based on Handwriting
19. Parziale, A., Senatore, R., Della Cioppa, A., Marcelli, A.: Cartesian genetic programming for diagnosis of Parkinson disease through handwriting analysis: Performance vs. interpretability issues. Artificial Intelligence in Medicine **111**, 101984 (2021)

20. Goel, N., Khanna, A., Gupta, D., Gupta, N.: Detection of Parkinson's disease using machine learning techniques for voice and handwriting features. International Conference on Innovative Computing and Communications, pp. 631–643
21. Ali, L., Zhu, C., Zhao, H., Zhang, Z., Liu, Y.: An integrated system for unbiased parkinson's disease detection from handwritten drawings. In: Zhang, J.F., Chen, C.M., Chu, S.C., Kountchev, R. (eds) Advances in Intelligent Systems and Computing. Smart Innovation, Systems and Technologies, **268** (2022). Springer, Singapore. https://doi.org/10.1007/978-981-16-8048-9_1
22. Seedat, N., Aharonson, V., Schlesinger, I.: Automated Machine Vision Enabled Detection of Movement Disorders from Hand Drawn Spirals (2020). arXiv:2006.12121v1cs.CV
23. Al-Wahishi, A., Belal, N., Ghanem, N.: Diagnosis of Parkinson's Disease by Deep Learning Techniques Using Handwriting Dataset. SIRS 2020: Advances in Signal Processing and Intelligent Recognition Systems, pp. 131–143
24. Pereira, C.R., et al.: Handwritten dynamics assessment through convolutional neural networks: an application to Parkinson's disease identification. Artificial Intelligence in Med. **87**, 67–77 (2018)
25. Diaz, M., Ferrer, M.A., Impedovo, D., Pirlo, G., Vessio, G.: Dynamically enhanced static handwriting representation for Parkinson's disease detection. Pattern Recognition Letters **128**, 204–210 (2019)
26. Moetesum, M., Siddiqi, I., Vincent, N., Cloppet, F.: Assessing visual attributes of handwriting for prediction of neurological disorders—a case study on Parkinson's disease. Pattern Recogn. Lett. **121**, 19–27 (2019)
27. Ali, L., Zhu, C., Golilarz, N.A., Javeed, A., Zhou, M., Liu, Y.: Reliable Parkinson's disease detection by analyzing handwritten drawings: construction of an unbiased cascaded learning system based on feature selection and adaptive boosting model. Digital Object Identifier https://doi.org/10.1109/ACCESS.2019.2932037

Security and Communications

A Multi-service Adaptive LTE Uplink Scheduler Based on Reservation for Machine -to- Machine Communication

Heba Nashaat[1](✉), Mohamed Abd Elgwad[1], Rawya Rizk[1], and Aboul Ella Hassanien[2]

[1] Department of Electrical Engineering, Port-Said University, Port-Said 42523, Egypt
{hebanashaat,mohamed_algwad,r.rizk}@eng.psu.edu.eg

[2] Faculty of Computers and Information, Cairo University, Giza, Egypt

Abstract. The importance of Machine-to-Machine (M2M) communications in the wide area communication networks set various challenges to cellular technologies such as the Long-Term Evolution (LTE) standard. The massive number of M2M devices generates small bursts of infrequent data packets with a wide range of delay requirements. Therefore, the main problems of M2M communications are resource allocation and satisfaction with the Quality of Service (QoS) of devices. In addition, it makes a massive race to existing Human-to-Human (H2H) devices for getting radio resources, affecting the performance of H2H communications. So, the scheduling procedure is essential to allocate physical resources for suitable users according to a specific algorithm. The Proposed Uplink Scheduler based on Reservation Adaption of Resources (UP_RAR) testing has been made for two scenarios and compared with different kinds of famous packet schedulers. In general, the (UP_RAR) algorithm maximizes resource utilization as it increases throughput for H2H for both scenarios. At the same time, the delay has been improved for the second scenario. Also, the performance is increased with an acceptable limit for the first scenario. Conversely, it had satisfied the QoS for M2M, and the traffic of M2M will not be affected.

Keywords: M2M · LTE · Uplink scheduling

1 Introduction

The Internet of Things (IoT) links smart objects to the Internet. It can enable an exchange of data, which was never available before, and securely bring users information. IoT can be characterized by interconnecting identifying objects [1]. The ecosystem for smart objects, which can connect to the Internet to cooperate and create new services for citizens and industrial applications, is called IoT, and the interconnection of smart objects, is called Machine-to-Machine (M2M) communication. M2M communication is an emerging technology that deals with IoT networking [1]. Typically, in IoT/M2M applications, M2M devices monitor events, which are relayed through a wireless communication network infrastructure to an application that translates the monitored event into meaningful information to make collective decisions with limited or no human intervention [2].

A. Hassanien et al. (Eds.): AISI 2023, LNDECT 184, pp. 413–431, 2023.
https://doi.org/10.1007/978-3-031-43247-7_37

Examples include smart transportation, environment monitoring, smart grid, smart healthcare, and fleet management M2M communication refers to automated applications involving devices that transmit and automatically collect data like temperature, humidity, speed, position, and heartbeat from a remote source without, or with only limited, human intervention through a public network infrastructure as Wi-Fi, WiMAX, UMTS, and LTE. Those applications generally use many low-cost devices, in the same area, with various resource constraints such as limited power and processing and storage resources [3].

The LTE standard is one of the most hopeful wireless network technologies for developing M2M communication because of its high data rates, low latency, high flexibility, and low cost. However, LTE was initially designed to support only Human-to-Human (H2H) communication, which includes services such as VoIP, media streaming, gaming, File Transfer Protocol (FTP), e-mail, and web. Compared with H2H communication, M2M communication has quite different characteristics. Amongst these characteristics:

1. Infrequent transmission at regular intervals of a small amount of data
2. Delay tolerance (milliseconds to tens of minutes)
3. The primary traffic is uplinked [4].

Because of differences between M2M and H2H communications, unexpected problems can occur when introducing M2M communication in the LTE network. For instance, when many M2M devices are connected in the same area, this may lead to a shortage of radio resources and reduced H2H traffic performance. Therefore, the main problems of M2M communications are resource allocation and satisfaction of Quality of Service (QoS). The addition of an enormous number of M2M devices into the cellular networks poses massive competition to existing H2H devices for getting radio resources, thereby affecting the performance of H2H communications. Presently, LTE is optimized for Human-to-Human (H2H) and Human-to-Machine (H2M) communications [1]. The characteristics of H2H/H2M communication differ from those of M2M in terms of traffic volume. Therefore, supporting M2M communication in LTE is a significant challenge to telecom network operators.

2 Uplink Packet Scheduler for LTE Network

The scheduling is performed in two channels (downlink and uplink) separately (one for the downlink and another one for uplink) in the Medium Access Control (MAC) layer of the base station. The scheduling of the two channels is performed periodically at TTI (Transmission Time Interval) of 1 ms, and the scheduler decides which Resource Blocks (RBs) (the smallest unit that the scheduler can allocate to a device) are allocated to which device. This decision aims to meet the requirements of device resources and get the best global system performance. An important characteristic distinguishing the downlink scheduler from the uplink scheduler is that the uplink has the constraint that all the RBs allocated to a single device should be continuous concerning frequency.

Scheduling is a procedure that allocates physical resources for users according to an algorithm in each TTI. The efficient allocation of radio resources between M2M and H2H applications is one of the biggest challenges. Because of global coverage and global connectivity, cellular networks can play a significant role in deploying M2M applications [1]. The frequency resources corresponding to the RB can be occupied for a period of TTI once one RB is allocated to a user [5]. When having data to send, a user first sends an uplink scheduling request to a base station and reports its cache, which can make the base station scheduler acquire necessary data and make an accurate decision about scheduling. Then the scheduler sends scheduling results to UE. When receiving the permissible scheduling message, UE sends data as required in the scheduling results, which specify the transmission format, the available resources, and the transmission time. Besides, LTE uplink scheduling should satisfy the constraint that all RBs assigned to a service request are contiguous in the frequency domain.

The radio access technology that LTE uplink uses is - SC-FDMA (Single Carrier - Frequency Division Multiple Access) – while OFDMA (Orthogonal FDMA) technology (used in LTE downlink), SC-FDMA the radio spectrum is divided into nearly perfect mutually orthogonal subcarriers. The transmissions compete for a share in the set of orthogonal (intra-cell interference-free) subcarriers. The total bandwidth allocated to a single UE depends on the available resources, the quality of the radio link, and the terminal's transmit tower. The main idea of packet scheduling in LTE networks is the possibility of scheduling users in two dimensions: time and frequency.

The bandwidth (BW) available for resource management is divided into subcarriers of 15 kHz. Twelve consecutive subcarriers clustered to form what it refers to as a resource block, with a bandwidth of 180 kHz. Each subcarrier consists of seven symbols. Resource elements are the intersection of one subcarrier with one symbol, meaning each resource block contains 84 Resource elements. It is considered the smallest discrete part of the frame and contains a single compound value representing data from a physical channel or signal.

In the time dimension, the access to the resource blocks is organized in time slots of 0.5 ms. Two slots of 0.5 ms form a TTI (subframe), the smallest scheduling unit in LTE. The frame in LTE consists of ten subframes. The scheduler can assign M resource blocks over the active flows in each TTI. The M resource blocks can vary from six resource blocks, which will give the 1.4 MHz, to 110 resource blocks, which will give the 20 MHz. The base station takes scheduling decisions, called eNodeB in LTE, for each TTI and potentially based on channel quality feedback provided by the (UEs). The overall architecture of the frame is shown in Fig. 1.

The packet scheduler decides which users are served and how many resource blocks are assigned to each selected user. The transmit power applied by any given UE is equally distributed over the assigned resource blocks. Therefore, a higher assigned number of resource blocks implies a lower transmit power per resource block. This affects the signal-to-interference-plus-noise ratio (SINR) experienced at the eNodeB. Therefore, the data rate a user can realize depends on the number of assigned resource blocks and SINR experienced per resource block, which determines the applied MCS (modulation and coding scheme). MCS is the term used to describe the combination of the radio carrier Modulation scheme and the Coding Scheme used when transmitting data where Each MCS has an associated data rate.

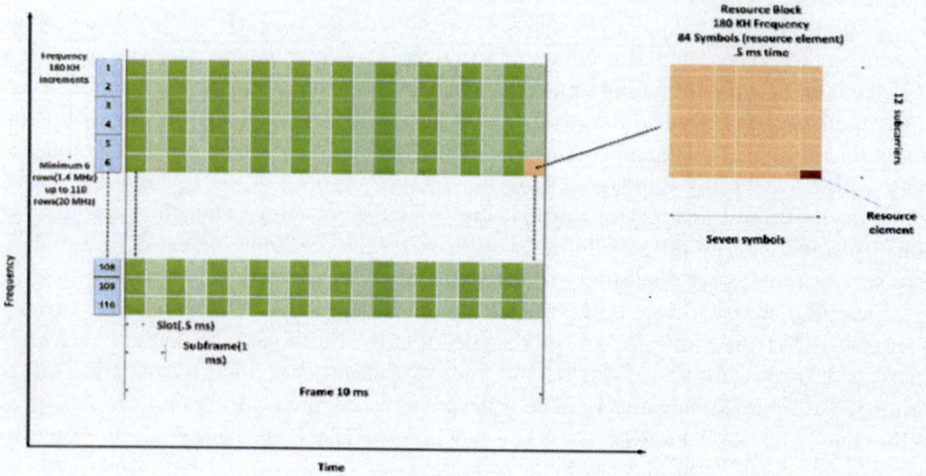

Fig. 1. Frame structure.

EPS Bearer is a pipeline through which data traffic flows within EPS (Evolved Packet-switched System). EPS Bearer is a virtual connection between UE and a PDN-GW(Packet Data Network Gateway), which provide transport service with specific QoS attributes. Sometimes multiple applications may run simultaneously in a UE, each with different QoS requirements. For example, a UE can be involved in a VoIP call while at the same time browsing a web page or downloading an FTP file. VoIP has more strict requirements for QoS regarding delay and jitter than web browsing and FTP, as it requires a much lower packet loss rate. Therefore, different bearers are structured within EPS to support multiple QoS requirements, each associated with a QoS. Generally, bearers can be classified into two categories based on the nature of the QoS they provide:

- Minimum Guaranteed Bit Rate (GBR) bearers that can be used for applications such as VoIP. These have an associated GBR value for which dedicated transmission resources are permanently allocated at bearer establishment modification. Bit rates more than the GBR may be allowed for a GBR bearer if resources are available. In such cases, a Maximum Bit Rate (MBR) parameter, which can also be associated with a GBR bearer, sets an upper limit on the bit rate expected from a GBR bearer.
- Non-GBR bearers do not guarantee any particular bit rate. These can be used for applications such as web browsing or FTP transfer. For these bearers, no bandwidth resources are allocated permanently to the bearer.

Allocation of RBs is made per UE and not per bearer or flow. The UE makes the task of assigning the scheduled RBs to bearers by an uplink control function. Schedulers are engaged in every TTI.

Channel-dependent scheduling is possible in LTE, where the Modulation and Coding Scheme (MCS) determines the bit rate in an allocated RB. There are several ways in which the eNodeB can deduce the channel condition. For example, CQI information transmitted by the UE can be directly mapped to particular MCS schemes. Alternatively, a UE can send an SRS, which can be used in estimating channel quality over all channels in a frame. Based on the inferred channel conditions, the eNodeB decides upon an MCS scheme for each UE. HARQ operation in the uplink is asynchronous, as opposed to the synchronous HARQ employed in the downlink. Accordingly, RB assignments occur at predefined times between the UE and eNodeB (retransmission must retransmit after exactly 8 TTIs). If a grant is not provided for the UE for retransmission, then a non-adaptive HARQ retransmission would be attempted without signaling. This requires no changes being made at the UE to attributes such as modulation order, code rate, and resource allocation in the lifetime of a HARQ process. Constraining the uplink scheduler in terms of the total number of grants to be assigned per UE is the physical RB availability, the quality of the radio link, and the transmit power budget.

RB assignments made to a single UE must be contiguous to maintain a low Peak-to-Average Power Ratio (PAPR) in SC-FDMA. The transmit power applied by any given UE is equally distributed over the assigned RBs. (so, a larger RB assignment for UE results in a lower transmit power per resource block.) Once the assignments have been decided, they are signaled by the eNodeB to the UEs on the Physical Downlink Control Channel (PDCCH) in the downlink frame. This naturally implies that the capacity of the PDCCH limits the number of UEs that can be scheduled in the uplink. The task of the uplink scheduler is to decide upon the number of RBs, their location, their MCS scheme, and the power with which each UE with traffic (be it data or control) will transmit. Once the above is decided upon, the eNodeB relays this information to the UE through a scheduling grant.

The UE decides upon which of its bearers gets served by how many RBs in the grant it is given. Once a UE receives its grant, it utilizes its uplink rate control function to assign the granted RBs among its radio bearers. RRC controls this function by giving each bearer a priority and a prioritized bit rate (PBR). The UE serves its radio bearer(s) as follows.

1) First, all the radio bearer(s) are served in decreasing priority up to their PBR, then
2) If any remaining resources are in the assigned grant, all the radio bearer(s) are served in decreasing priority until the grant is fully utilized.
3) Any remaining (unutilized) RBs are padded. If all PBRs are set to 0, the first step is bypassed, and the bearers are served in their strict priority order. Meanwhile, if more than one radio bearer has the same priority, the bearers will be served equally. Note that by limiting a UE's total grant, an eNodeB can ensure that the UE's AMBR (Aggregate Maximum Bit Rate), which is used for is used non-GBR flows, is not exceeded. The LTE uplink data transmission is shown in Fig. 2

Some literature introduces techniques used to improve scheduling that can be categorized into three main methods for LTE scheduling. The first is based on the grouping process; M2M devices are grouped into multiple clusters according to their QoS characteristics and requirements combinations. Then, each research chooses a combination of two or more QoS to achieve the best results.

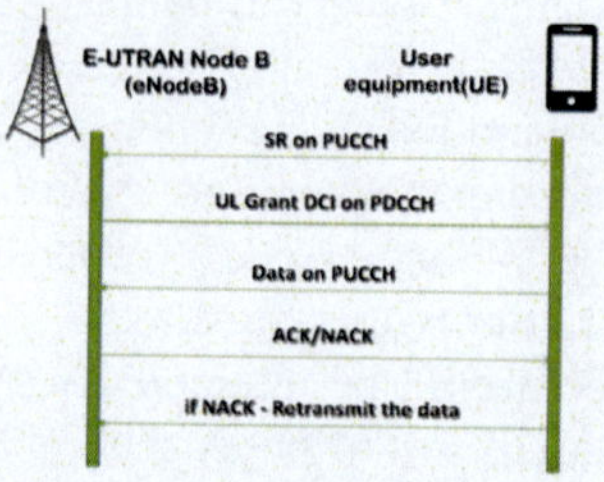

Fig. 2. LTE uplink data transmission.

In [2], the authors propose a scheduler whose devices are grouped into dynamically created classes according to their QoS requirements. The QoS requirements of the devices are placed into a class of two parameters: The arrival rate of packets and the maximum tolerable jitter. In [6], the call admission control scheme for the M2M communications in LTE is proposed. Moreover, the author supports the dynamic creation of the QoS classes according to the presence of the M2M devices, of which the existing classes cannot capture traffic characteristics. In [7], the author expressed the QoS requirement for an MTC device can be expressed as the interval of the data creation and the delay constraint. It manages the MTC devices group-based (a set of MTC devices with the exact QoS requirement grouped as a cluster). In [1], the proposed solution uses group-based scheduling to satisfy the QoS n requirements. This approach groups the devices into classes according to their deadline.

Furthermore, to control the impact of M2M on H2H, in the scheduling, it is given higher priority to the H2H devices than to the M2M devices in the same class. Although the priority of devices increases with the decrease of their deadlines, the problem of starvation may occur for the delay of the device tolerable in a congestion scenario. The second method considers the packet age information. In [8], the authors propose a scheduler to satisfy the QoS requirements. This scheduler prioritizes the devices with a critical deadline and more data transmission. Instead of assuming fixed capacity RBs, the author shows how scheduling devices for a variable number of RBs based on the size of their queues can increase efficiency and propose a new metric for the efficiency measurement of an LTE packet. In [9], a new delay-sensitive scheduler is proposed. It increases the probability of satisfying the delay budget of packets by exploiting Packet Age data explicitly reported by devices. A new MAC Control Element represents the age of the oldest pending packet in the device buffers sent by devices in parallel with the existing Buffer Status Report that contains volume and priority data. LTE protocol enhanced to allow a device to inform the eNodeB of the age of the oldest packet residing in its transmission buffers.

The last method divides resources between M2M and H2H communications. In [10], the scheduler controls the impact of M2M communication on H2H by limiting a percentage maximum of resources allocated for M2M devices. After separating resources, the scheduler prioritizes the devices with less delay tolerance, which results in no guarantee of fair allocation to devices more delay tolerable. In [11], an extension in paper[5], the author tries to satisfy the QoS requirements to ensure fairness in resource allocation and control the congestion caused by M2M devices. The author argues that the proposed approach can control the impact of M2M communication on H2H, ensure fairness and thus avoid the problem of starvation and meet the QoS requirements of applications.

In contrast to other scheduling algorithms, no classes of devices need to be formed. Two LTE uplink scheduling was introduced as in [12]. The first scheduler prioritizes devices with better channel conditions, and the second prioritizes devices less delay tolerant. Both schedulers proposed do not ensure fair allocation of resources and do not address the H2H traffic.

3 Proposed LTE Uplink Scheduler based on Reservation Adaption of Resources for M2M

An intelligent scheduler is proposed, dividing resources between M2M and H2H traffic flows. Feedback is always needed for current traffic information for each device. The combination of IoT and H2H applications will generate the problems of network resource starvation, so the need to satisfy the QoS requirements and control the effect of H2H traffic performance is needed. Because Event-Driven devices are delay-sensitive, there is a problem with the satisfaction of QoS requirements of these even if they have high priorities in scheduling. The level of satisfying QoS requirements for Event-Driven devices is the key to changing the maximum amount of M2M devices selected for each TTI. The standard QoS classes changed across time as in earlier times, the standard QoS classes were only nine classes with QCI from (1 to 9) and proposed in [13], but the number of standard QoS classes has been raised to cover all types of traffic, and that was proposed in [14].

LTE standard QoS classes are extended in UP_RAR to include the M2M traffic. In UP_RAR, One class is used for delay-sensitive traffic flows of Event-Driven applications with a high priority. The other classes are used for the traffic flow of Time-Driven applications. Table 1 describes the new QoS extensions for M2M traffic flows and 3GPP-standardized QCIs for LTE traffic flows. H2H can be classified into one of the standard QoS classes, defined according to their requirements.

The proposed UP_RAR contains four steps, as in Fig. 3. First, the suggested RBs calculated for H2H by finding the suggested RBs for H2H from the device (u) at time TTI (t):

$$\widehat{SR}_H(u, t) = \frac{BS(u, t) \times RB^{avg}}{BS^{avg}(u, t-1)} \tag{1}$$

where BS is the current size of the buffered data, RB^{avg} is the average number of RBs allocated to the device, and BS^{avg} is the average size of buffered data. The functions RB^{avg} and BS^{avg} are calculated using an Exponential Moving Average (EMA) with its constant based on the number of TTIs.

The maximum number of M2M devices selected can be calculated as follows:

$$U_M^{max}(t) = \max\left(\left\lfloor \frac{\Omega|RB|}{RB_M^{min}} \right\rfloor, \left\lfloor \frac{|RB| - \widehat{SR}_H(t)}{RB_M^{min}} \right\rfloor \right) \tag{2}$$

where RB_M^{min} is the minimum number of resources for each M2M device, and the percentage of resources ($0 \leq \Omega \leq 1$) ensures that M2M traffic flows not causing starvation for H2H. Simultaneously, this percentage was updated to support multiple M2M QoS.

Table 1. QCIS for LTE and new QoS extensions for M2M traffic flows

QCI	Type	Priority level	Delay budget	Error loss Rate	Example Services
1	H2H GBR	2	100 ms	10^{-2}	Conversational voice
2		4	150 ms,	10^{-3}	Conversational video (live streaming)
3		3	50 ms	10^{-3}	Real-time gaming, v2x messages Electricity distribution - medium voltage Process automation - monitoring
44		5	300 ms	10^{-6}	Non-conversational video (buffered streaming)
65		0.7	75 ms	10^{-2}	Mission critical user plane push to talk voice
66		2	100 ms	10^{-2}	Non-mission-critical user plane push to talk voice
67		1.5	100 ms	10^{-3}	Mission critical video user plane
75		2.5	50 ms	10^{-2}	V2x messages
82		1.9	10 ms	10^{-4}	Discrete automation (small packets)
83		2.2	10 ms	10^{-4}	Discrete automation (big packets)
84		2.4	30 ms	10^{-5}	Intelligent transport systems
85		2.1	5 ms	10^{-5}	Electricity distribution- high voltage
5	H2H Non-GBR	1	100 ms	10^{-6}	signaling
6		6	300 ms	10^{-6}	Video (buffered streaming) tcp-based (e.g., www, e-mail, chat, ftp, p2p file sharing, progressive video, etc.)
7		7	100 ms	10^{-3}	Voice, video (live streaming) interactive gaming
8		8	300 ms	10^{-6}	Video (buffered streaming) tcp-based (e.g., www, e-mail, chat, ftp, p2p file

(*continued*)

Table 1. (*continued*)

QCI	Type	Priority level	Delay budget	Error loss Rate	Example Services
9		9			Sharing, progressive video, etc.)
69		0.5	60 ms	10^{-6}	Mission critical delay-sensitive signaling (e.g., mc-ptt signaling, mc video signaling)
70		5.5	200 ms	10^{-6}	Mission critical data (e.g. Example services are the same as qci 6/8/9)
79		6.5	50 ms	10^{-2}	V2x messages
80		6.8	10 ms	10^{-6}	Low latency embb applications (tcp/udp-based); Augmented reality
10	M2M GBR	.5	50 ms	10^{-6}	Event-driven Emergency alerting
11	M2M Non-GBR	10	50 ms	10^{-2}	Time-driven Mobile streaming
12		11	100 ms	10^{-2}	Time-driven Mobile pos (instant pay in a vending machine and instant pay in intelligent toll)
13		12	150 ms	10^{-2}	Time-driven Regular monitoring(electricity transmission and switch monitoring in smart grid, monitoring in intelligent cargo, water quality monitoring in environment protect, street light control in public safety)
14		13	300 ms	10^{-2}	
15		14	400 ms	10^{-2}	
16		15	500 ms	10^{-2}	Time-driven Smart metering like (grid metering, smart water metering, smart gas metering; accounts of intelligent bus card uploading; accounts of mobile pay uploading)
17		16	1 s	10^{-2}	

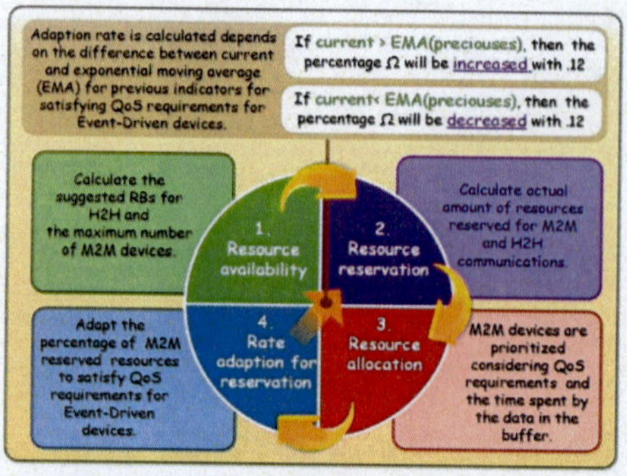

Fig. 3. The proposed LTE Uplink Scheduler Operation.

The second step is dividing the overall RBs into two consecutive groups regarding frequency. One for each type of communication (H2H and M2M) and with a size equal to its demand. The actual amount of resources reserved for M2M communication (AR_M) is formulated as:

$$AR_M(t) = RB_M^{min} \times \min\left(|u_m|, U_M^{max}(t)\right) \tag{3}$$

where u_m is the set of M2M devices requesting resources. And the actual amount of resources reserved for H2H communication (AR_H) is calculated as:

$$AR_H(t) = |RB| - AR_M(t) \tag{4}$$

In the third step, the resources are allocated to suitable users. M2M devices prioritized according to the utility function M_M^{TD} as in Eq. (5).

$$M_M^{TD}(u,t) = (1-\omega)\left(1 - \frac{T^{AVG}(u,t-1)}{\max_{n \epsilon u_M} T^{AVG}(n,t-1)}\right) + \omega\left(1 - \frac{\Delta D(u)}{\max_{n \epsilon u_M} \Delta D(n)}\right) \tag{5}$$

where $\Delta D(u)$ measures the M2M device that is not close to satisfying its QoS requirements and is equal to the difference between the maximum delay and the time spent by the data in the buffer.

Constant $\omega(0 \leq \omega \leq 1)$ indicates the weight to ensure fairness (first component of Eq. (5)) and to satisfy the QoS requirements (the second component of Eq. (5)) in prioritizing the devices. And $T^{AVG}(u, t-1)$ is an EMA of the throughput of device u until TTI t.

UP_RAR tries to balance between satisfying H2H traffic flow requirements and event-based M2M deadline. To maximize the utilization of the available network bandwidth utilization while increasing individual flows' satisfaction in proportion to their requested QoS requirements. Therefore, a new indicator is calculated $A(k, t)$ as a function to estimate the state of satisfying QoS requirements for event-driven devices. The function $A(k, t)$ can be formulated as:

$$A(k,t) = \left(\frac{1}{k}\right)\left(\begin{array}{c} \theta \sum_{i=0}^{i=k} \frac{BS(k,t) - \min_{k \in u_{me}} BS(k,t)}{\max_{k \in u_{me}} BS(k,t) - \min_{k \in u_{me}} BS(k,t)} + \\ (1-\theta) \sum_{i=0}^{i=k} \frac{\Delta D(k,t) - \min_{k \in u_{me}} \Delta D(k,t)}{\max_{k \in u_{me}} \Delta D(k,t) - \min_{k \in u_{me}} \Delta D(k,t)} \end{array}\right) \tag{6}$$

where K is the number of event-based M2M that has not been allocated yet and constant $\theta(0 \leq \theta \leq 1)$ indicates the weight of the total buffer size (first component of the Eq. (6)) and to satisfy the QoS requirements (the second component of the Eq. (6)).

In the final step, depending on the exponential moving average for previous indicators and current indicator, the adaption rate is calculated, and the percentage of M2M traffic flows allocated resources is updated according to the following equation

$$\Omega_{new} = \begin{cases} \Omega_{old} + .12 if A(\mathrm{k, t}) > A(\mathrm{k, t} - 1) \\ \Omega_{old} - .12 if A(\mathrm{k, t}) < A(\mathrm{k, t} - 1) \end{cases} \quad (7)$$

Subject to $.36 \leq \Omega_{new} \leq .6$
where the constrained to Ω_{new} as it must not exceed (0.6) or be less than (0.12).

4 Evaluation and Experimental Results

4.1 Simulation Environments

In the following, the simulation model will be described. The simulation utilizes the network simulator package NS-3 simulator. In this paper, several modifications in the source code of the NS-3 simulator have been made to simulate UP_RAR and other uplink schedulers.

First Scenario Simulation Configuration
This simulation scenario is assumed as in [15]. A single base station is considered where the devices are fixed and uniformed distributed. The scenario consisted of 30 H2H devices: ten devices for VoIP, ten for video, and ten for Constant Bit Rate (CBR) FTP. For M2M devices: 1/3 of devices for Event-Driven, and 2/3 of devices for Time-Driven. Moreover, the bandwidth will be set to 5 MHz (the default bandwidth) or 25 RBs. The modeling of the H2H traffic flows was based on the works of [15] (VoIP and CBR FTP) and [16] (Video). These three traffic flows were chosen to cover the various QoS classes in the LTE standard. A Poisson process modeled the burst transmissions of Event-Driven devices with rate $\gamma = 0.02$ packets/TTI [17]. The transmission interval of each Time-Driven device was uniformly distributed between 50 and 550 ms. The two categories of M2M applications have a packet size of 125 bytes. Table 2 summarizes the main parameters used in the simulations [18].

The number of resources required per M2M device ($(RB_M^{min} = 3)$) chosen to be able to transmit all buffered data in a single TTI in a channel with good quality. The waiting time after the request is denied established as a maximum of 10% ($\sigma = 0.1$) of the value of ΔD. The value of the parameter ω of Eq. (4) chosen is equal to 0.95 because it performs better in satisfying the QoS requirements of M2M devices.

Table 2. Parameters of the simulation for the first scenario.

General parameters			
Simulation time		3 s (3000 TTIs)	
Runs		30	
Distribution of devices		Fixed and uniform within 1400	
Number of base stations		1	
Bandwidth		5 MHz (25 RBs)	
Number of M2M devices		1/3 for Event-Driven, 2/3 for Time-Driven	
Number of H2H devices		10 VoIP, 10 Video, 10 CBR FTP	
Traffic type		Packet Size	Inter-arrival Time
H2H	Video	1200 bytes	75 ms
	VoIP	40 bytes	20 ms
	FTP	256 bytes	16.625 ms
M2M	Event-Driven	125 bytes	Poisson Process, = 50 ms
	Time-Driven	125 bytes	Uniformly distributed [50, 550] ms
General parameters of the scheduler			
$r_M^{min} = 3$		$\omega = 0.95$	$\sigma = 0.1$
Constant of EMA used in Equation			99 TTIs
Parameters of the dynamic scheduler			
$\theta = .2$			
$\Omega = 0.36, 0.48, 0.6$ and is changed dynamically			
The constant of EMA for congestion level			2 TTIs

Second Scenario Simulation Configuration

This simulation scenario is assumed as in [19]. A single base station also will be used here but with a bandwidth changed to 10 MHz or 50 RBs instead of 25 RBs. The scenario consisted of 30 H2H devices: ten for VoIP, ten for Video, and ten for Constant Bit Rate (CBR) FTP. For M2M devices: 60 devices will use video applications. The video application used here is based on EvalVid, a toolset for evaluating video quality over a real one [20, 21]. The well-known video sequence Akiyo will use. The akiyo video sequence duration is 12 s. The total simulation time was set to 30 s. Each scenario of the evaluation was repeated 50 times. The 60 devices will be distributed to the new eight classes of M2M, according to Table 3. The simulation will be for 30 s. Table 3 contains the main parameters of the second scenario.

Table 3. Parameters of the simulation for the second scenario

Antenna model	Isotropic antenna model
eNB TX Power	46 dBm
UE TX Power	23 dBm
Bandwidth	50 RBs (10 MHz)
AMC scheme	PiroEW2010
RLC mode	UM (buffer size: 10 MB)
Path loss model	COST 231
Fading loss model	Pedestrian EPA 3 km/h
User mobility model	Steady-state random waypoint Min speed: 0.8 m/s / Max speed: 0.83 m/s Min pause: 0s/ Max pause: 0.1 s Rectangle: 424m x 424m
Simulation time 30 s	Simulation time 30 s
Simulation runs 50	Simulation runs 50

4.2 Simulation Results

In the first scenario, the results for three schedulers: The First is Proportional Fair (PF), which is the scheduler guaranteed fair allocation of resources between the devices [3]. It also does not differentiate the types of communication (H2H and M2M), and it does not consider QoS and buffer size. Second, the Dynamic scheduler was developed in [5], where resources are divided between H2H and M2M. Still, the dynamic scheduler is according to the number of Event-Driven that has not been allocated yet where there are three states for congestion, and the number of M2M resources changed according to them. Finally, UP_RAR is the scheduler that has been proposed.

The three metrics used to evaluate the performance of the schedulers were the average throughput, the average of all end-to-end delays for all received packets of the flow, and the jitter parameter.

For H2H, the throughput is increased significantly with the proposed UP_RAR algorithm compared to a dynamic scheduler, and both are higher than the PF scheduler, as presented in Fig. 4. However, due to the increase in the throughput, the end-to-end delay is also increased for both dynamic and UP_RAR compared to the PF scheduler. In delay, the jitter for dynamic and proposed UP_RAR is very close for a few users, but they are more than the PF scheduler, as in Fig. 5 and Fig. 6.

M2M throughput is the same for all schedulers as in Fig. 7. The delay for the proposed UP_RAR scheduler is increased more than the dynamic and PF scheduler, as in Fig. 8. As opposed to delaying, the jitter for dynamic is the least and proposed UP_RAR more little as dynamic, but both are less than PF in Fig. 9.

The second scenario compares PF, Proposed UP_RAR scheduler, Strategic Genetic Algorithm (StGA), and Round Robin (RR). As in the first scenario, the three metrics (throughput, delay, and jitter) are presented considering the variation of M2M users.

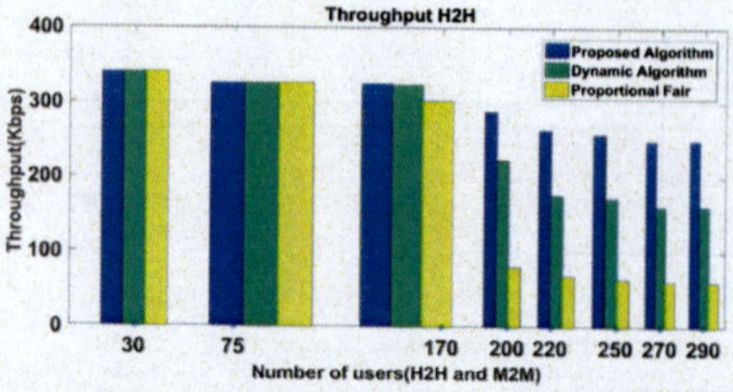

Fig. 4. The throughput for H2H devices versus the total number of UEs (Scenario 1).

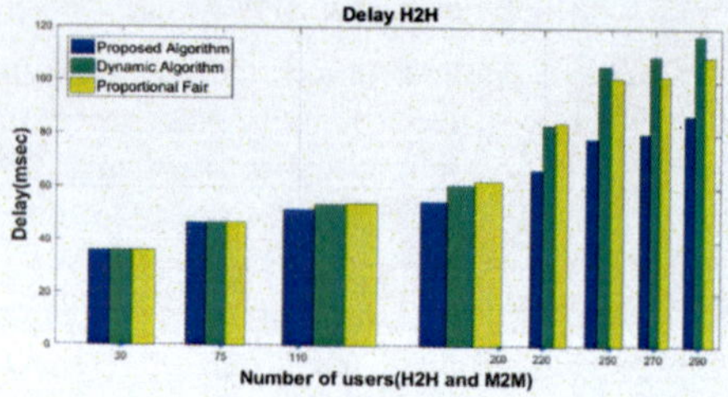

Fig. 5. The average of all end-to-end delays for received packets for H2H devices versus the total number of UEs (Scenario 1).

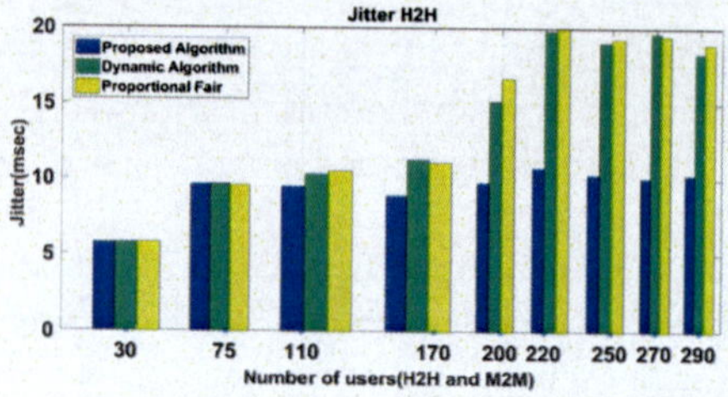

Fig. 6. The jitter received for H2H devices versus the number of users (Scenario 1).

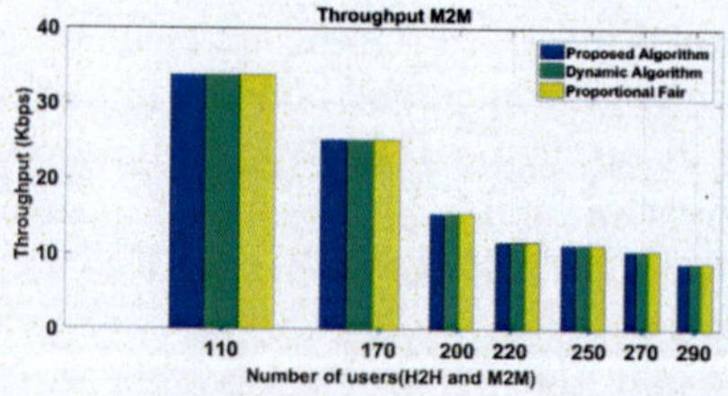

Fig. 7. The throughput for M2M devices versus the total number of UEs (Scenario 1).

In Figs. 10 and 11, The throughput and delay for four schedulers are simulated for H2H devices. The proposed UP_RAR scheduler is significantly better than all other schedulers as throughput increases, and delay is less than all other schedulers. However, Jitter is also increased, as in Fig. 12.

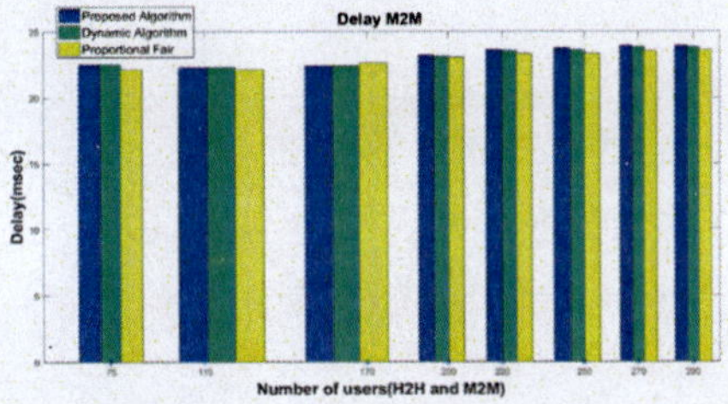

Fig. 8. The average of all end-to-end delays for received packets for M2M devices versus the total number of UEs (Scenario 1).

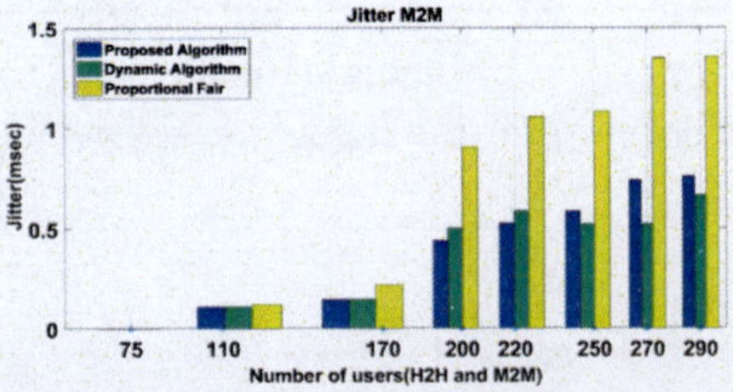

Fig. 9. The jitter received for M2M devices versus the total number of users (Scenario 1).

For M2M devices, the proposed UP_RAR scheduler's throughput and delay are better than all other schedulers as throughput will be the highest in Fig. 13, and delay is the least in Fig. 14. In addition, the jitter for the proposed UP_RAR scheduler is better than StGA and RR schedulers, but the PF scheduler is still the best, as in Fig. 15.

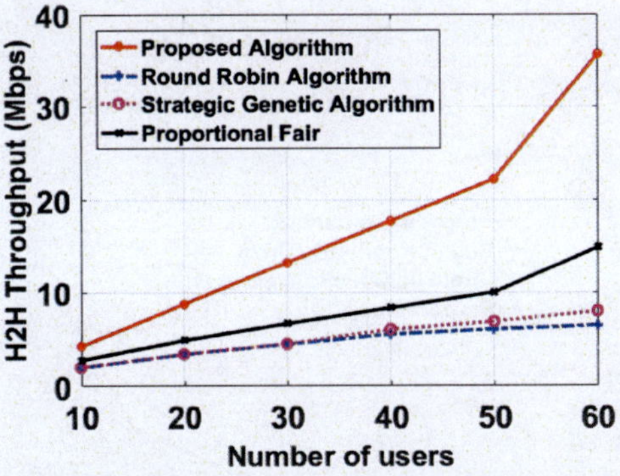

Fig. 10. The throughput for H2H devices versus the total number of UEs (Scenario 2).

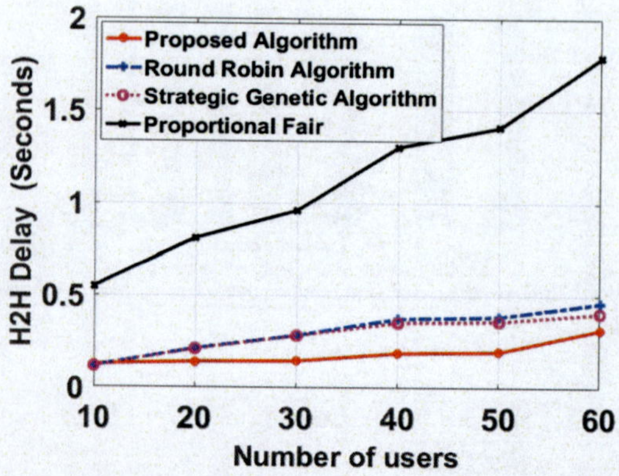

Fig. 11. The average of all end-to-end delays for received packets for H2H devices versus the total number of UEs (Scenario 2).

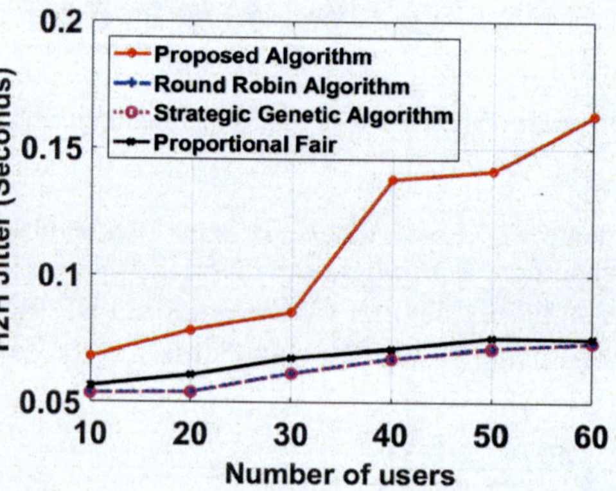

Fig. 12. The jitter received for H2H devices versus the number of users (Scenario 2).

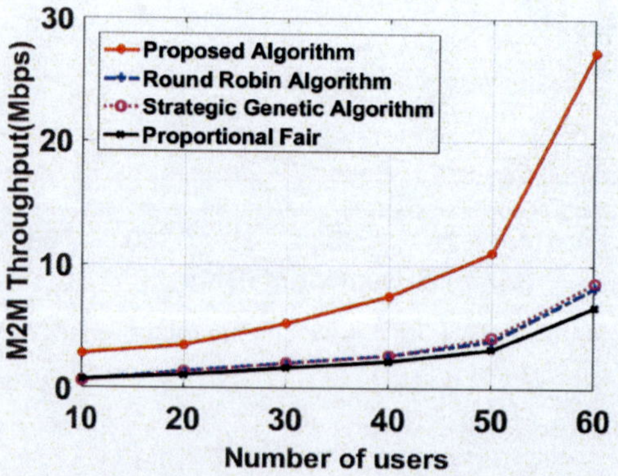

Fig. 13. The throughput for M2M devices versus the total number of UEs (Scenario 2).

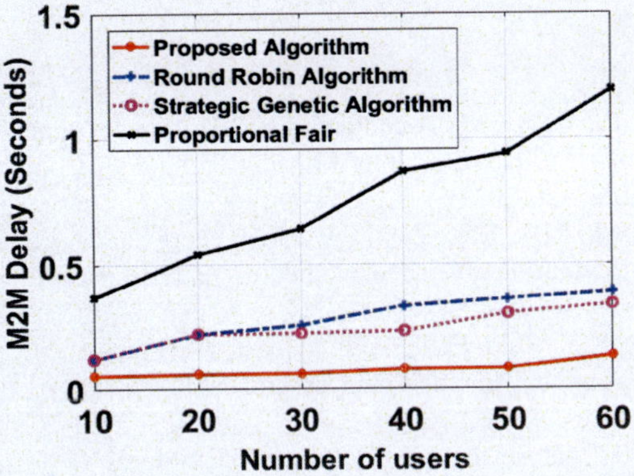

Fig. 14. The average of all end-to-end delays for received packets for M2M devices versus the total number of UEs (Scenario 2).

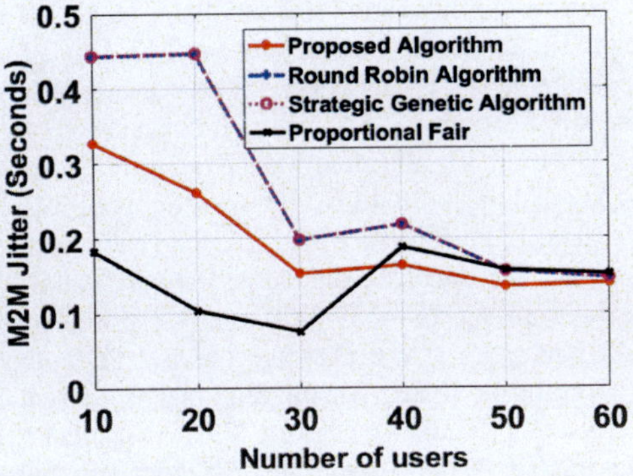

Fig. 15. The jitter received for M2M devices versus the number of users (Scenario 2).

5 Conclusion

For the first scenario, the proposed scheduler maximizes resource utilization as it succeeds in increasing the throughput for H2H traffic while keeping the throughput for M2M devices with no changes. Therefore, the increase in End-to-End delay and jitter for H2H will be acceptable compared to a massive increase in throughput. For M2M, the delay will also increase in H2H throughput, increasing the number of bytes, but the delay for M2M will also be accepted. On the other hand, the jitter will improve for M2M devices. In the second scenario, the proposed scheduler maximizes resource utilization as it succeeds in increasing the throughput for H2H traffic and M2M traffic and decreasing delay for both H2H and M2M devices. The only disadvantage will be the H2H jitter as it increases while the M2M jitter will be improved.

References

1. Sathya, V., Zhang, L., Yavuz, M.: A comparative measurement study of commercial WLAN and 5G LAN systems. In: 2022 IEEE 96th Vehicular Technology Conference (VTC2022-Fall), London, United Kingdom (2022)
2. Abo Hashish, S.M.M., Rizk, R.Y., Zaki, F.W.: Joint energy and spectral efficient power allocation for long term evolution-advanced. Computers and Electrical Engineering, Elsevier, **72**, pp. 828–845 (2018)
3. Magdy, S.M., Rizk, R.Y., Zaki, F.W.: Optimal multi-relay placement for LTE-A. In: Proc. of 11th International Computer Engineering Conference (ICENCO), pp. 195–200, Cairo, Egypt (2015)
4. Sahu, M.: Delay jitter analysis for uplink traffic in LTE systems. In: 11th International Conference on Communication Systems & Networks (COMSNETS), Bengaluru, India, pp. 504–506 (2019)
5. Maia, A.M., de Castro, M.F., Vieira, D.: A dynamic LTE uplink packet scheduler for machine-to-machine communication. In: IEEE 25th Annual International Symposium on Personal, Indoor, and Mobile Radio Communication (PIMRC), pp. 1609–1614 (2014)
6. Edemacu, K., Bulega, T.: Resource sharing between M2M and H2H traffic under Time-controlled scheduling scheme in LTE networks. In: proceeding of IEEE, Telecommunication Systems Services and Applications (TSSA), Kuta, Indonesia (2014)
7. Nashaat, H., Refaat, O., Zaki, F., Shaalan, I.E.: Dragonfly-based joint delay/energy LTE downlink scheduling algorithm. Journal IEEE Access, **8**, pp. 35392–35402 (2020)
8. Afrin, N., Brown, J., Khan, J.: Performance analysis of an enhanced delay sensitive lte uplink scheduler for m2m traffic. In: Telecommunication Networks and Applications Conference (ATNAC), 2013 Australasian, pp. 154–159 (2013)
9. Afrin, N., Brown, J., Khan, J.Y.: A packet age based LTE uplink packet scheduler for M2M traffic. In: proceeding of IEEE, 7th International Conference on Signal Processing and Communication Systems (ICSPCS), Carrara, VIC, Australia (2013)
10. Abdalla, I., Venkatesan, S.: A QoE preserving M2M-aware hybrid scheduler for LTE uplink. In: International Conference on Selected Topics in Mobile and Wireless Networking (MoWNeT), Montreal, QC, Canada, pp 127–132 (2013)
11. Maia, A.M., Vieira, D., Castro, M., Ghamri-Doudane, Y.: A mechanism for uplink packet scheduler in LTE network in the context of machine-to-machine communication. In: 2014 IEEE Global Communications Conference, pp. 2776–2782 (2014)
12. Lioumpas, A.S., Alexiou, A.: Uplink scheduling for machine-to-machine communications in LTE-based cellular systems. In: IEEE GLOBECOM Workshops (GC Wkshps), Houston, TX, USA (2011)
13. Beshley, H., Beshley, M., Medvetskyi, M., Pyrih, J.: QoS-aware optimal radio resource allocation method for machine-type communications in 5G LTE and beyond cellular networks. Wirel. Commun. Mob. Comput. **2021**, 1–18 (2021)
14. Kumar, N., Ahmad, A.: Quality of service-aware adaptive radio resource management based on deep federated Q-learning for multi-access edge computing in beyond 5G cloud-radio access network. Transactions on Emerging Telecommunications Technologies (2023)
15. Salah, M., Ali, N., Taha, A.-E., Hassanein, H.: Evaluating uplink schedulers in lte in mixed traffic environments. In: IEEE International Conference on Communications (ICC), pp. 1–5 (2011)
16. Dimitrova, D.C., van den Berg, H., Litjens, R., Heijenk, G.: Scheduling strategies for LTE uplink with flow behavior analysis. In: Fourth ERCIM workshop on e-mobility, Lulea Tekniska Universitet, Lulea, Sweden, pp. 15–26 (2010)

17. Gotsis, A.G., Lioumpas, A.S., Alexiou, A.: Analytical modelling and performance evaluation of realistic time-controlled m2m scheduling over LTE cellular networks. Trans. Emerging Telecommun. Technologies **24**(4), 378–388 (2013)
18. 3GPP. Analysis on traffic model and characteristics for mtc and text proposal. 3GPP, *Technical Report R1–120056* (2012)
19. Moses, L., Karthikeyan, S.: Strategic genetic algorithm for resource allocation in LTE uplink netwoks. In: Proceedings of the Second International Conference on Computing Methodologies and Communication (ICCMC), *IEEE*, pp. 436–441 (2018)
20. Salah, A., Abdel-Atty, H.M., Rizk, R.: Joint channel assignment and power allocation based on maximum concurrent multicommodity flow in cognitive radio networks. Wireless Communication and Mobile Computing (2018)
21. Nashaat, H., Rizk, R.: Smart prediction for seamless mobility in f-hmipv6 based on location based services. China Communications, Springer **15**(4), 192–209 (2018)

Optimized Helical Antenna for Wireless Application at 2.4 GHz Using Gray Wolf Optimizer

Habiba Hassan Nabwy[1](✉), Shereen Mostafa Ali[1], Maira Mohamed Khafagy[1], Salma Ahmed El-Tohamy[1], Basmala Samir Mohamed[1], Saif Eldeen Ehab Mohamed[1], and Ahmed Magdy Mohamed[2]

[1] Electrical Engineering Department, High Institute of Engineering and Technology, Fifth Settlement, Cairo, Egypt
habiba.abo3rb@gmail.com

[2] Electrical Engineering Department, Faculty of Engineering, Suez Canal University, Ismailia, Egypt
ahmed.m.1986@ieee.org

Abstract. This work introduces an AI-optimized helical antenna designed for wireless communication systems operating at 2.4 GHz. The antenna aims to achieve high gain with the fewest number of turns and smallest radius length while improving return loss and radiation-pattern characteristics. MATLAB is utilized for antenna design and simulation, resulting in a peak gain of 27.599 dBi. The Gray Wolf Optimizer (GWO) algorithm is employed for optimization, demonstrating significant enhancements in return loss and gain. Comparative analysis with previous studies highlights the superiority of this helical antenna design in terms of dimensions, parameters, and results. The integration of artificial intelligence algorithms enables the attainment of high gain with minimal size and turns, making it well-suited for wireless communication systems.

Keywords: AI · Helical Antenna · axial mode · Antenna Optimization

1 Introduction

Optimization is aimed at achieving the optimal performance of complex electromagnetic devices, which must be competitive with respect to high efficiency, in addition to reasonable costs and servicing capability [1]. This process involves selecting the objective functions and doing so by selecting the appropriate design parameters, which are usually contradictory. Several objectives must be attained during the process of optimization in order to achieve the optimum solution. Since these goals often contradict each other, there can be no one solution best suited to all of the goals under consideration [2]. The use of a suitable method for the problem is essential for a correct technical solution. For instance, there are proven technical solutions to a number of problems, which were achieved following several years of experimenting and testing. In other cases, it might

A. Hassanien et al. (Eds.): AISI 2023, LNDECT 184, pp. 432–443, 2023.
https://doi.org/10.1007/978-3-031-43247-7_38

be useful to come up with new designs that improve on the well-known default results, although they may be not global minimums. Nonetheless, conditions in which deterministic work well cannot be achieved, and therefore stochastic methods can be a good option such as Genetic Algorithms. The GWO is a very well-built algorithm capable of providing impressive results in comparison with the famous metaheuristics. The GWO is a new meta-heuristic, GWO refers to [3]. The Optimization technique of Grey Wolf is affected by grey wolves (Canis lupus). The inspiration for the GWO Algorithm is the hierarchies of leader's management and the hunt mechanisms of grey wolves in nature [4]. GWO is used for enhancement of the performance of the helical antenna, which was simulated and optimized using MATLAB to meet the necessary objectives, like their deployment in such Wireless Communication Systems applications, like Wi-Fi, at affordable Costs [5]. Helical antennas have a history, and since their inception have attracted a lot of research and developments for over 6 decades [6]. The helical Antenna consists of a hybrid of 2 simple radiation elements: The Dipole Antenna and the Loop Antenna. Helical antennae have widely been used in the past 5 decades as simple, practical radiating elements because of their noteworthy, unique characteristics. The precise analysis is extremely complex, so parameters such as gain, radiation pattern, input impedance, and the axial ratio of the helix are verified using experimental methods, proximity analysis techniques, and numerical analysis [7].

In this work, Our helical antenna is optimized using GWO operating at the S-band which is suitable for applications of Wireless Communication Systems [8].

2 System Model

The system model for the given project involves designing and optimizing a helical antenna using the Optimization technique of the Grey Wolf (G.W.O) algorithm as shown in Figure 1. The first step in the process is to calculate the parameters of the helical antenna, including the pitch angle, radius, number of turns, and wire diameter. Once the antenna parameters have been calculated, they are simulated using MATLAB to predict the antenna's radiation pattern, gain, and impedance bandwidth. The next step involves optimizing the antenna parameters using the algorithm of GWO, for which the metaheuristic algorithm is dependent on the hunting behavior of grey wolves. The GWO algorithm aims to optimize the antenna's parameters, which results in better radiation characteristics. The process of optimization involves running the algorithm on the simulated antenna parameters until the desired optimization criteria are met. After the antenna parameters have been optimized, the antenna is re-simulated using the optimized parameters to compare the results with the initial simulation. The comparison aims to determine the effectiveness of the G.W.O algorithm in optimizing the antenna's parameters. The comparison will include Radiation-pattern, Gain, and Impedance bandwidth. Finally, the results of the initial simulation and the optimized simulation are compared to determine the effectiveness of the G.W.O algorithm in optimizing the antenna's parameters. The comparison will help identify any improvements made to the antenna's performance. The system model is summarized in a flowchart, which illustrates the process of designing and optimizing the helical antenna using the G.W.O algorithm.

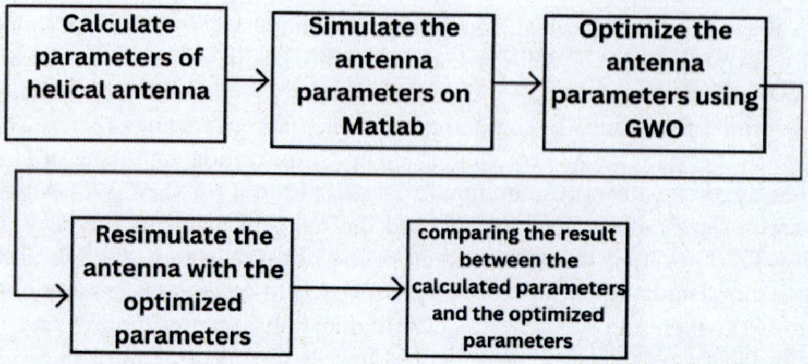

Fig. 1. The process of designing and optimizing a helical antenna using the Grey Wolf Optimization algorithm.

2.1 Helical Antenna Design

One of the primary considerations in antenna design is the frequency of operation, which determines the wavelength of the signal and the physical size of the antenna. The size of the antenna is proportional to the wavelength, and it is critical to ensure that the size is suitable for the application. For instance, in mobile phone antennas, the size of the antenna must be small enough to fit within the device, while also providing adequate performance. A helical antenna is selected in this paper as it serves a circular Polarization. This quality should be taken into account when installing the antenna in an area with many obstacles leading to extra signal reflections. Helical antennas with axial mode also offer several benefits over patch antennas, profitable. This antenna is preferred because of its wide bandwidth also high gain, and it can radiate the signal in a directional pattern [9]. The antenna has two main operating modes; normal (Broadside radiation) and axial (end-fire radiation) [10]. If the antenna has a diminutive diameter according to wavelength, it works in normal mode, but if the circumference of the helix is wider in the order of wavelength, it works in the axial mode, as a directional antenna [11]. In comparison to normal mode helical antenna that emits Omni-directional radiation. The antenna is designed at resonant frequency = 2.4 GHz. Figure 2, demonstrates the configuration of the desired helical antenna. It consists of conducting slot attached to a plate that is used as a ground plane. This structure was designed in MATLAB.

The wavelength (λ) of the desired antenna can be computed using the equation:

$$\lambda = \frac{c}{f}. \tag{1}$$

where C is the speed of light with the value of the speed of light with the value of (3810^8) m/s. while f is the chosen frequency of 2.45 GHz, the calculated λ is (122.449) mm For axial mode helical antenna, the helix circumference, C is equal to one wave- length [12]. Axial mode helical antenna performs well if the circumference of helix is in the following range

$$\frac{3}{4}\lambda \le c \le \frac{4}{3}\lambda. \tag{2}$$

The proposed antenna is coaxially fed with a circular ground. Key the helix's dimensions like radius, circumference, the distance between turns, height, and angle of inclination. Table. 1, contains each parameter that affects the performance of the helical antenna. The parameters acquired by calculation are based on the rules of helical antennas. The total number of turns of helix N was 2.8 turns, while the pitch angle α was calculated from the following equation:

$$\tan^{-1}\alpha = \frac{s}{c} \tag{3}$$

The antenna is fed with a coaxial wire with a circular ground. The input impedance of 50 Ω can be done by involving an impedance transformer in the feed or by modifying the location of the wire from the coaxial to the helix torsion. Through the formulation of the optimizing process, the physical parameter has been diverse to test the impact on the performance of the antenna [13]. The empirical formulas used to calculate the helical antenna parameters.

$$2a = \frac{\lambda}{100}. \tag{4}$$

where a is the wire radius.

$$R = \frac{\lambda}{2\pi} label. \tag{5}$$

R is the helix radius

$$D_0 = \frac{15NC^2}{\lambda^3}. \tag{6}$$

where D_0 is the directivity, N is the number of turns, S is the spacing between the turns, C is the helix circumference, and λ is the wavelength [14].

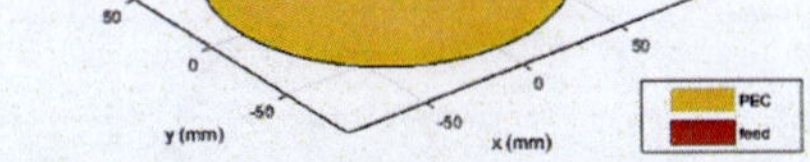

Fig. 2. Design of the proposed helical antenna.

2.2 Gray Wolf Optimization Algorithm

The Optimization technique of the gray wolf is an advanced metaheuristic, GWO refers to as once mentioned before Grey Wolf Optimizer that is inspired by grey wolves (Canis

Table 1. Optimum parameters of helix antenna

Optimum Parameters	Value(mm)
The radius of the Helix (R)	50
Total Axial Length (L)	NS
Width of Wire (W)	1.8
Spacing between Turns(S)	35
Number of Turns (N)	2.8
Circumference of Helix (C)	$2\pi r$
Ground Plane (G)	75
pitch angle (α)	14°

lupus) [15]. Figure 3 shows the inspiration for the GWO Algorithm is the hierarchies of leadership and the hunt mechanisms of the gray wolf in nature [16]. The four species of the gray wolf, the Alphas, the Betas, the Deltas, and the Omegas, were used to simulate the hierarchies of leadership. Furthermore, it implements three major arranged ways of hunting: searching and chasing the prey, circling around prey, and attacking prey [17, 18]. The common communicating attitude of grey wolves in group hunting is very interesting and unique. The main phases of a grey wolf hunt, based on Muro et al. [16] are:

1- Tracking, chasing, and approaching the prey
2- Pursuing, encircling, and harassing the prey until it stops moving
3- Attack the prey

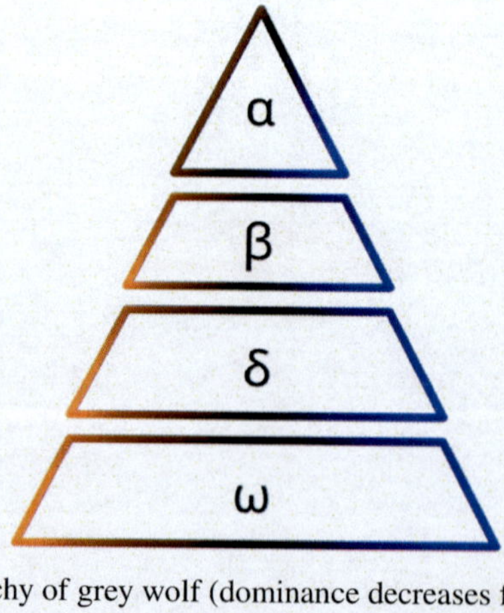

Fig. 3. Hierarchy of grey wolf (dominance decreases from top down).

2.2.1 Mathematical Model and Algorithm

The mathematical models for the social hierarchy, tracking, surrounding, and attacking prey are presented. Some points should be recorded to explain how GWO is theoretically capable of resolving optimization problems.

The Grey Wolf Optimization (GWO) algorithm is a nature-inspired optimization algorithm that is based on the social hierarchy and hunting behavior of grey wolves. The algorithm has gained popularity in recent years due to its effectiveness in solving a wide range of optimization problems.

One of the key features of the GWO algorithm is the encircling technique, which involves surrounding the current solution with a neighborhood that is formed like a circle and can be expanded to higher dimensions as a hypersphere. The search for better solutions is guided by two chance parameters, A and C, which determine the size and shape of the hypersphere.

The algorithm uses a hunting technique to identify the most promising regions for further exploration. This enables the algorithm to balance between exploration and exploitation, which is achieved through the adaptive values of the parameters (a and A). The adaptive nature of these parameters enables the algorithm to shift seamlessly between exploration and exploitation phases, depending on the progress of the search. As the algorithm converges on a solution, the value of A reduces, and the iterative process divides equally between exploration and exploitation. This enables the algorithm to gradually shift from exploration to exploitation and converge on the best possible solution.

The simplicity of the algorithm is reflected in the fact that only a pair of primary parameters (a and C) demand adjustment within GWO. This makes the algorithm easy to implement and requires minimal parameter tuning.

Overall, the GWO algorithm has proven to be effective in solving a wide range of optimization problems, and its various techniques, such as the encircling technique and hunting technique, enable it to balance between exploration and exploitation, adapt to changing conditions, and converge on the best possible solution.

The GWO algorithm terminates when the termination criteria are met as shown in Figure 4, below the GWO pseudo code and the flow chart diagram of GWO.

Algorithm 1 Grey Wolf Optimizer

Assume the grey wolf inhabitants Y_i for $i = 1, 2, \ldots, n$
Assume $\alpha, \beta, \delta, z, Z,$ and H
Determine the capability of each seeker
Y_α = the preferable seeker
Y_β = the second preferable seeker
Y_δ = the third preferable seeker
$k = 0$
while k = Ultimate number of repetitions
  for each seeker do
    Upgrade the place of the present seeker
  end for
  Upgrade $z, Z,$ and H
  Determine the capability of the whole seeker
  Upgrade Y_α, Y_β, and Y_δ
  $k = k + 1$
end while
return Y_α

And it is a good opportunity to apply it in the field of helical antenna testing.

GWO can improve antenna performance by changing the antenna design dimensions as follows:

After optimization, the number of turns changed from (3 turns) to (2.8320 turns). The radius changed from before optimization (46.875 mm) to after optimization (50 mm). The width before optimization changed from 1.25 mm to 1.7604 mm after optimization. The effect on the antenna result is to change its parameters, such as a Return Loss of −16.6475 dB before optimization to −57.3018 dB After optimization. Directivity of 7.36 dBi Before optimization, to 8.4 dBi after optimization. Out of the design, the algorithm of GWO was evaluated against twenty-nine (29) bench-mark consequences. The first twenty-three (23) bench-mark consequences are classic consequences used by a lot of seekers and researchers [19]. In spite of their clarity, we chose those consequences for a test so that their results can be compared with those of current metaheuristics. The benchmark consequence used are minimized consequences and can be split up into four groups: Unimodal, multimodal, and complex standard-dimensional consequences. A detailed description of the composite benchmark

The consequence can be found in the 2005 Technical Report of CEC [20]. The seventeenth function is the chosen function in this design as it serves the multiple inputs method in the seventeenth function with a maximum iteration is 1000 and seekers is 100.

3 Result Analysis and Simulation

The design of the helical antenna was simulated and implemented using MATLAB. The parameters involved in the simulation of the antenna were adjusted, such as turn's number and the radius of the ground, as the number of turns was 3 turns, the radius of ground was 93.75 mm and the width of the antenna was 1.25 mm, then to optimize the

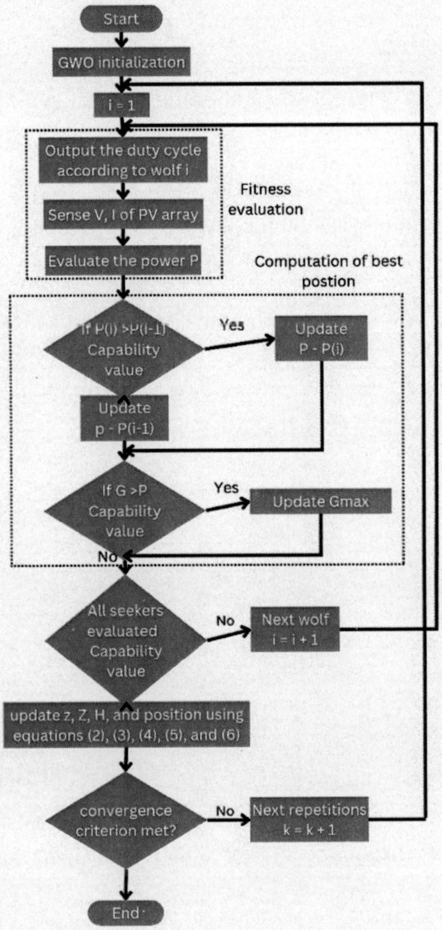

Fig. 4. The process of designing and optimizing a helical antenna using the Optimization of Grey Wolf (G.W.O) algorithm.

antenna's performance at 2.4 GHz. The parameters are adjusted to 2.83 turns, 50 mm, and 1.76 mm for the number of turns, the radius of the ground, and the width of the antenna, respectively. The simulation results show that the use of the G.W.O technique led to a significant improvement in the accomplishment of the antenna, particularly with respect to the return loss beside the gain, while the return loss is improved significantly from (−16 dB to −58 dB) with a −42 dB difference and the directivity improved from (7.36 dB to 8.52 dBi) with a 1.16 dBi difference. The antenna achieved a peak gain of 27.599 dB, and the radiation pattern showed high directivity in the desired direction. Overall, the simulation outcomes refer to the AI-optimized helical antennas that are appropriated to wireless communication systems.

3.1 Reflection Confection

To validate the simulation findings, the submitted helical antenna prototype was done. The simulated results of the Reflection Coefficient were performed via MATLAB. The suggested helical antenna's simulated S11 results vs frequency is shown in Fig. 5. At resonant frequency 2.4 GHz, the impedance bandwidth simulated for $S11 < 10$ dB is -16. A satisfactory agreement was obtained, which improved the simulated results of the mentioned helical antenna's reflection coefficient.

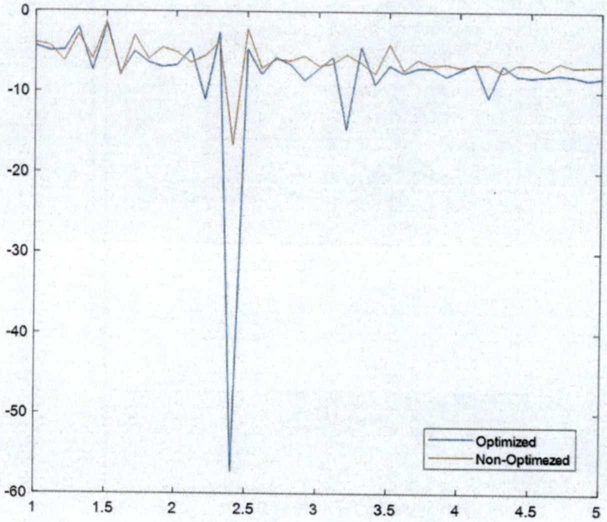

Fig. 5. simulated results of reflection coefficient result for the Optimized VS Non-Optimized for the proposed antenna.

3.2 Radiation Pattern

The radiation patterns of the mentioned Helical antenna upon the resonance frequency studied 2.4 GHz. Figure 6 clarifies the 3D radiation patterns. This gives the value of Directivity of 7.36 dBi for Non-Optimized, to 8.4 dBi for the Optimized proposed helical antenna. The far-field 2D radiation patterns have been simulated for the mentioned helical antenna is illustrated as two main planes: Elevation-plane or E-plane (x-z plane,), azimuth- plane or H-plane (x-y plane, =0) at frequency 2.4 GHz. From Fig 7 and Fig. 8. That should be noted that the after-optimization simulated outcomes are different in comparison with the before-optimization outcomes' simulation.

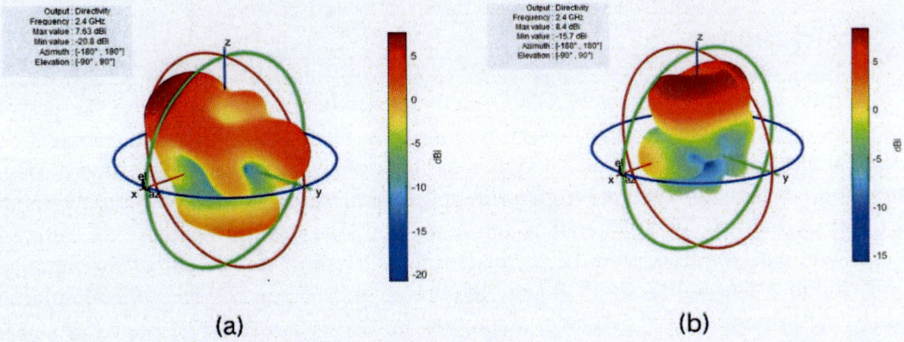

Fig. 6. 3D radiation patterns of the proposed helical antenna, (a) non-optimized, (b) optimized.

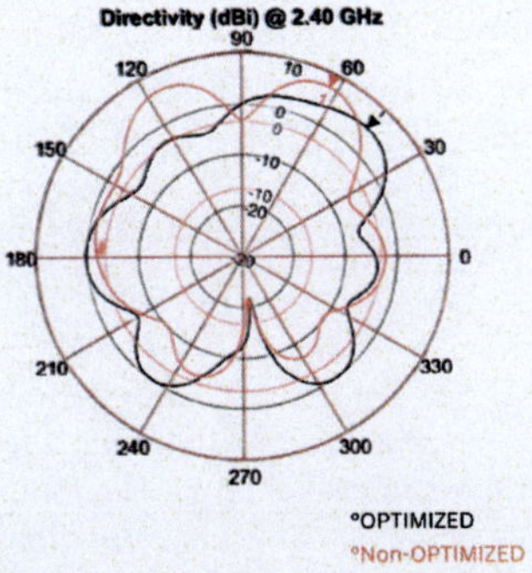

Fig. 7. The 2D plot for the radiation pattern in "E-plane" for the Optimized and Non-Optimized of the proposed helical antenna.

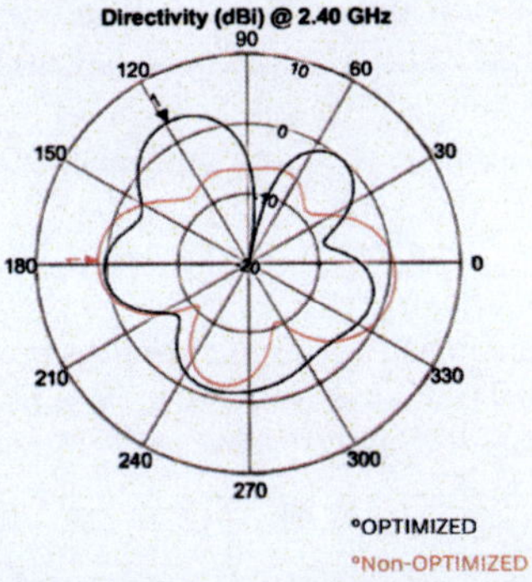

Fig. 8. The 2D plot for the radiation pattern in "H-plane" for the Optimized and Non-Optimized of the proposed helical antenna.

4 Conclusion

Within this work, the impact of GWO methods has been investigated on the performance of helical antennas in airborne environments. This is introductory research about a helical antenna optimized for 2.4 GHz's applications. The design is clarified through the outcomes that can generate higher directional patterns with high output power/gain as optimized with GWO rather than the non-optimized design. Known that different parameters can affect antenna performance. Coil's radius, the height of the antenna, turns' number. But with a slight change of parameters. The outcomes of the simulation clarify the importance of the GWO technique and led to a significant improvement in the antenna's performance, particularly in terms of return loss and gain, (−16 dB to −58 dB) with a −42 dB and (7.36 dB to 8.52 dBi) with a 1.16 dBi difference respectively. AI-optimized antenna using GWO gives a better performance with extremely high results. Out of the results, it is clear now that GWO was capable of providing a great comprehensive result with comparison to famously known different try and error techniques and that the algorithm of GWO executes efficiently in an unrevealed and difficult search space.

References

1. Wolpert, D.H., Macready, W.G.: No free lunch theorems for optimization. IEEE Trans. Evol. Comput. **1**(1), 67–82 (1997). https://doi.org/10.1109/4235.585893
2. Optimization by simulated annealing — science (no date). https://www.science.org/doi/ https://doi.org/10.1126/science.220.4598.671
3. Author links open overlay panelSeyedali Mirjalili a et al.: Grey Wolf optimizer, Advances in Engineering Software. Elsevier (2014). https://www.sciencedirect.com/science/article/abs/pii/S0965997813001853
4. PDF] grey wolf optimizer — semantic scholar (no date). https://www.semanticscholar.org/paper/Grey-Wolf-Optimizer-Mirjalili-Mirjalili/695bd9bb89df717227357a4ee8d15843a11f3609
5. Mirjalili, S., Mirjalili, S.M., Lewis, A.: Grey wolf optimizer. Adv. Eng. Softw. **69**, 46–61 (2014). https://doi.org/10.1016/j.advengsoft.2013.12.007
6. ISSN 1064–2269, Journal of Communications Technology and Electronics, **65**(8), 943–949. Pleiades Publishing, Inc. (2020)
7. Salbani, M.F., et al.: Helical antenna design for wireless power transmission: a preliminary study. 2011 IEEE International Conference on System Engineering and Technology [Preprint] (2011). https://doi.org/10.1109/icsengt.2011.5993448
8. Ping, X., et al.: Design of a high gain axial-mode helical antenna with a loaded plate. 2013 International Workshop on Microwave and Millimeter Wave Circuits and System Technology [Preprint] (2013). https://doi.org/10.1109/mmwcst.2013.6814581
9. Abdullah, S., Syed Hassan, S.I.: Design small size of high frequency (HF) helical antenna. 2009 5th International Colloquium on Signal Processing and Its Applications [Preprint] (2009). https://doi.org/10.1109/cspa.2009.5069229
10. Lenz, S.T.: Alan Agresti (2013): Categorical Data Analysis. Statistical Papers, **57**(3), 849–850 (2015). https://doi.org/10.1007/s00362-015-0733- 8
11. Pandey, A.K., Pathak, S.K.: Numerical and computational analysis of radiation characteristics of dielectric loaded helical antenna. International Journal of RF and Microwave Computer-Aided Eng. **31**(9) (2021). https://doi.org/10.1002/mmce.22756

12. Cha, S.H., et al.: A low-profile spiral termination helical antenna. 2017 International Symposium on Antennas and Propagation (ISAP) [Preprint] (2017). https://doi.org/10.1109/isanp.2017.8228817
13. Stutzman, W.L.: Antenna Theory and Design. John Wiley and Sons (2012)
14. Kreisselmier, G.: Indirect method for adaptive control. 1980 19th IEEE Conference on Decision and Control including the Symposium on Adaptive Processes, Albuquerque, NM, USA, pp. 594-598 (1980). https://doi.org/10.1109/CDC.1980.271865
15. Particle swarm optimization — IEEE conference publication - IEEE xplore (no date). https://ieeexplore.ieee.org/abstract/document/488968
16. Muro, C., et al.: Wolf-pack (canis lupus) hunting strategies emerge from simple rules in computational simulations: Semantic scholar, Behavioural Processes (1970)
17. Author links open overlay panelC. Muro a et al.: Wolf-pack (canislupus) hunting strategies emerge from simple rules in computational simulations, Behavioural Processes. Elsevier (2011)
18. Poli, R., Kennedy, J., Blackwell, T.: [PDF] particle swarm optimization: Semantic scholar, Swarm Intelligence (1995)
19. On benchmarking functions for genetic algorithms . Taylor & Francis
20. Liang, J., Qu, B., Suganthan, P.N., Das, S.: Novel Composition Test Functions for Numerical Global Optimization

Latin Square and Artificial Intelligence Cryptography for Blockchain and Centralized Systems

Abdelrahman Desoky[1,2], Hany Ammar[3], Gamal Fahmy[3], Shaker El-Sappagh[3], Abdeltawab Hendawi[4], and Sameh H. Basha[5,6](✉)

[1] Department of Computer Science, Claflin University, Orangeburg, SC, USA
desoky@desoky.com
[2] Scired: Scientific Research and Development, Orangeburg, SC, USA
[3] Faculty of Computer Science and Engineering, Galala University, Suez, Egypt
{hany.ammar,gamal.fahmy,Shaker.elsappagh}@gu.edu.eg
[4] Computer Science Department, University of Rhode Island, Kingston, USA
hendawi@uri.edu
[5] Faculty of Science, Cairo University, Giza 12613, Egypt
Samehbasha@cu.edu.eg, Samehbasha@gu.edu.eg
[6] Faculty of Science, Galala University, Suez 43511, Egypt

Abstract. The recent advances in cryptanalysis techniques and the leakage of information about the cryptosystem used are major threats to information systems. An adversary may succeed in decrypting ciphertexts, while users of a particular cryptosystem unknowingly continue to use the compromised cryptosystem. In this paper, we present a novel cryptosystem based on Latin Square and AI/ML for Blockchain and Covert Communications. This cryptosystem is capable of operating in quadruple modes keyless, symmetric, asymmetric, and hybrid encryption to Cipher In Cipher, and hence we call it CipherInCipher. Unlike all contemporary techniques including obscurity, CipherInCipher is a public-based approach that does not depend on the secrecy of any of its related components. It attains a high level of security that protects private information not only by having strong ciphertext but also by preventing an adversary from obtaining the actual ciphertext. The presented cryptanalysis validation demonstrates the robust capabilities of achieving the cryptographic goal.

Keywords: Cryptography · Cryptosystem · Cipher · Ciphertext · Security · Secure Communications · Covert Communications · Blockchain

1 Introduction and Related Work

Cryptography or cryptology is the science and art of concealing information, data, or both in an illegible format called ciphertext [1]. A ciphertext must be reversible back to its plaintext by only an authorized user via an operation called decryption [2]. Practically, the cryptographic goal is to avert an adversary from decrypting a ciphertext by generating

A. Hassanien et al. (Eds.): AISI 2023, LNDECT 184, pp. 444–455, 2023.
https://doi.org/10.1007/978-3-031-43247-7_39

a strong ciphertext. This can be done by utilizing a strong algorithm that may include techniques and mathematical operations, etc., to generate a resilient ciphertext [3]. On the contrary, cryptanalysis is the science and art of contributing to breaching a cryptosystem to eventually decrypt its ciphertext [4]. This may be achieved, even if a cryptosystem is unknown, via studying whatever is available to start from such as ciphertext, some predicted or leaked information about a cryptosystem used [5, 6].

Contemporary cryptography approaches are mainly categorized into two types, symmetrical and asymmetrical key cryptography [7]. Symmetric key cryptography primarily refers to particular encryption techniques in which a legitimate communicating party shares the same secret key for encrypting and decrypting [7]. Asymmetric key cryptography such as a public key cryptosystem uses pair of keys: a shared public key and an unshared private key [7]. Unlike symmetrical key encryption, a public key is shared and distributed publicly for anyone to use which resolved the key distribution problem [7].

Some additional concerns of contemporary cryptography approaches may be summarized as follows. Most contemporary cryptography approaches have been used for a long time and this by itself is risky regardless of the name and type of any cryptosystem [7]. This is because every cryptosystem has its own life cycle time. Theoretically, the longer it is used the higher probability of being breached because adversaries and researchers will have a good long time to contribute to breaching a cryptosystem [7]. In addition, over time scientific research, applications, tools, software applications, computer systems, scientists, experts, and adversaries become more advanced and can contribute to breaching a secure cryptosystem. Yet, all users of contemporary cryptography approaches must use cryptographic keys regardless of the type of cryptosystem used. Moreover, the increasing number of phishing attacks and losing key incidents became additional threats and vulnerabilities to contemporary cryptosystems too [8].

The above issues led us to develop and present in this paper a novel cryptosystem based on Latin Square (LS) to Cipher In Cipher (CipherInCipher). Our proposed cryptosystem is for blockchain and covert communications capable of operating in quadruple modes keyless, symmetric, asymmetric, and hybrid encryption. Unlike all contemporary techniques including obscurity, CipherInCipher is a public-based approach that does not depend on the secrecy of any of its related components. It attains a high level of security that protects private information not only by having strong ciphertext but also by preventing an adversary from obtaining the actual ciphertext. The presented validation demonstrates the robust capabilities of achieving the cryptographic goal.

The list of contributions of this short paper can be summarized as follows: 1- Presenting a novel concept of CipherInCipher that makes the real ciphertext hid-den; 2- Presenting the architecture design and the algorithm of the CipherInCipher cryptosystem; and 3- Presenting the implementation examples and cryptanalysis validation of CipherInCipher.

The remainder of this paper is organized as follows. Section 2 presents the CipherInCipher architecture. Section 3 demonstrates the implementation and the general validation of the proposed architecture and its security. Finally, Sect. 4 concludes the paper and its future work.

2 CipherInCipher

To illustrate CipherInCipher, let us consider the following scenario. Bob and Alice work for a high-tech company, which involves a high level of security to protect sensitive information about new inventions and new products because of recent cyberattacks. They considered the worst-case scenario where an adversary may be capable of decrypting all contemporary ciphertexts. Therefore, the CipherInCipher cryptosystem is selected to be utilized for this mission. This is because CipherInCipher is resiliently secure under the unfortunate assumption that an adversary can break all contemporary cryptography algorithms. CipherInCipher attains a high level of security that protects private information not only by having strong ciphertext but also by making the actual ciphertext invisible to an adversary. When using CipherInCipher, the only full ciphertext that is normally visible to an adversary is a false ciphertext. This false ciphertext is either an unwanted cyphertext that conceals non-sensitive information or a fake ciphertext, but it is not the pure actual ciphertext. However, when using CipherInCipher a pure actual ciphertext is shredded into different sizes and it either embeds it into one relatively large false ciphertext or multiple relatively large false ciphertexts.

Note that the shredding and embedding procedures are done in such a sophisticated way based on a relatively large Latin Square (LS) to make it more secure and to be extremely hard, if not impossible, for an adversary to reveal a plaintext [9].

Concisely, the CipherInCipher algorithm is achieved through three main phases as follows. First, CipherInCipher generates false ciphertext(s) by either legitimate ciphertext of unwanted plaintext or fake ciphertext. Second, it encrypts the intended message via symmetric, asymmetric, or hybrid encryption, as per user choice. Third, it embeds the encrypted message from Phase 2 into the generated ciphertext from Phase 1. Note that the embedding procedure is done according to a predetermined algorithm based on a relatively large LS with a minimum value of N = 11 to make it much more secure [9].

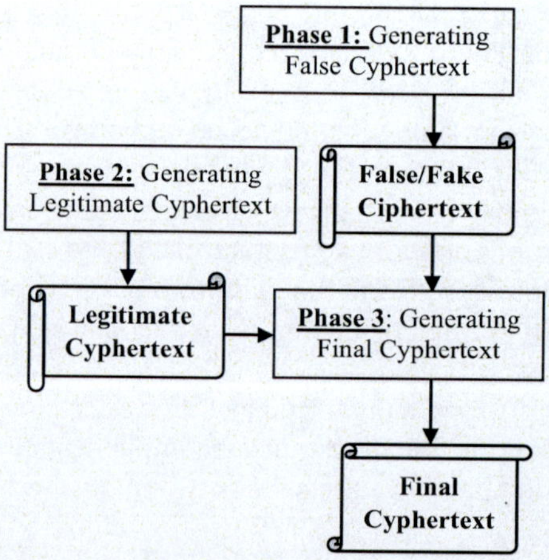

Fig. 1. Illustrates an overview of the CipherInCipher System and its phases.

It is worth noting that CipherInCipher gives legitimate users the option to operate in quadruple modes keyless, symmetric, asymmetric, and hybrid encryption by utilizing a contemporary cryptography algorithm to secure communications, data, etc., as per users' choice in configuring its setup. One may say if we use a contemporary cryptography algorithm, why do we need to use CipherInCipher? The answer in short, as stated early CipherInCipher makes an actual ciphertext invisible to an adversary, as demonstrated and discussed in detail in the following paragraphs.

The scenario of Bob and Alice in Sect. 2 demonstrates how the CipherInCipher paradigm can be utilized. This subsection presents a high-level overview of the overall core architecture of the CipherInCipher system and how it achieves the security goal. We then present the detailed components of the architecture in the following subsections. Note that the implementation of the CipherInCipher system may differ from one implementation to another. Therefore, the following presents an overview of the main core options of the CipherInCipher algorithm and its architecture. Architecturally, the CipherInCipher system consists of three main phases, as shown in Fig. 1. These three main phases are as follows:

Phase 1: CipherInCipher-based system generates false ciphertext by creating either a legitimate ciphertext of unwanted message/plaintext or fake ciphertext, as shown in Fig. 1, which will be described and discussed in this section below and the next sections. After performing ciphering of the fake text, we use an AI model to assure the robustness of the resulting ciphertext.

Phase 2: CipherInCipher encrypts the intended message via symmetric, asymmetric, or hybrid encryption, as per user choice and its setup, as shown in Fig. 1. This is done by using a contemporary cryptographic algorithm. This phase will be discussed further in more detail next. After performing ciphering of the sensitive text, we use an AI model to assure the robustness of the resulting ciphertext.

Phase 3: CipherInCipher embeds the encrypted message from Phase 2 into the generated ciphertext from Phase 1. Note that the embedding procedure is done according to a predetermined algorithm based on a relatively large LS with a minimum value of N = 11 to make it much more secure, as shown, mainly, in Fig. 1 along with all other figures and tables in this paper. This phase is described further in subsection D. After performing ciphering of the final text, we use the AI model to assure the robustness of the resulting final ciphertext.

Phase 3 embeds a legitimate ciphertext that is generated from Phase 2 into a false ciphertext that is generated from Phase 1. In other words, it embeds the encrypted message from Phase 2 into the generated false ciphertext from Phase 1. Note that a Latin Square matrix with a size of N is an N X N matrix that has a core property that each symbol S occurs only once in each row and column [9]. The embedding procedure is done according to a predetermined algorithm based on a relatively large LS with a minimum value of N = 11 to elevate its security. When LS uses a minimum value of N = 11 or greater than 11, will be an extremely large number of different unique LS as if it is a virtually infinite number of different unique LS [9]. Each one of the unique LS will be utilized as a security key for the embedding procedure in Phase 3. In this embedding procedure (Phase 3), CipherInCipher will select only one LS, from a virtually endless number of different LS, to embed the legitimate ciphertext according to the selected LS. It is worth noting that the security key (the selected LS) is unknown even to legitimate users because the CipherInCipher system securely computes the LS/security key internally without sharing it with anyone. This makes CipherInCipher a keyless cryptographic algorithm, despite giving the users an option to be partially involved in the key procedure generation or selection if users chose to operate under such an option (setup). While the CipherInCipher algorithm is publicly known to all adversaries, it assures the security of its ciphertext from any illegitimate deciphering and prevents any negative leakage.

For example, the CipherInCipher system ensures the secrecy of its security key(s) and prevents an adversary to conclude the LS/security key or any other information that may lead to breaking the CipherInCipher system via resilient mathematical computations (e.g., LS, no-reuse keys, rotating procedures used, etc.), as will be shown in this section and the upcoming sections as well too. Note that the CipherInCipher system does not allow reusing an LS/security key once it is used. Furthermore, there are several ways to select a secure LS which plays the role of a security key in the CipherInCipher system, which will be described and discussed in this section below and the next sections.

In this paper, due to space constraints, this section will discuss just a few concepts and examples in general of what and how the CipherInCipher system and its implementation may utilize, including the selection of secure LS, to achieve the security and cryptographic goal, as follows. For instance, how is CipherInCipher generates and utilizes an LS/security key securely and internally? CipherInCipher system is a flexible system that may utilize the following concepts, some of them, or more than them. This confirms the fact that the CipherInCipher system may differ from one implementation to another.

Keyless: CipherInCipher is a public algorithm which means, it is known to all adversaries. However, it assures the security of its ciphertext from any illegitimate deciphering and prevents any negative leakage. CipherInCipher system is a keyless system. Therefore, users neither hold keys nor use keys by themselves. However, if users want to use their keys, CipherInCipher can do that too, but it is not recommended due to the concern of losing a key, key leakage, stealing a key, key exchange, or similar issues.

Latin Square (LS) Security Key: CipherInCipher system internally and securely generates keys to be utilized without sharing them with anyone. So, users have no clue about any key used. For instance, one type of key that is internally, securely, and randomly generated and utilized is the Latin Square (LS) with a minimum of $N = 11$ to elevate its security. It selects a specific LS to assign it to a group of users to use only one time. So, there is no reuse of any internal keys nor will be shared with users (e.g., LS, any other keys, etc.). CipherInCipher system will mark it to keep track of the used LS to avoid reusing keys. This includes other virtual keys such as the order or selection of mathematical operations utilized will be changed or rotated etc. For example, the key for computing and generating the sequence of all LS is unknown to all users, and will be no reuse of such key as well too. Mathematically and logically, the matrix of Latin Square (LS) will be discussed in more detail according to its utilization in the CipherInCipher system in the validation section (Sect. 3) [9].

Other Mathematical Operations Based on LS Matrix: Concerning the mathematical operations, the CipherInCipher algorithm and its implementation can utilize a very sophisticated mathematical operation consisting of a group/combination of multiple mathematical operations that can be resiliently secure even when it is publicly known to an adversary, which is the case in CipherInCipher. Furthermore, a group/combination can be built using simple and cheap mathematical operations like addition, subtraction, multiplication, swap, shifting, and other operations, etc.

However, the sequence order for a set of mathematical operations is based also on the LS matrix, which makes it in a dynamic order rather than just a static sequence. For example, when CipherInCipher utilizes an LS-based matrix, CipherInCipher may select a row or column to apply a specific sequence order for a set of mathematical operations which will differ from one raw to another and one column to another too, according to properties of LS or/and partial LS utilized. Each cell of a raw or column in the LS matrix conveys a specific mathematical operation while that operation is unique in the entire raw and column according to LS property. Each time of communication will utilize a new row or column. Note that the CipherInCipher system will keep track of the sequence used for a particular group of users and it will differ each time used and from one group of users to another. The utilization of the LS-Based matrix will ensure the achievement of the security goal by changing the selected raw or column used each time via predetermined protocol internally randomly without sharing it with anyone. So, the CipherInCipher system will know/compute such sequence utilized but it will not be shared with anyone.

105fb3JYNa3uCUW6X510gPxCK68JTJXd/hn2XSFEFfJ1v7r
LA+WzNSu9S38FjS81Zt01epcbrrtQcq8w6gxFPiUL6Ymmbr
y8+/rOlZOxNGNvM7A0EUUwmcx2dC4/noetl4A8KygBfZ0I
IH2aBiQlsIxGEp69weAoYnki5G1gKdI4Ay2ETjkNjXl/nuO1
aAs5yfrO7X+LgZxvdbwcUpbFDHU4mTkTdcSk9MSxkLkae
0wokhqXMPp2c17zrMcSmimKbZsILmTi6oNNj7VopSdE8k
eYlHY/AQDOmwxarcbvlOr+a6g1ykS6Cx5jtTPztW1YXiNe
LGvoS+w9UByBdc4IAjRMQMTBVWDLXqdfD7OoRt/Ff8L
and+ujFwei4TqS9S9P4lUqTPxqLzp048Isxp+zAToKw2sjQ/T
6Zab7cZILDAC

Fig. 2. Illustrates the output of Phase 1, which is the altered false cyphertext. This is produced after encrypting random unwanted plaintext.

hNYyDQwaB4dDqFhcfDW9O5tMJsyP53b9VcCx/OwStRY

Fig. 3. Illustrates the output of Phase 2, which is the legitimate cyphertext.

3 Implementation and Security Validation

This section aims to demonstrate the implementation validation and feasibility of the CipherInCipher system and its distinct robustness capability of achieving the security and cryptographic goal, as follows.

105fb3JhNYyDQwW6XaB4gPxCKdDqFhXd/hn2XSFcfDW9
O5tMA+JsNyu9S38FjSP53b9VcCcbrrtQcq8wStRYPiUL6Ym
mbry8+/rOlZOxNGNvM7A0EUUwmcx2dC4/noetl4A8KygB
fZ0IIH2aBiQlsIxGEp69weAoYnki5G1gKdI4Ay2ETjkNjXl/n
uO1aAs5yfrO7X+LgZxvdbwcUpbFDHU4mTkTdcSk9MSxk
Lkae0wokhqXMPp2c17zrMcSmimKbZsILmTi6oNNj7VopSd
E8keYlHY/AQDOmwxarcbvlOr+a6g1ykS6Cx5jtTPztW1YXi
NeLGvoS+w9UByBdc4IAjRMQMTBVWDLXqdfD7OoRt/F
f8Land+ujFwei4TqS9S9P4lUqTPxqLzp048Isxp+zAToKw2sj
Q/T6Zab7cZILDAC

Fig. 4. Illustrates the output of Phase 3, which is the final cyphertext of the CipherInCipher system that conceals the actual legitimate cyphertext.

3.1 Implementation

To avoid abstraction, this section demonstrates and discusses a possible implementation example and illustrates synthetic samples. The CipherInCipher system is flexible, and it can differ from one implementation to another and from one group of users to another. Thus, it can be built to make it harder for adversaries to attack the CipherInCipher system, as previously discussed. However, because of space constraints and for the sake of making it easier to understand and follow up with it, this section demonstrates easy examples as follows.

Sample of the Output of Phase 1: CipherInCipher system generates false ciphertext by creating either a legitimate ciphertext of unwanted plaintext or fake ciphertext, to avoid repetition, as explained before and shown in Fig. 2, Fig. 1, Table 1, and Table 2, along with online tools used in [10, 11].

Sample of the Output of Phase 2: CipherInCipher encrypts the intended message via using a contemporary cryptographic such as symmetric, asymmetric, or hybrid encryption, as per user choice and its setup. This is as shown in Fig. 3, and Fig. 1, along with online tools used [10, 11].

Sample of the Output of Phase 3: CipherInCipher embeds the encrypted message from Phase 2 into the generated ciphertext from Phase 1. Note that the embedding procedure is done according to a predetermined algorithm based on a relatively large Latin Square (LS) with a minimum value of $N = 11$ to make it more secure, as shown in Figs. 4, 1, and Table 1, and 2, along with online tools used [10, 11]. Note that Latin Square (LS) is elaborated on in more detail next.

Table 1. Shows the selected Latin Square used.

7	3	5	9	2	1	8	4	10	11	6
1	8	10	3	7	6	2	9	4	5	11
9	5	7	11	4	3	10	6	1	2	8
3	10	1	5	9	8	4	11	6	7	2
8	4	6	10	3	2	9	5	11	1	7
11	7	9	2	6	5	1	8	3	4	10
10	6	8	1	5	4	11	7	2	3	9
5	1	3	7	11	10	6	2	8	9	4
6	2	4	8	1	11	7	3	9	10	5
4	11	2	6	10	9	5	1	7	8	3
2	9	11	4	8	7	3	10	5	6	1

3.2 Mathematical Analysis and Proofs of Utilizing Latin Square

This subsection aims to show the mathematical complexity of how CipherInCipher takes advantage of the hard problem of computing a relatively big Latin Square (LS) (e.g., n = > 11).

Definition: An $n \times n$ matrix is a Latin Square (LS) only if it consists of n sets (e.g., symbols, numbers, letters, a combination of all, etc.) arranged in such a way that no orthogonal (row or column) contains the same value more than one time [9]. Note that there are special cases of LS such as Latin Rectangle and Partial LS that retain the same properties of LS [9].

How Many Latin Squares Are There? Mathematically, there is no easy computation method for the number L_n of $n \times n$ Latin squares with symbols 1, 2, ... , n, as up to date, see Eq. 1. The most known upper and lower bounds assumed for large n are still vague. However, van Lint and Wilson stated a classical result that is never fully confirmed and is still fuzzy [9]. This classic result is that:

$$\prod_{k=1}^{n} (k!)^{\frac{n}{k}} \geq L_n \geq \frac{(n!)^{2n}}{n^{n^2}} \tag{1}$$

Shao and Wei reached a simple and explicit equation for calculating the number of Latin Squares [12]. However, it is still a hard problem to compute because of the exponential increase in the number of terms. Equation 2 is for the number L_n of $n \times n$ Latin squares is [9].

$$L_n = n! \sum_{A \in B_n} (-1)^{\sigma_0(A)} \binom{per\ A}{n} \tag{2}$$

where B_n is the set of all $n \times n$ $\{0, 1\}$ matrices, $\sigma_0(A)$ is the number of zero entries in matrix A, and per(A) is the permanent of matrix A [9].

It is worth noting that the On-Line Encyclopedia of Integer Sequences (OEIS) was developed and maintained by Neil Sloane during his research at AT&T Labs. Then, it was transferred the intellectual property and hosting of the OEIS to the OEIS Foundation in 2009 [13, 14]. Table 2 shows all known exact values of all Latin squares of size n, as referenced via OEIS. It is observed that the numbers increase exceptionally rapidly. For instance, the number of all Latin Squares registered a dramatic big jump between n and $n + 1$, especially starting when n is equal to 3, 4, 5, 6, and so on, as shown in Table 2. It is also observed that the number of all Latin Squares holds a relatively very large number from n = 6 and so on. Once n reaches a value of 11, the number of all Latin squares becomes a virtual infinity number, as shown in Table 2. For each n, the number of Latin squares altogether (sequence A002860 in the OEIS), as shown in Table 2 [9, 13, 14].

The usage of cell sequence in an LS can be via a column-based, row-based, combination of both rows and columns together, and/or random-based, as predetermined. This will add more strength to the use of LS. Yet, if the usage level is just rows and columns, then, each LS will be used twice. Thus, the above large number will be multiplied by 2 which will create an extremely larger number than its original one. In addition, CipherInCipher can construct LS from any random size with a minimum $n = 11$ along with a random set of values that satisfy the properties of LS. Moreover, special cases of LS such as Latin Rectangle and Partial LS can also be utilized by the CipherInCipher system which also will add more strength to it. Therefore, the prediction, calculation, brute force, and similar ways to attack are invisible to be used against CipherInCipher because the virtual infinity number of LS makes it extremely difficult for an adversary to deal with. For example, the task of searching or predicting a key among a virtually endless number of keys (e.g., LS used).

This makes LS with a relatively big n as at least $n = 11$ to be very promising in the field of cryptography in general and particularly in the presented cryptosystem in this paper, as shown in Table 2.

3.3 Centralized, Decentralized, and Hybrid System

CipherInCipher system can utilize a centralized, decentralized, blockchain, or hybrid (combination of centralized and decentralized) [15–19] keyless management system by taking advantage of LS proprieties, e.g., a large number of LS for $n = 11$, *etc.*, to securely handle an LS-based keyless system, as discussed before [9, 12].

3.4 Artificial Intelligence and Machine Learning Cryptanalysis-Based

CipherInCipher exploits Artificial Intelligence and Machine Learning (AI/ML) techniques to play the role of Cryptanalysis on the generated ciphertext to ensure security [20–22]. In short, due to space limitations, CipherInCipher inspects its generated ciphertext in all phases to approve or disapprove the use of its outputs (e.g. false ciphertext, real ciphertext, and final ciphertext). In case of CipherInCipher system disapproves its ciphertext of any phases, CipherInCipher will redo the phase of the disapproved ciphertext to regenerate another ciphertext that will pass its inspection. To be robust against possible

Table 2. Shows all Latin Square of size n (sequences a002860 in the OEIS) increases by increasing the value of N.

n	All Latin squares of size *n* (Sequence A002860 in the OEIS)
1	1
2	2
3	12
4	576
5	161, 280
6	812, 851, 200
7	61, 479, 419, 904, 000
8	108, 776, 032, 459, 082, 956, 800
9	5, 524, 751, 496, 156, 892, 842, 531, 225, 600
10	9, 982, 437, 658, 213, 039, 871, 725, 064, 756, 920, 320, 000
11	776, 966, 836, 171, 770, 144, 107, 444, 346, 734, 230, 682, 311, 065, 600, 000

adversarial attacks, the CipherInCipher system attacks its own generated cyphertext to ensure it is secure before sending it to the legitimate recipient. As a result, CipherInCipher takes precautionary steps to prevent failures, and this will be done intelligently and automatically using advanced AI/ML techniques such as residual connections and gated linear units.

4 Conclusion

In this paper, we presented the architecture and validation of a novel cryptosystem, namely CipherInCipher, based on a relatively large Latin Square for blockchain and covert communications demonstrating the robust capabilities of achieving the cryptographic and security goal. Unlike all contemporary techniques including obscurity, CipherInCipher is a public-based approach that does not depend on the secrecy of any of its related components. It attains a high level of security that protects private information, data, or both not only by generating strong ciphertext but also by preventing an adversary from even obtaining the actual ciphertext by concealing it in a false ciphertext. CipherInCipher gives legitimate users the option to operate in triple modes symmetric, asymmetric, and hybrid encryption for the actual ciphertext. In addition, the CipherInCipher system is keyless for the main ciphertext that conceals an actual ciphertext. We presented the architecture, implementation, and validation of the approach to illustrate this promising approach for securing information and data whether in the static or in the transmission stage.

The future work of this research should focus on applying the proposed approach to many different domains such as the medical and financial application domains. These applications would use different sets of data types such as graphs, images, and audio files including big data sets and different application systems. This then could detect limitations that need to be addressed to use the CipherInCipher approach in a wide variety of industrial applications.

Acknowledgment. We would like to express our thanks to the well-known cryptographer Dr. Taher Elgamal for the great generosity of his support, effort, and time in discussing and advising in building the CipherInCipher system. This work of this paper is fully supported and funded by The Scired: Scientific Research and Development under TR/Project 0115202201.

Compliance with Ethical Standards

- There are no conflicts of interest.
  - The Research does not involve Human Participants and/or Animals.
  - There is no need for Informed consent.

Competing Interests

- There are no financial conflicts of interest.
  - There are no financial gains.

Research Data Policy and Data Availability Statements. All data, information, etc., are available in the public domain for anyone to use as a reference.

References

1. The National Institute of Standards and Technology (NIST). https://csrc.nist.gov/glossary/term/cryptography. Accessed 22 Oct 2022
2. Wade, M.I., Gill, T.: The Iso-ElGamal cryptographic scheme. In: 2022 IEEE International IOT, Electronics and Mechatronics Conference (IEMTRONICS). IEEE (2022)
3. Habib, H.B., Hussein, W.A., Abdul-Rahman, A.Kh.: A hybrid cryptosystem based on latin square and the modified BB84 quantum key distribution. Tikrit J. Pure Sci. **27**(4), 100–103 (2022)
4. The National Institute of Standards and Technology (NIST). https://csrc.nist.gov/glossary/term/cryptanalysis. Accessed 22 Oct 2022
5. The National Institute of Standards and Technology (NIST), NIST Special Publication 800-57 Part 2 Revision 1. https://nvlpubs.nist.gov/nistpubs/SpecialPublications/NIST.SP.800-57pt2r1.pdf. Accessed 22 2022
6. Munir, N., et al.: Cryptanalysis of encryption scheme based on compound coupled logistic map and anti-codifying technique for secure data transmission. Optik **267**, 169628 (2022)
7. Koblitz, N.: A Course in Number Theory and Cryptography, 2nd edn, pp. 54–76. Springer, New York (1994). https://doi.org/10.1007/978-1-4419-8592-7
8. The Department of Homeland Security: Cybersecurıty & Infrastructure Securıty Agency, Cybersecurity Attacks, Incident Reporting. https://www.cisa.gov/stopransomware/general-information. Accessed 23 Oct 2022
9. Keedwell, A.D., Dénes, J.: Latin Squares and Their Applications. Elsevier, Amsterdam (2015)
10. An Online Encryption. https://www.online-toolz.com. Accessed 31 Aug 2022
11. An Online Encryption. https://www.dcode.fr. Accessed 31 Aug 2022
12. Shao, J.Y.: A formula for the number of Latin squares. Discrete Math. **110**(1–3), 293–296 (1992)
13. An On-Line Encyclopedia of Integer Sequences (OEIS). https://oeis.org/wiki/Welcome. Accessed 31 Aug 2022

14. An On-Line Encyclopedia of Integer Sequences (OEIS). https://oeis.org/A002860. Accessed 31 Aug 2022
15. Liu, Y., et al.: A semi-centralized trust management model based on blockchain for data exchange in iot system. IEEE Trans. Serv. Comput. (2022)
16. Stennikov, V., et al.: Coordinated management of centralized and distributed generation in an integrated energy system using a multi-agent approach. Appl. Energy **309**, 118487 (2022)
17. Kandi, M.A., et al.: A decentralized blockchain-based key management protocol for heterogeneous and dynamic IoT devices. Comput. Commun. **191**, 11–25 (2022)
18. Pelletier, A.: Centralization vs. Decentralization: Why You Need A Hybrid Model More Now Than Ever. https://perkuto.com/blog/centralization-vs-decentralization-why-you-need-a-hybrid model more now than ever/. Accessed 23 Oct 2022
19. Islam, N., et al.: A blockchain-enabled distributed advanced metering ınfrastructure secure communication (BC-AMI). Appl. Sci. **12**(14), 7274 (2022)
20. Kim, H., et al.: Quantum artificial ıntelligence on cryptanalysis. Cryptology ePrint Archive (2023)
21. Liao, M., et al.: Deep-learning-based ciphertext-only attack on optical double random phase encryption. Opto Electron. Adv. **4**(5), 05200016 (2021)
22. Desoky, A.: Noiseless Steganography: The Key to Covert Communications. Information Security Publisher/Taylor & Francis Group, 18 November 2012–19 April 2016

Egyptian Intention Behavior to Use Bitcoin Based on the Unified Theory of Acceptance and Use of Technology

Lamiaa Mostafa[1](✉) and Sara Beshir[2]

[1] Business Information System Department, Arab Academy for Science and Technology and Maritime Transport, Alexandria, Egypt
Lamiaa.mostafa31@aast.edu, Lamiaa.mostafa31@gmail.com
[2] Egypt-Japan University of Science and Technology, Alexandria, Egypt
sara.beshir@ejust.edu.eg

Abstract. After COVID-19, people's financial activities are digitalized. Bitcoin is the first decentralized digital currency that attracts people from all over the world to use it. The study discovers the Egyptian behavioral intention to use Bitcoin through using an extended unified theory of acceptance and use of technology (UTUAT). A structural equation modeling approach using Analysis of Moment Structures (AMOS) was employed to confirm the reliability of the model, data collected from a sample of 3761 employees working in an Egyptian private university. The statistical tests used are the reliability test, chi-square test, and person correlation test. The results show that the performance expectancy, effort expectancy, habit, and social influence have a high correlation value to the behavior intention to use Bitcoin while, facilitating conditions, hedonic motivation, and price value have a very high correlation to the behavior intention to use Bitcoin. Implications of the findings and recommendations for further studies were discussed.

Keywords: Bitcoin · Cryptocurrencies · Bitcoin Exchange · Digital Banking · Unified Theory of Acceptance and Use of Technology (UTUAT)

1 Introduction

Cryptocurrencies have become essential for e-commerce platforms as they rely on a peer-to-peer tool to meet economic requirements. Bitcoin, often referred to as the currency of the Internet, is a decentralized digital monetary unit that operates without the involvement of any government or central bank. Its unique transaction structure allows for digital money transfers without the need for a third party. Bitcoin utilizes cryptographic protocols and a network of clients to mint, trade, store, and transfer Bitcoins. It has gained significant popularity, with widespread usage in various online services such as booking hotels, shopping for furniture, and purchasing video game consoles. Bitcoin's value has experienced significant fluctuations over time, and its economy surpasses that of some smaller countries [9].

A. Hassanien et al. (Eds.): AISI 2023, LNDECT 184, pp. 456–470, 2023.
https://doi.org/10.1007/978-3-031-43247-7_40

Cryptocurrencies depend on the transaction between people through commercial sites. Bitcoin is identified as the future currency of the Internet, world's first decentralized digital monetary unit [1, 27]. Bitcoin is used in different stores and services, also Bitcoin ATMs are accessible. Currently, the Bitcoin economy is now compared with smaller countries' economies [27]. Bitcoin provides several advantages such as faster transaction speed and lower transaction fees, but at the same time, many people resist to adopt especially in developing countries such as Egypt. People are afraid to use Bitcoin because of risks and uncertainties. Bitcoin reliability and validity are still ambiguous [9, 10].

While Bitcoin offers advantages such as lower transaction fees and faster speeds, many individuals remain hesitant to adopt it due to risks and uncertainties. Concerns about the reliability, validity, and ambiguous status of Bitcoin in several nations contribute to this resistance. Cyber-attacks and privacy issues within the network are additional concerns. Understanding the root causes of this opposition is crucial for promoting wider acceptance and usage of Bitcoin.

The Unified Theories of Acceptance and Use of Technology (UTAUT) model and the Technology Acceptance Model (TAM) provide valuable frameworks for studying user behavior and acceptance of technology. In the context of cryptocurrencies, such as Bitcoin, understanding user acceptance and adoption is crucial. By applying the UTAUT and TAM models to the cryptocurrency domain, researchers gain insights into the factors influencing user acceptance and usage behavior [21–28]. Variables such as performance expectancy, effort expectancy, social influence, and facilitating conditions are examined to identify drivers and barriers to cryptocurrency adoption. These models contribute to a deeper understanding of the complexities surrounding user acceptance and usage of cryptocurrencies, guiding the development and implementation of cryptocurrency technologies.

The primary goal of the research is to understand the Egyptian opinion towards Bitcoin and whether they will use Bitcoin in financial transactions or not. This paper has the lead to provide a summary of people's opinions on such an important topic which is the bitcoin and the cryptocurrencies.

The rest of this paper is organized as follows. Section 2 discusses previous studies in bitcoin and bitcoin in Egypt, research methodology, and the formulation of the hypothesis are described in Sect. 3; Sect. 4 includes statistical analysis and results. Section 5 is the paper's conclusion and future work.

2 Previous Studies

2.1 Bitcoin Previous Studies

In 2009, open-source software known as the Bitcoin system (or Bitcoin), It is a decentralized consensus mechanism. Each transaction is authenticated by the miners. As a result, the Bitcoin system is more secure and maintains its key principle of "Maintain trust in an untrusted environment" without the requirement for a third party to be trusted as payment. Transaction fees are collected by miners for the transactions they verify [5]. When Bitcoin first debuted, anyone could become a miner, but as the user base grew, mining become more difficult and needed specialized equipment. Large warehouses with inexpensive electricity are preferred by miners [6].

According to [7] users are categorized into 6 types, (1) active investor: a more than two-transaction that exclusively sends bitcoin in transactions worth more than USD $2,000, (2) passive investor: an investor who has made more than two transactions but only receives bitcoin in transactions worth more than $100 and neither sends nor receives bitcoin in transactions worth more than $100, (3) currency user: uses more than one transaction, sends, and receives transactions, and sends transactions with a value of less than US$2000, (4) tester: only does one transaction totaling less than $100 USD, (5) Miner: a person who has ever mined bitcoin, as determined by a person who receives newly created bitcoin, and (6) hybrid user: Every other user who is not classified.

The market structure of cryptocurrencies is fundamentally different, even though they share many characteristics with conventional financial markets like equity and foreign exchange. Bitcoin is a speculative investment asset with trading available around-the-clock that also acts as an electronic medium of exchange [8]. The President of the European Central Bank referred to cryptocurrencies as highly speculative assets [9]. The advantages of Bitcoin are listed as follows: ease of use, decentralization, and universal availability, while in contradiction the drawbacks are lack of knowledge, volatility, and low government regulation.

Bitcoin ease of use in which to access the wallet, you need a device that supports the internet (a cell phone, laptop, desktop). You can use cryptocurrencies by using your wallet [10].

Decentralization in which networks are shared among all users, each computer mining node is a part of this framework, and cryptocurrencies lack centralized management. By doing this, it made sure that the central authority cannot impose rules on cryptocurrency owners [11]. Universal Availability: if you want to open a bank account, they need from you many documents, if there are any errors or missing documents, they refuse the request to open an account also it is difficult to o access money in various locations [11].

Lack of knowledge of Bitcoin which many users are exposed to hackers since they are unaware of best practices or how to use cryptocurrencies. Because cryptocurrency technology is somewhat complex, people should educate themselves before utilizing or investing [12]. Volatility is slowing down the adoption of cryptocurrencies is the high volatility of cryptocurrencies. Unknown cryptos might be used by con artists, and in a shaky environment, users might ultimately lose money [10]. Government regulations are few and not clear [12].

The Egyptian government has played a significant role in motivating users to adopt and use Bitcoin, albeit with certain regulatory restrictions. In recent years, the government has taken steps to encourage digital payment systems and embrace financial technology innovations, including cryptocurrencies [13]. The Central Bank of Egypt (CBE) has recognized the potential benefits of blockchain technology and cryptocurrencies while expressing caution about their unregulated nature [13]. The government has also shown interest in exploring the possibility of issuing a digital currency backed by the state [3]. These efforts reflect the government's recognition of the growing importance of digital currencies and its willingness to engage with the evolving financial landscape. However, it is important to note that the government has imposed restrictions on the use of cryptocurrencies for fear of money laundering and illicit activities [4]. While the government's role in motivating users to embrace Bitcoin has been a mixed approach of cautious exploration and regulatory control, its recognition of the potential benefits demonstrates a willingness to adapt to changing financial dynamics.

The Egyptian government has made several attempts to motivate Egyptian citizens to use cryptocurrencies. However, in 2018, Egypt's Dar al-Ifta, the top Islamic law-maker in the country, issued a religious decree prohibiting bitcoin business transactions under Islamic law, citing concerns about potential harm to central financial systems, national security, and the financing of terrorism [13]. Despite this warning, the Central Bank of Egypt announced in 2019 that it was developing a draft law that would outlaw the creation, trading, or marketing of cryptocurrencies without a license [14]. The government has also taken steps to support the digital transformation of the banking and financial sector. In 2020, the Egyptian Parliament passed the Central Bank and Banking Sector Law No. 194, introducing technology and digital tools such as electronic money, digital check settlement, fintech, and regulatory technology to facilitate the sector's digital transformation [15]. These measures reflect the government's efforts to navigate the evolving landscape of cryptocurrencies and embrace digital innovations while addressing concerns related to regulation, security, and financial stability.

2.2 Consumer Intention Behavior Previous Studies

The Unified Theories of Acceptance and Use of Technology (UTAUT) model provides valuable insights into the adoption and usage of technology, including Bitcoin [16]. According to the UTAUT model, user acceptance and usage behavior are influenced by four key factors: performance expectancy, effort expectancy, social influence, and facilitating conditions. In the context of Bitcoin, performance expectancy refers to the user's belief in the benefits and advantages of using Bitcoin, such as faster and cheaper transactions [17]. Effort expectancy relates to the perceived ease of use and user-friendly nature of the Bitcoin system [17]. Social influence considers the impact of peers, family, and influential individuals on a user's decision to adopt Bitcoin [17]. Facilitating conditions encompass the availability of resources, infrastructure, and technical support necessary for Bitcoin usage [17]. By considering these factors, the UTAUT model can help researchers and practitioners better understand and predict user behavior when it comes to Bitcoin adoption and usage.

UTAUT has four constructs influencing the behavioral intention in using technology: performance expectancy, effort expectancy, social influence, and facilitating condition [16].

Researchers in [17] enhanced the previous model by adding hedonic motivation, service price value, and habit, known as the UTAUT2 model. UTAUT2 variables proved that they had a significant impact on the behavioral intention to adopt a particular technology [18, 19].

Perceived risk and consumer innovation have been used to assess their impact on blockchain and cryptocurrency adoption in the UAE. Currency acceptance factors in the UAE. Results had shown that there is a global acceptance difference and similarities regarding perceived usefulness, perceived trust, social influence, and perceived ease of use [21]. In Malaysia [22]. They examined the user intention to use Bitcoin by the UTAUT model and the findings of a pilot study confirmed that the measurement items and constructs of the proposed model were reliable and valid. In the same context,

Authors in [23] used the UTAUT model to test the behavior intention to use Bitcoin in Korea. They concluded that performance expectancy can be influenced by the strategy design of Bitcoin service. Perceived security is a very important behavior while using cryptocurrency.

Examining the Gulf area in [24], the study investigated the factors that influence the adoption of cryptocurrencies for remittance transactions in the United Arab Emirates (UAE). Variables such as behavioral intention (BI) and perceived risk (PR), and consumer innovation (CI) are examined. The conclusions propose UAE remittance market and UAE cryptocurrency industry focus on customer behavior acceptance.

TAM can also be used to examine user intention behavior such as researchers in [27] who proposed a framework to discover the antecedents of behavioral intention to use cryptocurrencies in the USA, Extending the technology acceptance model (TAM) with used and 397 individuals surveyed. The results indicated that perceived risk negatively predicted the behavioral intention to use Bitcoin. Whereas, social influence, perceived usefulness, perceived ease of use, and attitude positively predicted the behavioral intention. Additionally, trust was identified as a crucial factor that significantly influenced the acceptance of telemedicine services and TAM is used in the investigation of acceptance of telemedicine services [28].

3 Research Methodology

The paper aims to understand Egyptian behavior toward using Bitcoin. The researcher used performance expectancy, effort expectancy, social influence, and facilitating condition, hedonic motivation, service price value, and habit selected from the literature.

This research aims to significantly identify the factors (and their interdependence relationships) that determine the behavior of Egyptian users towards Bitcoin. The research conducted an online survey among the employees working in a private Egyptian University. 3761 employees filled out the survey. The questionnaire involves all study variables and three demographic information. Twenty-seven questions are placed in the questionnaire, and the Likert-type five-level scales are employed. The options for each question using points 1 to 5 represent extremely disagree, disagree, neutral, agree, and extremely agree, respectively as shown in Table 1.

Table 1. Respondents' Profile

Attribute	Category	Frequency	Percent
Gender	Male	1170	31.1%
	Female	2591	68.8%
Job title	Administrative	3011	80.1%
	Management	750	19.9%

(*continued*)

Table 1. (*continued*)

Attribute	Category	Frequency	Percent
Income	Less than 5,000	1051	27.9%
	5001–10,000	1918	50.9%
	More than 10,000	792	21.2%
Total	Total	3761	

The second part of the questionnaire contains five sets of questions related to UTAUT variables proposed by [16] and has eight constructs influencing the behavioral intention in using technology which is listed as follows: performance expectancy, effort expectancy, social influence, and facilitating condition. UTUAT2 holds the following variables: hedonic motivation, service price value, and habit [17] which are represented in Table 2.

Table 2. Research Model Factors

Attribute	Definition
Performance Expectancy	It is defined as individual's belief that Bitcoin can enhance the financial transactions [21–26]
Effort Expectancy	The ease of using Bitcoin helps to minimize travel and movement effort [21–26]
Social Influence	The degree to which people around the user perceive Bitcoin is significant [21–26]
Facilitating Condition	To what extent Bitcoin can facilitate the transactions [21–26]
Hedonic Motivation	HM translate user's happiness and enjoyment when using Bitcoin in the learning process [21–26]
Service price value	The cost and the price to use Bitcoin [21–26]
Habit	The process in which the Bitcoin functions became an ordinary action done by users [21–26]

The second part of the questionnaire contains eight sets of questions related to UTAUT variables and UTAUT 2 proposed by [16, 17] as shown in Table 3.

After defining the factors and their associated elements, hypotheses are formulated to understand the relationships between factors. Table 4 summarizes the research hypotheses we used in the study.

Table 3. Research Model Factors and the items

Factor	Encoding	Statements
Performance Expectancy (PE) [21–26]	PE1	I believe that using Bitcoin is useful
	PE2	I believe that the Bitcoin provide faster service
	PE3	I think Bitcoin is useful to make it faster and easier
	PE4	The interaction with Bitcoin is clear and understandable
Effort Expectancy (EE) [21–26]	EE1	I feel that I will be an active member of the society Bitcoin
	EE2	I feel happy when I use Bitcoin
	EE3	I will use all the features of Bitcoin
	EE4	I will miss Bitcoin if for any reason I can not use it
Social Influence (SI) [21–26]	SI1	Most of my friends will use Bitcoin
	SI2	Most of my friends motivate me to use Bitcoin
	SI3	My family will use Bitcoin
	SI4	My family members motivate me to use Bitcoin
Facilitating Condition (FC) [21–26]	FC1	I have the resources necessary to use Bitcoin
	FC2	I have the knowledge necessary to use Bitcoin
	FC3	Bitcoin are compatible with other technologies I use
	FC4	I can get help from others when I have difficulties using Bitcoin
Hedonic Motivation (HM) [21–26]	HM1	Learning the use of Bitcoin will be easy for me
	HM2	I will enjoy using Bitcoin
	HM3	I believe that the use of Bitcoin would be enjoyable Process

(*continued*)

Table 3. (*continued*)

Factor	Encoding	Statements
	HM4	I believe that the use of Bitcoin would be a fun process
Service price value (PV) [21–26]	PV1	Bitcoin are a good value for the money
	PV2	Bitcoin are reasonably priced
Habit	H1	I believe that Bitcoin usage will be a habit in my life
	H2	I will use Bitcoin as a regular basis
Behavioral Intention (BI) [21–26]	BI1	Given the opportunity, I will use Bitcoin
	BI2	I am willing to use Bitcoin in the near future
	BI3	I am open to use Bitcoin
	BI4	I intend to use Bitcoin in the future

Table 4. Research Hypothesis

#	Hypothesis
H1-PE	Performance Expectancy has a significant effect on Behavior intention to use Bitcoin
H2-EE	Effort Expectancy has a significant effect on Behavior intention to use Bitcoin
H3-SI	Social Influence has a significant effect on Behavior intention to use Bitcoin
H4 FC	Facility Conditions has a significant effect on Behavior intention to use Bitcoin
H5 HM	Hedonic Motivation has a significant effect on Behavior intention to use Bitcoin
H6-SPV	Service price value has a significant effect on Behavior intention to use Bitcoin
H7-H	Habit has a significant effect on Behavior intention to use Bitcoin

4 Statistical Analysis and Results

The research type is quantitative, and the proposed model was constructed based on a literature review. The data collection is a survey that was employed using an online google form. The target population was expressed as Egyptian users. Two statistical software, Statistical Package for the Social Sciences (SPSS) 25.0 and AMOS 23.0, AMOS is a statistical software that functions as an adjunct module for SPSS. Its primary purpose is to facilitate various analytical techniques such as Structural Equation Modeling, path analysis, and confirmatory factor analysis. Were used to measure the proposed model variables' relationships. The following tests are described descriptive analysis, reliability test, Chi-square test, and Pearson Correlation.

4.1 Reliability Test

Reliability tests the consistency of the data; it is measured by the Cronbach's alpha coefficient. Table 5 presents the results of the reliability test. According to [20], the value of the Cronbach's alpha coefficient must be greater than 0.7.

Table 5. Reliability Test

Latent Variable	Item	Factor Loading	Composite Reliability CR	Cronbach's alpha	Average Variance Extracted AVE
Performance Expectancy	PE1	0.810	0.784	0.805	0.806
	PE2	0.773			
	PE3	0.721			
	PE4	0.705			
Effort Expectancy	EE1	0.836	0.736	0.799	0.790
	EE2	0.802			
	EE3	0.702			
	EE4	0.709			
Social Influence	SI1	0.996	0.915	0.941	0.963
	SI2	0.987			
	SI3	0.900			
	SI4	0.919			
Facilitating Conditions	FC1	0.881	0.893	0.823	0.877
	FC2	0.825			
	FC3	0.808			
	FC4	0.894			
Hedonic Motivation	HM1	0.972	0.918	0.927	0.918
	HM2	0.915			
	HM3	0.956			
	HM4	0.922			
Price Value	PV1	0.981	0.955	0.971	0.987
	PV2	0.988			
Habit	H1	0.712	0.788	0.776	0.753
	H2	0.731			

(continued)

Table 5. (*continued*)

Latent Variable	Item	Factor Loading	Composite Reliability CR	Cronbach's alpha	Average Variance Extracted AVE
Behavior intention to use Bitcoin	BI1	0.853	0.873	0.811	0.859
	BI2	0.896			
	BI3	0.800			

4.2 Chi-Square Test

Chi-square tests whether the dependent and independent variables are associated and whether this association is significant [25]. Chi-square results show that all association with the independent variable BI is significant as shown in Table 6.

Table 6. Chi-square Test

Association	Chi-square value	Asymptotic significance	Result
PE-> BI	1.525	0.000 <0.01	Significant
EE-> BI	1.207	0.000 <0.01	Significant
SI-> BI	1.351	0.000 <0.01	Significant
FC-> BI	1.126	0.000 <0.01	Significant
HM-> BI	1.316	0.000 <0.01	Significant
PV-> BI	1.223	0.000 <0.01	Significant
H-> BI	802	0.000 <0.01	Significant

4.3 Pearson Correlation Test

Pearson correlation test defines the extent to which the increase or decrease in the factors used in the model results in a change (increase or decrease) in the dependent factor with which it is correlated [25]. Thus, the results are represented in Table 7 and Fig. 1 shows the research model variables and Pearson correlation value.

Based on [25], if value of Pearson Coefficient value is between (0–0.2) is a very weak correlation, between (0.2–0.4) represents a weak correlation, between (0.4–0.6) is a reasonable correlation; between (0.6–0.8) is high correlation and finally (0.8–1) is very high correlation. Performance expectancy, effort expectancy, habit, and social influence have a high correlation value to the behavior intention to use Bitcoin while, facilitating conditions, hedonic motivation, and price value have a very high correlation to the Behavior intention to use Bitcoin. Based on Fig. 1, the results have shown that the hypothesis is accepted.

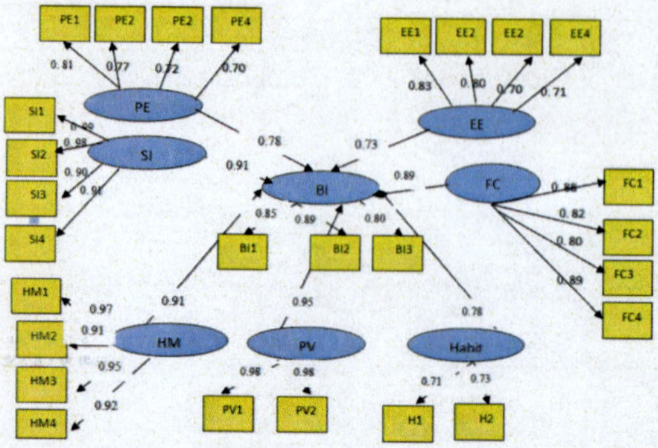

Fig. 1. Research Model variables and Pearson Correlation Values.

Table 7. Person Correlation Test Results

Hypotheses	Pearson's correlation coefficient	P-value
H1(BI ← PE)	0.738	$p < 0.01$
H2 (BI ← EE)	0.771	$p < 0.01$
H3 (BI ← SI)	0.681	$p < 0.01$
H4 (BI ← FC)	0.829	$p < 0.01$
H5 (BI ← HM)	0.823	$p < 0.01$
H6 (BI ← PV)	0.913	$p < 0.01$
H7 (BI ← H)	0.742	$p < 0.01$

4.3.1 Performance Expectancy has a Significant Effect on Behavior Intention to Use Bitcoin

This hypothesis was supported. The findings indicated that the performance expectancy of using Bitcoin is a strong indicator of users' intention to continuously use them in the future. These findings have been supported also by [21, 22, 30] found that perceived usefulness is a major determinant in the usage of Bitcoin. Similarly, [23] suggested that increasing awareness and education about cryptocurrency could enhance its adoption in the country.

4.3.2 Effort Expectancy has a Significant Effect on Behavior Intention to Use Bitcoin

This hypothesis was supported. Our results showed that users' behavior intention largely depends on the effort expectancy which agrees with the results in [23, 30] that also suggests the importance of cryptocurrency system security. Our findings underscore the importance of effort expectancy in the user intention to use Bitcoin, which it is critical for Bitcoin providers. It is also important for these companies to ask users for suggestions that allow these companies to introduce more services on these apps that are important for these users.

4.3.3 Social Influence has a Significant Effect on Behavior Intention to Use Bitcoin

This hypothesis was supported. Our findings showed that social influence is an important element that affects the users' intention to use Bitcoin. Authors in [21] emphasize the importance of social influence in affecting and developing digital currency frameworks, and in the same context [27] reached the same findings when examining users' intention to use Bitcoin.

4.3.4 Facility Conditions has a Significant Effect on Behavior Intention to Use Bitcoin

This hypothesis was supported. Our findings showed that social influence is an important element that affects the users' intention to use Bitcoin. Authors in [21, 27, 30] suggest that Bitcoin companies should focus on providing easy-to-use and secure platforms while emphasizing the benefits of using Bitcoin and provide facilitating conditions in the usage and the dealing process in Bitcoin, also [27] emphasis the reduction of risk and increase user adoption.

4.3.5 Hedonic Motivation has a Significant Effect on Behavior Intention to Use Bitcoin

This hypothesis was supported. Our findings showed that hedonic motivation is an important factor that affects users' intention to use Bitcoin. The feeling of fun and joy when interacting and dealing with Bitcoin is very important in user behavior. These findings are similar to the findings of [25] which showed that hedonic motivation is an important factor that affects users' intention to use bitcoin. Also, our findings are consistent with [30] that argued that user intention is influenced by hedonic motivation. Thus, based on our results and the support in the extant literature, we recommend that cryptocurrency providers should continuously monitor the intention of the users. This can be done through surveys and effective customer service.

4.3.6 Service Price Value has a Significant Effect on Behavior Intention to Use Bitcoin

As expected, the hypothesis was supported since our findings confirmed the significant effect of service price on user intentions to use Bitcoin. The relationship between the price and the service affects the customer behavior as stated previously in different research [22–24]. Bitcoin providers should study the factors that affect users' behavior to ensure their use and continuous intention.

4.3.7 Habit has a Significant Effect on Behavior Intention to Use Bitcoin

This hypothesis was supported. The findings indicated that user Habit affects the user intention behavior. These findings have been supported in the previous literature. For instance, [25] found that Hedonic motivation is important in users' willingness to use Bitcoin. Thus, based on these findings, the bitcoin providers that have succeeded to build a strong trust with their customers should rely on this in building positive attitudes towards usage of bitcoin.

4.4 Discussion of Results

The paper examines the intended behavior of Egyptians to use Bitcoin in financial transactions. The results of the research are important for the following reasons. Firstly, the paper contributes theoretically to the usage of Bitcoin in Egypt. Mainly, it focuses on performance expectancy, effort expectancy, social influence, facilitating conditions, hedonic motivation, price value, and habit. The research can emphasize research in Egypt that focuses on cryptocurrency, bitcoin, and digital banking.

Results of the analysis show that performance expectancy, effort expectancy, habit and social influence have a high correlation value to the behavior intention to use bitcoin while, facilitating conditions, hedonic motivation and price value have a very high correlation to the behavior intention to use bitcoin. Egyptian are going to enjoy and be pleased when they use Bitcoin on the same perspective the people around them motivate the employees to use Bitcoin also the price of the services available to them has a great focus to them. Research limitations are summarized as follows. First, this research sample is limited to 3761. A larger sample size is required to understand Egyptian perception to use bitcoin, and a comparative study should provide very interesting results such as testing the variables on European countries. In addition, this study's results and model can be tested in Gulf countries that have experiences with Bitcoin. The quantitative approach in the research is limited to the number of collecting data through an online questionnaire. Future research can adopt qualitative research such as interviews to provide more understanding of the intention of Egyptian to use Bitcoin. Finally, the research did not examine how the demographic and cultural dimensions affect Egyptian use of bitcoin such as gender, job, and income.

5 Conclusion and Future Work

This paper examines the Egyptian intention to use Bitcoin in financial transactions. 3761 employees working at a private university in Egypt filled out the survey. Results of the analysis show that facilitating conditions, Hedonic Motivation, and Price value are a very high correlation with significant behavior intention to use Bitcoin. Performance expectancy, effort expectancy, facilitating condition, and habit have a high correlation significance on the Egyptian user's intention to use Bitcoin. Research limitations are summarized as follows. First, this research sample is limited to 3761. A larger sample size is required to understand Egyptian users use Bitcoin, and a comparative study should provide very interesting results such as testing the variables on European countries.

In addition, this study's results and model can be tested in Gulf countries that have experiences with Bitcoin. The quantitative approach in the research is limited to the number of collecting data through an online questionnaire. Future research can adopt qualitative research such as interviews to provide more understanding of the intention of

Egyptian users to use Bitcoin. Finally, the research did not examine how the demographic and cultural dimensions affect Egyptian opinion to use Bitcoin such as age, gender, job, and income.

References

1. Nerurkar, P., Patel, D., Busnel, Y., Ludinard, R., Kumari, S., Khan, M.: Dissecting bitcoin blockchain: empirical analysis of bitcoin network (2009–2020). J. Netw. Comput. Appl. **177** (2021)
2. Park, S., Im, S., Seol, Y., Paek, J.: Nodes in the bitcoin network: comparative measurement study and survey. IEEE Access **7**, 57009–57022 (2019)
3. Feng, Q., He, D., Zeadally, S., Khan, M., Kumar, N.: A survey on privacy protection in blockchain system. J. Netw. Comput. Appl. **126**, 45–58 (2019)
4. Wang, L., Shen, X., Li, J., Shao, J., Yang, Y.: Cryptographic primitives in blockchains. J. Netw. Comput. Appl. **127**, 43–58 (2019)
5. Aggarwal, S., Chaudhary, R., Aujla, G.S., Kumar, N., Choo, K.-K.R., Zomaya, A.Y.: Blockchain for smart communities: applications, challenges and opportunities. J. Netw. Comput. Appl. **144**, 13–48 (2019)
6. Baur, D., Hong, K., Lee, A.: Bitcoin: medium of exchange or speculative assets? J. Int. Financ. Mark. Inst. Money, 54177–189 (2018)
7. Yermack, D.: Bitcoin, innovation, financial instruments, and big data. In: Handbook of Digital Currency, pp. 31–43 (2015)
8. Conghui, C., Lanlan, L., Zhao, N.: Fear sentiment, uncertainty, and bitcoin price dynamics: the case of COVID-19. Emerg. Mark. Finance Trade **56**(10), 2298–2309 (2020). https://doi.org/10.1080/1540496X.2020.1787150
9. Rana, R., Tricase, R., Cesare, D.: Blockchain technology for a sustainable agri-food supply chain. Br. Food J. (2021)
10. Alqassem, I., Rahwan, I., Svetinovic, D.: The anti-social system properties: bitcoin network data analysis. IEEE Trans. Syst. Man Cybern. Syst. **50**(1), 21–31 (2020). https://doi.org/10.1109/TSMC.2018.2883678
11. Sousa, P., Resende, J., Martins, R., Antunes, L.: The case for blockchain in IoT identity management. J. Enterp. Inf. Manag. (2020)
12. Kant, N.: Blockchain: a strategic resource to attain and sustain competitive advantage. Int. J. Innov. Sci. (2021)
13. Central Bank of Egypt: A warning statement (2018). Press Release
14. Religion Decree No 4205: Central Bank of Egypt, A Warning Statement, The Status of Transactions in Bitcoins and other Cryptocurrencies under Islamic Law, Egypt's Dar al-Ifta (2017)
15. Egypt Independent: CBE to ban issuance of cryptocurrency without licenses (2019)
16. Venkatesh, V., Morris, M., Davis, G., Davis, F.: User acceptance of information technology: toward a unified view. MIS Q. **27**(3), 425–478 (2003). https://doi.org/10.2307/30036540
17. Venkatesh, V., Thong, J., Xu, X.: Consumer acceptance and use of information technology: extending the unified theory of acceptance and use of technology. MIS Q. **36**(1), 157–178 (2012). https://doi.org/10.2307/41410412
18. Mehta, A., Morris, N., Swinnerton, B., Homer, M.: The influence of values on E-learning adoption. Comput. Educ. **141**, 103617 (2019). https://doi.org/10.1016/j.compedu.2019.103617
19. Pérez, G., Hurtado, M., Sánchez-Martí, A.: Towards smart learning spaces in Catalan schools: teachers' perceptions of change. Learn. Environ. Res., 1–17 (2021)

20. Cortina, J.: What is coefficient alpha? An examination of theory and applications. J. Appl. Psychol. **78**, 98–104 (1993). [e-journal]. https://doi.org/10.1037/0021-9010.78.1.98
21. Saif Almuraqab, N.: Predicting determinants of the intention to use digital currency in the UAE: an empirical study. Electron. J. Inf. Syst. Dev. Ctries. **86**(3), 1–12 (2020). https://doi.org/10.1002/isd2.12125
22. Yeong, Y., Kalid, K., Sugathan, S.: Cryptocurrency acceptance: a case of Malaysia. Int. J. Eng. Adv. Technol. **8**(5), 28–38 (2019). https://doi.org/10.35940/IJEAT.E1004.0585C19
23. Hwang, Y., Moon, Y.: A quantitative model of determinants of use behavior for the cryptocurrency system in terms of security concerns and risks. J. Theor. Appl. Inf. Technol. **97**(17), 4623–4635 (2019)
24. Jegerson, D., Khan, M., Mertzanis, C.: Adoption of cryptocurrencies for remittances in the UAE: the mediation effect of consumer innovation. Eur. J. Innov. Manag. Emerald (2022)
25. Mostafa, L., Essawi, N.: The effect of social media on electronic-purchase decisions during the COVID-19 in Egypt. Financ. Bus. Res. Portsaid 3(23) (2022)
26. Mostafa, L.: Student intention behavior to use smart campus in Egyptian University. J. Commer. Environ. Stud. **2**(23) (2023). Suez Canal University
27. Koksalmis, G.H., Arpacı, İ., Koksalmis, E.: Predicting the intention to use bitcoin: an extension of technology acceptance model (TAM) with perceived risk theory. In: Al-Emran, M., Shaalan, K. (eds.) Recent Innovations in Artificial Intelligence and Smart Applications. SCI, vol. 1061, pp. 105–120. Springer, Cham (2022). https://doi.org/10.1007/978-3-031-14748-7_6
28. Kamal, S.M., Shafiq, M., Kakria, P.: Investigating acceptance of telemedicine services through an extended technology acceptance model (TAM). Technol. Soc. **60**, 101212 (2020)
29. Nair, A.: Egypt Considering Issuing Digital Currency, Says Central Bank Governor. CoinDesk (2021). https://www.coindesk.com/markets/2021/03/29/egypt-considering-issuing-digital-currency-says-central-bank-governor/. Accessed 9 May 2023
30. Purbandini, A., Hau, A.: Unified theory of acceptance and use of technology model for user acceptance analysis of Bitcoin. AIP Conf. Proc. **2329**(1), 050011 (2021). https://doi.org/10.1063/5.0042374

Image and Video Processing

Three-Dimensional Photometric Stereo Reconstruction By shadow Elimination

Lyes Abada(✉), Omar Chouaab Malki, Mohamed Mekkaoui, and Saliha Aouat

Artificial Intelligence and Data Science Department, USTHB,
16024 Bab-Ezzouar, Algiers, Algeria
{labada,saouat}@usthb.dz

Abstract. Computer vision is a very wide field, although, research is advancing rapidly. But there are still many incomplete areas of research. For this reason we are interested on 3D reconstruction field, which consists of creating a 3D object from one or more 2D images. In this paper, we propose a method to improve 3D reconstruction by photometric stereo (SP) based on the surface orientation. This approach aims to eliminate shadow in 2D images in order to perfect the accuracy and results of the 3D object. The 3D objects are used in many applications such as medical diagnosis, object recognition and augmented reality. Finally we show the obtained results before and after applied our improved on the Singular Value Decomposition (SVD) and Microfacet-based Photometric Stereo (MFPS) methods.

Keywords: Photometric stereo · 3D reconstruction · Shape From Shading · Singular Value Decomposition · Shadow elimination

1 Introduction

Computer vision is a very wide field, although the research is advancing rapidly. Especially the three-dimensional (3D) reconstruction of an object from a 2D image is the inverse problem of photography, which consists in obtaining a 3-dimensional representation of an object or a scene from one or several two-dimensional images and to recover the dimension lost during the process of image formation. For that, we propose in this paper a method to improve the quality and the precision of the 3D reconstruction. The proposed method consists in eliminating the shadow in the 2D images to improve the 3D object using the surface orientation to eliminate shadow in 2D images. Our project aims to eliminate the shadow in 2D images to reconstruct a 3D scene from one or more 2D images.

3D models are used in many fields, that's why we have to reconstruct these models from the real world. Among the 3D reconstruction methods we can find the Photometric Stereo (PS) [1–3], which allows to compute the 3D object from

A. Hassanien et al. (Eds.): AISI 2023, LNDECT 184, pp. 473–481, 2023.
https://doi.org/10.1007/978-3-031-43247-7_41

several images taken by the same position of the camera with different light source positions, it is based on the photometric information contained in each object point.

Photometric stereo (PS) based on another domain which is the Shape From Shading (SFS) [4,5]. SFS is a particular case of the PS which uses only one image. SFS is considered as an ill-posed problem, since theoretically it is not possible to determine a unique solution of a single equation with three unknowns (x,y,z) [6].

Several techniques have been used to solve photometric stereo such as machine learning [3,7], meta heuristics [1,2,4], etc. The PS can be classified into two categories according to the light sources positions. It is calibrated if the light sources positions are known, otherwise it is considered uncalibrated [8,9]. Several works consider the PS as calibrated, since we can determine the positions of the light sources with a calibration object [10]. Photometric reconstruction uses the basic notions of a field called Image Formation (IF) [11], which is the reverse operation of the SFS [12,13]. The basic equation of the image formation (Eq. 1) is used to determine the relationship between the 3D object and the 2D image.

$$E = \frac{\alpha}{4}(\frac{p}{f})^2 I \cos^4 \theta L \tag{1}$$

Such as:

- E : is the illumination of a pixel (we can consider it as grayscale),
- L : is the luminance of the apparent surface,
- I : is the intensity of the light source,
- The internal parameters of the camera:
  - θ : the angle between the projection line and the optical axis,
  - p : the lens diameter,
  - f : the focal length,

Certain constraints must be imposed on light sources, surface type, and camera model [14] which allow to simplify the image formation equation (Eq. 1) in order to be able to use it in photometric 3D reconstruction. Most of the works in this field use the following constraints in order to obtain Eq. 2:

- Lambertian surface,
- Light source located at the infinity and parallel with the z-axis,
- Camera with a parallel projection model,

Based on these constraints, the image formation equation can be simplified as follows [5,15]:

$$E = \rho N.S \tag{2}$$

This equation is used in the SFS, it gives the relation between the 3D object given by the normal vector N (x, y, z) and the 2D image given by the pixel illumination E. Theoretically, Eq. 2 does not have a unique solution, (three unknowns with one equation). Photometric Stereo is the field that uses the same equation many times such that the camera position is the same but only the light sources positions are different. By using m images we can obtain a system of m equations, with the same normal vector but with different gray scales E_i and light sources S_i. We can obtain the following equations system [10]:

$$E_i = \rho N S_i \quad \backslash i = 1..m \tag{3}$$

With this equations system (Eq. 3), we have three possible cases:

- $m < 3$: the system has no unique solution,
- $m = 3$: the system has a unique exact solution,
- $m > 3$: theoretically the system has no exact solution,

We can use three images to get a unique solution, but if only one source is affected by the shadow, the result will be completely wrong [16]. So it is necessary to use more than three equations (more than three images) and determine an optimal solution as a minimization problem.

We can solve the above system of equations with linear algebra using a matrix representation. Such that E is a matrix of size [m,1] containing the gray scales of the pixel, S is a matrix of size [m,3] containing the light sources directions, and $N = (x, y, z)^T$ is the normal vector. we can obtain the following equation (Eq. 4):

$$E_{[m,1]} = \rho S_{[m,3]} N_{[3,1]} \quad //with\ m\ is\ the\ number\ of\ images \tag{4}$$

Since the matrix S is not square, it is not possible to compute its inverse matrix, on the other hand we can compute the pseudo-inverse matrix S^{-1} with Singular Value Decomposition (SVD) [17]. We can obtain the following equation [18]:

$$N_{[3,1]} = \frac{1}{\rho} S^{-1}_{[3,m]} E_{[m,1]} \tag{5}$$

The solution given by this equation is relative to the positions of the light sources, when we multiply all gray levels by the same factor the result remains the same, so we can replace the albedo ($1/\rho$) with 1 to simplify our system (ρ can be easily determined thereafter).

In the next section, we will present our proposal to improve the 3D reconstruction by photometric stereo with the shadow elimination.

2 3D Reconstruction by Shadow Elimination

In order to achieve a photometric stereo solution which consists in search three-dimensional shape of a single view, we propose a method based on the orientation of the surface, in order to find a threshold to eliminate certain gray levels in different positions of the light source of each pixel, then we exploit its thresholds to improve the result of 3D reconstruction techniques.

In order to validate our proposition, we will use two existing techniques. The mathematical solution SVD [19], and the Microfacet-based technique (MFPS) [20]. In this part, we will see the calculation techniques used to have the normal matrix of all the pixels of a 3D object, in order to carry out our work. For some square matrices, we can compute their inverses using eigenvalue decomposition. But to do this transformation, the matrix must be square and diagonalizable [19].

The singular value decomposition is similar to the eigenvalue decomposition, it works in any m x n matrix. This technique can be used to solve the photometric stereo equations system to get the 3D object normals matrix to generate the 3D object [19]. The SVD method is applied to Lambertian surfaces only, such as, the "albedo" reflectance has a constant value. In order to apply our proposals on non-Lambertian surface objects, we use another technique which is based on a general isotropic reflection, where, the reflectance is given by a function called BRDF [20] which depends on the three factors (the position of the camera, the position of the light source and the normal vector). The simplified equation of the image formation becomes as follows:

$$E = B(V, N, S)N.S \tag{6}$$

Such that, V represents the camera position and B represents the BRDF function (Bidirectional Reflectance Distribution Function). The solution of this formulation is given by the a minimization problem form. For more details, please see Sect. 3 of the Chen et al. [20] work.

3 Nos Contributions

Shadow is a region where light cannot reach the surface due to an obstacle [21]. The regions that will be considered as shadows are dark areas (low intensities) as shown in Fig. 1. The pixels affected by the shadow influence the quality of the result during the creation of the 3D object, for that, we propose two methods to eliminate the pixels that represent a shadow, in order to improve the quality of the 3D reconstruction.

4 Shading Elimination Using Surface Orientation

Our proposal is based on the surface orientation in order to eliminate the shadow. According to results we had in previous works [4,5], we found that the pixels affected by the shadow (see the Fig. 1) form an angle between the normal pixel

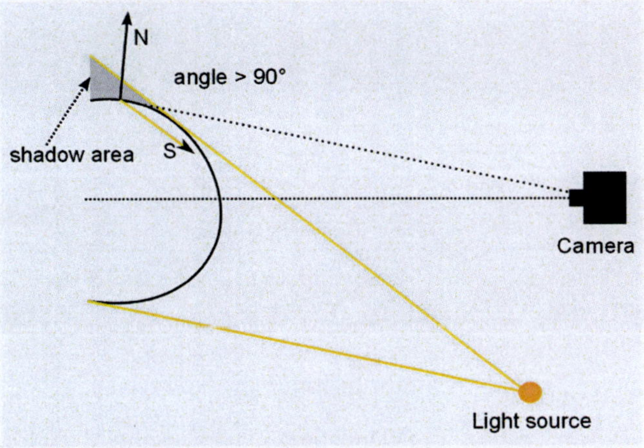

Fig. 1. Shadow and light.

vector and the light source vector greater or equal to 90°. In this case the pixel is opposed to the light source, which will consequently have a very low gray scale. Our proposal consists in eliminating certain gray scale lower than a certain threshold of each pixel. This last generates an angle between the normal vector and the light source vector which tends towards 90°. To solve this problem, we have processed all the pixels by computing for each pixel the mean angular error (MAE) between the normal vector and the all light source positions.

Our method consists in computing the normal vector, the first time using all the light sources, then we calculate the angle between this normal and each light source. In order to eliminate the light sources that generate the shadow, we set a threshold on the gray scale.

We will test all thresholds between min_threshold equal to 80° and max_threshold equal to 95°. The second step consists to compute a new value of the normal using the remaining light sources (not eliminated) . We repeat this operation until the stabilization of the values of the normals. This method eliminates the gray scales affected by shadow to obtain improved normal values for each pixel.

5 Experiments

In order to validate our method, we use the "DiLiGenT" [22] dataset. It is a photometric stereo dataset that contains 10 different objects (as shown in Fig. 2). For each object, we have 96 PNG images of 16-bit depth with a size of 612×512 taken under 96 different light source directions.

In this section, we present the results of the proposed method, which is based on the elimination of the shadow on two methods for improving the normal fields. The first with the SVD mathematical technique, then with the MFPS technique using a threshold specific for each pixel then a fixed threshold for all the pixels.

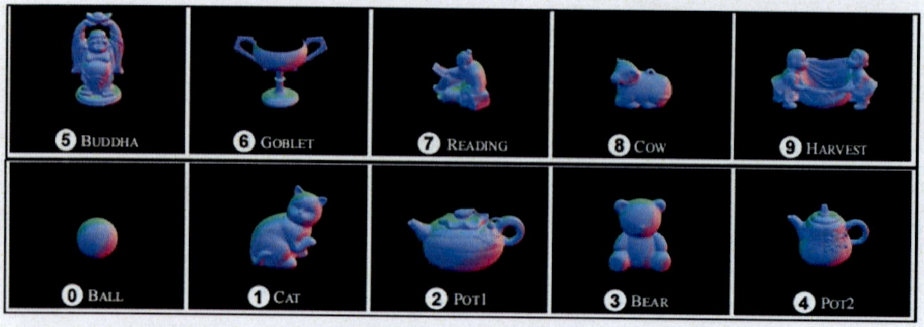

Fig. 2. Stereo-photometric dataset.

Table 1. SVD method improvement

Objects	Before SVD	After SVD
bear	8.3887	7.5575
cat	8.4132	7.9035
goblet	18.4999	17.4481
pot1	8.8944	8.2743
reading	19.8031	18.836
Ball	4.0962	3.7097
Moyenne	11.3492	10.6215

Table 2. MFPS method improvement

Objects	Before MFPS	After MFPS
bear	10.3807	9.5038
cat	7.1237	6.6160
goblet	12.1096	11.5326
pot1	7.3890	7.2953
reading	17.0776	17.0812
Ball	3.8112	2.7357
Moyenne	9.6486	9.12

From the results of Table 1, we notice that with the use of the precise thresholds for each pixel (technique 1), we see that there is an improvement for each object, the better it was with the object Goblet, the result is improved by 1.05° on average of all pixels on average. Our method with the calculation technique (SVD) gives an average improvement of all the objects. In Table 2, we changed the calculation technique to the MFPS technique and we see that there is an improvement for each object. The best is given with the Ball object. According to the results of Tables 1 and 2, we notice that with using of specific thresholds

Table 3. 3D object generation

2D images	3D Object	2D images	3D Object

for each pixel, we see that there is an improvement for each object, the best one was with the Goblet object. The results are improved by 1.05° on average. The results of our proposed method with the SVD technique are improved. We also tested on the second MFPS technique. We see that there is an improvement for each object. The best was with the Ball object where the result is improved by 1.01° on average of all pixels. Our method with the MFPS technique gives a significant improvement on average of all objects.

6 Conclusion

In this paper, we have used surface orientation in an effort to find the best threshold to eliminate certain shade-affected gray levels. This method gave us better results, using the 2 calculation methods "SVD" and "MFPS", comparing the results obtained using the two calculation techniques, we had a good improvement.

In the perspectives of this work, we will propose an improvement of the obtained results using Deep Learning. As well as the generalization of our problem from single-view photometric reconstruction to multi-view reconstruction.

References

1. Abada, L., Aouat, S.: Improved photometric stereo based on local search. Multimed. Tools Appl. **81**(21), 31181–31195 (2022)
2. Abada, L., Bennaceur, M., Boudjenana, A.A., Aouat, S.: Using pso metaheuristic to solve photometric 3d reconstruction. In: 2022 7th International Conference on Image and Signal Processing and Their Applications (ISPA), pp. 1–6 (2022). IEEE
3. Abada, L., Aouat, S., Elfani, R., Zamoum, Y.H.: Three-dimensional photometric reconstruction of a single view using machine learning techniques. In: 2022 7th International Conference on Image and Signal Processing and Their Applications (ISPA), pp. 1–6 (2022). IEEE
4. Abada, L., Aouat, S.: Tabu search to solve the shape from shading ambiguity. Int. J. Artif. Intell. Tools **24**(05), 1550035 (2015)
5. Abada, L., Aouat, S.: Improved shape from shading without initial information. Front. Comp. Sci. **11**(2), 320–331 (2017)
6. Abada, L., Aouat, S.: Shape from shading with and without boundary conditions. In: Chen, L., Kapoor, S., Bhatia, R. (eds.) Intelligent Systems for Science and Information. SCI, vol. 542, pp. 369–387. Springer, Cham (2014). https://doi.org/10.1007/978-3-319-04702-7_21
7. Bourahla, O.E.F., Debbah, A.M., Abada, L., Aouat, S.: Shape from shading: a non-iterative method using neural networks. In: Abraham, A., Hanne, T., Castillo, O., Gandhi, N., Nogueira Rios, T., Hong, T.-P. (eds.) HIS 2020. AISC, vol. 1375, pp. 759–768. Springer, Cham (2021). https://doi.org/10.1007/978-3-030-73050-5_74
8. Chen, G., Waechter, M., Shi, B., Wong, K.-Y.K., Matsushita, Y.: What is learned in deep uncalibrated photometric stereo? In: European Conference on Computer Vision (2020)
9. Chen, G., Han, K., Shi, B., Matsushita, Y., Wong, K.-Y.K.: Self-calibrating deep photometric stereo networks. In: Proceedings of the IEEE Conference on Computer Vision and Pattern Recognition, pp. 8739–8747 (2019)
10. Cho, D., Matsushita, Y., Tai, Y.-W., Kweon, I.S.: Semi-calibrated photometric stereo. IEEE Trans. Pattern Anal. Mach. Intell. **42**(1), 232–245 (2018)
11. Abada, L., Aouat, S.: Solving the perspective shape from shading problem using a new integration method. In: 2013 Science and Information Conference, pp. 416–422 (2013). IEEE
12. Abada, L., Aouat, S.: Facial shape-from-shading using features detection method. Int. J. Adv. Intell. Paradigms **8**(1), 3–19 (2016)
13. Abada, L., Aouat, S.: Enhanced variational approach by fast-marching to recover a 3d shape from shading. In: Proceedings of the 20th International Database Engineering & Applications Symposium, pp. 156–163 (2016)
14. Ouyang, W., Wang, R., Xu, B.: Fabric pilling measurement using three-dimensional image. J. Electron. Imag. **22**(4), 043031 (2013)
15. Rouy, E., Tourin, A.: A viscosity solutions approach to shape-from-shading. SIAM J. Numer. Anal. **29**(3), 867–884 (1992)
16. Cho, D., Matsushita, Y., Tai, Y.-W., Kweon, I.: Photometric stereo under non-uniform light intensities and exposures. In: Leibe, B., Matas, J., Sebe, N., Welling, M. (eds.) ECCV 2016. LNCS, vol. 9906, pp. 170–186. Springer, Cham (2016). https://doi.org/10.1007/978-3-319-46475-6_11
17. Shi, B., Wu, Z., Mo, Z., Duan, D., Yeung, S.-K., Tan, P.: A benchmark dataset and evaluation for non-lambertian and uncalibrated photometric stereo. In: Proceedings of the IEEE Conference on Computer Vision and Pattern Recognition, pp. 3707–3716 (2016)

18. Hayakawa, H.: Photometric stereo under a light source with arbitrary motion. JOSA A **11**(11), 3079–3089 (1994)
19. Abdi, H.: Singular value decomposition (svd) and generalized singular value decomposition. Encyclopedia of measurement and statistics, pp. 907–912 (2007)
20. Chen, L., Zheng, Y., Shi, B., Subpa-Asa, A., Sato, I.: A microfacet-based model for photometric stereo with general isotropic reflectance. IEEE Trans. Pattern Anal. Mach. Intell. **43**(1), 48–61 (2019)
21. Almoussa, N.: Variational retinex and shadow removal. The Mathematics Department, UCLA (2008)
22. Shi, B., Mo, Z., Wu, Z., Duan, D., Yeung, S.-K., Tan, P.: A benchmark dataset and evaluation for non-lambertian and uncalibrated photometric stereo. IEEE Trans. Pattern Anal. Mach. Intell. **41**(2), 271–284 (2019). https://doi.org/10.1109/TPAMI.2018.2799222

Check for updates

Machine Learning for Fog Computing: Perceiving QoE for IoT Applications

Eman Ahmed(✉), Heba Nashaat, and Rawya Rizk

Computer and Control Department, Faculty of Engineering, Port-Said University, Port-Said, Egypt
{Eman_Ahmed,hebanashaat,r.rizk}@eng.psu.edu.eg

Abstract. Since the Internet of Things (IoT) is a rapidly growing industry, so does the range of IoT applications; IoT transforms every aspect of our lives, from smart homes and cities to healthcare and agriculture. With the increasing popularity of IoT applications in many different areas, the round-trip delay of data processing in the cloud affects the user's quality perception. A portable operating system distribution, high environmental consistency, and resource isolation are all features of the lightweight application virtualization technology known as containerization. Mainstream cloud service providers for automated application administration have widely incorporated container technologies into distributed system architectures. However, the distributed nature of cloud and edge computing environments and workloads significantly increases the complexity of orchestration systems. Therefore, the Quality of Experience (QoE) is required for Internet of Things (IoT) applications. Machine Learning (ML) algorithms are employed by container orchestration systems for modeling, predicting the multi-dimensional performance metrics, and improving the overall QoE. The main objective of this paper is to provide a comprehensive review of existing machine learning enablement of container orchestration. Moreover, we discuss and classify the ML implementation challenges in IoT applications and enlist the applications of machine learning algorithms with IoT.

Keywords: QoE · IoT · ML

1 Introduction

Technological advances such as edge computing, low-power wireless networks, microservices, serverless applications, containerization, and machine learning capabilities drive such growth. As the number of connected devices grows, so does the data generated. This considerable data amount presents opportunities and challenges for service providers and consumers. On the one hand, the generated data can be used to gain valuable insights into consumer behavior and optimize infrastructure resource utilization. On the other hand, managing and real-time analyzing this large amount of data can affect the overall application Quality of Experience(QoE). Fog computing complements cloud services to serve users at the network edge, extending the cloud to be closer to the things that produce and act on Internet of Things (IoT) data.

A. Hassanien et al. (Eds.): AISI 2023, LNDECT 184, pp. 482–496, 2023.
https://doi.org/10.1007/978-3-031-43247-7_42

Fog computing performs low latency processing, while latency-tolerant and large-scale tasks that need more computational capabilities than that offered by fog nodes can still be efficiently processed by the cloud [1]. Since user-centric IoT applications/services directly affect the user's life. Moreover, some play a critical role in the user's daily activities, such as environmental monitoring systems, e-healthcare applications, and transportation monitoring. QoE reflects user needs, preferences, and expectations regarding the application. It indicates the perceived quality of the applications, which in turn helps determine if there is a need to improve the data processing time, quality of the network, and proper resource utilization [2]. Containerization enables the creation and automated management of lightweight, portable, and scalable software applications.

Container orchestration is critical in handling the automation of deployment, maintenance, scheduling, and autoscaling of containerized applications. However, the dynamic and heterogeneous nature of cloud and edge computing environments and workloads complicate orchestration systems duties and, thus, the quality of experience of IoT applications evaluation. Container orchestration systems employ Machine Learning (ML) algorithms for application, workload, and infrastructure performance prediction based on multi-dimensional performance metrics. Such insights could further improve applications' overall QoE. This paper presents a comprehensive literature review of existing ML enablement of container orchestration from two perspectives: IoT user-centric application QoE and infrastructure resources utilization optimization. Also, the paper discusses the ML implementation challenges in IoT applications. Further, we classify the challenges of IoT QoE preserving into three main groups. Finally, applications of machine learning algorithms with IOT are depicted.

The remainder of the paper is organized as follows: Sect. 2 discusses the IOT applications in fog computing. Section 3 categorizes the applications of machine learning for perceiving IOT. Finally, the conclusion and future work are drawn out in Sect. 4.

2 QoE of IoT Applications in the Fog Computing Environment

To enable IoT applications with low latency response requirements, fog computing intends to provide Cloud-like services at the network edge. Application deployment in Fog is problematic due to computing instances' hierarchical, distributed, and heterogeneous nature. The placement of applications is a challenge that the variety of user expectations and IoT device features could improve. System services' QoE can be improved by allocating applications to compatible Fog instances to user expectations.

2.1 IoT Application Architecture

The microservice architecture (MSA) pattern has emerged as an alternative to monolithic architectures and applications [3], which are difficult to maintain and evolve due to the high coupling among their components. Microservice architecture is based on building an application as a set of small interconnected services which communicate through light protocols [4]. The microservice architecture (MSA) is proposed to split single-component applications into multiple loosely coupled and self-contained microservice components [5]. Each microservice unit can be deployed and operated independently for different functionalities and business objectives. By decomposing applications into lightweight and independent microservice units, MSA has significantly reduced application development and deployment costs and complexity. Nonetheless, the growing number of components and dynamic inter-dependencies between microservices in MSA raises the problem of load distribution and resource management at the infrastructure level. Many works have used microservice architecture for applying and presenting different IoT applications in variant sectors as air quality monitoring and alerting in smart ports [6], in which they propose a reusable microservice architecture standardized through the use of the Web of Things paradigm and with high efficiency in real-time data processing, supported by complex event processing techniques, Fire evacuation planner application, and intelligent city platform [7].

The main characteristics of microservices architecture, IoT applications, containerization, and Fog computing paradigm fit perfectly together. Figure 1 lists these characteristics and groups them to show their relationships. Microservices architecture with identipent deployment ability and scalability provides advantages of service reusability and distributed application integration with different technology stacks. The IoT ecosystem's complexity includes its interoperability, mobility, ubiquitous, and heterogeneous nature. MSA addresses such complexity by breaking down applications into manageable and reusable services, making microservices serverless, scalable, and easy to deploy in diverse locations.

Regarding these capabilities, container-based microservices can accelerate flexibility to changes, reduce time to market, and collaborate among multiple development teams from a deployment and maintenance perspective. Lightweight containers with considerably lower startup times enable quick deployment and migration, which result in higher

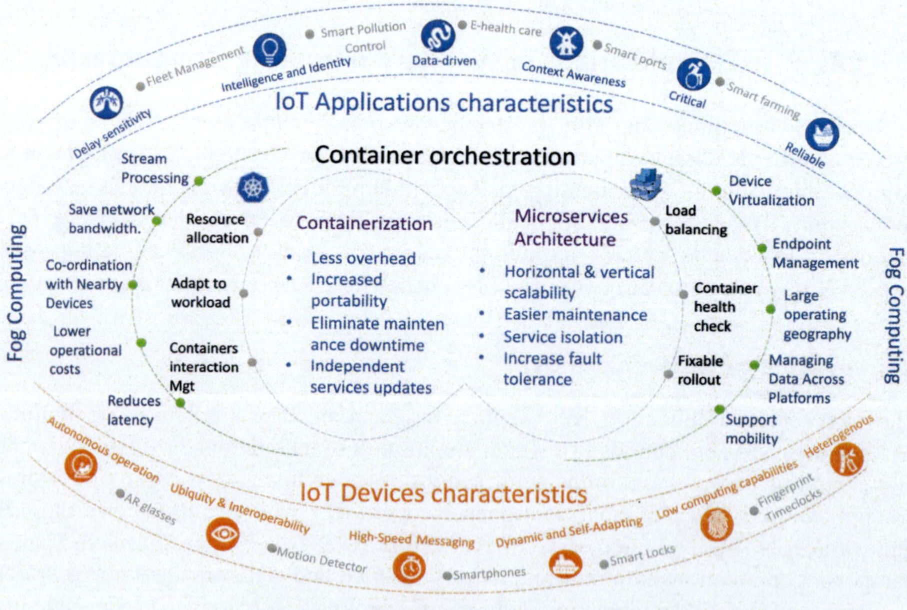

Fig. 1. IoT and Containers Microservices

service availability and reliability under dynamic conditions. The decomposition of the application into small independent units allows only the updated or newly developed microservices to be re-deployed and just the performance-affected microservices to be scaled within each deployment cycle. Thus, MSA meets large and complex IoT software systems' scalability, maintainability, extensibility, and interoperability requirements [9]. As a result, MSA is increasingly adopted for IoT application development in many areas, such as smart ports, smart cities, and intelligent healthcare [6, 7, 10–12].

2.2 Container-Based IoT Microservices on the Fog Computing Environment

Container orchestration is a container lifecycle's fine-grained and automated management process, including resource allocation, deployment, autoscaling, health monitoring, migration, load balance, security, and network configuration. A refined and robust container orchestration system is critical in controlling overall resource utilization, energy efficiency, and application performance for cloud service providers, which have to handle hundreds or thousands of containers simultaneously. Under the surge of cloud workloads regarding resource demands, bandwidth consumption, and quality of service (QoS) requirements, the traditional cloud computing environment is extended to fog and edge infrastructures that are close to ending users with extra computational power [7]. Consequently, this also requires current container orchestration systems to be further enhanced in response to the rising resource heterogeneity, application distribution, workload diversity, and security requirements across hybrid cloud infrastructures.

Sending a large amount of data to the cloud would result in higher network congestion and larger latency values, lowering the overall QoE rank of the user-centric IoT applications. Fortunately, Fog computing emerged to overcome such challenges, enabling the data processing closer to the end devices that generate the data, decreasing data volume sent towards the cloud. Also, fog provides the required computing resources for latency-sensitive services to be placed closer to the end user. In contrast, the rest of the service could still use cloud resources. Fog computing resources give such agility because it's empowered by containerized deployment, dynamic composition, and load-balancing across federated multi-fog and multi-cloud environments with the support of other cloud-native technologies like container orchestrators (i.e., Kubernetes, Docker Swarm) and service mesh technologies (i.e., Istio, Linkerd). Container orchestration plays an automatic role in operational effort management required to run containerized workloads and services, including the container's lifecycle, resource provisioning, deployment, scheduling, scaling (up and down), networking, load balancing, and more. Container orchestration empowers cloud service providers to decide how containerized applications are configured, deployed, and maintained under cloud/edge/fog computing environments. It targets automatic application deployment and dynamic configuration adjustment at runtime for diverse workloads.

2.3 Quality of Experience Evaluation in IoT Applications

QoS plays a vital role in maintaining services with acceptable criteria for IoT systems. Most prior studies attempted to exploit QoS metrics for various purposes of controls to improve user experiences like optimizing network delay, network quality, and energy consumption. Many types of research have shown that designing control policies using QoS metrics alone is insufficient to capture actual user experiences. QoS metrics can be seen as system factors that impact user experiences. In addition, human factors can affect user experiences [12]: user profiles, tendencies, history, characteristics, gender, and age. Context factors such as location, places, and application-related factors (e.g., live streaming) are other non-system factors. Therefore, more than optimizing the system based on QoS metrics is required. Many recent works focused on studies of quality factors that have an impact on user experiences [13], where a multi-dimensional QoE model was proposed, which includes three main Influence Factors (IFs)groups: Environment runtime context, application usage, and user expectations while considering the QoS violation as feedback. Much research focuses on evaluating IoT QoE from different perspectives. The wide image of evaluating QoE in the IoT ecosystem could be summarized in [14], which considers what QoE IFs should be at each layer of the IoT ecosystem. Then quantifying or measuring these IFs with predefined metrics, as shown in Table 1.

Table 1. QoE IFs at each layer of the IoT ecosystem and the related parameters

Level	Influence factors	Metrics
User level, including Context	Gender, age, social characteristics, education, user location, background, usage history, living environments, job, and personal interest	User engagement metrics include user profile and background records, number of downloads, average visit time, screen views per visit, retention rate, and user event tracking
Application level	IFs are associated with the contents of applications: – video-related Ifs: video codec, resolution, video types, voice codec, – voice-related Ifs: voice signal impairment, call setup success rate, blocking call rate, average call setup time – Web-related Ifs: applications, download, and fetch time	Metrics are associated with Content: – video-related metrics from a system perspective video codec, frame rate and interval, compression ratio, buffer size, chunk size, and resolution – voice-related metrics: Call setup time, codec, Mean Opinion Score (MOS), and Perceptual Evaluation of Speech Quality (PESQ)

(*continued*)

Table 1. (*continued*)

Level	Influence factors	Metrics
Service level	Including: Device-Related IFs: device registration, device discovery, and device management – Data-Related IFs: data semantic, data repository and management, data subscription, and notifications – System-Related Ifs: in the Service level are application management, group management, and communication (connectivity)	Including: – Device-related metrics: registration failure rate, registration time, discovery time, number of errors in the discovery process, and ease of usage – Data-related metrics: notification failure rate, number of subscriptions, notification success rate, notification time, transaction statistics, and service costs – System-related metrics: ease of usage, number of connectivity failures, number of hand-over connectivity, number of security breaches
Control or Network level	Network quality	The network-level QoS metrics are jitter, delay, packet loss rate, data rate and volume, reliability, congestion indicators, busy period, and round trip time
Sensing level	Channel quality, including bandwidth, signal strength, bit error rate, dropping probability, power consumption QoD (Quality of data)IFs: data accuracy, truthfulness, and completeness – QoI (Quality of information) refers to the amount of helpful information for a decision or action	Parameters include bandwidth, subcarriers, modulation schemes, coding, and transmission power

However, recent research in Artificial Intelligences (AI) allows us to exploit novel ML techniques for predicting or estimating QoE. One of the well-known QoE indicators is the Mean Opinion Score (MOS). Which has two different implementations: one is subjective, and the second is objective. Subjective MOS is suitable for non-real-time applications like multimedia, video streaming, and voice/speech transmission. In contrast, the objective MOS is suitable for real-time application or dynamic systems evaluation where service quality continuously changes. MOS is quantified into five levels of quality experiences from 1 to 5 (bad, poor, fair, reasonable, and excellent). Subjective MOS has many different methods for measurement, including Absolute Category Rating (ACR), Double-Stimulus Impairment Scale (DSIS), Double- Stimulus Continuous Quality Scale (DSCQS), Subjective Assessment Methodology for Video Quality (SAMVIQ), Degradation Category Rating (DCR) and others. They commonly measure QoE in multimedia applications like video streaming and voice/speech transmission. As shown in Table 1, each QoE level IFs can formulate the associated MOS objective, such as the Peak Signal to Noise Ratio (PSNR), Structural Similarity (SSIM), and Video QualityMetrics (VQM), which the content-related parameters could formulate.

2.4 Machine Learning Enablement of IoT

ML allows the IoT applications to gain insights and knowledge from tradenames sensed data volume. It also gives IoT the ability of business intelligence and proactive readiness. The primary goal of IoT is to perceive what is happening in our surroundings and allow automation of decision-making through intelligent methods, which will mimic the decisions made by humans. IoT became closer to their client's needs, perceptions, emotions, feelings, and ideas. To do so, IoT has to be as intelligent as required. Thus, understanding and explaining the data produced by millions of objects is the bargaining power of IoT application sustainability. Machine learning provides Embedded Intelligence (EI) to enable IoT to achieve its objectives. EI is the fusion of product and intelligence to achieve better automation, efficiency, productivity, and connectivity. In brief, Machine learning gives a glance view of IoT application providers' applications, from development and deployment performance to application quality of experience. IoT is ubiquitous, so availability and mobility handling are a must in such applications. Machine learning is based on mathematical models that can predict essential data patterns. ML has gained popularity over the past decades, and it has numerous use cases, including but not limited to image recognition, text-to-speech, trend and forecasting, and recommendation systems.

As shown in Fig. 2, three main steps are to be handled as pre-steps of the suitable machine learning model per the requirement and the mission's objective. The main three steps are described as follows:

1. Data characterization: based on the available dataset type, which determines the ML model:
   a. Supervised: where the target (desired output) in the dataset is known, considering that dataset could contain multiple targets (labels).
   b. Unsupervised: where the target is unknown. The dataset here contains only the features with no labels.
   c. Reinforcement learning: the learning algorithm collects the dataset by exploring the environment. Take actions in an environment to maximize cumulative reward by rewarding desired behaviors and punishing undesired ones. The algorithm tries to optimize a decision process toward a goal.
2. Data provisioning:
   a. Offline (batch): where the model should be trained on all the examples provided during the training phase. The prediction result from the training is then used in

production, and no more learning occurs. Other methods are variants of batch learning to accept the data in small chunks. Mini-batch techniques still require the entire data set at training time but "consume" data by small subsets. Stochastic methods consume one element at a time and pick this element according to a probability distribution.

b. Online (real-time): Unlike batch learning, an online learning technique can progressively provide new training examples and change its representations accordingly, even in production. An online learning algorithm takes the current best predictor as an additional input. Recursive algorithms used to apply the new predictor depend on the current best predictor and all the previous examples (already learned). Where in, or previous examples can be expressed using an aggregation function. In this case, we do not store these data elements but rather the aggregation function and the information needed for its computation (including its previous result).

3. Training mode selection:
   a. Instance: compares new data points to the training dataset. For example, a new point joins the dataset. A distance metric or sometimes a similar measure is calculated (Distance metrics include the Euclidian distance, the Hamming distance, etc.), then apply the strategy of "voting," and averaging is to check the data point going to the data set. Voting means that the algorithm's estimation will equal the most common value closest to the new data (according to the chosen distance metric). Averaging means giving as our estimation the average value of the property of interest of closest neighbors.
   b. Model: A model based-learning algorithm generalizes from the training examples. The mathematical representation that the algorithm obtains from this generalization process is the model it will use to estimate new data.

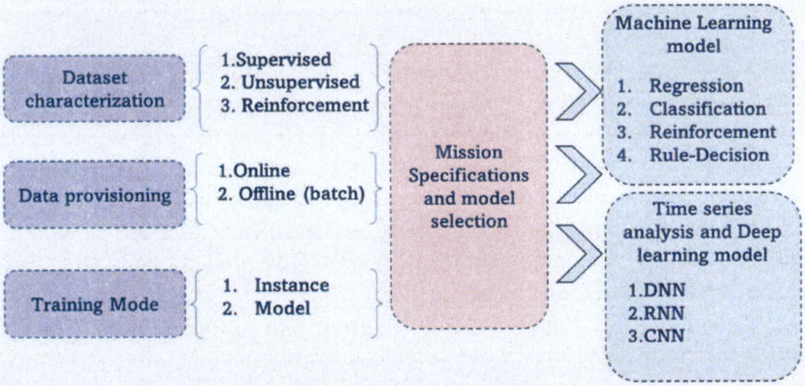

Fig. 2. Machine learning selection process

3 Machine Learning Enablement for Perceiving IoT Applications QoE

The QoE model has emerged as a potential method for estimating real-world satisfaction levels with services. A comprehensive design approach that considers the limits of several QoE metrics is required to meet the needs of IoT applications and services. Additionally, IoT services might function in areas with scarce resources. As a result, efficient management of services and system resources is crucial for supporting QoE.

As discussed, The quality of experience objective assessment computes a metric as a function of QoS parameters [15–17]. For instance, The most commonly used QoE objective model, POLQA [18], is a function of different parameters, including lost packets, time-frequency mapping, time alignment, alignment with a calibrated listening level, compressed loudness scaling, and frequency warping. However, Objective assessment models in offline mode make it insufficient for online/QoE real-time prediction. Besides, it needs to apply some critical feature engineering tasks as a prerequisite step before ML model training. Applying feature selection or dimension reduction to overcome the consequences of high-dimensional data or too large many features might reduce ML prediction power. Recently intelligent QoE models adapted to use the power of advanced data analysis to serve online/real-time QoE prediction requirements. ML plays a master role in IoT QoE estimation and prediction. Many researchers attempted to solve real-time QoE prediction based on relationships between QoS and QoE metrics, using Deep Nuaral Network (DNN), Conventional Nuaral Network (CNN), and Recurrent NuaraL Network (RNN), not only overcome feature selection but also to take advantages of its computing capabilities.

QoE prediction mechanisms help in full image awareness of application ecosystem performance, thus achieving higher satisfaction. It is required feedback information from the system, service, application, and user metrics for predicting a new QoE. Classical ML techniques, e.g., Naive Bayes, Support Vector Machine (SVM), K-Nearest Neighbors (KNN), Decision Tree (DT), Random Forest (RF), and Neural Network (NN), have been studied and proposed to predict QoE value for various systems. For instance, some of these works are explored and summarized in Table 2, including the proposed work of [19–21]. F. Laiche [19] uses social context factors and user engagement as subjective factors to structure a user QoE evaluation model. Applying feature engineering to the prediction model. A QoE management approach was proposed in software-defined network/multi-access edge computing that implements the proposed QoE prediction model to optimize the video delivery transmission chain.

While, Authors of [20] proposed monitoring web applications to classify QoE influence factors into four categories: system factors, human factors, context factors, and social-behavioral factors using supervised ML algorithms like decision trees, k-nearest neighbors, and support vector machines. Finally, software-defined network/multi-access edge computing introduced a QoE management approach that implements the proposed QoE prediction model to optimize the video delivery transmission chain. Laiche [21] used KNN, RF & DT in social context factors and user engagement as subjective factors to structure a user QoE evaluation model. Analyze social context factors and user engagement characteristics and investigate their correlation with QoE. A deep neural network (DNN) is developed to learn the relationships between the network parameters and the subjective QoE scores. The experimental results show that the proposed approach can effectively remove most features irrelevant to QoE prediction.

Table 2. Samarizaton of implementing Machine learning in IoT applications

Ref	Applied ML	ML Group	Data Prov	Training mode	Objective	QoE level	Feature Engineering	App
[19]	KNN, RF, and DT	Supervised: **Classification**	**Offline**	model	QoE Prediction	User level	Applied	Multimedia
[20]	DT, KNN, SVM	Supervised: **Classification**	**Offline**	model	Optimize the video delivery transmission chain	Network level	Applied	Multimedia
[21]	KNN, RF & DT	**Classification**	**Offline**	Instance	Maximizing QoE	User and application level	Applied	Multimedia
[22]	RL	**Classification**	**Offline**	model	Classify web **traffic**	Network level	Applied	Web browsing
[23]	Deep RL	**Classification**	Real-time	model	Real-time RSU cache content Optimizing update	User and application level	Applied	Multimedia
[24]	Deep RL	Supervised: **Classification**	Real-time	model	Maximizing QoE	Application level	Applied	Multimedia
[25]	RL, Deep Q-Learning	Regression	Real time	Instance	Maximizing QoE	Application level	Applied	Multimedia
[26]	RFand LSTM	Supervised: **Classification** &time series analysis	Real time	Instance	QoE Estimation	Network level	Applied	Multimedia
[27]	LSTM-CNN, & RL	Time series analysis	Real-time	Instance	QoE Estimation	User and application level	Applied by Model	
[28]	LSTM, DNN & IRNN	Time series analysis	Real-time	Instance	Real-time Evaluation of QoE	User and Network level	Applied by Model	Multimedia
[29]	NN, Bayesian & RT	Supervised: **Classification** and regression	**Offline**	model	QoE Prediction	Network and User level	Applied	Multimedia
[30]	XG Boost)	Supervised: **Classification**	**Offline**	model	QoE Estimation	Network level	Applied	Multimedia

(*continued*)

Table 2. (*continued*)

Ref	Applied ML	ML Group	Data Prov	Training mode	Objective	QoE level	Feature Engineering	App
[31]	CNN, RNN	Supervised: Classification	Offline	model	Detecting anomalies on video transmission	Network level	Applied by Model	Multimedia
[32]	DNN	Supervised: Classification	Offline	model	QoE Prediction	Network level	Applied by Model	Multimedia
[33]	DNN	Classification	Offline	model	QoE Prediction	User level	Applied	Multimedia
[34]	DNN	Classification	Offline	model	Maximizing QoE	Application level	Applied	Multimedia
[35]	Generic	Generic	Offline	model	Application performance prediction	Application and network level	Applied	Multimedia

Several researchers [22–25] used Reinforcement Learning (RL) as their machine learning model for prediction using the reward concept. Authors of [22] proposed a semi-supervised learning method to classify the websites into three classes (light, average and heavy) that group pages by their complexity, like the number and size of page elements. These classes define the MOS predictor used by the control loop using RL. C. Song in [23] proposes a QoE-driven edge caching method to make the cached content hit the file access requests sent by moving vehicles on the road as often as possible to reduce the delay caused by downloading from the cloud server and improve the users' QoE using file class popularity distribution. Deep reinforcement learning was used to address the QoE-driven RSU cache update issue effectively Deeplive was proposed in [24], which maximizes QoE through Deep Reinforcement Learning (DRL), so it does not rely on fixed rules. To accelerate the training process of Deeplive, we propose optimization, including window completion with historical data and quickstart with the rate-based algorithm. Authors of [25] proposed a QoE-oriented rate adaptation framework based on enhanced Deep Q-Learning. Using the subjective QoE model as the reward function in RL so that the strategy can converge toward maximizing the subjective QoE score.

Besides, Long Short Term Memory (LSTM) has been used by researchers to fulfill time series analysis tasks with the support of Deep Learning capabilities, including Recurrent Neural Networks (RNN) and Convolutional Neural Networks (CNN). ReCLive [26] is a machine-learning method for live video detection and QoE measurement based on network-level behavioral characteristics. Also, they propose a method that uses the chunk-based features collected from the network flows to estimate QoE metrics for live streaming in terms of resolution using a random forest classifier and predict buffer stalls using a statistical model. [27] The QoE estimation approach was proposed based on three metrics: perceived video quality, buffering time, and video quality using DNN to stream video data to users while maintaining the QoE adaptively. Authors [28] propose IRNN, a hybrid network model that integrates a deep neural network (DNN) and an Improved Recurrent Neural Network (RNN) for representation learning of view-level QoE evaluation. Then a graph-based ranking algorithm was applied to conduct user behavior analysis.

In [29], the authors propose a Bayesian Network model using Nural Network to predict the re-buffering ratio. Finally, Capturing hidden context information variable models boost performance for all QoE-related measures. [30] proposed an estimation method according to current network conditions using an ML model. The estimation method uses Internet Control Message Protocol (ICMP) probing to estimate QoE for users consuming video on demand. M. Lopez [31] proposed a classifier to detect anomalies at the current time instant and predict them at the next immediate instant using network parameters (jitter, delay, bandwidth, and loss packets) to tag the received streams with the QoE errors: columns, blur, ghost, chrominance, blackness, color bleeding, and black pixel. The proposed classifier is based on a combination of a Convolutional Neural Network (CNN), Recurrent Neural Network (RNN), and Gaussian Process (GP) classifier. Visualizing the effects obtained from different variations of the network and video by extending the capabilities of CNN, especially since CNN's natural domain is image processing. X. Tao [32] used a deep neural network (DNN) to extract the relationships between the network parameters and four subjective QoE scores. The proposed DNN is a data-driven objective QoE prediction approach for mobile video transmission to predict user QoE scores. The experimental results show that the proposed approach can effectively remove most features irrelevant to QoE prediction.

Moreover, the authors of [33] propose a learning-based QoE prediction approach for mobile video transmission was proposed. 4 QoE metrics were: elicited visual quality, loading quality, stalling quality, and overall quality. Each QoE metric is related to only some of the 89 network parameters. Application and network level Influence factors were used in [34], who used the DNN to train the QoE models based on diverse observations related to network bandwidth, actually received bitrate, and segment size, corresponding to video segments received in the past. The overall QoE is obtained as a linear combination of various input parameters such as perceived video quality, buffering time, and smoothness of the video session. Similarly, the authors of [35] propose a module built on the SDN controller that uses this information to correlate network-level metrics with application conditions. This module will learn how the network features affect the application condition. This information is then used to make decisions about network resources for the application. The framework is structured into three main modules: data collection and aggregation, network/application learning's and network management with prediction. The proposed framework is intended to investigate link types and their effects on end-to-end path selection for application flows in SDN. The researcher provided either offline or online QoE objective assessments. The applied ML and DL models are classified as the classical ML approaches in Fig. 3.

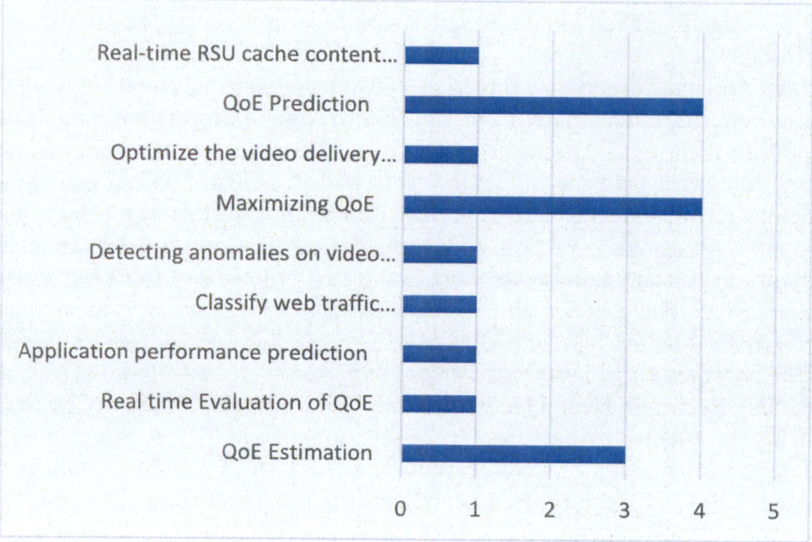

Fig. 3. Machine learning objectives for researchers

4 Conclusion

In conclusion, the paper showed that many works have been introduced to handle QoE prediction in an Offline mode which cannot satisfy the time sensitivity and mission-critical nature of IoT applications. Based on insights obtained from this study, both DT and reinforcement learning shows the best MOS estimation. Also, using DNN, RNN, and CNN offline is mainly for implementing feature engineering and extracting the most significant parameters that affect the overall QoE. To the best of our knowledge, most recent proposals study the QoE from one level, such as user, network, application, or, at most, the combination of the two. Thus the overall QoE prediction still needs to be covered, which could steer the future direction of our research.

References

1. Pham, X., Huh, E.: Towards task scheduling in a cloud-fogcomputing system. In: 2016 18th Asia-Pacific Network Operations and Management Symposium (APNOMS), Kanazawa, Japan, pp. 1–4 (2016)
2. Lin, Y., Shen, H.: Cloud fog: towards high quality of experience in cloud gaming. In: 2015 44th International Conference on Parallel Processing, Beijing, China, pp. 500–509 (2015)
3. Zdun, U., Navarro, E., Leymann, F.: Ensuring and assessing architecture conformance to microservice decomposition patterns. In: Maximilien, M., Vallecillo, A., Wang, J., Oriol, M. (eds.) ICSOC 2017. LNCS, vol. 10601, pp. 411–429. Springer, Cham (2017). https://doi.org/10.1007/978-3-319-69035-3_29
4. Richardson, C., Smith, F.: Microservices: From Design to Deployment. NGINX Inc., San Francisco (2016). https://www.nginx.com/resources/library/designing-deploying-microservices/

5. Kecskemeti, G., Marosi, A., Kertesz, A.: The ENTICE approach to decompose monolithic services into microservices. In: Proceedings of the 2016 International Conference on High Performance Computing Simulation, pp. 591–596 (2016)
6. Guadalupe, O., et al.: A microservice architecture for real-time IoT data processing: a reusable Web of things approach for smart ports. Comput. Stand. Interfaces (2021)
7. Krylovskiy, A., Jahn, M., Patti, E.: Designing a smart city Internet of Things platform with microservice architecture. In: 2015 3rd International Conference on Future Internet of Things and Cloud, Rome, Italy, pp. 25–30 (2015)
8. Pallewatta, S., Kostakos, V., Buyya, R.: Placement of microservices-based IoT applications in fog computing: a taxonomy and future directions
9. Razzaq, A.: A systematic review on software architectures for IoT systems and future direction to the adoption of microservices architecture. SN Comput. Sci. **1**, 1–30 (2020)
10. Bugshan, N., Khalil, I., Moustafa, N., Rahman, M.: Privacy-preserving microservices in industrial Internet of Things driven smart applications. IEEE Internet Things J. (2021)
11. De Iasio, A., Futno, A., Goglia, L., Zimeo, E.: A microservices platform for monitoring and analysis of iot traffic data in smart cities. In: Proceedings of the 2019 IEEE International Conference on Big Data (BigData), pp. 5223–5232. IEEE (2019)
12. Nguyen, L., et al.: IoHT-MBA: an internet of healthcare things (IoHT) platform based on microservice and brokerless architecture. Int. J. Adv. Comput. Sci. Appl. **12** (2021)
13. Nashaat, H., Ahmed, E., Rizk, R.: IoT application placement algorithm based on multi-dimensional QoE prioritization model in fog computing environment. IEEE Access **8**, 111253–111264 (2020)
14. Saovapakhiran, B., Naruephiphat, W., Charnsripinyo, C., Baydere, S., Özdemir, S.: QoE-driven IoT architecture: a comprehensive review on system and resource management. IEEE Access **10**, 84579–84621 (2022)
15. Guo, B., Zhang, D., Yu, Z., Liang, Y., Wang, Z., Zhou, X.: From the Internet of Things to embedded intelligence. World Wide Web **16**, 399–420 (2013)
16. Guo, B., Zhang, D., Wang, Z.: Living with Internet of Things: the emergence of embedded intelligence. In: Proceedings of the2011 International Conference on Internet of Things and 4th International Conference on Cyber, Physical and Social Computing, Dalian, China, pp. 297–304 (2011)
17. Mattern, F., Floerkemeier, C.: From the internet of computers to the Internet of Things. In: Sachs, K., Petrov, I., Guerrero, P. (eds.) From Active Data Management to Event-Based Systems and More. LNCS, vol. 6462, pp. 242–259. Springer, Heidelberg (2010). https://doi.org/10.1007/978-3-642-17226-7_15
18. ITU. ITU-T P.863 Recommendation: Perceptual objective listening quality prediction (2021). Accessed 2018
19. Laiche, F., Letaifa, A., Elloumi, I., Aguili, T.: When machine learning algorithms meet user engagement parameters to predict video QoE. Wirel. Pers. Commun. **116**(3), 2723–2741 (2021)
20. Laiche, F., Letaifa, A., Aguili, T.: QoE-aware trafc monitoring based on user behavior in video streaming services. Concurr. Comput. Pract. Exp. (2021)
21. Fatima, L., Asma, L., Imene, E., Taoufik, A.: When machine learning algorithms meet user engagement parameters to predict video QoE. Wirel. Pers. Commun. (2021)
22. Moura, H., Macedo, D., Vieira, M.: Wireless control using reinforcement learning for practical web QoE. Comput. Commun. **154**, 331–346 (2020)
23. Song, C., Xu, W., Wu, T., Yu, S., Zeng, P., Zhang, N.: QoE-driven edge caching in vehicle networks based on deep reinforcement learning. IEEE Trans. Veh. Technol. **70**(6), 5286–5295 (2021)

24. Tian, Z., Zhao, L., Nie, L., Chen, P., Chen, S.: Deeplive: QoE optimization for live video streaming through deep reinforcement learning. In: 2019 IEEE 25th International Conference on Parallel and Distributed Systems (ICPADS), Tianjin, China, pp. 827–831 (2019)
25. Liu, J., Tao, X., Lu, J.: QoE-oriented rate adaptation for DASH with enhanced deep Q-learning. IEEE Access **7**, 8454–8469 (2019)
26. Madanapalli, S., Mathai, A., Gharakheili, H., Sivaraman, V.: ReCLive: real-time classification and QoE inference of live video streaming services. In: Proceedings of the IEEE/ACM 29th International Symposium on Quality Service (IWQOS), p. 17 (2021)
27. Lekharu, A., Moulii, K., Sur, A., Sarkar, A.: Deep learning based prediction model for adaptive video streaming. In: Proceedings of the International Conference on Communication Systems & Networks (COMSNETS) (2020)
28. Yue, T., Wang, H., Cheng, S., Shao, J.: Deep learning based QoE evaluation for internet video. Neurocomputing **386**, 179–190 (2020)
29. Vasilev, V., Leguay, J., Paris, S., Maggi, L., Debbah, M.: Predicting QoE factors with machine learning. In: Proc. IEEE International Conference on Communications (ICC), 1–6 (2018)
30. Miranda, G., Macedo, D., Marquez-Barja, J.: A QoE inference method for DASH video using ICMP probing. In: Proceedings of the 16th International Conference on Network and Service Management (CNSM), p. 15 (2020)
31. Lopez-Martin, M., Carro, B., Lloret, J., Egea, S., Sanchez-Esguevillas, A.: Deep learning model for multimedia quality of experience prediction based on network ow packets. IEEE Commun. Mag. **56**(9), 110–117 (2018)
32. Tao, X., Duan, Y., Xu, M., Meng, Z., Lu, J.: Learning QoE of mobile video transmission with deep neural network: a data-driven approach. IEEE J. Sel. Areas Commun. **37**(6), 13371348 (2019)
33. Xiaoming, T., Yiping, D., Mai, X., Zhishen, M., Jianhua, L.: Learning QoE of mobile video transmission with deep neural network: a data-driven approach. IEEE J. Sel. Areas Commun., 1, 10 (2019)
34. Lekharu, A., Moulii, K.Y., Sur, A., Sarkar, A.: Deep learning based prediction model for adaptive video streaming. In: 2020 International Conference on COMmunication Systems & NETworkS (COMSNETS), Bengaluru, India, pp. 152–159 (2020)
35. Wang, L., Delaney, D.: QoE oriented cognitive network based on machine learning and SDN. In: 2019 IEEE 11th International Conference on Communication Software and Networks (ICCSN), Chongqing, China, pp. 678–681 (2019)

Stress Detection Using Horror Video Game and Machine Learning Algorithms

Ahmed Khattab, Hager Mostafa, Merrna Ibrahim, Karim Samer, Yasmeen Abdelaty, and Gehad Ismail Sayed(✉)

School of Computer Science, Canadian International College (CIC), New Cairo, Cairo, Egypt
Gehad_Sayed@cic-cairo.com

Abstract. Psychology and neuroscience research have both shown an interest in the ability to recognise people's levels of stress. Human behaviour, however, is complicated and subject to large variations depending on a variety of variables. Self-reporting and observation are two common traditional ways of identifying human behaviour, although they can be subjective, biased, and time-consuming. A stress detection model based on a video game and machine learning algorithms is proposed in this research. Game engines can offer a regulated and engaging setting for spotting human behaviour. Age, gender, and various frequency and time domain variables connected to the recorded audio through the game are just a few of the features that are extracted. The effectiveness of the feed-forward neural network, support vector machine, and random forest algorithms is examined. The proposed stress detection model is a very promising model, according to the experimental results. Overall, it achieved 96.2% accuracy, 96.4% precision, 96.2% sensitivity, and a 96.1% f1-score.

Keywords: Machine Learning · Human Behavior Detection · Video Games · Spectral Features · Stress Detection

1 Introduction

Researchers can employ human behaviour detection to better understand how people think, feel, and act in diverse circumstances when researching mental health [1]. This is crucial because behaviour analysis can show the underlying causes of mental health issues [2]. Mental health issues can have a big impact on a person's behaviours. Additionally, human behaviour detection offers important new insights into the actual lives of people with mental health issues. Researchers can gain a better understanding of the views, needs, and difficulties of people with mental health illnesses by monitoring and analysing their behaviour [3]. Using this information, better therapies and support systems can be created for those who suffer from these conditions.

The exponential increase in the volume of data generated by humans in various forms, such as social media, online transactions, and other digital activities, has given machine learning more significance for understanding human behaviour in contemporary times [4]. While traditional approaches like surveys and interviews have a limited scope and

A. Hassanien et al. (Eds.): AISI 2023, LNDECT 184, pp. 497–507, 2023.
https://doi.org/10.1007/978-3-031-43247-7_43

accuracy, machine learning can analyse vast volumes of data and discover patterns and trends that would be difficult or impossible for people to manually identify. Insights into human behaviour, such as societal trends, consumer preferences, and issues with public health, can be gained via machine learning [5]. It has developed into an essential tool for studying human behavior in a variety of sectors because of its ability to handle and interpret enormous volumes of data [6].

Game engines are a helpful tool for machine learning-based human behavior detection. This is due to that they make it possible to create interactive, immersive virtual worlds that can be used to deliberately and safely raise stress levels [7]. When using game engines to simulate stressful situations, such as public speaking [8] or emergencies, researchers can create custom stress levels based on participant preferences and levels of stress tolerance. The incorporation of numerous sensors, such as heart rate monitors and EEGs, enables the collection of a significant amount of physiological and behavioral data while performing a stressful task [9]. In order to produce more accurate predictions and insights, machine learning algorithms can be used to analyse this data and look for patterns and changes that might indicate signs of stress. Additionally, custom stress management plans that will help people better manage their stress can be made using the behavioural data gathered by game engines, such as speech patterns and reaction times [10].

This paper proposes an approach model for investigating human behaviour under stressful conditions, utilizing a controlled and immersive virtual environment provided by a horror video game. The proposed model employs machine learning algorithms to analyze the audio features of the player during stress-triggering events, to classify the player's state as either stressed or not stressed based on the spectral features of the audio. It consists of three main phases. The first phase is data acquisition, the second phase is data preparation and the third phase is classification. In data acquisition, several voice features are obtained from the video game. Then, in data preparation, data normalization followed by random oversampling is applied to the collected dataset. Finally, the prepossessed data are used to feed a machine learning algorithm. To best the best of our knowledge, there is few work consider using video games for stress detection. The main contribution of this paper can be summarized in the following:

a) This paper introduces the use of horror video games and machine learning algorithms to detect stress in individuals.
b) The proposed model can provide a unique and engaging means of inducing stress in study participants, without exposing them to real-world stressful situations that may be unethical or impractical to simulate.
c) The study findings open up new possibilities for stress detection and monitoring across different domains.
d) The combination of horror video games and machine learning algorithms can provide potential applications in healthcare, psychology, and entertainment industries.

The rest of the paper is organized as follows: Sect. 2 summarizes the previous studies related to human behaviour detection. Section 3 discusses in detail the overall proposed stress detection model. The analysis of the conducted results is shown in Sect. 4. Finally, the main finding of the paper with the future directions is presented in Sect. 5.

2 Literature Survey

According to recent studies [11], physiological and conventional methods have been widely used to identify human behaviour and emotions. The exponential growth of data has made machine-learning algorithms more useful and efficient than ever for studying human behaviour [12]. This approach may result in analysis that is more accurate and scalable, making it possible to find trends and insights that might be challenging to find using conventional methods. Machine learning algorithms have consequently developed into a potentially helpful tool for researching human motivation and emotion. The use of video games to track human behaviour is a growing academic field [13]. While this is the case, not many studies have demonstrated their potential. The authors of a recent study in [9] used a virtual reality-based game along with classification models to identify individuals who were under stress when speaking in front of an audience. In this study, three categorization models were used to conduct the analysis. The findings of this study demonstrated that random forest classification produced the most accurate results based on the audio properties of amplitude envelope, root-mean-square, and Mel-frequency cepstral coefficients. Similar to this, a study was conducted in the area of mental health [14] that utilized VR technology in exposure therapy to assist people in overcoming their phobias in a secure environment. Virtual reality video games could fundamentally alter how patients receive mental health therapy if they are used in this setting. The authors conducted an investigation in [15] on the use of neural network models to analyze and detect stress in individuals using an audio dataset. The dataset that the researchers used contained seven distinct emotions: surprise, anger, disgust, disgust, neutral, and melancholy. On this dataset, the researchers trained and tested a variety of classification models, including convolutional neural networks, to see how well they could detect stress. The models' 76.08% accuracy shows that these techniques are capable of identifying emotional states from audio inputs. This method significantly expands the body of knowledge on the use of machine learning techniques for emotion recognition and opens up new avenues for further study in the area.

The authors of another study in [16] evaluated the effectiveness of examining two spectral features with mel-frequency cepstral coefficients in classifying stressed speech using a neural network model based on the Levenberg-Marquardt algorithm. The two spectral characteristics outperformed the MFCCs in the study's preliminary findings, identifying stressed speech with an accuracy rate of 83% compared to 75% for the MFCC features. These results suggest that spectral data can be combined with machine learning techniques to improve the accuracy of stressed speech detection, which has important implications for fields like mental health research and speech therapy. The majority of these studies had to create or use their own datasets because there aren't many trials in this field, which brings us to the important finding from the earlier studies. Each of the experiments discussed and reviewed in this section also demonstrates the potential of using video games and neural network models to analyze speech and spectral data and accurately identify stress levels and emotional states. Consequently, the proposed model will consider using a neural network classifier.

3 Proposed Stress Detection Model

The proposed model described utilizes a unique and innovative approach to stress detection by leveraging the immersive experience of a first-person psychological horror game. By constructing a stressful, realistic environment that gives players the ability to alter the course of the story during gameplay. The model employs a mic input analysis technique during gameplay that enables audio feature extraction in order to conduct stress detection. The microphone records and extracts audio features that are then analyzed along with the player's in-game choices to give insight into their stress levels. The microphone is triggered by particular game events. This model offers an approach for stress detection that may provide more precise and significant results by fusing audio feature analysis with gameplay choices. Additionally, the proposed model aims to produce its own dataset in order to better assess the model's efficacy given the paucity of research in this area. The proposed model presents a promising avenue for stress detection and has important implications for fields such as mental health research, speech therapy, and game design. The overall proposed model consists of three main phases These phases are data acquisition, data preprocessing, and classification. Figure 1 shows a block diagram illustrating the main phases of the model. The diagram provides a high-level overview of the model's workflow, emphasizing the sequential nature of the process.

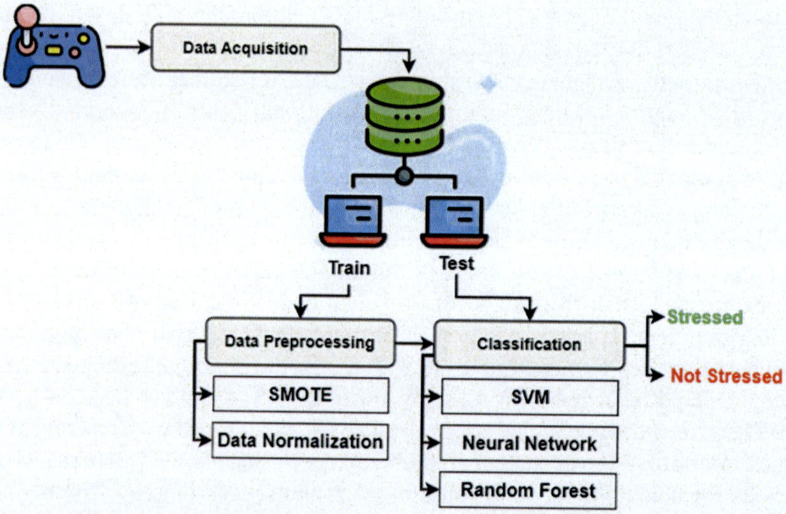

Fig. 1. The block diagram of the proposed stress detection model.

During the data acquisition phase, the video game is used to extract a number of voice features. In the proposed video game, a horror survival game, players can assume the role of a nurse who accepted a job in a remote mansion. While attending to the elderly patient, the nurse begins to notice strange occurrences inside and outside the mansion. It was developed with Unreal Engine 5 and placed a strong emphasis on realism to boost player interest. The feature extraction process involves separating pertinent acoustic features from the audio data produced by the microphone following audio capture triggering events. The spectral centroid, spectral flatness, spectral roll-off, spectral crest, spectral difference, root mean square, peak energy, and zero crossing rate are some examples of these acoustic features. The required duration time to complete the game, the decisions players make while playing, and the type of ending are additional features.

The study used a method known as random oversampling using Synthetic Minority Oversampling Technique (SMOTE) to address data imbalance in the second phase, which is data preparation. By using this method, it was possible to balance the dataset and provide the model with more representative data by adding fictitious data points. Normalization was also performed to ensure that all features were on a comparable scale and to enhance the accuracy of the model. The third stage of the stress detection model involves feeding the preprocessed data into three distinct classification models: random forest, support vector machine, and feed-forward neural network. Entropy, Gini, and log loss are three different criteria that are applied to random forests. Similar to RBF, linear, and polynomial kernel modes, SVM also uses three different kernel modes. Based on features extracted from the video game and additional features related to age, gender, and game performance, all three of the adopted models are used to categorize a player's behavior as stressed or not stressed. Table 1 shows the description of each adopted feature collected from the game.

Table 1. Features Description.

Feature	Description
Age	Participant's age
Gender	Participant's gender either male and female
Duration	The time the participant takes to finish the game
Decision 1	It is a binary attribute, the player either selects the first decision or not
Decision 2	It is a binary attribute, the player either selects the second decision or not
Decision 3	It is a binary attribute, the player either selects the third decision or not
Ending Type	This game has two possible endings either escaping from the house or death; 1 for reaching the goal, and 2 if the player died
Spectral Centroid	A measure of the amplitude at the center of the spectrum of the signal distribution
Spectral Flatness	The estimation of the uniformity of signal energy distribution in the frequency domain
Spectral Rolloff	The frequency below which a specified percentage of the total spectral energy
Spectral Crest	Is an indication of the peakiness of the spectrum
Spectral Difference	Measures the difference in spectral magnitude between two consecutive frames of a signal

(continued)

Table 1. (*continued*)

Feature	Description
Root Mean Square	Measures the average loudness of an audio track within a window of roughly 300 ms
Peak Energy	Measures the highest energy value in a frame of a signal
Zero Crossing Rate	Is the rate of sign-changes of the signal during the frame

4 Experimental Results and Discussion

Multiple experiments are conducted in this paper. All experiments have been performed on the same PC. Also, it should be mentioned that the reported results are reported on an average of ten independent runs. The primary objective of the first experiment in Fig. 2 is to demonstrate the imbalance issue with the gathered dataset. As can be seen, the adopted dataset has a sample imbalance issue, with the stressed class having more samples than the unstressed class. Therefore, this issue can be solved by using the random over-sampling method.

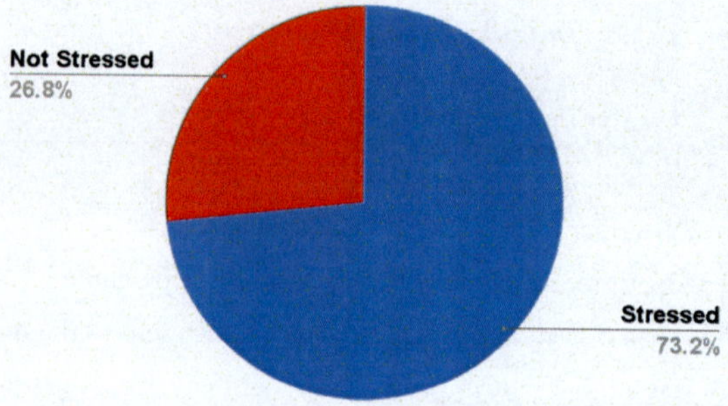

Fig. 2. The class distribution of the collected dataset.

The second experiment aims to compare the performance of the proposed model before and after adopting each part of the preprocessing phase including data normalization and data oversampling method. Moreover, it aims to identify the best parameter settings of each adopted machine learning algorithm. Table 2 shows the performance of using different classifiers with different parameter settings of the original dataset before preprocessing. These classifiers are random forest (RF), support vector machine (SVM), and feed-forward neural network (NN). As can be seen from this table that random forest with log loss criterion obtained the best results compared with Gini and entropy criterion. Also, it can be observed that SVM with polynomial obtained the best results compared with the RBF and linear kernels. From this experiment, the best parameter setting for each adopted classifier is determined and it will be further used in the upcoming experiments.

Figure 3 compares the performance of the proposed model before and after applying the random over-sampling technique to address the class imbalance issue in the dataset.

Table 2. The performance of different classifiers with different parameters settings on the original dataset

Classifier	Accuracy (%)	Precision (%)	Sensitivity (%)	F1-Score (%)
RF-gini	0.91	0.91	0.91	0.906
RF-entropy	0.922	0.921	0.922	0.919
RF-logloss	0.925	0.925	0.925	0.922
SV-linear	0.784	0.774	0.784	0.777
SVM-poly	0.833	0.838	0.833	0.813
SVM-RBF	0.815	0.808	0.815	0.799
NN	0.934	0.934	0.934	0.934

The results indicate that random over-sampling slightly improved the accuracy of the model.

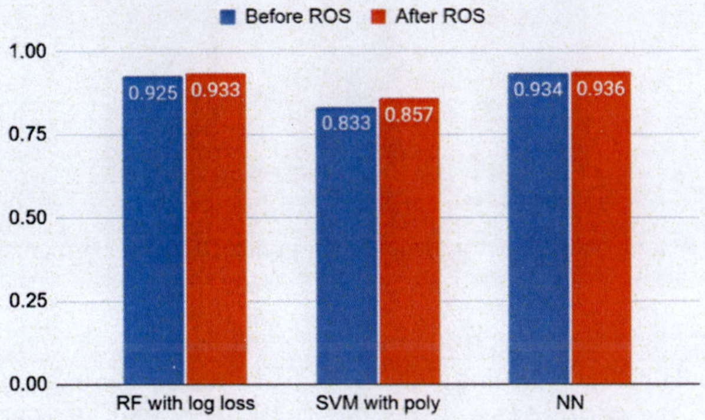

Fig. 3. Accuracy result before and after applying ROS.

Table 3 compares the performance of the best parameter setting of the adopted classifiers, random forest with log loss criterion, SVM with the polynomial kernel, and feed-forward neural network. The conducted results in this experiment consider applying ROS of the preprocessing without normalization part. As can be observed the feed-forward neural network classifier obtained the best result compared with the other two classifiers.

To show the significance of adopting both preprocessing parts, another experiment is conducted in Fig. 4. In this figure, both normalization and ROS are considered, where the accuracy results are reported before and after adopting these parts. As can be seen, there is a significant improvement in the conducted results. The accuracy increases from 92% to 95% for the random forest with log loss criterion, from 83% to 89% for SVM with the polynomial kernel, and finally from 93% to 96% for the feed-forward neural network. Moreover, it can be observed that the combination of these techniques resulted

Table 3. The results of RF, SVM, and NN after applying ROS

Classifier	Accuracy (%)	Precision (%)	Sensitivity (%)	F1-Score (%)
RF-logloss	0.933	0.933	0.933	0.933
SVM-poly	0.857	0.881	0.857	0.855
NN	0.936	0.942	0.936	0.936

in a more balanced dataset and better feature scaling, leading to an overall improvement in the classification performance.

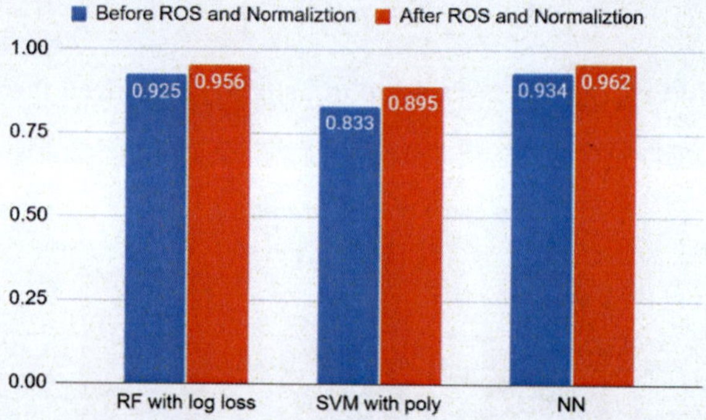

Fig. 4. Accuracy result before and after applying ROS and normalization.

For further evaluation of the performance of the overall stress detection model based on video games and machine learning algorithms, five other measurements are employed. Table 4 compares the performance of the three classifiers with their best-performing parameters settings after applying both random over-sampling and normalization on the dataset in terms of precision, sensitivity, and f1-score. As can be seen, the feed-forward neural network classifier was the best-performing classifier compared to the random forest and support vector machine.

Table 4. The results of RF, SVM, and NN after applying ROS and normalization

Classifier	Accuracy (%)	Precision (%)	Sensitivity (%)	F1-Score (%)
RF-log loss	0.956	0.956	0.956	0.956
SVM-poly	0.895	0.911	0.895	0.894
NN	0.962	0.964	0.962	0.962

Figure 5 shows the receiver operating characteristics (ROC) curve with the area under the curve obtained from the feed-forward neural network. ROC depicts the performance of a binary classification model, with the x-axis showing the false positive rate and the y-axis showing the true positive rate. As can be seen, the proposed model using a NN almost reach the maximum value.

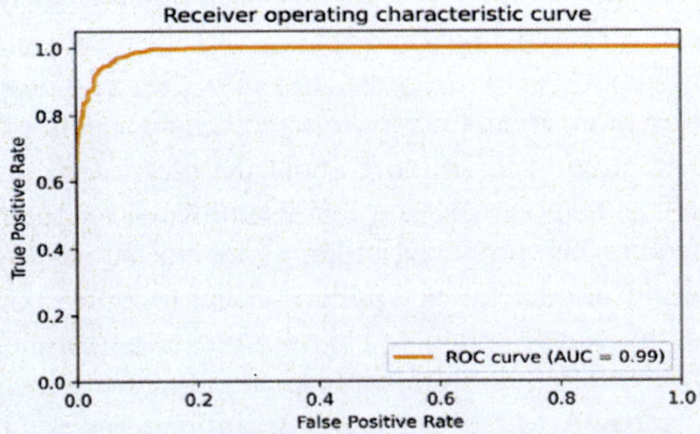

Fig. 5. Receiver operating characteristic curve of the proposed model

Furthermore, the confusion matrix is shown in Fig. 6. The confusion matrix is used to represent the performance of a machine learning model in a binary classification task, where the predicted classes are either "stressed" or "not stressed". The true positive and true negative values represent the correct predictions. As can be observed, the proposed model can significantly distinguish between stress and not-stressed classes.

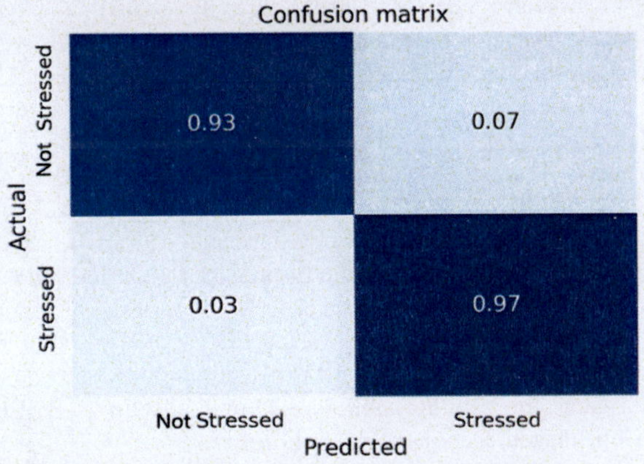

Fig. 6. The confusion matrix of the proposed model

5 Conclusion

In this paper, a model for stress detection based on video games and machine learning algorithms is proposed. On a real dataset, the proposed model is examined. This dataset combines a variety of voice-related features and other features related to the player's choices in the video game. Three well-known machine learning algorithms—random forest, support vector machine, and feed-forward neural network—are used and evaluated. The robustness and dependability of the proposed model were demonstrated by the experimental results. It achieved an overall accuracy of 96% with a feed-forward neural network, 89% with a support vector machine, and 95% with a random forest. The results can also be used to learn more about the physiological and psychological causes of stress. The results can also be used to learn more about the physiological and psychological mechanisms of the stress response, laying the groundwork for future studies on stress management and therapy. The proposed video game can be used to detect additional human emotions, and this model can be applied further to more complicated datasets.

Further investigations can be conducted to improve the performance of the proposed model. Additionally, investigating more sophisticated feature selection techniques can increase the model's capacity to recognize significant patterns and boost classification precision.

References

1. Aung, M.H., Matthews, M., Choudhury, T.: Sensing behavioural symptoms of mental health and delivering personalized interventions using mobile technologies. Depress. Anxiety **34**(7), 603–609 (2017)
2. Jun, E.R., Kim, S.H., Cho, Y.J., Kim, Y.A., Lee, J.Y.: The influence of negative mental health on the health behavior and the mortality risk: analysis of Korean longitudinal study of aging from 2006 to 2014. Korean J. Fam. Med. **40**(5), 297–306 (2019)
3. Wittchen, H.U., et al.: The need for a behavioural science focus in research on mental health and mental disorders. Int. J. Methods Psychiatr. Res. **23**, 28–40 (2014)
4. Zhu, J., et al.: Machine learning human behavior detection mechanism based on Python architecture. Mathematics **10**(17), 1–31 (2022)
5. Qiu, J., et al.: A survey of machine learning for big data processing. EURASIP J. Adv. Signal Process. **2016**(1), 1–16 (2016)
6. Cavus, N., et al.: A systematic literature review on the application of machine-learning models in behavioral assessment of autism spectrum disorder. J. Pers. Med. **11**(4), 299 (2021)
7. Moriarty, C.: Learning human behavior from observation for gaming applications. Electronic Theses and Dissertations, 2004-2019, pp. 1–132 (2007)
8. Dillon Arushi, R., Teoh, A.N.: Real-time stress detection model and voice analysis: an integrated VR-based game for training public speaking skills. In: 2021 IEEE Conference on Games (CoG), Copenhagen, Denmark, pp. 1–4 (2021)
9. Yu, Z., Wang, Z.: Human Behavior Analysis: Sensing and Understanding, pp. X-271. Springer, Singapore (2020). https://doi.org/10.1007/978-981-15-2109-6
10. Vuong, Q., Ho, M., La, V., Lê, T., Nguyen, T., Nguyen, M.: A multinational data set of game players' behaviors in a virtual world and environmental perceptions. Data Intell. **3**(4), 606–630 (2021)
11. Dzedzickis, A., Kaklauskas, A., Bucinskas, V.: Human emotion recognition: review of sensors and methods. Sensors **20**(3), 592 (2020)
12. Siam, A.I., Soliman, N.F., Algarni, A.D., Abd El-Samie, F.E., Sedik, A.: Deploying machine learning techniques for human emotion detection. Comput. Intell. Neurosci. **2022**, 1–16 (2022)
13. Shute, V., Ventura, M.: Measuring and Supporting Learning in Games: Stealth Assessment, pp. 1–83. MIT Press, Cambridge (2013)

14. Meyerbröker, K., Morina, N.: The use of virtual reality in assessment and treatment of anxiety and related disorders. Clin. Psychol. Psychother. **28**(3), 466–476 (2021)
15. Divyashree, P., Yadav, A.G., Jayadev, N., Chidaravalli, S.: Stress and anxiety detection through speech recognition using deep neural network. Int. J. Innov. Res. Technol. **8**(11), 730–734 (2022)
16. Sanaullah, M., Chowdhury, M.H.: Neural network based classification of stressed speech using nonlinear spectral and cepstral features. In: 2014 IEEE 12th International New Circuits and Systems Conference (NEWCAS), Trois-Rivieres, QC, Canada, pp. 33–36 (2014)

Lightweight Mask Wearing Detection Algorithm Based on Improved YOLOv5

Xiaodong Su[1,2], Hongyan Xu[1,2](✉), Yurong Zhang[1,2], Jiayuan Zhao[1,2], Fengchun Zhang[1,2], and Xu Chen[1,2]

[1] Harbin University of Commerce, Harbin, China
2902512892@qq.com
[2] Heilongjiang Key Laboratory of E-Commerce and Intelligent Information Processing, Harbin, China

Abstract. To prevent the spread of the epidemic, a lightweight mask-wearing detection algorithm based on the modified YOLOv5 is proposed. The three categories of mask-wearing, mask-less, and non-standard mask-wearing were studied, and images of faces covered with hands and clothing were added as interference items to make the study more realistic. Using YOLOv5 as the base model, the modified deformable convolution is introduced into the CSP-Darknet53 module of the backbone network, which effectively enhances the model's ability to extract key information such as faces and masks and improves the detection accuracy of the model. In addition, image defogging based on dark channel prior is introduced to make the model adapt to detect whether pedestrians wear a mask correctly in special weather. Finally, the knowledge distillation method is used to perform light-weight compression on the model to solve the problem of slow detection speed in practical applications. Training and testing on the self-made T-Mask dataset, the mAP of the model reaches 93.6%, which is about 3% points higher than the original YOLOv5m algorithm, and the number of parameters is reduced to 1/3 of the original one, which has a higher practical value.

Keywords: Mask wearing detection · YOLOv5 · Deformable Convolution · Image dehazing · Knowledge distillation

1 Introduction

Novel coronavirus (2019-nCoV) is the third coronavirus pandemic facing mankind. After the human are infected with the novel coronavirus, it can cause symptoms such as fever and pneumonia. On 13 March 2020, the WHO declared COVID-19 a pandemic. Research [1] shows that the new coronavirus is mainly transmitted by droplets and contact. Wang Rui et al. [2] pointed out that masks, as one of the most convenient and effective physical barrier methods, are used to block the spread of pathogens by droplets. Wearing a mask can not only prevent the spread of pathogens between people, but also reduce the risk of human inhaling pathogens, which has the effect of two-way isolation and protection. Studies [3–5] have shown that wearing masks by people infected with diseases can protect

A. Hassanien et al. (Eds.): AISI 2023, LNDECT 184, pp. 508–520, 2023.
https://doi.org/10.1007/978-3-031-43247-7_44

other people and reduce the incidence of infectious diseases. At present, many countries have regarded wearing masks as an important measure to prevent and control epidemic diseases transmitted through the respiratory tract [6]. Currently, manual monitoring is still the main monitoring method for wearing masks in public places such as schools and stations. However, manual monitoring is not only difficult to efficiently identify mask-wearing in a crowd, but also may have difficulties such as monitoring human infection and insufficient manpower to monitor it. Therefore, using computer vision to build an intelligent mask wearing detection model is of great significance and research value for the current epidemic prevention and control [7].

Currently, a number of scholars have started to investigate mask wearing detection algorithms. For example, Niu Zuodong et al. [8] realized mask wearing detection based on the RetinaFace face recognition algorithm by optimizing the loss function and increasing the research method of the attention mechanism. Wu Peishu et al. [9] adopted the Im-Res2Net-101 feature extractor, combined the Res2Net module and deep residual network, and used the enhanced path aggregation network En-PAN for feature fusion to improve the robustness and generalization ability of the model. Kumar Akhil et al. [10] designed a tiny-YOLOv4w network by integrating a spatial pyramid pooling (SPP) module and an additional YOLO detection layer, and tested it on a self-made mask detection dataset consisting of more than 50,000 images and verify. Zhang Xin et al. combined YOLOv3's backbone network DarkNet-53 with Inception-v4 ideas, and introduced a feature pyramid pooling structure to improve detection accuracy [11]. The above algorithms all have the following drawbacks: they only detect two classes of faces, those with and without masks, and ignore the possibility that irregular mask wearing is common in real life. There is still room for improvement in the accuracy of model detection when deploying applications in real-time detection tasks. Further research is needed on issues such as large target differences, complex scenes, and special weather affecting model detection in natural environments. In response to the above issues, in this paper, we propose an improved mask detection algorithm based on YOLOv5, which detects three categories: wearing a mask, not wearing a mask, and wearing an irregular mask, using a self-made T-Mask dataset. The aim is to improve the detection accuracy of the algorithm on the basis of ensuring real-time detection.

2 Theory

YOLO (You Only Look Once) [12] is a high-performance general-purpose object detection model. YOLOv1 innovatively uses a single-stage model structure to perform both tasks of object classification and localization simultaneously. The subsequent YOLOv2 [13] and YOLOv3 [14] achieved a significant improvement in the speed and accuracy of object detection and accelerated the implementation of object detection in the industry. The lightweight YOLOv4 model [10] even can be trained on an ordinary GPU (1080Ti). From YOLOv1 to the present, the YOLO series has evolved to YOLOv5. Compared with YOLOv4, YOLOv5 is more flexible. To some extent, the YOLOv5 model is the SOTA (state-of-the-art) of all known YOLO implementations. It is available in 4 sizes, namely YOLOv5s, YOLOv5m, YOLOv5l, and YOLOv5x. The accuracy of the 4 versions of the model increases sequentially with model size, distinguished by the number of bottlenecks. The channel and layer control factors are used to realize the version change, which can be chosen according to the application scenario with the correct size model. In this paper, we fully consider the real-time and accuracy of model detection so that it can be better applied to practical detection tasks, and therefore YOLOv5m is chosen as the base model for the experiment.

2.1 YOLOv5m Model Structure

As shown in Fig. 1, the YOLOv5m model structure mainly consists of four basic components: input, backbone network, Neck network and output. The input end mainly completes two tasks: (1) Adaptive anchor box calculation. Cluster analysis is performed on the target real boxes in the dataset to obtain suitable initial anchor boxes. A set of suitable initial anchor boxes can accelerate network convergence; (2) Perform preprocessing operations on the batch images transmitted to the network. In addition to random crop, zoom, flip and HSV augmentation, the image preconditioning operation employed by YOLOv5 introduces mixup and copy-paste data augmentation. The mixup can reduce the impact of noisy samples on the model and improve the stability of model training. The problem of non-equilibrium distribution of the number of categories in the dataset is solved by the copy-paste. When the model is good enough, the benefit of the mixup on model improvement becomes very small, while the copy-paste can still play a stabilizing role.

YOLOv5 uses the improved CSPDarkNet53 [15] as the backbone network to extract features from the input image. First, it introduces the Focus module, which divides each 2×2 adjacent pixels into a patch. Then the pixels at the same position in each patch are spliced to obtain four independent feature layers, and then these four feature layers are stacked. At this time, the channel dimension is enlarged to four times the original dimension, and the width and height dimensions are reduced to 1/4 of the original dimension. The Focus operation cleverly utilizes the reset tensor dimension method, which enables downsampling without loss of image information, while reducing computational cost and cuda memory usage. Delete part of the Conv module in BottleneckCSP and then obtain the C3 module. YOLOv5 contains two types of C3 modules, which are divided into C3_T and C3_F depending on whether the Bottleneck in the module is set to a residual edge or not. The Neck network is located behind the backbone network, which can further improve the diversity and robustness of features and better utilize the image features extracted by the backbone network. YOLOv5 uses the SPP (spatial pyramid pooling) module, the FPN (feature pyramid network) module, and the PAN (path aggregation network) module in the Neck network to complete the transfer and fusion of information. The output layer mainly uses the previously extracted features to make a final prediction on the objects in the image. In addition, YOLOv5 uses a weighted NMS method to filter bounding boxes to enhance the detection ability of occluded and overlapping targets.

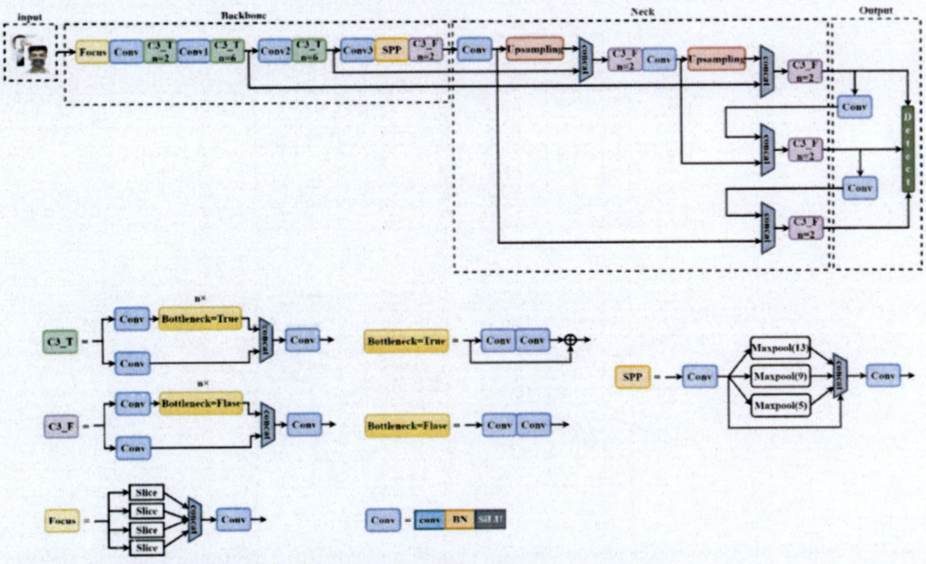

Fig. 1. YOLOv5m model structure

2.2 Deformable Convolution

The traditional rectangular convolution kernel has a good feature extraction effect on approximately rectangular objects (such as cars, tables, TVs, etc.). However, for objects with large scale and shape changes or irregular outlines, some sampling points of the convolution kernel will inevitably fall in the background area, thereby introducing background noise, affecting the feature extraction of the target, and reducing the detection accuracy of the model. For mask detection tasks, considering the variability of camera's shooting angles and human behaviors, using deformable convolution (DCNv1) [16] can better capture face information and feature extraction, and is more conducive to the model to detect whether pedestrians wear masks properly.

Deformable convolutions (Eq. 2) introduce the ability to learn geometric deformations of the target space. It does not change the computational operation of conventional convolution (Eq. 1), but adds a learnable parameter, namely the 2D offset, so that the sampling region can be freely shaped, which makes the network more sensitive to the shape of the object. Taking the 3 × 3 convolution kernel as an example, Fig. 2 shows the position distribution of traditional convolution and deformable convolution sampling points. Among them, Fig. 2(a) represents the sampling points of the traditional convolution kernel. Figure 2(b) represents the sampling points of the convolution kernel introducing a random offset vector. Figure 2(c) and Fig. 2(d) are the special cases of deformable convolution, which introduces convolution kernel sampling points of regular offset vectors. And show that deformable convolutions can be used as a special case of scale variation, scale variation and rotation variation.

$$y(P_0) = \sum_{P_n \in R} W(P_n) \times x(P_0 + P_n) \tag{1}$$

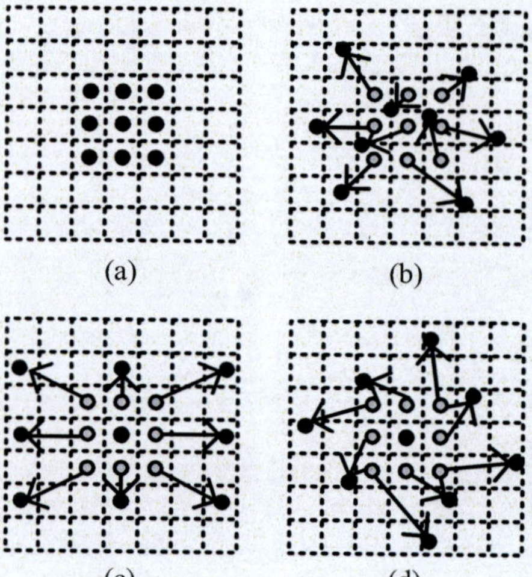

Fig. 2. Comparison of location distribution of sampling points. (a) Traditional sampling points. (b) Randomly offset sampling points. (c) Specific scale offset sampling points. (d) Specific rotation offset sampling points.

$$y(P_0) = \sum\nolimits_{P_n \in R} W(P_n) \times x(P_0 + P_n + \Delta P_n) \quad (2)$$

where y is the output feature matrix; x is the input feature matrix; R defines the size of the receptive field, covering the entire input feature matrix; P_n is the pixel point at any position on R; W is the weight matrix of the sampling value; $W(P_n)$ is the weight on the position of pixel point P_n; ΔP_n is the position offset of pixel point P_n, which is usually not an integer, and the value of the sample point after offset is determined by bilinear interpolation.

3 Experiment

3.1 Backbone Network Optimization Design

To improve the model's global feature extraction ability for masks and faces, we designed a network structure D-YOLOv5m based on deformable convolution. Among it, the backbone network CSPDarkNet53 introduces improved deformable convolution (DCNv2) [17], and the Neck network further improves the diversity and robustness of the features extracted by the backbone network. Deformable convolution associates the position coordinates of the sampling point with the specific sampling value, so that the sampling position offset can be learned and updated along with the model. As shown in Table 1, through a series of comparative experiments, we found that using DCNv2 in the shallow layer of the backbone network does not improve the detection accuracy of the model significantly. The deeper the model is, the better the detection accuracy of the model can be improved by using DCNv2 to improve it. Therefore, it is finally determined that DCNv2 is used to improve the SPP module, which is shown in Fig. 3.

Table 1. Comparative test using DCNv2 improved model.

Backbone Network Module	mAP% (with_mask)	mAP% (without_mask)	mAP% (mask_weared_ incorrect)	mAP%
Convert Conv1 to DCNv2	92.7	88.6	95.9	92.4
Convert Conv2 to DCNv2	93.2	90	97.8	93.7
Convert Conv3 to DCNv2	93.6	90.8	97.3	93.9
Convert SPP to DCNv2	94	91.6	97.1	94.2

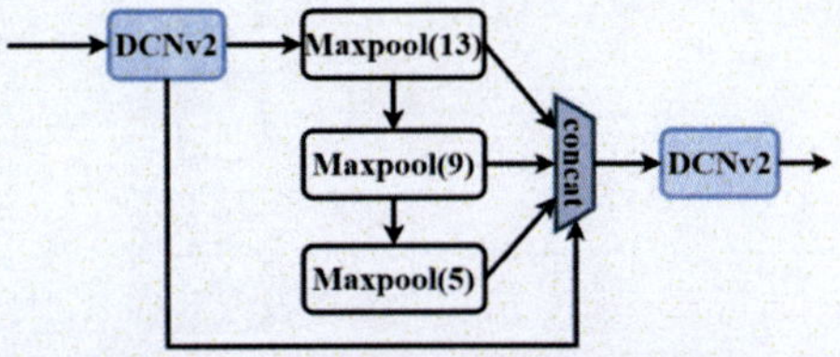

Fig. 3. Improved SPP structure

3.2 Improved Deformable Convolution

The improved deformable convolution (DCNv2) [17] is introduced into the backbone network CSPDarkNet53 for optimization and improvement. Applying traditional convolution to objects with large scale and shape changes or irregular contours may introduce background noise and affect feature extraction operations. However, the use of a simple sampling offset in DCNv1 will cause the sampling points to be biased towards weaker target features. Sampling points that are initially located in and around the center of the feature-rich target will be shifted to the edges of the target, affecting the object detection effect. DCNv2 (Eq. 3) sets a learnable modulation factor $\Delta m_k \in [0, 1]$ on the basis of DCNv1, which can make the offset point to the position with rich target features. Therefore, compared with DCNv1 and traditional convolution, DCNv2 has better learning ability of deformable targets, not only can obtain more effective information, but also the information obtained will be more accurate.

$$y(P_o) = \sum\nolimits_{P_n \in R} W(P_n) \times x(P_o + P_n + \Delta P_n)\Delta m_k \tag{3}$$

3.3 Model Lightweight Design

To meet the needs of practical applications, the algorithm finally deployed on the terminal usually has high requirements for computational efficiency, so we perform knowledge distillation [18] on the D-YOLOv5m algorithm to obtain the L-YOLOv5m model. We use an offline distillation method, using pre-trained D-YOLOv5m as the teacher network and YOLOv5s as the student network. In knowledge distillation, only the student network is

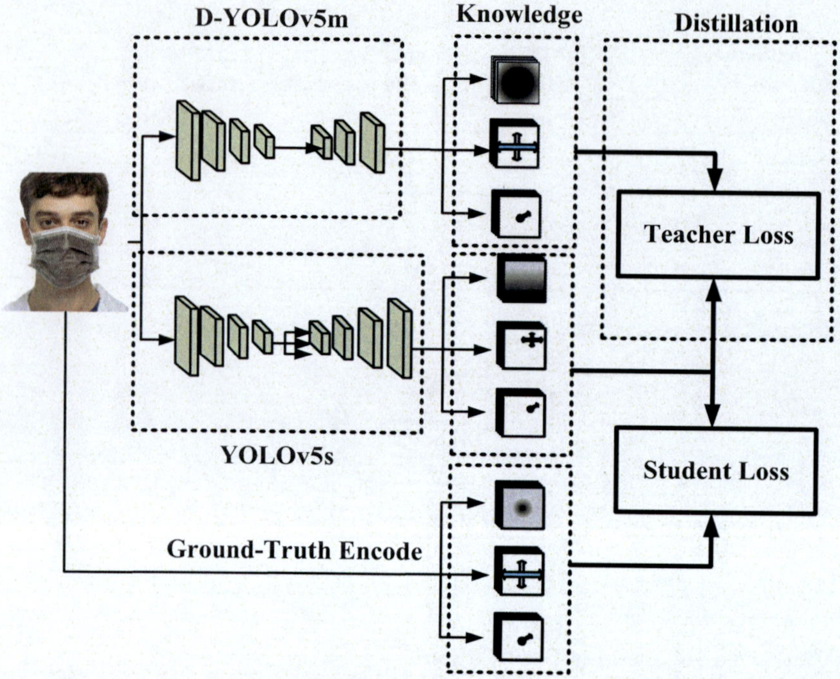

Fig. 4. Distillation mechanism

backpropagated and the parameter iteration of the teacher network is no longer performed (Fig. 4).

The logits output after D-YOLOv5m training are distilled, and the class prediction probability distribution obtained after the softmax layer is used as the soft targets. Meanwhile, the class prediction probability distribution obtained from the logits output of the student network after distillation and softmax layer will be used as the soft prediction, and the loss L_{soft} between the two is calculated.

$$L_{soft} = -\sum_{j}^{N} p_j^T \log\left(q_j^T\right) \tag{4}$$

where p_j^T is the value output by the softmax of the teacher network in the jth class when the temperature is T; q_j^T is the value output by the softmax of the student network in the jth class when the temperature is T, and N is the total label quantity.

In addition, the results obtained by training the teacher network will also have a certain error rate. Using ground truth can effectively reduce the possibility of errors being transmitted to the student network. Therefore, using the ground truth to train the student network and obtain the loss L_{hard}, at this time $T = 1$.

$$L_{hard} = -\sum_{j}^{N} c_j \log\left(q_j^T\right) \tag{5}$$

The loss function L_{soft} and L_{hard} are added together as the final loss function L.

$$L = \alpha L_{soft} + \beta L_{hard} \tag{6}$$

where α and β are used to adjust the contribution of teacher loss and student loss. The larger α is, the more emphasis is placed on learning the latent knowledge of the teacher network. The larger β is, the more inclined to learn the knowledge of ground truth. In this experiment, we did not pay too much attention to the influence of α and β on the distillation network, and set them to 1.

3.4 Image Defogging

To avoid the influence of weather factors on the model detection effect, we introduce the image defogging algorithm based on the dark channel prior (DCP) [19] to enhance the dataset. DCP proposes an effective and fast dark-channel prior theory to calculate the transmittance and atmospheric light values of an image. The imaging model of the image under foggy conditions is as follows.

$$I(x) = J(x)t(x) + A(1 - t(x)) \tag{7}$$

where x represents the position of the pixel; $I(x)$ is the acquired haze image; $J(x)$ is the reconstructed haze-free image at a single pixel x and A represents the ambient light value of the atmosphere. $t(x) \in [0, 1]$ is the transmittance.

$$J^{\text{dark}}(x) = \min_{y\in\Omega(x)}\left(\min_{c\in\{r,g,b\}} J^c(y)\right) \rightarrow 0 \tag{8}$$

This formula is the core of the dark channel prior theory, which means that there are always some pixels in the non-sky region whose value of at least one color channel is almost zero. $J^{\text{dark}}(x)$ represents the value of a certain pixel; $J^c(y)$ represents each channel of the color image; $\Omega(x)$ is a rectangular block centered on pixel x and y represents the value in the $\Omega(x)$ window area.

After normalizing the atmospheric light A, the dielectric function is approximated using a minimum filter, which is then refined using a soft-matting method to remove artifacts, resulting in a map of the corresponding transmittance distribution. The haze-free image is restored as follows.

$$J(x) = \frac{I(x) - A}{max(t(x), t_0)} + A \tag{9}$$

A parameter t_0 is introduced into the formula, which is usually a very small value to ensure that the denominator will not be smaller than 0, thus preventing the transmittance from being smaller than 0 and causing the formula to lose its meaning.

Since the number of special weather images in the self-made dataset T-Mask is very small, this method does not significantly improve the accuracy of the model. However, after defogging the images under special weather conditions, the trained model is used to detect the wearing of face mask on the images before and after defogging. As shown in Fig. 5, it can be proved that this processing method has greatly improved the effect of model detection.

Fig. 5. Comparison of image detection before and after defogging

4 Experimental Results

4.1 Main Dataset Introduction, Evaluation Metric

Currently, the official mask detection datasets are mostly divided into two categories: wearing masks and not wearing masks. In real life, it is common to wear masks irregularly. Therefore, in this paper, we integrate the mask detection dataset officially released by Paddle [20], the MAFA (Multi-Attribute Face Annotation) dataset, and the IMFD (the Incorrectly Masked Face Dataset) dataset [21] and use the IMFD dataset to expand the class of mask-wearing irregularities. A total of 5862 images were screened from the public dataset and divided into a training set and validation set according to the 8:2 ratio. Then use the image annotation tool labeling to manually label the dataset and define the dataset labels into three categories: wearing a mask (with_mask), not wearing a mask (without_mask) and wearing an irregular mask (mask_weared_incorrect). According torelevant epidemic prevention regulations, a face whose nose and mouth are safely covered by a mask is defined as a mask-wearing category. A face that completely exposes the entire face and is covered by a breathing mask, towel, clothing, etc. is defined as not wearing a mask. A face with a mask on the face but not completely covering the nose or mouth is defined as being in the category of irregular mask wearing. As shown

Fig. 6. Sample diagram of dataset category

in Fig. 6, the first row is the with_mask class, the second row is the without_mask class, and the third row is the mask_weared_incorrect class. The distribution of samples from different categories in the dataset is shown in Table 2.

Table 2. Number distribution of different types of images in the data set.

Category	Training set	Validation set
with_mask	3934	1576
without_mask	4214	1164
mask_weared_incorrect	2003	618
total	10151	3358

In this experiment, mAP (mean average precision) is used as the measurement index, which is commonly used to measure the accuracy in object detection.

$$AP = \frac{1}{11}\sum\nolimits_{r\in(0,0.1\ldots1)} P_{\text{interp}\,(r)} \tag{10}$$

$$mAP = sum(AP)/N \tag{11}$$

where $P_{\text{interp}\,(r)}$ is the interpolation precision; AP is the 11-point interpolation average precision of this class, and the AP of the three classes are summed and averaged to obtain the mAP of the experiment.

4.2 Analysis of Results

The experiments in this paper are trained and tested on the self-made T-Mask mask wearing detection dataset. The experimental results are shown in Fig. 7. The first column shows the experimental results for D-YOLOv5m and the second column shows the results for L-YOLOv5m. The figure shows that model D-YOLOv5m achieves a precision of 91.6%, while L-YOLOv5m is slightly less accurate than D-YOLOv5m at 91.3%. The recall value of models D-YOLOv5m and L-YOLOv5m has reached 99%. Combined with Table 3, it can be seen that the mAP value of the D-YOLOv5m model has reached 94.2%, and the mAP value of the lightweight model L-YOLOv5m has reached 93.6%, which is about 3 percentage points higher than the original YOLOv5m, while the number of parameters is reduced to 1/3 of the original.

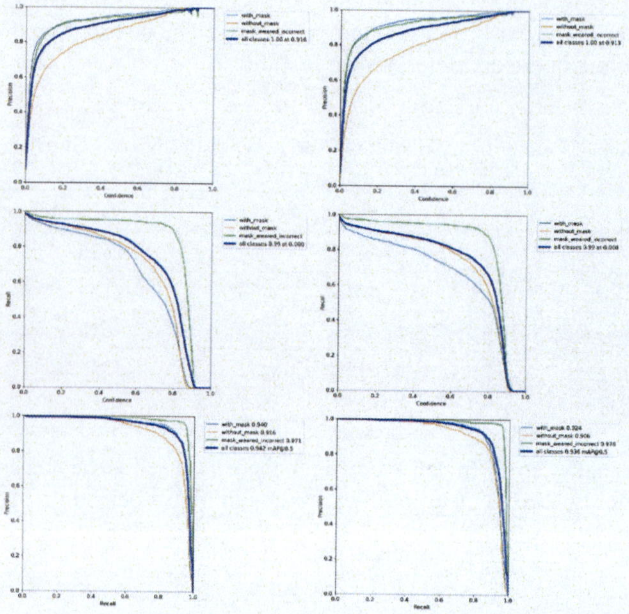

Fig. 7. Comparison of training results between D-YOLOV5m and L-YOLOv5m

Table 3. mAP comparison of each model

Model	with_mask %	without_mask %	mask_weared_incorrect %	Total %	Parameters 10^6
YOLOv5s	87.7	84.3	92.6	88.2	7.1
YOLOv5m	90.9	87	94.3	90.7	21.06
D-YOLOv5m	94	91.6	97.1	94.2	21.12
L-YOLOv5m	92.4	90.6	97.8	93.6	7.1

The detection effect of this experimental model is shown in Fig. 8. The first row is an example of the detection results without and with the mask on irregularly. It can be seen that the model in this paper still performs well in detecting pictures where the face is covered with a towel or clothing. The second row is an example of the detection results in a real scene, which shows that this model performs well in detecting whether a face far or close to the camera is wearing a mask.

Fig. 8. The model detection effect of this paper

5 Conclusion

This experimental model is based on the classical YOLOv5m model. Considering the real-time performance and accuracy of the model, YOLOv5m was chosen as the base model for the experiments to be optimized and improved. The modified deformable convolutional DCNv2 is introduced into the backbone network CSPDarkNet53 module. The introduction of deformable convolutional DCNv2 improves the model's ability to adapt to the deformation of the target space, allowing the model to focus on learning the salient features of the target while reducing the interference from background noise. The optimized D-YOLOv5m is obtained via knowledge distillation to obtain the lighter network L-YOLOv5m. Second, an image defogging operation based on dark channel priors is introduced to facilitate the model to adapt to the detection task under special conditions. This experiment is trained and tested on the self-made T-Mask dataset. The T-Mask dataset includes three categories: wearing a mask, not wearing a mask, and wearing an irregular mask, which makes the experimental study more realistic. Experimental results show that the mAP value of the model L-YOLOv5m on the T-Mask dataset reaches 93.6%, which is about 3 percentage points higher than the original YOLOv5m model, while the number of parameters is reduced to 1/3 of the original. Experimental results demonstrate that this model can be applied to pedestrian mask wearing detection tasks in most scenarios. The design in this paper will be of greater application value in future practical applications. With the lightweight design, the algorithm in this paper is more suitable for edge devices.

Acknowledgement. This work is supported by Heilongjiang Provincial Natural Science Foundation of China (No. LH2022F035) and 2022 Harbin University of Commerce's the support program project of faculty "Innovation" project (No. XL0068) and Harbin University of Commerce Graduate Innovative Research Project (No. YJSCX2022-743HSD).

References

1. Ni, X., Xing, Y., Suo, J., et al.: Characteristics and functions of microbial aerosols in medical institutions. Chin. J. Hosp. Infect. **30**(8), 1183–1190 (2020)
2. Wang, R., Zeng, Q., Liu, H.: Meta-analysis of the relationship between medical staff wearing masks and infections of respiratory infectious diseases. J. Environ. Health **29**(3), 269–270 (2012)
3. Davies, A., Thompson, K.-A., Giri, K., et al.: Testing the efficacy of homemade masks: would they protect in an influenza pandemic? Disaster Med. Public Health Prep. **7**(4), 413–418 (2013)
4. Macintyre, C.R., Cauchemez, S., Dwyer, D.E., et al.: Face mask use and control of respiratory virus transmission in households. Emerg. Infect. Dis. **15**(2), 233 (2009)
5. Kux, L.: Department of health and human services. Food and Drug Administration [docket no FDA-2012-D-0146]: guidance for industry on irritable bowel syndrome-clinical evaluation of drugs for treatment: availability Fed Regist, vol. 77, pp. 32124–32125 (2012)
6. Henry, S., Manuguerra, J.-C., Escourolle, D., et al.: National plan for the prevention and control "influenza pandemic". La Revue du Praticien **58**(15), 1687–1693 (2008)
7. Suzhen, C., Dongsen, W., Xing, C., et al.: Research on Chinese residents wearing masks for protective behavior during the COVID-19 epidemic. Environ. Sci. Res. **33**(7), 1649–1658 (2020)
8. Zuodong, N., Tao, Q., Handong, L., et al.: Improved RetinaFace mask wearing detection algorithm in natural scenes. Comput. Eng. Appl. **56**(12), 1–7 (2020)
9. Wu, P., Li, H., Zeng, N., et al.: FMD-Yolo: an efficient face mask detection method for COVID-19 prevention and control in public. Image Vis. Comput. **117**, 104341 (2022)
10. Kumar, A., Kalia, A., Sharma, A., et al.: A hybrid tiny YOLO v4-SPP module based improved face mask detection vision system. J. Ambient Intell. Humaniz. Comput., 1–14 (2021)
11. Xin, Z., Rui, L., Weiqi, L.: Mask wearing detection algorithm based on improved YOLOV3. Comput. Eng. Des. **43**(05), 1319–1326 (2022)
12. Redmon, J., Divvala, S., Girshick, R., et al.: You only look once: Unified, real-time object detection. In: Proceedings of the IEEE Conference on Computer Vision and Pattern Recognition (2016)
13. Redmon, J., Farhadi, A.: YOLO9000: better, faster, stronger. In: Proceedings of the IEEE Conference on Computer Vision and Pattern Recognition (2017)
14. Redmon, J., Farhadi, A.: Yolov3: an incremental improvement. arXiv preprint arXiv:180402767 (2018)
15. Wang, C.-Y., Liao, H.-Y.M., Wu, Y.-H., et al.: CSPNet: a new backbone that can enhance learning capability of CNN. In: Proceedings of the IEEE/CVF Conference on Computer Vision and Pattern Recognition Workshops (2020)
16. Dai, J., Qi, H., Xiong, Y., et al.: Deformable convolutional networks. In: Proceedings of the IEEE International Conference on Computer Vision (2017)
17. Zhu, X., Hu, H., Lin, S., et al.: Deformable ConvNets V2: more deformable, better results. In: Proceedings of the IEEE/CVF Conference on Computer Vision and Pattern Recognition (2019)
18. Gou, J., Yu, B., Maybank, S.J., et al.: Knowledge distillation: a survey. Int. J. Comput. Vis. **129**(6), 1789–1819 (2021)
19. He, K., Sun, J., Tang, X.: Single image haze removal using dark channel prior. IEEE Trans. Pattern Anal. Mach. Intell. **33**(12), 2341–2353 (2010)
20. https://github.com/PaddlePaddle/PaddleHub
21. Cabani, A., Hammoudi, K., Benhabiles, H., Melkemi, M.: MaskedFace-Net - a dataset of correctly/incorrectly masked face images in the context of COVID-19. Smart Health (2020). ISSN 2352-6483. https://doi.org/10.1016/j.smhl.2020.100144

Facial Expression Recognition Based on Fused Features and Support Vector Machine

Zhi-jie Zhao and Ming-yang Jiao(✉)

School of Computer and Information Engineering, Harbin University of Commerce, Harbin 150028, Heilongjiang, China
1151317344@qq.com

Abstract. Traditional facial expression recognition methods often suffer from low feature dimension and low recognition rates. These methods typically extract features limited to commonly used image features, while overlooking the semantic information of facial expressions. In this paper, we proposed an innovative facial expression recognition method that incorporates a novel calculation method for facial landmark features. These features are combined with Histogram of Oriented Gradients (HOG) and wavelet features using weighted concatenation to form the final feature vector. Support Vector Machines (SVM) are employed for classification. The proposed method achieves an impressive accuracy of 93.4% on the JAFFE dataset.

Keywords: Facial Expression Recognition · Facial Landmark Feature · Feature Weighting · SVM

1 Introduction

Facial expressions are an essential component of human emotional communication, playing a crucial role in conveying emotions and intentions. The facial expressions of the face can mainly be categorized into seven types: angry, disgust, fear, happy, sad, surprise, and neutral. (Fig. 1 illustrates the facial expressions among them) [1]. Facial expression recognition, as a significant research direction in computer vision and machine learning, aims to develop automated systems that can accurately recognize and interpret the emotional information conveyed in facial expressions [2].

Despite significant progress in the field of facial expression recognition, there are still challenges and difficulties. Firstly, there are differences and diversities in facial expression expressions among different individuals, including cultural and individual variations. These individual differences add complexity and difficulty to facial expression recognition. Secondly, facial expressions can vary greatly under different contexts and environments, such as changes in lighting conditions and poses, posing challenges to the robustness of recognition algorithms. Additionally, imbalanced datasets and a lack of annotated data also limit the performance of facial expression recognition algorithms.

To address these challenges, researchers have proposed various facial expression recognition methods, including feature-based approaches, deep learning methods, and

A. Hassanien et al. (Eds.): AISI 2023, LNDECT 184, pp. 521–531, 2023.
https://doi.org/10.1007/978-3-031-43247-7_45

multi-modal fusion approaches [3]. These methods have made progress in improving the performance of facial expression recognition. However, there is still room for improvement, especially in more complex tasks such as emotion recognition, multi-angle, and multi-person emotion analysis.

Therefore, this paper aims to investigate and explore the key techniques and methods for facial expression recognition to enhance its performance and robustness in practical applications. Our research is primarily based on the Japanese Female Facial Expression (JAFFE) dataset [4], considering a variety of feature extraction methods. Finally, a Support Vector Machine (SVM) [5] is employed to classify the extracted feature vectors.

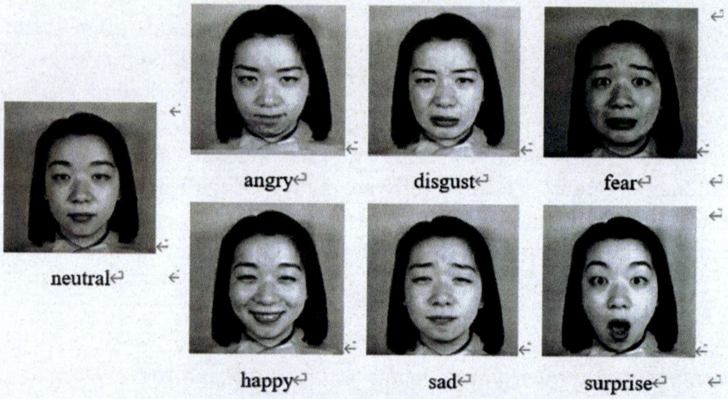

Fig. 1. Seven facial expressions of human face.

2 Related Work

In this section, we provide an overview of the key research work and advancements in the field of facial expression recognition.

Early facial expression recognition methods mainly involved extracting manually designed features from facial images. These features encompassed geometric characteristics, such as the distance between facial landmarks, as well as texture-based attributes, like Local Binary Patterns (LBP) [6] and Histogram of Oriented Gradients (HOG) [7]. Typically, classifiers such as Support Vector Machines (SVM) and decision trees were utilized to classify these handcrafted features.

With the advent of deep learning, convolutional neural networks (CNNs) [8] have made significant strides in facial expression recognition. Architectures such as VGGNet [9], ResNet [10], and InceptionNet [11], which are based on CNNs, have demonstrated the ability to automatically learn discriminative features from raw facial images. These deep learning models excel at capturing both low-level features, such as edges and textures, and high-level semantic information, thereby significantly enhancing recognition accuracy. In recent years, there has been a growing interest in multi-modal facial expression recognition, which involves combining facial visual information with other modalities such as audio and physiological signals.

Another pivotal aspect in facial expression recognition is the availability of appropriate datasets for training and evaluation purposes. Several benchmark datasets have been developed, including CK+ (Extended Cohn-Kanade) [12], JAFFE (Japanese Female Facial Expression) [13], and FER2013 (Facial Expression Recognition 2013) [13]. These datasets offer diverse facial expressions captured under varying conditions, empowering researchers to develop and compare different recognition methods.

Moreover, deep generative models, such as variational autoencoders (VAEs) [14] and Generative Adversarial Networks (GANs) [15], have exhibited promise in generating synthetic facial expressions and augmenting limited training data. These generative models enable researchers to overcome challenges posed by data scarcity and enhance the generalization capabilities of facial expression recognition models [16]. To summarize, facial expression recognition has made remarkable progress, transitioning from traditional handcrafted feature methods to deep learning approaches. The availability of benchmark datasets, coupled with the exploration of multi-modal fusion techniques, has significantly enriched this field.

In this paper, we continue to employ handcrafted feature extractors and propose a facial landmark feature extraction method that combines Histogram of Oriented Gradients (HOG) and image wavelet features [17]. We employ feature weighting to concatenate different feature vectors into a complete feature vector [18], which is subsequently fed into an SVM classifier for facial expression classification. Through rigorous experimentation, our proposed method effectively leverages the strengths of image features and facial landmark features, ultimately achieving substantial accuracy in facial expression classification.

3 Implementation

3.1 Pre-processing

Facial alignment plays a crucial role in facial expression recognition tasks [19]. It ensures consistent localization of facial landmarks in images, enabling accurate and reliable analysis. By eliminating pose variations, facial alignment standardizes facial appearance, allowing effective comparison of facial features. It enhances feature extraction and localization capabilities, aiding in the recognition of subtle changes and fine details in facial expressions. Additionally, facial alignment reduces the impact of external factors such as lighting, improving the robustness of expression recognition algorithms. Overall, facial alignment plays a vital role in achieving accurate and reliable facial expression analysis by standardizing facial appearance, eliminating pose variations, and enhancing feature extraction. It enhances the robustness and generalization of facial expression recognition models.

In this paper, we utilize the dlib library [20] to perform facial bounding box and landmark detection on the faces present in the image. The resulting 68 facial landmarks encompass key points in regions such as the eyes, eyebrows, nose, and mouth. Subsequently, we employ the Procrustes analysis Alignment method to align the faces. Procrustes analysis is a commonly used alignment technique that employs geometric transformations on facial images to ensure consistency in the positions of the key points across all facial images.

By employing Procrustes analysis face alignment, we can eliminate variations in pose and scale among different faces, thereby maintaining consistent spatial positioning of the key points. This allows for better comparison and analysis of facial expressions, accurate feature extraction, and reliable classification and recognition. In this study, we align the detected faces in the image to a size of 256 * 256 pixels. Figure 2 illustrates the process of face alignment, where Fig. 2a represents the input original image, Fig. 2b depicts the facial bounding boxes and corresponding key point positions detected using the dlib library from the original image, and Fig. 2c shows the resulting face image after the face alignment operation.

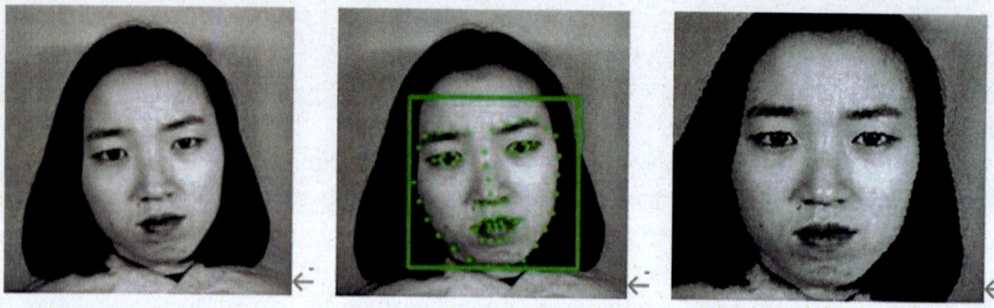

a b c

Fig. 2. The process of face alignment.

3.2 Hog Feature Extraction

Histogram of Oriented Gradients (HOG) is a commonly used image feature descriptor widely applied in computer vision tasks such as facial expression recognition [7]. It captures local gradient information by dividing the image into small cells and computing histograms of gradient orientations within each cell. In facial expression recognition, HOG features have been successfully utilized to extract discriminative facial texture patterns. By analyzing the distribution of local gradient orientations, HOG can capture important facial cues related to facial expressions, such as edges, corners, and texture variations.

In this study, we selected image HOG features as one of the features for facial expression recognition. The input to the HOG feature extractor is the aligned face image, which is the result of face alignment. Since the JAFFE dataset provides grayscale images, no additional grayscale conversion is required. Subsequently, we performed gamma correction [21] (as shown in Eq. 1. In the equation, x and y represent the coordinates of the current pixel, I represent obtaining the pixel value at the specified position in the image, gamma is a variable parameter, and in this paper, gamma is set to 0.8.) and color space normalization on the pre-processed image to adjust image contrast and reduce the impact of local shadows and lighting variations on expression prediction.

$$I(x, y) = I(x, y)^{gamma} \tag{1}$$

We divided the input image into 256 16 × 16 cells and computed the gradient histograms for each cell. The angle interval for computing the gradient histogram was set to 20°, representing each cell's gradient histogram as an array of length 9. Since image gradients are highly sensitive to overall illumination, we aimed to make the feature descriptor invariant to lighting changes. To achieve this, we normalized the gradient histograms of each cell (as shown in Eq. 2, n represents the length of HOG feature vector). We selected a block size of 2 × 2 for normalization, and then concatenated the histograms of each block to form the final feature representation. This setting resulted in HOG feature vectors with a length of 8100. The process of extracting HOG feature maps from facial images is illustrated in Fig. 3. Figure 3a represents the input original image, Fig. 3b shows the image after gamma correction, and Fig. 3c displays the HOG feature map.

$$F_{HOG} = \frac{[x_1, x_2, \ldots x_n]}{\sqrt{x_1^2 + x_2^2 + \ldots + x_n^2}} \tag{2}$$

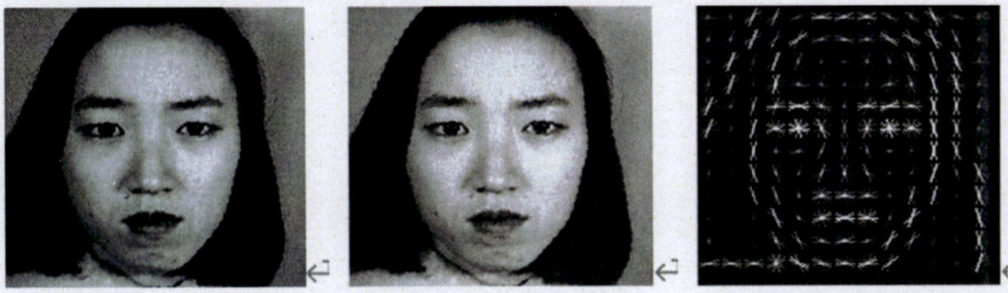

a b c

Fig. 3. HOG feature extraction.

3.3 Wavelet Feature Extraction

Wavelet features are commonly used image features in image recognition, capturing local intensity variations [17]. In facial expression analysis, wavelet features can effectively capture facial texture patterns, such as edges and corners, which are crucial for differentiating between different expressions. Wavelet features are computed by convolving a set of wavelet prototype filters with the image at different scales and positions. The obtained responses are then used to construct feature vectors that represent the presence or absence of specific facial patterns. By analyzing these feature vectors, classifiers can learn to differentiate between different facial expressions.

HAAR wavelets can extract features at different frequencies through multi-scale wavelet transforms, enabling them to capture various details and structural information in facial images, thus enhancing feature richness and dimension [22]. HAAR wavelets have a relatively fast computation speed (as shown in Eq. 3). Due to their rectangular waveform basis functions, the computation process can be efficiently implemented using techniques like fast integral computation. Moreover, HAAR wavelet features exhibit robustness to lighting variations and noise. They achieve this by extracting features from local regions and combining the features of multiple small regions, reducing the influence of global lighting and noise on the overall image and improving the stability and accuracy of facial expression recognition. In this paper, we employ HAAR wavelets to extract wavelet features.

$$x + y = z \tag{3}$$

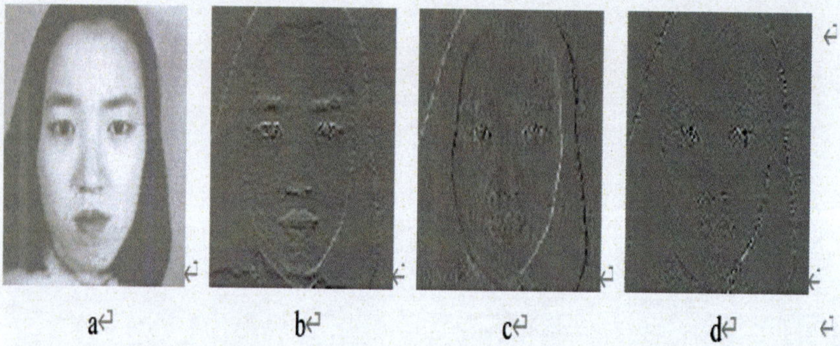

Fig. 4. Haar wavelet feature maps.

Figure 4 presents the Haar wavelet feature maps, where Fig. 4a represents the low-frequency feature map, Fig. 4b corresponds to the high-frequency horizontal feature map, Fig. 4c depicts the high-frequency vertical feature map, and Fig. 4d illustrates the high-frequency diagonal feature map.

To fully represent the details of local facial regions, we first divide the image into 8 × 8 blocks (as shown in Fig. 5). Then, we extract Haar wavelet features from each block individually, removing the low-frequency components from the wavelet features. Next, we perform statistical analysis on the high-frequency wavelet coefficients, considering values within the range [−450, 450] as valid. We divide this range into 10 equal intervals and count the number of coefficients falling into each sub-interval. In this study, the length of the wavelet feature vector for each image is 1920.

3.4 Key-Points Feature Extraction

The Delaunay triangulation algorithm [23] is commonly used in face recognition tasks because it employs facial landmarks that correspond to the facial feature points, each carrying semantic information. In this study, we also adopt the Delaunay triangulation to represent the topological features of facial expressions.

First, we utilize the dlib library to detect 68 facial landmarks, also known as facial feature points. These landmarks are crucial for understanding and analyzing facial expressions. Subsequently, we construct Delaunay triangles using these points. Delaunay triangulation is a method for creating triangles from a set of points such that no point is

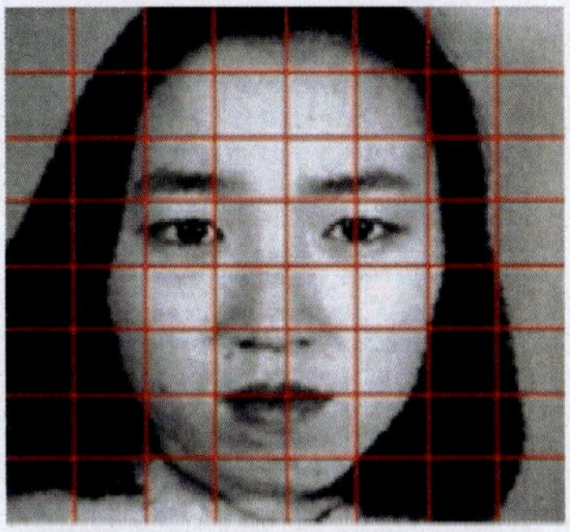

Fig. 5. The division of facial image.

inside the circumcircle of any triangle. Once we obtain the Delaunay triangles, various features can be extracted. In this study, we consider the edge lengths of each triangle, the area (as shown in Eq. 4, the lengths of the three sides are represented by a, b and c respectively), and the three internal angles (as shown in Eq. 5, A represents one of the interior angles of the triangle) as elements of the facial keypoint feature vector. These features capture significant geometric information about facial expressions and provide insights into facial muscle movements and shape changes associated with different expressions. This feature vector, as a concise and informative representation of facial expressions, can be utilized for further facial expression classification.

$$S = \sqrt{p(p-a)(p-b)(p-c)}$$
$$s.t. \quad p = \frac{a+b+c}{2} \tag{4}$$

$$A = \arccos((b^2 + c^2 - a^2)/(2bc)) \tag{5}$$

By employing the dlib library for facial landmarks detection, constructing Delaunay triangles, and extracting features such as triangle edge lengths, areas, and internal angles, we can effectively represent the topological features of facial expressions. This allows us to obtain a topological feature vector that accurately represents facial expressions for further facial expression classification. Figure 6 shows the result of Delaunay triangulation.

3.5 A Subsection Sample

In this paper, we extracted three types of features from facial images, namely HOG, wavelet, and facial keypoint topological features. However, the importance of each feature in facial expression recognition varies. Therefore, for feature vector fusion, we adopted a weighted fusion approach to concatenate the feature vectors (as shown in the Eq. 6, X represents the feature vector, w represents the weight of each feature). This weighted concatenation of feature vectors allows for complementary feature expression capabilities between different features, thereby enhancing the discriminative power of

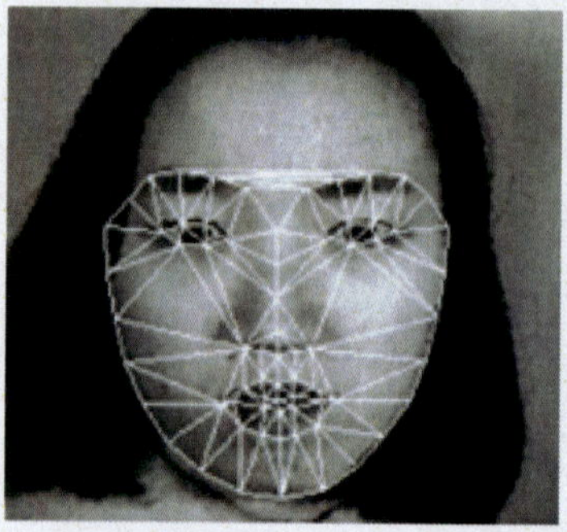

Fig. 6. The result of Delaunay triangulation.

the feature vector. In this paper, we assigned weights to the three features as follows: HOG: 0.3, wavelet: 0.2, keypoints: 0.5.

$$F_{fusion} = [w_1 * X_1,\ w_2 * X_2,\ w_3 * X_3]$$
$$s.t.\ \begin{cases} w_1 + w_2 + w_3 = 1 \\ 0 < w_1, w_2, w_3 < 1 \end{cases} \tag{6}$$

3.6 Classification

Support Vector Machines (SVM) is a powerful supervised machine learning algorithm that has gained widespread application in various fields such as pattern recognition, image classification, and natural language processing [5]. SVM is highly acclaimed for its ability to efficiently handle both linear and non-linear classification tasks. The fundamental principle of SVM is to search for an optimal hyperplane that maximally separates different classes in the input data, enabling robust and accurate classification.

n scenarios where the data is not linearly separable, SVM utilizes a technique known as the kernel trick. The kernel trick maps the input data into a higher-dimensional feature space, where the data becomes linearly separable. This mapping is implicitly performed, avoiding the computational burden of explicit data transformation. SVM then constructs a hyperplane in this high-dimensional space to effectively classify the data.

A key advantage of SVM is its capability to handle high-dimensional data with a relatively small number of samples. It achieves this by utilizing a subset of the training data known as support vectors, which define the key points for the hyperplane. By focusing on these support vectors, SVM can efficiently generalize to unseen data points.

In this paper, we adopt the Radial Basis Function kernel (RBF) (as shown in the Eq. 7, γ is a hyper parameter, x and y are feature vectors) [24] and employ SVM to tackle the multi-class facial expression recognition task.

$$K(x, y) = e^{-\gamma \|x-y\|^2} \tag{7}$$

4 Experiment and Analysis

After extracting feature vectors for the three types of features, we employ a weighted fusion approach to concatenate the feature vectors. Subsequently, SVM is utilized to classify the concatenated feature vectors.

Dataset: mThe Japanese Female Facial Expression (JAFFE) database contains 213 samples of posed expressions from 10 Japanese females. Each person has 3–4 images with each of six basic facial expressions and one image with a neutral expression. We divided the dataset into training and test set, consisting of 144 and 69 images, respectively.

In the experimental section, we conducted two main sets of experiments primarily on the JAFFE dataset. In the first experiment, we compared the impact of incorporating the facial keypoint topological features proposed in this paper on the accuracy of the test set. The results are presented in Table 1. In the second experiment, we compared the effect of using weighted concatenation during the feature fusion step on the test accuracy. The results are shown in Table 2.

Table 1. Results of the first experiment.

Facial keypoint topological features	Recognition accuracy
w/o	91.3%
w/	92.6%

From Table 1, we can observe that the inclusion of facial landmark topological features resulted in a recognition accuracy improvement of 1.3% compared to the absence of this feature. This indicates that facial landmark topological features effectively capture the semantic information of the face, thereby enhancing the recognition accuracy in facial expression recognition tasks.

Table 2. Results of the second experiment.

Feature weighting	Recognition accuracy
w/o	92.6%
w/o	93.4%

From Table 2, it is evident that weighted feature fusion does contribute to an improvement in recognition accuracy. By assigning a higher weight to the facial landmark topological feature that possess clearer semantic information, and then concatenating them with the other two feature vectors, using such a vector for classification leads to a 0.8% increase in accuracy. This result is 3.55% higher than Wang's [25] and 0.6% higher than Shervin Minaee's [26].

5 Conclusion

The key contribution of this research lies in the introduction of facial landmark topological features, which effectively capture the semantic information of facial expressions. By incorporating these features into the fusion process and assigning higher weights to the more explicit facial landmark topological features, significant improvements in classification performance are achieved.

The achieved accuracy on the JAFFE dataset validates the effectiveness of the fused feature vector and SVM classifier. This research holds significant importance in advancing facial expression recognition technology and has broad prospects for applications in human-computer interaction, emotion analysis, and facial behavior understanding.

In future studies, we will explore different feature extraction algorithms and combine classification algorithms with state-of-the-art large models to strive for greater breakthroughs in facial expression recognition tasks.

References

1. Li, S., Deng, W.: Deep facial expression recognition: a survey. IEEE Trans. Affect. Comput. **13**(3), 1195–1215 (2022)
2. Mellouk, W., Handouzi, W.: Facial emotion recognition using deep learning: review and insights. Procedia Comput. Sci. **175**, 689–694 (2020)
3. Zhang, J., Yin, Z., Chen, P., Nichele, S.: Emotion recognition using multi-modal data and machine learning techniques: a tutorial and review. Inf. Fus. **59**, 103–126 (2020)
4. Lyons, M.J.: "Excavating AI" re-excavated: debunking a fallacious account of the JAFFE dataset. http://arxiv.org/abs/2107.13998 (2021)
5. Salcedo-Sanz, S., Rojo-Álvarez, J.L., Martínez-Ramón, M., Camps-Valls, G.: Support vector machines in engineering: an overview. Wiley Interdiscip. Rev. Data Min. Knowl. Discov. **4**(3), 234–267 (2014)
6. Shan, C., Gong, S., McOwan, P.W.: Facial expression recognition based on local binary patterns: a comprehensive study. Image Vis. Comput. **27**(6), 803–816 (2009)
7. Dalal, N., Triggs, B.: Histograms of oriented gradients for human detection. In: 2005 IEEE Computer Society Conference on Computer Vision and Pattern Recognition (CVPR 2005), San Diego, CA, USA, vol. 1, pp. 886–893. IEEE (2005)
8. Li, Z., Liu, F., Yang, W., Peng, S., Zhou, J.: A survey of convolutional neural networks: analysis, applications, and prospects. IEEE Trans. Neural Netw. Learn. Syst. **33**, 6999–7019 (2022)
9. Simonyan, K., Zisserman, A.: Very deep convolutional networks for large-scale image recognition. http://arxiv.org/abs/1409.1556 (2015)
10. He, K., Zhang, X., Ren, S., Sun, J.: Deep residual learning for image recognition. In: 2016 IEEE Conference on Computer Vision and Pattern Recognition (CVPR), Las Vegas, NV, USA, pp. 770–778. IEEE (2016)
11. Szegedy, C., et al.: Going deeper with convolutions. In: 2015 IEEE Conference on Computer Vision and Pattern Recognition (CVPR), Boston, MA, USA, pp. 1–9. IEEE (2015)
12. Lucey, P., Cohn, J.F., Kanade, T., Saragih, J., Ambadar, Z., Matthews, I.: The extended Cohn-Kanade dataset (CK+): a complete dataset for action unit and emotion-specified expression. In: 2010 IEEE Computer Society Conference on Computer Vision and Pattern Recognition – Workshops, San Francisco, CA, USA, pp. 94–101. IEEE (2010)

13. Zahara, L., Musa, P., Wibowo, E.P., Karim, I., Musa, S.B.: The facial emotion recognition (FER-2013) dataset for prediction system of micro-expressions face using the convolutional neural network (CNN) algorithm based Raspberry Pi. In: 2020 Fifth International Conference on Informatics and Computing (ICIC), Gorontalo, Indonesia, pp. 1–9. IEEE (2020)
14. Doersch, C.: Tutorial on variational autoencoders. http://arxiv.org/abs/1606.05908 (2021)
15. Goodfellow, I., et al.: Generative adversarial networks. Commun. ACM **63**(11), 139–144 (2020)
16. Yang, L., Tian, Y., Song, Y., Yang, N., Mam K., Xie, L.: A novel feature separation model exchange-GAN for facial expression recognition. Knowl. Based Syst. **204**, 106217 (2020)
17. Liu, C., Wechsler, H.: Gabor feature based classification using the enhanced fisher linear discriminant model for face recognition. IEEE Trans. Image Process. Publ. IEEE Signal Process. Soc. **11**(4), 467–476 (2002)
18. Qiao, Z., Zhang, Q., Dong, Y., Yang, J.-J.: Application of SVM based on genetic algorithm in classification of cataract fundus images. In: 2017 IEEE International Conference on Imaging Systems and Techniques (IST), Beijing, pp. 1–5. IEEE (2017)
19. Zhang, K., Zhang, Z., Li, Z., Qiao, Y.: Joint face detection and alignment using multitask cascaded convolutional networks. IEEE Signal Process. Lett. **23**(10), 1499–1503 (2016)
20. Amos, B., Ludwiczuk, B., Satyanarayanan, M.: OpenFace: a general-purpose face recognition library with mobile applications. CMU Sch. Comput. Sci. **6**(2), 1–20 (2016)
21. Huang, S.-C., Cheng, F.-C., Chiu, Y.-S.: Efficient contrast enhancement using adaptive gamma correction with weighting distribution. IEEE Trans. Image Process. **22**(3), 1032–1041 (2013)
22. Lepik, Ü., Hein, H.: Haar wavelets. In: Lepik, Ü., Hein, H. (eds.) Haar Wavelets. Mathematical Engineering, pp. 7–20. Springer, Cham (2014). https://doi.org/10.1007/978-3-319-04295-4_2
23. Lee, D.T., Schachter, B.J.: Two algorithms for constructing a Delaunay triangulation. Int. J. Comput. Inf. Sci. **9**(3), 219–242 (1980)
24. Soman, K., Sathiya, A., Suganthi, N.: Classification of stress of automobile drivers using radial basis function kernel support vector machine. In: International Conference on Information Communication and Embedded Systems (ICICES2014), Chennai, India, pp. 1–5. IEEE (2014)
25. Zhang, L., Wang, D., Chen, Y., Sun, H., Han, X.: Facial expression recognition based on GMRF and KNN algorithm. Comput. Appl. Softw. **37**(10), 214–219 (2020). (in Chinese)
26. Minaee, S., Minaei, M., Abdolrashidi, A.: Deep-emotion: facial expression recognition using attentional convolutional network. Sensors **21**(9), 3046 (2021)

![Check for updates]

Research on Multimodal Retrieval Method Based on Commodity Images

Xinghai Zhang[1,2], Ming Chen[1,2], Mingyang Jiao[1,2], and Zhijie zhao[1,2](✉)

[1] Harbin University of Commerce, Harbin 150028, China
zhijiezzz@aliyun.com

[2] Heilongjiang Provincial Key Laboratory of Electronic Commerce and Information Processing, Harbin 150028, China

Abstract. With the rapid development of science and technology, online shopping has gradually changed the way people buy goods because of its good quality and fast circulation. Therefore, online shopping has become one of the important modern shopping modes. With the development of the Internet, there are more and more channels to find product images and accompanying descriptive texts. How to retrieve the required goods accurately and efficiently is one of the important problems to be solved urgently. The existing traditional image retrieval methods mostly focus on the semantic information of the modality itself, and learn the binary code through quantization to find the similarity between the modalities. The heterogeneity of different types of modalities needs further discussion. Therefore, it is difficult to explain the low average accuracy of commodity image retrieval. Different from the traditional linear regression model retrieval method, this study uses a polynomial regression model method, which focuses on directly using polynomial regression to guide hash learning. The results show that the average accuracy of multimodal commodity retrieval using different methods to extract commodity image features is obviously better than that of traditional linear regression model, and polynomial regression model is the main factor to guide hash learning. When the code length is 128 bits, the average accuracy of LBP and TF_IDF is the highest. By analyzing the reasons why the average retrieval accuracy can be improved between different modes with the same semantics, the target commodity image can be searched to the greatest extent, saving more space cost and time cost.

Keywords: commodity image · Multimodal · Image features · Hash

1 Introduction

In recent years, with the rapid development of science and technology and the rapid popularization of the Internet, people can find commodity pictures [1] and accompanying words through various channels. However, in the face of a large number and variety of commodity pictures, the retrieval results often have the problems of low efficiency and longtime consumption. Because the commodity image [2] is different from the traditional landscape images and people images, the commodity image itself has the

A. Hassanien et al. (Eds.): AISI 2023, LNDECT 184, pp. 532–541, 2023.
https://doi.org/10.1007/978-3-031-43247-7_46

characteristics of commodity in the center of the picture and simple background. Therefore, how to quickly and accurately retrieve the required product images is the main problem to be solved urgently. Different from the single-mode data that only describes the information contained in the data unilaterally, multimodal retrieval [3] has become a new hot research direction because it can describe the information contained in the data more comprehensively, retrieve commodity images quickly and improve the retrieval efficiency in a targeted manner.

Commodity image retrieval is an image retrieval based on the characteristics of commodity images and the needs of online shopping in the field of e-commerce. Commodity image retrieval methods are mainly divided into content-based commodity image retrieval methods and deep learning-based commodity image retrieval methods. Content-based commodity image retrieval extracts the visual features of the bottom layer [5] of the commodity image and carries out commodity image retrieval by calculating the similarity between the feature vectors of the test commodity image and each vector in the query feature matrix. The method of commodity image retrieval based on deep learning generally uses a neural network [6] and other models to extract the features of commodity images and then realizes image retrieval. The retrieval of commodity images can only be aimed at the underlying features of a single image, and it is difficult to fully reflect the real characteristics of commodity images. At present, for the research of finding the most similar commodity image [7], scholars only use linear regression models, matrix decomposition [8], and other methods, so they pay insufficient attention to the practical attributes and relative attributes of commodity images.

Multi-modal commodity retrieval is a method to retrieve images. By combining the semantic information of labels with the study of the underlying features, a more comprehensive idea of multi-modal commodity retrieval is constructed. Multi-modal hash retrieval plays an important role in multi-modal retrieval because of its characteristics of eliminating the heterogeneous gap [9] and high efficiency. Therefore, this study uses multimodal hash retrieval to learn the hash transformation of different modes, uses polynomial regression model to guide hash learning, maps the features of different modes to the same Hamming space [11], and then realizes similarity measurement in Hamming space [12], which improves the effective time of retrieval, and makes reasonable explanations for different kinds of commodity image retrieval, so as to put forward effective suggestions and strategies for improving the average precision of commodity retrieval.

2 Framework Structure of Multimodal Retrieval Method

2.1 The Significance of Commodity Image Retrieval

Commodity image retrieval is to search and match commodities by analyzing and identifying commodity images. The significance of commodity image retrieval is to improve the user experience at first. Commodity image retrieval allows users to find the needed commodities intuitively and quickly, avoiding the complexity and misleading brought by traditional text description, which can greatly improve user satisfaction. Secondly, commodity image retrieval can promote sales. Commodity image retrieval enables merchants to provide customers with more accurate search services, so that users' real needs can be met as much as possible, thus enhancing brand trust and consumer loyalty. Finally, commodity image retrieval can reduce labor costs. Modern image retrieval mainly relies on image processing, pattern recognition, artificial intelligence and computer vision, and realizes image retrieval by using the characteristics of color, shape, texture and spatial relationship of the image itself. Because commodity images themselves contain a lot of data information, these information can not be described by text-based image retrieval technology. Therefore, modern commodity image retrieval technology can replace the manual marking process, which not only saves time, but also ensures the integrity and accuracy of commodity image information. It can greatly reduce labor costs and time expenses. Commodity image retrieval has great commercial value. As people's requirements for shopping experience are not improved, it will become an indispensable technical means in the e-commerce industry in the future.

2.2 Feature Extraction

The so-called feature extraction means that in machine learning, pattern recognition and image processing, the computer reads the information of the image itself or non-redundant derived values, etc., and makes necessary preparations for commodity image retrieval. This study is aimed at how to extract the features of commodity images and texts. Because of the particularity of commodity images, this study adopts three methods to extract image features, namely local binary pattern (LBP), directional gradient histogram (HOG) and one-hot. The method of extracting different features from commodity images, so as to compare the difference of average accuracy.

2.3 Feature Fusion

In the work of multimodal retrieval [13], the retrieval performance is uneven due to the poor robustness of individual feature matrix and different anti-interference ability. Feature fusion is to combine the strengths of different sub-features to achieve the purpose of complementing each other's weaknesses. In this study, we try to use mosaic method to realize feature fusion through many experiments, which makes the retrieval efficiency high and the operation efficiency fast. By fusing the features of different scales, users can retrieve the commodity images that match the input query better.

2.4 Similarity Measurement

In multimodal commodity image retrieval, it is necessary to judge the similarity of two samples by similarity measurement, and the common similarity measurement method is realized by finding the distance between two sample data. Because Hamming distance can be simply interpreted as the number of characters with different positions corresponding to two binary strings, and hash codes have only the particularity of two code values, Hamming distance is adopted to realize similarity measurement. Hamming distance can be simply expressed by the following formula:

$$d(x, y) = \sum x \oplus y \tag{1}$$

Where x and y represent n-bit binary codes and exclusive OR operation.

2.5 Evaluating Indicator

A common performance index is average accuracy (mAP). Given a query and a list of search results, the average accuracy (AP) is defined as:

$$AP = \frac{1}{N}\sum_{r=1}^{R} P(r)\partial(r) \tag{2}$$

where n is the number of relevant instances in the retrieved data set, indicating the first r instances retrieved, and if the retrieved instances are true, it is = 1; Otherwise, = 0. Then the average precision is the average precision. The higher the average precision, the better the retrieval effect. R is set to 100 here (Fig. 1).

2.6 Model Structure Diagram

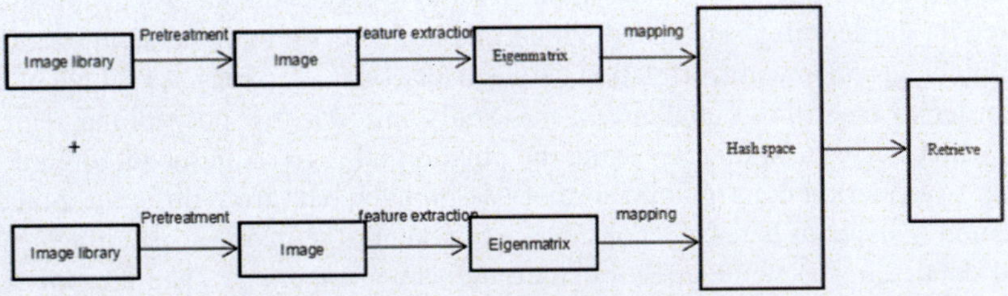

Fig. 1. Research Framework of Multimodal Retrieval

3 Matrix Processing of Polynomial Regression Model Based on Hash Transformation

3.1 Advantages and Ideas of the Linear Regression Model

Linear regression is a commonly used modeling method, and it is also a method chosen by many scholars in their research. In this method, because the independent variable is continuous or discrete and the regression curve is linear. Linear regression uses the best fitting straight line (regression line) to establish the relationship between the dependent variable y and one or more independent variables X. It is expressed by the equation Y = a + b*X + e, where A is the intercept, B is the slope of a straight line and E is the error term. This equation can predict the target value from the known prediction variable S. As shown in Fig. 2.

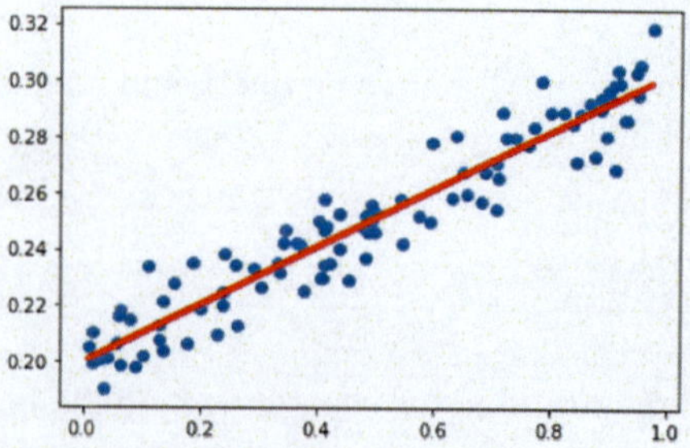

Fig. 2. Linear Regression Model 1

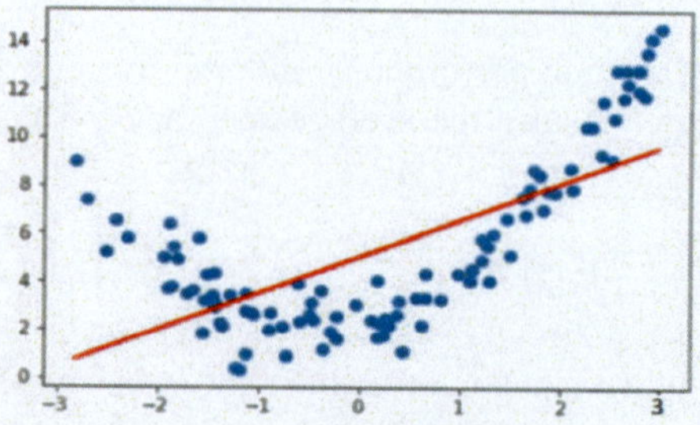

Fig. 3. Linear Regression Model 2

3.2 Problems and Improvement Ideas of the Linear Regression Model

Because the linear regression model is simple in thinking and easy to implement, it can be modeled quickly, and it is excellent in small data and simple relationships, and it is also the basis of many powerful nonlinear models. This is very beneficial to decision analysis. However, its shortcomings are also obvious, and it is difficult to model nonlinear data or polynomial regression with correlation between data characteristics. As shown in Fig. 3.

In order to meet these challenges, this study introduces a polynomial regression model, which focuses on directly using the polynomial regression model to guide hash learning. Considering the semantic similarity of related data from different modalities, this method transforms heterogeneous data into potential semantic space, in which multimodal data from the same category share the same semantics. Then the polynomial regression model is used to improve the shortcomings of the linear regression model. Therefore, the hash code quantized by the obtained representation is consistent with the semantics of the original data, so the multimodal retrieval task has a stronger discrimination ability. n the polynomial regression method, introducing higher power features (such as square and cubic) can effectively describe the nonlinear variation of data. The basic formula of the multiple regression model is:

$$h(x) = x_1\omega_1 + x_2\omega_2 + (x_1)^2\omega_3 + x_1x_2\omega_4 + (x_2)^2\omega_5 + b \quad (3)$$

Among them, $\mathrm{x_i}$ is the first characteristic attribute, ω_i represents all the weights as one vector. ***b*** represents the error scalar. The advantage of the polynomial regression model is that it can model quickly, and it is very effective for simple relationships with slightly larger data. The nonlinear regression model is very easy to understand and is beneficial to decision analysis. Able to express highly complex data well. In regression analysis, the linear fitting of linear regression will lead to poor results. If the data points in the scatter plot are found to be polynomial curves, polynomial regression can be considered for analysis. As shown in Fig. 4.

If higher-order terms are added, the complexity of the model will increase accordingly. With the increase in model complexity, the capacity of the model and the ability of fitting data increase. The hash code quantized by the obtained representation is consistent with the semantics of the original data, so it can have a stronger discrimination ability for multimodal retrieval tasks.

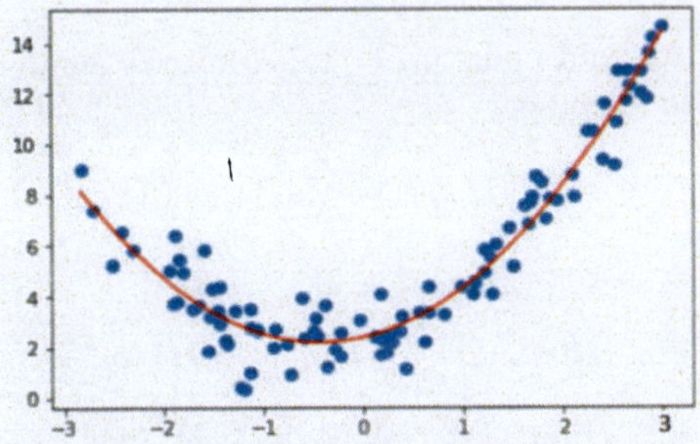

Fig. 4. Polynomial regression model

3.3 Mathematical Description

Because of the complexity of the commodity image data set, the use of a linear regression model will lead to the loss of semantic information, and finally, the retrieval effect will gradually decline with the increase of data volume. The polynomial regression model becomes an important regression model with the increase in model complexity. In this paper, linear regression is changed to a polynomial regression model on the basis of [15], and Frobenius norm is selected as the loss function, and its objective function is as follows:

$$L = \min_{\omega_i} \sum_{i=1}^{5} \|B - H(x)\|_F^2 + \|\omega_i\|_F^2 \tag{4}$$

$$H(x) = \omega_1\varphi_1(x) + \omega_2\varphi_2(x) + \omega_4\varphi_1(x)\varphi_2(x) + \omega_3(\varphi_1(x))^2 + \omega_5(\varphi_1(x))^2 \tag{5}$$

B represents a hash code, ω_i represents the corresponding weight.

Considering the polynomial regression model comprehensively, the overall objective function is constructed as follows:

$$\min_{U,B,Q,\mu} \|\varphi(x) - UB\|_F^2 + \beta L + \delta\|B - QY\|_F^2 + \gamma(\|U\|_F^2 + \|Q\|_F^2) \tag{6}$$

The objective function consists of four parts. The first item is the matrix decomposition of a multi-modal matrix, which decomposes the modes into a shared potential semantic space. The second term is used to perform a polynomial regression model. The third item is to guide hash learning through tags. The fourth item is to prevent over-fitting. Aiming at the NP-hard mixed integer optimization problem caused by the discreteness of hash codes, an efficient hash algorithm based on an augmented Lagrange multiplier is studied to directly learn discrete multimodal hash codes, and auxiliary variables are added to the hash codes to reduce the influence of impossible values on system performance on the premise of minimizing the loss of infeasible values, thus realizing efficient learning of multimodal hash codes. On this basis, a new hash code algorithm is proposed, which has the advantages of simplicity, high efficiency, and low storage overhead.

Step 1: Update B, V, E, and U, as shown in [15].

Step 2: fix U, V, T, Q, and update, i = 1,2,3,4,5, set the derivative of the overall objective function pair to 0, and get:

$$\omega_1 = (\varphi_1^T(x)\varphi_1(x) - 2\boldsymbol{I})^{-1}.\varphi_1^T(x).B \quad (7)$$

$$\begin{aligned}\omega_2 &= (\varphi_2^T(x)\varphi_2(x) - 2\boldsymbol{I})^{-1}.\varphi_2^T(x).B \\ \omega_3 &= (B(\varphi_1^2(x))^T - I)(\varphi_1^2(x)(\varphi_1^2(x))^T)^{-1}\end{aligned} \quad (8)$$

$$\omega_3 = (B(\varphi_1^2(x))^T - I)(\varphi_1^2(x)\left(\varphi_1^2(x)\right)^T)^{-1} \quad (9)$$

$$\omega_4 = (B(\varphi(x_1)\varphi(x_2)) - I)((\varphi(x_1)\varphi(x_2))(\varphi(x_1)\varphi(x_2))^T)^{-1} \quad (10)$$

$$\omega_5 = (B(\varphi_2^2(x))^T - I)(\varphi_2^2(x)\left(\varphi_2^2(x)\right)^T)^{-1} \quad (11)$$

Step 3: by fixing u, v, q, and updating t, get

$$T = \varphi_1(x)\varphi_2^T(x) * (\varphi_2(x)\varphi_2^T(x) + I)^{-1} \quad (12)$$

4 Experiments and Results

4.1 Data Set Introduction

In this research experiment, the data set of commodity images published by the Harbin University of Commerce is used, which contains 3800 images and related labels, including 38 categories of commodity images. Two types of image features are extracted from these images: HOG and LBP respectively, and the text is TF-IDF. In comparing commodity images, different feature values are extracted and compared under different hash code lengths. Considering the complexity of calculation and the existing experimental conditions, this study selected 20 categories from the data set (shoes, skirts, vest suspenders, shoulder bags, pendants, short-sleeved T-shirts, windbreakers, high heels, waistcoats, furs, business shoes, handbags, bracelets, shoulder bags, sweaters, bra sets, casual pants, caps, fishermen's hats and boots).

4.2 Experimental Result

The experiment compares the average precision of multimodal retrieval when the hash code length is 16 bits, 32 bits, 64 bits and 128 bits respectively. For the experimental results, this study repeated the calculation of each measurement method for 10 times, and then calculated the average value as the final experimental result. The longer the hash code is, the more semantic information the commodity image is mapped into the feature space, so the higher the average retrieval accuracy is. However, it is relatively difficult to detect the direction change of commodity images in the data set, and it is difficult to accurately express the semantic information of commodity retrieval, which leads to the inconspicuous improvement of the average accuracy of retrieval. Because the contour information in this data set is clear, the average accuracy is relatively improved. Therefore, the experimental results show that HOG and TF_IDF are the best methods for multimodal image retrieval when the hash code length is 128 bits.

4.2.1 Experimental Result

(1) When the hash code length is 16, 32, 64 and 128 respectively, when the feature extraction methods are LBP and TF_IDF, the comparison between the two methods for retrieving commodity images is made (Tables 1, 2 and 3).

Table 1. Average Accuracy of Retrieval Commodity Images

Search method	Self-built data set			
	16 bit	32 bit	64 bit	128 bit
ITQ	0.3748	0.3881	0.3954	0.4023
Our method	0.4537	0.4611	0.4703	**0.4782**

By comparing the above tables and charts, it can be found that the overall retrieval effect is better when the hash code length is 128 bits than when the hash code length is 32 bits. When multimodal retrieval is used and the code length is 128, the average accuracy is the highest.

(2) When the hash code length is 16, 32, 64 and 128 respectively, the features are extracted by HOG and TF_IDF, and the comparison between the two methods for retrieving commodity images is made.

Table 2. Average Accuracy of Retrieval Commodity Images

search method	Self-built data set			
	16 bit	32 bit	64 bit	128 bit
ITQ	0.7213	0.7346	0.7501	0.7698
Our method	0.8559	0.8607	0.8668	**0.8779**

By comparing the above tables and charts, it can be found that the overall retrieval effect is better when the hash code length is 128 bits than when the hash code length is 32 bits. When multimodal retrieval is used and the code length is 128, the average accuracy is the highest.

(3) When the hash code length is 16, 32, 64 and 128 respectively, the methods of extracting features are one-hot and TF_IDF, and the comparison between the two methods of retrieving commodity images is made.

By comparing the above tables and charts, it can be found that the overall retrieval effect is better when the hash code length is 128 bits than when the hash code length is 32 bits. When multimodal retrieval is used and the code length is 128, the average accuracy is the highest.

Table 3. Average Accuracy of Retrieval Commodity Images

Search method	Self-built data set			
	16 bit	32 bit	64 bit	128bit
ITQ	0.4183	0.4357	0.4481	0.4569
Our method	0.5312	0.5427	0.5529	**0.5642**

5 Conclusion

According to the research on multimodal commodity retrieval, if images can be combined with text information, the average accuracy of retrieval can be effectively improved. Aiming at the limitation of linear regression of multi-retrieval commodity images, this study uses the semantic information of the polynomial regression model to guide hash learning, which can greatly improve the average accuracy of retrieval. The experimental results show that: (1) Considering the low average accuracy of hash retrieval based on the linear regression model, the polynomial regression model is used to guide hash learning, which optimizes the system of multimodal retrieval of commodity images and builds a bridge between the model and commodity images, which greatly improves the average accuracy of multimodal commodity retrieval. (2) According to different methods of extracting low-level features, the advantages of this experiment are verified, and it is proved that different methods of extracting features lead to great differences in average accuracy.

Acknowledgment. This research was supported by the Industrialization Project of Harbin University of Commerce (22CZ04).

References

1. Xia, X., Li, X.: Analysis on the structural characteristics and driving factors of spatial correlation network of e-commerce development in China. Guizhou Soc. Sci. **2**, 132–140 (2021)
2. Zhang, H.: Research on Product Image Retrieval Method Based on Multimodal Data. Harbin University of Commerce, Harbin (2022)
3. Chen, F., Meng, Z.: Research on multimodal data retrieval based on hash algorithm. Comput. Sci. **46**(10), 49–54 (2019)
4. Qiao, B.: Research on Commodity Image Retrieval Method Based on Multimodal Technology. Harbin University of Commerce, Harbin (2021)
5. Lamia, T., Taesik, K., Hoon, J.Y.: Analysis of calcareous sand particle shape using fourier descriptor analysis. Int. J. Geo-Eng. **1** l(1) (2020)
6. Chen, C., Yang, R., Wang, C.: Research and realization of commodity image retrieval system based on deep learning. Parallel Architecture, Algorithm and Programming. PAAP, pp. 376–385 (2017)
7. Kumar, S., Udupa, R.: Learning hash functions for cross-view similarity search. In: Proceedings International Joint Conference on Artificial Intelligence, Barcelona, Spain, vol. 22, no.1, p. 1360 (2011)

8. Zhang, D., Wang, F., Si, L.: Composite hashing with multiple information sources. In: Proceeding of the International ACM SIGIR Conference on Research and Development in Information Retrieval (SIGIR), pp. 225–234 (2011)
9. Wang, L.: Research on Cross-Media Retrieval Algorithm Based on Discriminant Subspace Learning. Shandong Normal University, Jinan (2019)
10. Srivastava, N., Salakhutdinov, R.: Learning representations for multimodal data with deep belief nets. In: International Conference on Machine Learning Workshop, 79: 3 (2012)
11. Erin Liong, M., Lu, J., Wang, G., et al.: Deep hashing for compact binary codes learning. In: Proceedings of the 28th IEEE Conference on Computer Vision and Pattern Recognition, pp. 2475–2483 (2015)
12. Owen Christopher, G., Rudnicka Alicja, R.: Measuring retinal vessel tortuosity in 10-year-old children: validation of the Computer-Assisted Image Analysis of the Retina (CAIAR) program. Inves. Ophthalmology Visual Sci. (5) (2009)
13. Yuan, T.: Research on Image Retrieval Based on Hash Learning. Beijing University of Posts and Telecommunications, Beijing (2020)
14. Li, X., Guo, X.: Research on Commodity Image Detection and Recognition Method of Unmanned Supermarket Based on Deep Learning. Shenzhen University, Shenzhen (2020)
15. Lu, X.: Supervised Discrete Multimodal Hash Algorithm for Multimodal Retrieval. Shandong Normal University, Jinan (2021)

Enhancing Children's Fire Safety Skills with Machine Learning-Powered Virtual Reality Training System

Youssif Wagih Ali, Mahmoud Khaled Abdel-Basset, Mario Talaat Ghaly, Leqaa Ahmed Amin, Ahmed Abdullah Mousa, and Gehad Ismail Sayed(✉)

School of Computer Science, Canadian International College (CIC), New Cairo, Cairo, Egypt
Gehad_Sayed@cic-cairo.com

Abstract. Fires, regardless of whether they are small or huge, can have a profound effect on anyone, especially children. They can cause not only some serious physical harm but also could lead to long-term psychological trauma to them. The old-fashioned training ways, either tutorials or a real fire simulation cannot be suitable for children. In this paper, a new fire safety training system based on using virtual reality (VR) and machine learning for training children how to deal with fire is proposed. The proposed system consists of two main parts working in parallel, the first part is called the VR simulator, and the second part is called the Non-Player Character (NPC) recommendation system. Two well-known machine learning algorithms are used and evaluated. These algorithms are support vector machine (SVM) and neural networks. The experimental results revealed SVM with polynomial obtained the best result. The proposed system obtained overall accuracy of 90.8%, precision of 90.7%, recall of 90.8%, and f-score of 90.7%.

Keywords: Virtual Reality · Fire simulation · Recommendation system · Machine learning · Virtual Fire Drills

1 Introduction

According to the U.S Fire Administration, there are an average of 291 children per year from (2011 to 2020) under the age of 14 had gone because of a home fire, and out of 15,200 injured cases there are 8.9% of them are children [1]. The fire burns could leave permanent marks, which affect badly on the injured children either physically or psychologically. They should be taught how to survive if a fire occurred but there is no doubt that teaching children at the age of (6 to 15) is tough, especially in this significant issue. They are always getting bored because of the traditional ways, like the safety drill courses or watching tutorial videos. It is too hard to deliver the idea to those young kids through these methods. The study discovered that the use of immersive virtual reality serious games with DL-assisted learning can effectively improve fire safety training and research, specifically in high-rise fire evacuation scenarios [2]. So with the trendiest technology nowadays, which is VR, kids can now be taught the main instructions for dealing with these critical situations rather than ordinary fire drills. As when children

A. Hassanien et al. (Eds.): AISI 2023, LNDECT 184, pp. 542–551, 2023.
https://doi.org/10.1007/978-3-031-43247-7_47

have used VR, they have shown promising results in enhancing their preparedness and response skills without any harm to the user [3]. On the other hand, there are seriously dangerous sides to conventional real-fire simulation. Just like carbon dioxide, which is risky even in the most advanced adult training environment.

One of the main reasons fires because death is because people inhale these gases [4]. In traditional ways, accidents could occur, like a fire going out of control. Besides the psychological effect that could cause trauma to some children as they would face a real fire occurrence, now, it could be concluded that the traditional ways are not suitable for the children, both physically and mentally. So virtual reality technology was recommended as it could provide a secure and controlled environment for children to practice fire safety procedures, which can be difficult to simulate in real life [5]. As it's the safest, most effective, and most efficient approach to instructing children how to deal with fire in a safe and fun way. Also, they will be able to experience realistic fire and smoke simulations, providing them with a more immersive and engaging learning experience [6]. Concerning the child's physiology, it was preferred to do these drills in the form of a game and develop VR software to simulate different scenarios with an artificially intelligent (AI) non-player character voice trainer (NPC) to guide the children on what they should do in each case by using predefined scenarios to figure out the fire assessment and decide the best solution for the current situation. Every task the user achieves in a certain time gets him a point, so at the end of the simulation, the score is collected and pushed to the database. After several tries, the performance is evaluated based on the score that the child gains through the simulation. While the back part, namely the recommendation system, is used to guide the kid through the simulation. The main contribution of the paper can be summarized as follows:

1. A new virtual reality fire safety training system based on VR and machine learning algorithms is proposed.
2. Different fire scenarios are proposed for training a child with various sources of fire.
3. A comparative study between different machine learning algorithms with different parameter settings is reported.
4. Several evaluation metrics are adopted to evaluate the overall proposed recommendation system, such as accuracy, sensitivity, precision, f-score, and area under the curve.

The rest of the paper structure is organized as follows; Sect. 2 shows the earlier work on using VR for fire simulation, and Sect. 3 shows a detailed description of the adopted dataset. Section 4 introduces the overall proposed fire training simulation system based on VR and machine learning algorithms. Section 5 discusses the obtained results. Finally, in Sect. 6, the main finding of the paper with the future direction is proposed.

2 Literature Survey

The use of VR technology has increased significantly across several areas, including education and training. The study [7] talks about a VR training program designed to improve kids' fire safety and evacuation abilities in case of a fire. There are three training options available: investigate the scenario, reply to inquiries, and simulate evacuations.

The results showed that children's knowledge of fire safety and their capacity to evacuate during a fire emergency were both improved by the VR instruction. People's knowledge and proficiency with fire extinguishers can be improved by VR fire extinguisher training [8]. Training was available in "Tutorial Mode" and "Game Mode," respectively. The findings demonstrate that VR fire extinguisher training significantly increased participants' fire extinguisher knowledge and proficiency.

The authors conducted a thorough literature assessment on the use of VR for fire safety teaching in [9]. The goal of this study was to assess how effectively VR technology might imitate actual fire threats for educational reasons and to provide a starting point for future research on VR-based fire safety training. The investigation revealed the range of ways VR has been utilized for fire safety instruction, including simulation models, group immersive emergency drill training, and gamified views.

Another research in [5] examined a CAVE-VR game designed to prepare elementary school teachers for fires and found that it was effective for teaching and offered challenge, amusement, and mastery. With CAVE-VR systems, up to six people may work together, decreasing user isolation. Virtual reality training program for teaching kids how to flee during a fire [10]. Children between the ages of 8 and 12 were divided into two groups, one of which had training in a VR system and the other in a traditional classroom. The authors draw the conclusion that teaching children how to evacuate during a fire using virtual reality can be a helpful and enjoyable way. Children in [11] were instructed on the usage of fire extinguishers using a VR simulator. The study added to the expanding body of research showing that VR technology may be a beneficial training aid for disaster planning and response by showing that it can be used to teach children how to behave responsibly in emergency circumstances.

3 Dataset Description

The Fire Detection Dataset includes details on the condition of several game components in the virtual gaming environment. The data was created using in-game information from players, which was then gathered to create the dataset. A recommendation system for fire training is then studied using the dataset in virtual situations. The dataset includes a binary variable called "isRoadToDoorClear," indicating whether the path from the player's current location to the destination is clear (1 = clear, 2 = not clear). Additionally, there are 59 binary variables, "Element 0–58 Status," indicating whether each of the game elements is on fire or not (1 = on fire, 2 = not on fire). The elements are a category of flammable objects including desks, cardboard, chairs, and flasks, which have the potential to catch fire. The target variable of the dataset is "Sol," which represents the ID of the selected solution to extinguish the fires (0–3) acquired by the recommendation system. The dataset provides valuable information about the game element status and path clearance in a virtual game environment, which can be used to develop and test fire detection systems. There are various scenarios in which the user can express fire training, such as in our scene, where the recommendation system begins to give the user hits about what he should do and the fire type, and when the user starts the game, this dataset offers four proposed solutions to extinguish the fires, which are: First Solution (class 0) the user is informed about the type of fire and is provided with a recommendation on how to handle it based on the identified fire type. Second Solution (class 1) is based on the type of fire, the system recommends the appropriate fire extinguisher type and provides instructions to the user on how to use it effectively. Third Solution (class 2) If the fire has spread over a large area in the scene, the system recommends using the fire alarm, as it can help extinguish the fire and cover a larger area with its sparks. Fourth Solution (class 3) as a final option, if the fire has spread extensively throughout the scene, the system recommends evacuating the area by running to the door.

Figure 1 shows the class distribution of the adopted dataset. As can be observed, the adopted dataset suffers from an imbalance problem that can affect its performance. This problem will be tackled in the proposed model.

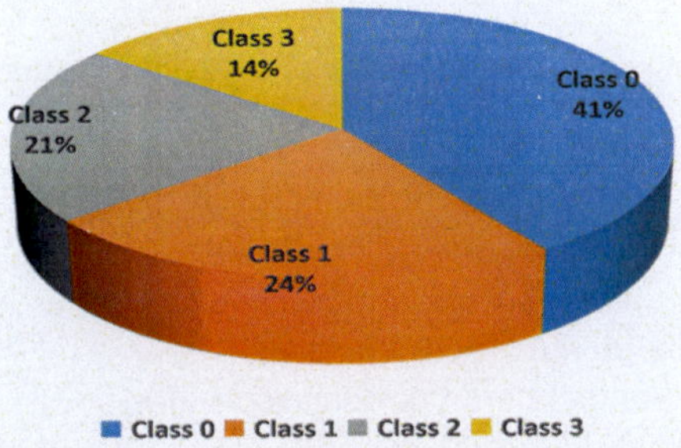

Fig. 1. The class distribution of the adopted dataset

4 Proposal Fire Training System

The proposed fire training system based on using VR and machine learning algorithms consists of two main parts, working in parallel. The front part namely (VR) simulator is used to handle the user interaction between the system and (NPC) which is responsible for guiding the user during the game. While the back part namely the recommendation system is used to guide the kid through the simulation. Figure 2 shows the block diagram of the proposed system. A detailed description of each part is stated in the upcoming sub-sections.

4.1 Front Part

This part appears for the child when starting the system. Once the child selects the scene, the NPC starts to work to give the user general instructions about the scene, identify the type of fire that is going to be faced (electrical, chemical, etc.), and keep providing him/her with instructions and recommended actions that came from the recommendation system till the end of the scene, as it's going to be clarified in the VR simulator part. The fire simulation would start after 10 s once the player starts playing. Then the fire would continue spreading after a specific interval of time. These solutions that were successfully achieved had a score that was counted at the end to evaluate the performance of the child and show his improvement along the way. The proposed VR simulator in the proposed

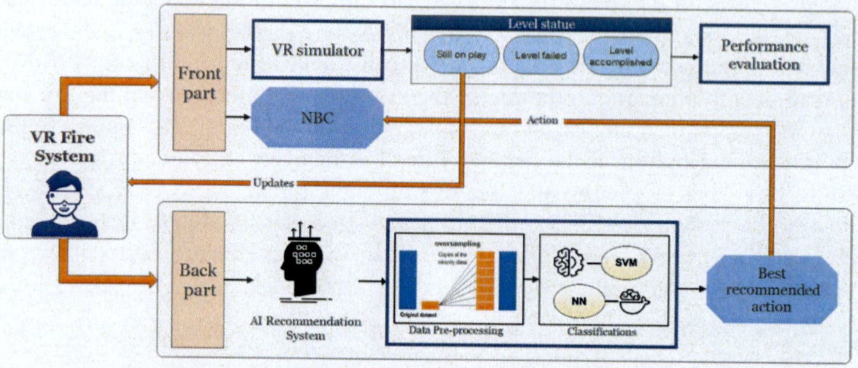

Fig. 2. The block diagram of the proposed fire training system

system is the scene that is simulated for the child. It could be at home, school, or any other place needed to perform the drills. The child can interact with the surrounding environment (fire extinguishers, fire sprinklers, power switches, etc.) and apply changes to it. It works in parallel with the NPC.

In the proposed system, there are two scenes developed: the first one is in the laboratory room of a school, while the second one is in the bedroom. In the first scenario, the user would be inside a laboratory that has desks, flasks, cupboards, and boards as flammable objects, a fire extinguisher and a fire alarm (fire sprinklers) as tools to put out the fire, and finally, an emergency exit door to evacuate the laboratory room. According to the fire type, the NPC would be informed of the recommended action from the recommendation system so the user could use the right tools at the right time. A small fire would start randomly at the beginning of the scene and start to increase and propagate as time passed. For each case in the scene, there is a preferred solution, like in the case of a small fire, where the child would grab a fire extinguisher and follow the instructions to use it. If the fire increases, he or she will pull the fire alarm to activate the fire sprinklers. Otherwise, the recommendation system will ask the child to evacuate the laboratory.

The second scene is a bedroom scene with flammable objects like a bed, a cupboard, and window curtains. While a fire extinguisher, a blanket, and a bucket of water are the tools that exist to douse the flame, and a door exists to exit the room. Similarly, in the above scenario, the recommendation system would inform the child through an NPC what to do according to the situation and the type of fire. If the fire is small, the recommended action would be to use a blanket to cut off the oxygen supply, a bucket of water, or a fire extinguisher (if available) to prevent it from spreading. But in case that fire is relatively large, the recommendation system would suggest turning on the house fire alarm. Finally, if it propagates on a large scale, it would be recommended to evacuate the room. Every solution has a score that is counted at the end of the drill to evaluate the child or the user's performance and see how it is improved after several attempts.

The NPC refers to non-player characters. The main role of it is to guide the user while in the scene, by giving him hints about what he should do through receiving the best action recommended in the current situation from the proposed recommendation system (the back part). While the level status describes the status of the level whether it's accomplished, failed, or still in play. If the level is still in play, that means there are still more reads being sent to the VR system to be processed in the recommendation system part, and this process will be repeated until the child reaches the end.

Finally, to evaluate the performance of the child through the VR simulator, Eq. 1 is applied to get the total score by multiplying the obtained score (S_{Obt}) and timing (T) by their respective weights (W) and adding them together. S_{Obt} Is calculated by summing up the scores of the solution used by the child in each scenario, divided by the number of scenarios.

$$Total\ Score = W_1 \times \frac{\sum_{i=1}^{n} S_{Obt}}{n} + W_2 \times \frac{\sum_{i=1}^{n} T}{n} \tag{1}$$

$$T = \frac{T_{opt}}{T_{obt}} \tag{2}$$

where n is the number of scenes, T_{opt} is the optimal timing, and T_{obt} is the obtained timing from the user.

W_1 And W_2 are the weights coefficients. That is used to balance the importance of the score and the timing. W_1 Set to 0.7 for the score and W_2 set to 0.3 for the timing, this is because the score is more important than the timing.

4.2 Back Part

The proposed recommendation system in the back part comprises two main phases: data pre-processing and classification. In the data preprocessing phase, a random oversampling technique is used to tackle the imbalance problem of the adopted dataset. Then, the processed data is used to feed a machine-learning algorithm. The output from this phase is a recommendation of the best action that the child can take in the current situation. As time passes, new reads are provided to the VR system to reprocess and suggest new, suitable suggestions.

The random oversampling technique has been used to deal with the imbalanced dataset issue as a data pre-processing technique, where one class of data significantly outweighs the others [12, 13]. In this technique, by randomly replicating instances from the minority class, the minority class is oversampled until the number of instances in the minority class is comparable to that of the majority class. This would help to enhance the accuracy of the machine learning model by reducing the bias towards the majority class and allowing the model to learn more from the minority class.

The pre-processed data are also passed through a classifier. In this phase, supported vector machines (SVM) with different kernel functions such as SVM-RBF (radial basis function), SVM-polynomial kernel, SVM-linear kernel, and neural networks (NN) are adopted. The performance of these algorithms is compared using several measurements. These measurements are accuracy, precision, recall, and F1-score, as it's going to be shown in the result part.

5 Experimental Results and Discussion

In this section, several experiments were conducted to evaluate the overall performance of the proposed VR fire training system. It should be mentioned that all the conducted experiments are implemented on the same PC with this spec; 16 RAM, Core I7, and 64-bit processor. Figure 3 compares the class distribution before and after applying the random oversampling method. As can be seen, the class imbalance problem is handled after applying ROS, where the class distribution percentage of the class 0 which consider the majority class moves from 40.6% to 25%. Additionally, after applying ROS, all four classes have equal distribution.

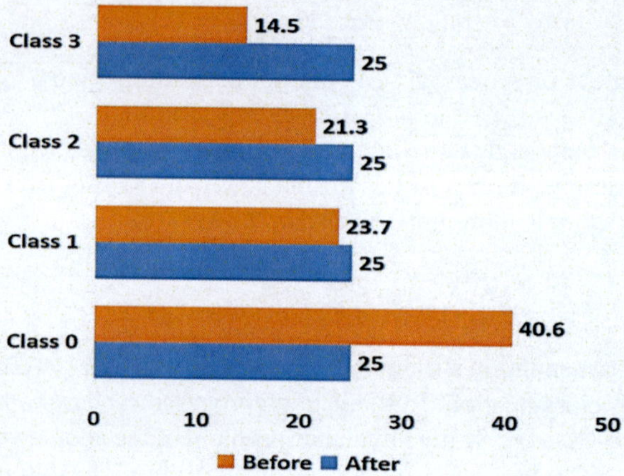

Fig. 3. The class distribution before and after applying ROS

A comparison of the obtained results of the proposed VR fire training system before and after using ROS in terms of accuracy, precision, recall, and f1-score is conducted in Tables 1 and 2, respectively. As can be observed from Table I, SVM with polynomial or shortly poly obtained the best accuracy results compared with SVM with linear kernel function and SVM with Radial Basis Function (RBF). Additionally, it can observe that the proposed VR fire training system with ROS obtained higher results compared with the obtained result before considering it in the proposed system. From the obtained results in Tables 1 and 2, it can be concluded that the original suffers from an imbalance problem, as the accuracy results jumped to around 3%.

Another experiment is conducted in Fig. 4. In this experiment the performance of the proposed VR fire training model is compared when using SVM with polynomial kernel function and when using neural network machine learning algorithm. It should be mentioned that in this experiment, ROS is considered. As can be observed, SVM with the polynomial kernel function obtained the best result compared with neural networks (NN).

Table 1. The performance of SVM with different kernels before using ROS

Classifier	Accuracy (%)	Precision (%)	Recall (%)	F1-Score (%)
SVM-RBF	89.06	88.98	89.06	88.92
SVM-Poly	89.11	89.05	89.11	88.92
SVM-Linear	87.77	87.61	87.77	87.56

Table 2. The performance of SVM with different kernels After ROS

Classifier	Accuracy (%)	Precision (%)	Recall (%)	F1-Score (%)
SVM-RBF	90.81	90.76	90.81	90.789
SVM-Poly	90.85	90.84	90.85	90.83
SVM-Linear	89.17	89.56	89.17	89.25

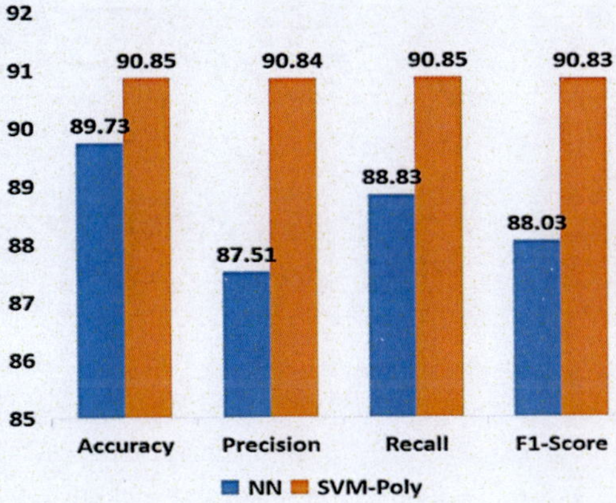

Fig. 4. SVM vs. NN in terms of accuracy, precision, recall, and f1-score.

Figure 5 shows the confusion matrix of the proposed VR fire training system using SVM with RBF kernel function. As can be observed from the diagonal of the matrix, which represents the true positives and true negatives. The obtained values are high in the diagonal which can indicate that the proposed system is performing well.

Finally, the evaluation performance of the children during the attempts in different intervals is illustrated in Fig. 6. This evaluation was conducted by calculating the score obtained by two children users in each attempt using Eq. 1 to determine whether the performance of this child is improved over time. As can be observed from this figure, the performance of this child is remarkably improved.

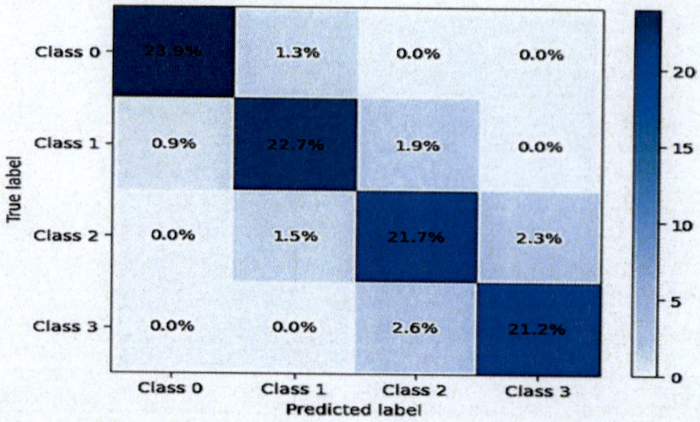

Fig. 5. Confusion matrix of the proposed VR fire training system

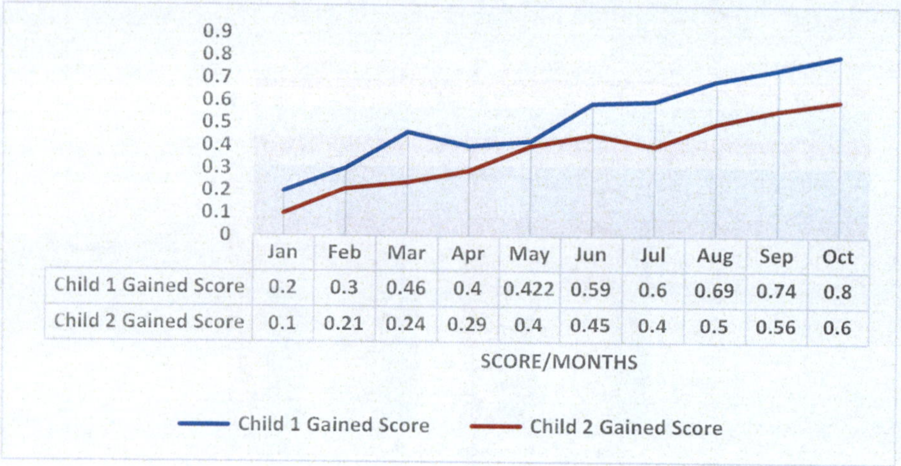

Fig. 6. Performance evaluation of a child

6 Conclusion

The number of children's deaths caused by fire increases; virtual reality fire training programs are becoming more crucial. To address this issue, this paper proposed a system that integrates a VR simulator and a Non-Player Character recommendation system to raise awareness and educate children about dealing with fire emergencies. In the VR simulator, two 3D environment scenes are developed. The NPC guides children on how to react to each scenario. By collecting scores from the children's actions and using machine learning algorithms such as SVM and neural networks, the system evaluates their performance. The experimental results demonstrated that employing ROS oversampling method in the proposed recommendation system oversampling improves the overall system's performance. It obtained an overall accuracy of 91%. Additionally, the results revealed that the overall proposed VR fire training system can be valuable for training a child how to deal with fire in different situations. The proposed system can be further considered in other scenarios. Also, the performance of the proposed recommendation system can be further improved by investigating different machine-learning algorithms.

References

1. U.S. Fire Administration. "Children and Fire." U.S. Fire Administration, 15 May 2020. https://www.usfa.fema.gov/statistics/deaths-injuries/children.html. Accessed 3 March 2023
2. Ko, Y.J., Kim, J.H., Shin, D.H.: Immersive virtual reality serious games with DL-assisted learning in high-rise fire evacuation on fire safety training and research. Front. Psychol. **13**, 786314 (2022)
3. Isleyen, E., Duzgun, H.S.: Use of virtual reality in underground roof fall hazard assessment and risk mitigation. Int. J. Min. Sci. Technol. **29**(4), 603–607 (2019)
4. Yaman, F.N., Kerkutluoglu, M., Guler, O., Hakkoymaz, H.: End-tidal carbon dioxide measurements in unintentional non-fire-related carbon monoxide poisoning. Iranian J. Toxicol. **16**(4), 267–274 (2022)
5. Mystakidis, S., et al.: Design, development, and evaluation of a virtual reality serious game for school fire preparedness training. Educ. Sci. **12**(4), 1–18 (2022)
6. Shaw, E., Roper, T., Nilsson, T., Lawson, G., Cobb, S.V.G., Miller, D.: The heat is on: exploring user behaviour in a multisensory virtual environment for fire evacuation. In: Proceedings of the 2019 CHI Conference on Human Factors in Computing Systems, New York, NY, USA, pp. 626:1–626:13 (2019)
7. Bhaskar, A., Reddy, V.K.: Development of virtual reality training for fire safety education. Int. J. Adv. Trends Comput. Sci. Eng. **9**(3), 3839–3843 (2020)
8. Lin, Y., Yu, J., Hung, Y.: The evaluation of virtual reality fire extinguisher training. Front. Psychol. **11**, 1–17 (2020)
9. Srivastava, L.P., Singh, A., Bhardwaj, R., Iqbal, Khanzode,V.: A systematic literature review on virtual reality-based fire-safety training in India. In: Singh, L.P., Bhardwaj, A., Iqbal, R., Khanzode, V. (Eds.) Productivity with Health, Safety, and Environment, HWWE 2019, Design Science and Innovation. Springer, Singapore, pp. 23-35 (2022)
10. Kurniawan, F.D., Wibowo, S.A., Yudhana, A.: Comparing the effectiveness of fire extinguisher virtual reality and video training. Virtual Reality **24**(2), 311–320 (2020)
11. Pietroforte, D., Ricci, L.: Virtual reality-based training platform for educating children to act correctly in fire emergency situations. In: ICERI2020 Proceedings, pp. 7461–7468 (2020)
12. Sayed, G.I., Khoriba, G., Haggag, M.H.: A novel chaotic equilibrium optimizer algorithm with S-shaped and V-shaped transfer functions for feature selection. J. Ambient Intell. Human. Comput. 1–26 (2022)
13. Sayed, G.I., Solyman, M., Hassanien, A.: A novel chaotic optimal foraging algorithm for unconstrained and constrained problems and its application in white blood cell segmentation. Neural Comput. Appl. **31**, 7633–7664 (2019)

Performance Improvement of Breast Cancer Diagnosis Using Artificial Intelligence and Image Processing Techniques

Rasha Sameh(✉), Basem E. Elnaghi, Atef Ghuneim, and Ahmed Magdy

Electrical Engineering Department, Faculty of Engineering, Suez Canal University, Ismailia, Egypt
rashasameh85@eng.suez.edu.eg

Abstract. One of the most deadly and terrible diseases in the world that kills women is breast cancer. The timely detection of breast cancer can significantly impact and potentially save the lives of numerous women. Radiologists and clinicians have turned to computer-aided diagnosis as the burden of diagnosing breast cancer has grown. Based on thermal imaging, machine learning, and deep learning are employed in this study to diagnose breast cancer. Three stages—preprocessing, features extraction, and artificial neural network classifier—are used to investigate the proposed technique. Additionally, four Deep Learning models—AlexNet, GoogleNet, SqueezNet, and ResNet18—are suggested to attain high performance and accurate breast cancer diagnosis. The Machine Learning algorithm achieves accuracy of 89,74%, sensitivity of 82,35%, and specificity of 95,45% in simulation results. Additionally, Deep Learning models, like AlexNet, achieve accuracy levels of 100%.

Keywords: Breast cancer · computer-aided diagnosis · Thermogram image · Artificial Neural Network · convolutional neural network

1 Introduction

One of the most frequently diagnosed cancers in women worldwide is breast cancer [1]. Globally, 2.3 million women (11.7%) received a breast cancer diagnosis in 2020, and 685 000 people died from the disease. By the conclusion of 2020, breast cancer is projected to become the most prevalent form of cancer worldwide, with approximately 7.8 million women living who were diagnosed within the past five years [2].

Breast cancer can be found using a variety of breast image modalities, including ultrasound, magnetic resonance imaging (MRI), mammography, thermography, and others. In contrast to thermal imaging, which is suitable for all patients, non-invasive, inclusive, painless, and safe, mammography is expensive, ineffectual, and not ideal for women with dense breasts. Mammography is considered the definitive method for early detection of breast cancer [3].

Thermal imaging of the breast shows that breast cancer tissue has a higher temperature than normal tissue, resulting in hotspots on the thermogram. In healthy breasts, the

A. Hassanien et al. (Eds.): AISI 2023, LNDECT 184, pp. 552–564, 2023.
https://doi.org/10.1007/978-3-031-43247-7_48

temperature distribution is symmetric [4]. Due to errors in interpreting and differentiating between healthy and diseased breast thermal images using the naked eye, doctors have turned to computer-aided diagnosis (CAD), smart systems, and computer vision techniques, leading to consistent and effective diagnostic performance [5].

2 Literature Survey of Research

CAD heavily relies on Artificial Intelligence (AI), specifically Machine Learning and Deep Learning techniques, to analyze the breast. These techniques analyze the breast through different methods. The three fundamental steps in any computer-assisted breast cancer detection system are pre-processing, feature extraction, and classification.

2.1 Step I: Pre-processing Process and Segmentation

The pre-processing stage of an image is used to improve the thermographic images' quality. Following the conversion of the RGB colour input image into intensity grayscale, the steps of the image pre-processing include image segmentation and image de-noising by a filter. Due to its significance, some study separates the image segmentation step from the pre-processing stage of the image. Image segmentation is the process of breaking an image up into distinct pieces to make analysis easier. Many image segmentation techniques, such as edge detection, thresholding, clustering, compression-based techniques, histogram-based techniques, etc., are presented in [6–12] either entirely manually or fully automatically [6].

The technique of segmenting the breast region involves dividing it from the rest of the body. The Region of Interest is the name of this area (ROI). The diagnosis will be accurate depending on how thoroughly this area is segmented. The ROI must be included Breast structure and all its tissues.

In [7], manual breast segmentation was performed using techniques that divide each breast into four quadrants and others that use an ellipse on a patient to isolate the ROI. Due to the lack of a distinct shape and visible breast boundaries in thermal images in [8], others were manually split.

The area of the tumor search can be narrowed down, and the time and effort needed for manual segmentation can be reduced with automatic segmentation and separation of the breast area from thermal pictures. Using the subtle edge detection operator and the gradient operator, [9] created a novel method for automatically segmenting the ROI of a breast thermogram. In [10], the ROI of the right and left breasts were extracted using a segmentation method that used four terminal lines to divide the breast region (upper, lower, left, and right). In [11], it was demonstrated that a suggested algorithm could successfully divide all different breast types (small, medium, large, asymmetric, and flat). In [12], it was suggested to use trained Convolutional and Deconvolutional Neural Networks (C-DCNN) to automatically segregate breast areas in thermal images.

2.2 Step II: Features Extraction

Feature extraction, one of the key procedures in CAD systems, is used to lessen the complexity of the classification process. The texture of thermogram photographs reflects changes in temperature. There are numerous ways to extract texture features, including statistical and structural information transformation. Tumors and certain tissue activities are some common characteristics of cancer that experts utilize to make an early diagnosis [13].

In [14], a Non-parametric t-test analysis, which gave the statistical and textural features differences between sick and healthy breasts, was used to choose the features. Texture and entropy features were used in [15] to extract 112 features. From 40 thermal images, 20 features were extracted by GLC matrices in a study [16].

The feature extraction process can be carried out either manually or mechanically. Deep learning (DL) is frequently employed in the context of autonomous extraction by training a dataset with multi-layer convolutional neural networks, a well-liked machine learning (ML) method [17].

2.3 Step III: Classification

After pre-processing and features extraction, classification is the last step. For categorization, ML and DL are employed. Multi-Layer Perceptron (MLP), K Means Clustering Algorithm, Support Vector Machine (SVM), Decision Trees (DT), Random Forests, K-Nearest Neighbors (KNN), Extreme Learning Machines (ELM), and Artificial Neural Network (ANN) are the most used ML algorithms.

Different approaches to classifying breast tumors were put forth in [18] by extracting properties using the Haralick and Zernike descriptors. DT, MLP, ELM, and Bayesian classifiers were applied. The accuracy of the MLP classifier achieved 76.01%. [19] employed the Backpropagation Neural Network Model (BPNN) classifier, which has good accuracy. In [20], an SVM classifier was used to classify 80 frontal breast pictures, 50 of which were normal and 30 of which were pathological. This approach's accuracy, sensitivity, and specificity were 91.25%, 93.3%, and 90%, respectively. In [21], significant machine learning classification models—KNN, DT, SVM, and RF—were employed to classify the photos and identify potential malignant images. In this investigation, the KNN classifier provides the best classification.

ELM classifier with a more significant activation function was utilized in [22]. One of the best methods for detecting breast cancer is the ELM classifier with "Tribas" activation function, which achieved an accuracy of 95.94%. In [23], two techniques: ELM and MLP, were evaluated for the thermal imaging-based early diagnosis of breast cancer. The outcomes demonstrated that the ELM classifier had the best accuracy. In [24], the K-Star classifier was used for dynamic thermography images (20 DIT images), obtaining high accuracy using a 10-fold cross-validation method. In [25], images captured using dynamic thermography were classified using the ANN. The system demonstrated high accuracy with negative predictive values of 99.1% and positive predictive values of 83.67%. Deep learning was used in [26] to categorize breast thermograms. Less than 190 photos from the dataset were used. Although the results were not satisfactory and not all aspects were accurately graded, it was nevertheless a significant step forward for deep learning.

The difference between ML and DL is that the DL doesn't need to extract features because it contains this process in its architecture. In DL, CNN is regarded as the most popular computer vision-based AI approach for identifying breast cancer. Due to their resilience and ease of use, CNN methods are more practical than those relying on texture and statistical data. In [27], a CNN model named "AlexCNNbreast.m" was used to classify images using the Gradient Vector Flow technique (GVF), which provided great accuracy. [28] suggested the CNN app as a classifier for dynamic and static images. This study's accuracy rates for the dynamic and static protocols were 95% and 98%, respectively. In [29], static and dynamic images were classified using four CNN models: ShuffleNetV2, DenseNet, ResNet101, and MobileNetV2 for breast cancer detection. The results showed that MobileNetV2 achieved high accuracy.

The performance of seven pre-trained CNN models, including ResNet-101, ResNet-50, Inception-v3, GoogLeNet, VGG-19, AlexNet, and VGG-16, was compared to that of the dataset in [30]. The VGG-16 had the highest specificity and balanced accuracy, both of which was 82.35%. Another comparison of the CNN designs resnet18, resnet34, resnet50, resnet152, vgg16, and vgg19 can be found in [31]. The outcomes were highly accurate. Conversely, ShuffleNet was used and altered in [32] by adding a convolutional layer and extra filters to enhance its performance and achieve higher accuracy.

[33] compares various CNN models, including Inception V4, Inception V3, and Inception MV4, based on various optimization techniques, including Root Mean Square Prop (RMSPROP), Adaptive Moment Estimation (ADAM), and Stochastic Gradient Descent (SGD). This DCNN Modified Inception MV4 is more accurate and faster than Inception V4. The study in [34] introduced the CNN DenseNet121 model that was used to extract the features for constructing a classifier. The performance of DenseNet121 was compared with six other pre-trained models (VGG16, VGG19, DenseNet169, Xception, and Inception) in terms of extracting features from 3-channel edge-prominent breast images. The results showed that the DenseNet121 model achieved impressive accuracy.

3 Proposed Method

Two totally autonomous systems for the early identification of breast cancer are suggested in this research. The three stages of a proposed system are, in order, pre-processing the images, feature extraction, and machine learning classifier. Second, to quickly and accurately diagnose the other automatic system, four CNN models are pre-trained to be used: AlexNet, SqueezNet, ResNet18, and GoogleNet.

This section demonstrates the dataset, the proposed methods, the pre-processing and segmentation step, the features extraction step, and various classification methods.

3.1 Dataset

In this paper, the two automatic systems are implemented based on a database of 190 thermal images (112 healthy and 78 sick) obtained from [35]. Figure 1 presents procedures for taking a thermal image of the breast in real work. The patient must be in an upright position with their hands raised. Images are taken with a camera made of Forward-looking infrared (FLIR). These images usually included the abdomen and breasts axilla, and neck areas.

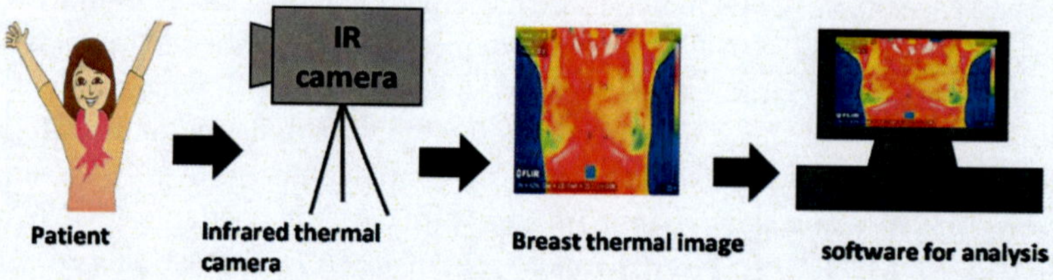

Fig. 1. The procedure for capturing thermal images.

3.2 The First Proposed Method

The pre-processing stages for thermal photos must be completed first. The proposed method then divides the breast into its left and right halves for analysis. Different features are taken from the image after it has been segmented. The ANN classifier receives the retrieved features as input to evaluate this new model as shown in Fig. 2.

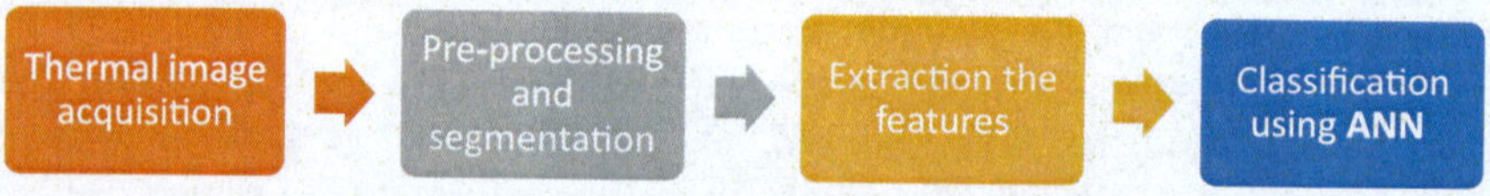

Fig. 2. The procedural framework for carrying out the core operations in the first proposed approach.

3.2.1 Image Pre-processing and Segmentation Stage

The image pre-processing method is used to enhance the thermographic image quality and increase the effectiveness of the features extraction phase from the Region of Interest (ROI) and classification phase. Image noise reduction is a crucial stage in the pre-processing process which comes after transforming the RGB input image into grayscale image where black pixels indicate the lowest temperatures and white pixels the highest.

To eliminate the impulsive noise brought on by the flaws in the thermal camera, the median filter is used for the original grey image. The Region of Interest (ROI) section is separated from other body parts in the proposed Algorithm (1) by dividing the image horizontally into two halves (the top part of the body and the ROI part), as illustrated in Fig. 3(a).

Algorithm 1: Proposed Image Pre-processing Approach

1: Read original thermal image X.

2: Convert X to grey scale image G.

3: Apply median filter to G.

4: Determine the height of G as T.

5: Define upper ratio U, initially set U=0.225, which may be changed.

according to the axilla position found in Algorithm 2.

6. Compute the upper part M = round (T*U).

7: Extract the ROI part G_L.

In ROI segmentation: The proposed segmentation algorithm is used to segment the left and right breasts from the ROI image based on the breast structure.

Step I: The ROI image is converted to a binary image using threshold value λ.

Step II: Breast points determination for the right breast (left part in the image) The axilla point is determined by the point at which the position (P) of the white pixels at the boundary increases and then decreases starting from the top left of the image as shown in Fig. 3(b). After the axilla, the maximum curvature point is determined by the pixel at the position where the white pixels decrease and then increase. The position of the white pixels increases again after the curvature point till it reaches a constant value which means that the breast ends. And the opposite will be done for the left breast (breast in the right part), starting from the top right.

The breast area in the grey version is formed by a horizontal cut at the axilla and a vertical cut through the center of the image. This proposed algorithm does not cut through the inframammary incision, but the cut is determined by the axilla not to lose a part of the breast because some images have the nipple at that point.

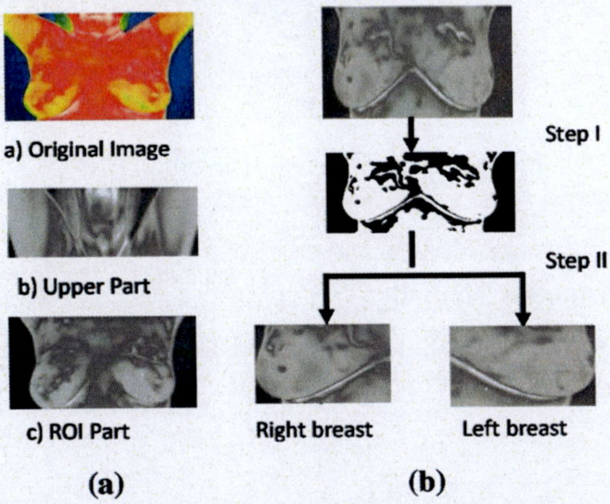

Fig. 3. (a) The results of pre-processing step, (b) The results of the segmentation step.

3.2.2 In the Feature Extraction Stage

Extraction of efficient features is a step required for asymmetry analysis. The segmented image of breast thermograms is used in this study to extract two different sorts of characteristics. The mean, variance, skewness, and kurtosis are examples of first-order statistical features. By counting the occurrences of various combinations of grey levels in an image or by distributing the grey levels among the area's pixels, texture characteristics are formed. It was calculated for each pixel-distance interval. Twenty-two features were taken from these textures (GLCM) [36]. These features are called Energy, Cluster Prominence Contrast, Cluster Shade, Correlation,.. Etc. [16]. So, the total extracted features are 26 features. Finally, the features will be passed to the ANN classifier.

3.2.3 In the Classification Stage

A typical Artificial neural network (ANN) is one of the ML methods containing millions of units that form an interconnected network that processes enormous amounts of information. In this study, ANN was used with one hidden layer. After data pre-processing and segmentation, 22 features were extracted using GLCM from each image. Four additional features were appended, variance, mean, skewness, and kurtosis, to obtain 26 features for each image. These features feed ANN (26 inputs) one hidden layer with 52 neurons and one output (healthy or sick), as shown in Fig. 4. 80% of the dataset (152 images) was trained along 1000 epochs to find optimal performance. After that, we tested 20% of the dataset we tested 38 images.

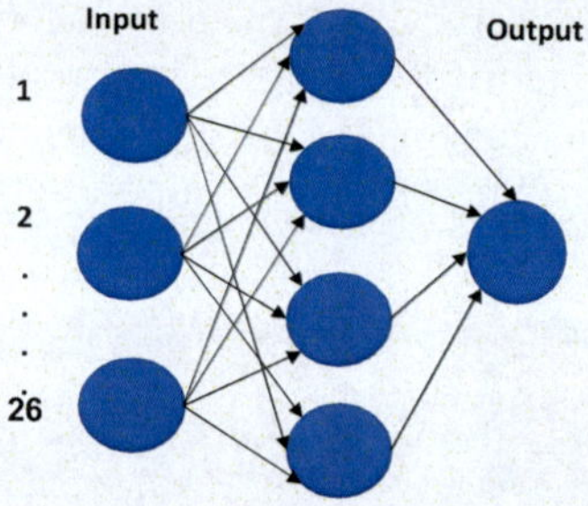

Fig. 4. Artificial Neural Network (ANN)

3.3 The Second Proposed Method

Because of the poor results for ML, we turned to DL to improve accuracy. To automate and enhance the accuracy of the proposed technique, four pre-trained CNN models were used after the pre-processing stage, as shown in Fig. 5.

3.3.1 In Image Pre-processing and Resizing

In this phase, the image is resized and converted from RGB to grayscale. Additionally, the thermal images are resized to a smaller dimension of 227 × 227 pixels to enhance computational efficiency.

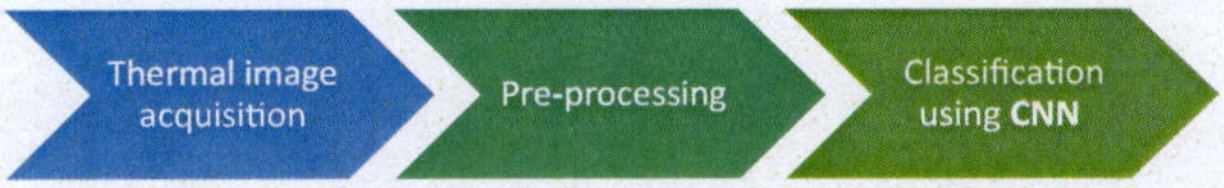

Fig. 5. The primary steps undertaken in the second proposed approach.

3.3.2 In Deep Learning Model for Classification

In this method, the CNN app was used as a classifier for thermal images. CNNs comprise layers that receive an input image, conduct mathematical operations, and generate predictions for the class or label probabilities at the output, as shown in Fig. 6. The types of layers are:

- Convolutional Layers (CONV): The Convolutional Layers extract local features from input data and improve spatial awareness for tasks like image recognition.
- Pooling Layers (POOL): The pooling layers reduce the spatial dimensions of the input data, helping to extract and retain important features while reducing computational complexity.
- Fully Connected Layers (FC): Fully Connected Layers make the final classification decision based on the extracted features. They enable the model to learn complex relationships and make accurate predictions.

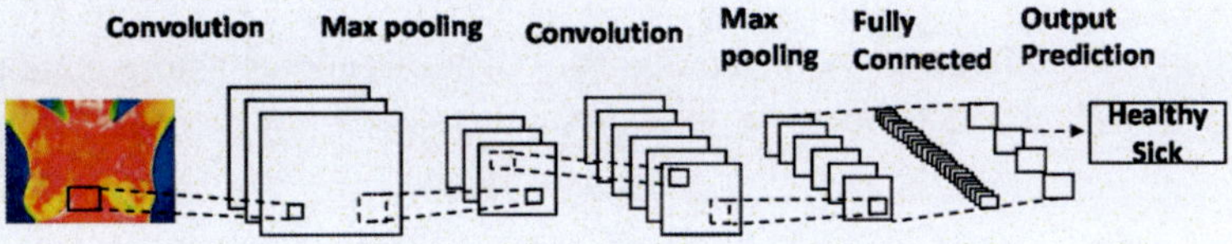

Fig. 6. The Convolutional Neural Network (CNN) architecture

In this study, we created a set of options for training CNN. Then, we loaded the dataset after pre-processing and split them randomly into 80% for training and 20% for testing. In this paper, we created a set of options for training CNN. Then, the pre-processed dataset was loaded and randomly divided into 80% for training and 20% for testing, following the approach used in prior studies to achieve optimal system performance. The accuracy of the network was then calculated.

4 Simulation Results and Discussion

This section presents the results of two classification methods used and several tests performed based on the DMR-IR database to evaluate the performance of the proposed approaches.

4.1 Evaluation of the Machine Learning

In the breast cancer detection method, the last step is classification, which entails both training and testing phases. Hence, in the first proposed method, we use the extracted features to analyze the thermal breast images as inputs into the classification algorithms. This paper utilizes ANN and Deep Learning Models classifiers, and their outcomes have been assessed and analyzed. A series of equations are employed to assess the experiments conducted for breast cancer detection, encompassing metrics such as accuracy, specificity, and sensitivity.

Accuracy: measures the proportion of cases that are correctly classified. It is computed by dividing the total number of correct predictions by the total number of predictions made. Eq. (1) is used for computation.

$$\text{Accuracy} = \frac{TP + TN}{TP + FP + FN + TN} \tag{1}$$

Specificity: is computed by considering the count of patients whose condition was incorrectly identified. Eq. (2) is used for computation.

$$\text{Specificity} = \frac{TN}{FP + TN} \tag{2}$$

Sensitivity: is determined by accurately estimating the number of patients with the condition. Eq. (3) is used for computation.

$$\text{Sensitivity} = \frac{TP}{TP + FN} \tag{3}$$

First, 190 frontal thermogram images are used, and extract 26 features for each thermal image. Next, the ANN classifier is utilized to differentiate between normal and abnormal breast tissue and give us TP = 10, TN = 23, FP = 3, FN = 3 that achieved accuracy 89,74%, sensitivity 82,35%, and specificity 95,45%.

4.2 Assessment of the Deep Learning Models' Performance

In the second method proposed, the performance of a deep convolutional neural network is assessed by employing four pre-trained CNN models. Table 1 presents a comparison of the evaluation outcomes of the pre-trained CNN models. Based on the findings presented in Table 1, it can be observed that the AlexNet model outperforms other CNN models in terms of performance on this particular dataset. AlexNet was used with different values of epochs, learning rates, and optimization methods. The proposed model, AlexNet, demonstrated exceptional performance with 100% accuracy, sensitivity, and specificity. It achieved these results within a training period of 6.49 minutes, utilizing 8 epochs, a learning rate of 1×10-4, and employing the Stochastic Gradient Descent with Momentum (SGDM) solver. The training progress of the model can be visualized in Fig. 7.

Table 1. Comparing the Performance Metrics of Various CNN Models

CNN model	Accuracy	Sensitivity	Specificity
AlexNet	100	100	100
GoogleNet	94,7	87.5	100
SqueezNet	84.2	62.5	100
ResNet18	97,4	100	95,5

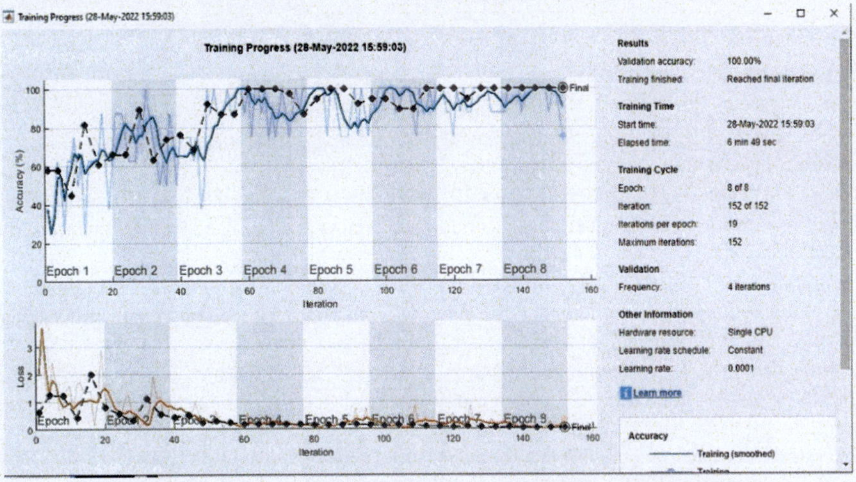

Fig. 7. Progress in Training the Proposed Deep Learning Model.

5 Conclusion

Breast cancer ranks among the top causes of death for women worldwide. Numerous research has used various techniques for breast segmentation and classification. Utilizing infrared technology, thermal imaging is an efficient diagnostic method for identifying breast cancer. In this study, we introduce two fully automated systems for breast cancer detection. The first system involves image processing to partition the breast, followed by feature extraction and classification using an ANN. The ANN achieved an accuracy of 89.74%, sensitivity of 82.35%, and specificity of 95.45%. In recent years, CNNs have emerged as a prominent and effective approach for breast cancer diagnosis. CNN-based CAD systems are preferred over texture and statistical feature-based systems due to their robustness and ease of implementation. In the second approach, the thermal image undergoes processing and classification using four pre-trained convolutional neural network (CNN) models, namely AlexNet, SqueezeNet, GoogleNet, and ResNet18. Compared to other pre-trained models, the Alexnet model achieved a fast and accurate diagnosis of breast cancer with accuracy, sensitivity and specificity of 100%. In order to improve efficiency, it will involve utilizing a significantly larger database. This expanded database will not only increase the diversity of the data but also provide a more comprehensive foundation for training and evaluation purposes. Additionally, cutting-edge machine learning and deep learning techniques will be employed to further refine the accuracy and performance of the breast cancer detection system. By leveraging advanced algorithms and models, it is anticipated that the system will exhibit enhanced precision, sensitivity, and specificity, leading to more reliable and efficient outcomes in the detection and diagnosis of breast cancer.

References

1. Mohamed, N.A.E.-R.: Breast cancer risk detection using digital infrared thermal images. Int. J. Bioinform. Biomed. Eng. (2015)
2. Sung, H., et al.: Global cancer statistics 2020: GLOBOCAN estimates of incidence and mortality worldwide for 36 cancers in 185 countries. Cancer J. Clinicians **71**, 209–249, February 2021
3. Sadoughi, F., Kazemy, Z., Hamedan, F., Owji, L., Rahmanikatigari, M., Azadboni, T.T.: Artificial intelligence methods for the diagnosis of breast cancer by image processing: a review. Breast Cancer Targets Therapy **10**, 219 (2018)
4. Zuluaga-Gomez, J., Zerhouni, N., Al Masry, Z., Devalland, C., Varnier, C.: A survey of breast cancer screening techniques: thermography and electrical impedance tomography. J. Med. Eng. Technol. **43**, 305–322 (2019)
5. Morales-Cervantes, A., et al. An automated method for the evaluation of breast cancer using infrared thermography. EXCLI J. **17**:Doc989 (2018). ISSN 1611-2156
6. Mahmoud, A.A., et al.: Medical image segmentation techniques, a literature review, and some novel trends. Menoufia J. Elec. Eng. Res. **27**, 23–58 (2018)
7. Bezerra, L.A., et al.: Infrared imaging for breast cancer detection with proper selection of properties: from acquisition protocol to numerical simulation. Multimodality Breast Imaging: Diagnosis and Treatment. SPIE Press, Bellingham, vol. 1, pp. 285–332 (2013)
8. Borchartt, T.B., Conci, A., Lima, R.C.F., Resmini, R., Sanchez, A.: Breast thermography from an image processing viewpoint: a survey. Sig. Process. **93**, 2785–2803 (2013)

9. Kapoor, P., et al.: Image segmentation and asymmetry analysis of breast thermograms for tumor detection. Int. J. Comput. Appl. **50** (2012)
10. Hossam, A., Harb, H.M., Abd El Kader, H.M.: Automatic image segmentation method for breast cancer analysis using thermography. JES J. Eng. Sci. **46**, 12–32 (2018)
11. Adel, M., Abdelhamid, B., El-Ramly, S.: Automatic image segmentation of breast thermograms. In: Proceedings of the 2018 7th International Conference on Bioinformatics and Biomedical Science – ICBBS, vol. 18 (2018)
12. Guan, S., Kamona, N., Loew, M.: Segmentation of thermal breast images using convolutional and deconvolutional neural networks. In: 2018 IEEE Applied Imagery Pattern Recognition Workshop (AIPR) (2018)
13. Guzman-Cabrera, R., et al.: Digital processing of thermographic images for medical applications (2016)
14. Usha, N., et al.: Feature selection and classification for analysis of breast thermograms. In: 2019 2nd International Conference on Signal Processing and Communication (ICSPC) (2019)
15. Dey, N., Rajinikanth, V., Hassanien, A.E.: An examination system to classify the breast thermal images into early/acute DCIS class. In: Proceedings of International Conference on Data Science and Applications (2021)
16. Milosevic, M., Jankovic, D., Peulic, A.: Thermography based breast cancer detection using texture features and minimum variance quantization. EXCLI J. **13**, 1204 (2014)
17. Wehle, H.-D.: Machine learning, deep learning and AI: what's the difference. Data Scientist Innovation Day, pp. 2–5 (2017)
18. Santana, M.A.d., et al.:Breast cancer diagnosis based on mammary thermography and extreme learning machines. Res. Biomed. Eng. **34**, 45–53 (2018)
19. Soliman, O.O., Sweilam, N.H., Shawky, D.M.: Automatic breast cancer detection using digital thermal images. In: 2018 9th Cairo International Biomedical Engineering Conference (CIBEC) (2018)
20. Karim, C.N., Mohamed, O., Ryad, T.: A new approach for breast abnormality detection based on thermography. Med. Technol. J. **2**, 245–254 (2018)
21. Mishra, V., Singh, Y., Rath, S.K.: Breast cancer detection from thermograms using feature extraction and machine learning techniques. In: 2019 IEEE 5th International Conference for Convergence in Technology (I2CT) (2019)
22. El-Soud, M.W.A., Eltoukhy, M.M.: Breast cancer detection in thermal images using extreme learning machine. J. Intell. Math Fuzzy Syst. **38**, 2673–2681, March 2020
23. AlFayez, F., El-Soud, M.W.A., Gaber, T.: Thermogram breast cancer detection: a comparative study of two machine learning techniques. Appl. Sci. **10**, 551 (2020)
24. Silva, L., Seixas, F., Fontes, C., Muchaluat-Saade, D., Conci, A.: A Computational method for breast abnormality detection using thermographs. In: 2020 IEEE 33rd International Symposium on Computer-Based Medical Systems (CBMS) (2020)
25. Ghobadi, H., Thainimit, S., Sugino, N., Gansawat, D., Zadeh, H.G.: Comparative accuracy of Digital Infra-red Thermal Imaging (DITI) in breast cancer diagnoses. J. Chem. Pharma. Res. (2016)
26. Mambou, S., Maresova, P., Krejcar, O., Selamat, A., Kuca, K.: Breast cancer detection using infrared thermal imaging and a deep learning model. Sensors **18**, 2799 (2018)
27. Tello-Mijares, S., Woo, F., Flores, F.: Breast cancer identification via thermography image segmentation with a gradient vector flow and a convolutional neural network. J. Healthcare Eng. **2019**, 1–13 (2019)
28. de Freitas Oliveira Baffa, M., Lattari, L.G.: Convolutional neural networks for static and dynamic breast infrared imaging classification. In: 2018 31st SIBGRAPI Conference on Graphics, Patterns and Images (SIBGRAPI) (2018)

29. Roslidar, R., Saddami, K., Arnia, F., Syukri, M., Munadi, K.: A study of fine-tuning CNN models based on thermal imaging for breast cancer classification. In: 2019 IEEE International Conference on Cybernetics and Computational Intelligence (CyberneticsCom) (2019)
30. Torres-Galvan, J.C., Guevara, E., Gonzalez, F.J.: Comparison of deep learning architectures for pre-screening of breast cancer thermograms. In: 2019 Photonics North (PN) (2019)
31. Fernández-Ovies, F.J., Santiago Alférez-Baquero, E., de Andrés-Galiana, E.J., Cernea, A., Fernández-Muñiz, Z., Fernández-Martínez, J.L.: Detection of breast cancer using infrared thermography and deep neural networks. In: Rojas, I., Valenzuela, O., Rojas, F., Ortuño, F. (eds.) Bioinformatics and Biomedical Engineering. IWBBIO 2019. Lecture Notes in Computer Science(), vol. 11466. Springer, Cham (2019). https://doi.org/10.1007/978-3-030-17935-9_46
32. Roslidar, R., et al.: BreaCNet: a high-accuracy breast thermogram classifier based on mobile convolutional neural network. Math. Biosci. Eng. **19**, 1304–1331 (2021)
33. Husaini, M.A.S.A., Habaebi, M.H., Gunawan, T.S., Islam, M.R., Elsheikh, E.A.A., Suliman, F.M.: Thermal-based early breast cancer detection using inception V3, inception V4 and modified inception MV4. Neural Comput. Appl. **34**, 333–348 (2021)
34. Dey, S., Roychoudhury, R., Malakar, S., Sarkar, R.: Screening of breast cancer from thermogram images by edge detection aided deep transfer learning model. Multimedia Tools Appl. **81**, 9331–9349 (2022)
35. Silva, L.F., et al.: A new database for breast research with infrared image. J. Med. Imaging Health Inform. **4**, 92–100 (2014)
36. Acharya, U.R., Ng, E.Y.K., Tan, J.-H., Sree, S.V.: Thermography based breast cancer detection using texture features and support vector machine. J. Med. Syst. **36**, 1503–1510 (2010)

Author Index

A. Hassanien et al. (Eds.): AISI 2023, LNDECT 184, pp. 565–567, 2023.
https://doi.org/10.1007/978-3-031-43247-7